Phytochemicals in Citrus

Applications in Functional Foods

FUNCTIONAL FOODS AND NUTRACEUTICALS SERIES

Series Editor
John Shi, Ph.D.
Guelph Food Research Center, Canada

Phytochemicals in Citrus: Applications in Functional Foods **(2017)**
Xingqian Ye

Food as Medicine: Functional Food Plants of Africa **(2016)**
Maurice M. Iwu

Chinese Dates: A Traditional Functional Food **(2016)**
Edited by Dongheng Liu, Ph.D., Xingqian Ye, Ph.D., and Yueming Jiang, Ph.D.

Functional Food Ingredients and Nutraceuticals: Processing Technologies, Second Edition **(2015)**
Edited by John Shi, Ph.D.

Marine Products for Healthcare: Functional and Bioactive Nutraceutical Compounds from the Ocean **(2009)**
Vazhiyil Venugopal, Ph.D.

Methods of Analysis for Functional Foods and Nutraceuticals, Second Edition **(2008)**
Edited by W. Jeffrey Hurst, Ph.D.

Handbook of Fermented Functional Foods, Second Edition **(2008)**
Edited by Edward R. Farnworth, Ph.D.

Functional Food Carbohydrates **(2007)**
Costas G. Biliaderis, Ph.D. and Marta S. Izydorczyk, Ph.D.

Dictionary of Nutraceuticals and Functional Foods **(2006)**
N. A. Michael Eskin, Ph.D. and Snait Tamir, Ph.D.

Handbook of Functional Lipids **(2006)**
Edited by Casimir C. Akoh, Ph.D.

Handbook of Functional Dairy Products **(2004)**
Edited by Collete Short and John O'Brien

Herbs, Botanicals, and Teas **(2002)**
Edited by G. Mazza, Ph.D. and B.D. Oomah, Ph.D.

Phytochemicals in Citrus

Applications in Functional Foods

Edited by
Xingqian Ye

CRC Press
Taylor & Francis Group
Boca Raton London New York

CRC Press is an imprint of the
Taylor & Francis Group, an **informa** business

CRC Press
Taylor & Francis Group
6000 Broken Sound Parkway NW, Suite 300
Boca Raton, FL 33487-2742

Printed on acid-free paper

International Standard Book Number-13: 978-1-4987-4272-6 (Hardback)

Library of Congress Cataloging-in-Publication Data

Names: Ye, Xingqian, editor.
Title: Phytochemicals in citrus : applications in functional foods / [edited by] Xingqian Ye.
Other titles: Functional foods & nutraceuticals series.
Description: Boca Raton : CRC Press/Taylor & Francis, 2017. | Series: Functional foods and nutraceuticals | Includes bibliographical references.
Identifiers: LCCN 2017011306 | ISBN 9781498742726 (hardback : alk. paper)
Subjects: | MESH: Citrus--chemistry | Phytochemicals | Flavonoids--therapeutic use | Functional Food | Food-Processing Industry
Classification: LCC QP144.C58 | NLM WB 430 | DDC 572/.2--dc23
LC record available at https://lccn.loc.gov/2017011306

Visit the Taylor & Francis Web site at
http://www.taylorandfrancis.com

and the CRC Press Web site at
http://www.crcpress.com

Contents

Series Preface

In recent years, consumers have paid more attention to their health led by the increasing belief that the food they eat contributes directly to it. The consumption of functional foods has emerged as a major consumer-driven trend, playing an outstanding role in the prevention of nutrition-related diseases and improving the physical and mental well-being of consumers. This trend is expected to continue, which increases the need for scientific information concerning all aspects of functional foods. The aim of the "Functional Foods and Nutraceutical" series is to provide a comprehensive source of up-to-date information on the latest developments in this field. Many bioactive components present in food, from both plant and animal sources, have been shown to be effective in disease prevention and health promotion. The series covers relevant aspects of chemistry, biochemistry, epidemiology, nutrigenomics and proteomics, engineering, formulation, and processing technologies as they relate to nutraceuticals, functional foods, and dietary supplements, as well as new product developments and research progress.

Extensive studies have confirmed that diets rich in fruits and vegetables are strongly associated with numerous health benefits and a lower risk of disease. Citrus fruits, one of the most popular fruits in world market today, have long been valued as part of a nutritious and tasty diet. Citrus fruits, ranging from sweet to sour in taste, include oranges, tangerines, grapefruits, limes, and lemons. Well-known for their vitamin C content, these fruits also provide energy, fiber, folate, and potassium. In addition, they are a good source of disease-fighting antioxidants such as flavonoids. The different flavors of citrus are among the preferred in the world, and it is increasingly evident that citrus fruits not only taste good, but are good for our health too. It is well established that citrus fruits and products are a rich source of vitamins, minerals, and dietary fiber (nonstarch polysaccharides) that are essential for normal growth, development, and overall nutritional well-being. Moreover, there is an increasing awareness that some biologically active, non-nutrient compounds found in citrus and other plants (phytochemicals) can also help reduce the risk of many chronic diseases. For instance, citrus has been shown to help keep your heart healthy, specifically oranges, which have been shown to reduce the risk of heart disease and arrhythmias.

The "Functional Foods and Nutraceutical" series is appropriate for academic use, aimed at providing a solid scientific reference for faculties and students in the fields of food science and technology, nutritional science, and pharmaceutical science. This series is also meant as a reference for food science professionals in either government or industry pursuing functional food research, food ingredient development, or research and development for food companies. Herein, readers will obtain a current and sound scientific fact base of information about functional food products and their latest developments. It is our hope that the scientific community will appreciate our endeavor in preparing this series and the impact it will have on furthering the knowledge of functional foods and nutraceuticals.

John Shi, PhD
Series Editor
Senior Research Scientist
Guelph Food Research Center
Agriculture and Agri-Food Canada

Preface

Citrus is the largest genus in the Rutaceae family native to tropical and subtropical regions, and now it is the most traded horticultural product in the world. Taxonomic identification is difficult because there are many spontaneous and commercial hybrids, but citrus can be generally classified into the following categories: sweet oranges, mandarins (tangerines, clementines), sour/bitter oranges, lemons, limes, grapefruits, pummelos, hybrids, citrons, and so on. They are mainly characterized by a rough and bright color that ranges from yellow to orange. Citrus fruits not only contain large amounts of vitamin C, but also a variety of beneficial nutrients and vitamins such as folate, thiamin, minerals, fiber, carotenoids, flavonoids, and limonoids. Vitamin C is a powerful antioxidant that protects the body from damaging free radicals. It is also required for the synthesis of collagen, which helps wounds heal and helps hold together blood vessels, tendons, ligaments, and bone. Folate is essential for cell division and DNA synthesis. Thiamin is a B vitamin, which is important in metabolism. Citrus fruits have the advantage of containing several different antioxidants. There is considerable evidence that citrus fruits have antioxidant and antimutagenic properties and positive associations with the health of bones, the cardiovascular system, and the immune system. For example, citrus flavonoids are strong antioxidants that can neutralize free radicals and may guard against heart disease. Studies show that citrus flavonoids can improve blood flow through the coronary arteries, reduce the ability of arteries to form blood clots and prevent the oxidation of LDL cholesterol, which is an initial step in the formation of arterial plaque. Citrus fruits, as a whole, are packed with beneficial antioxidants and essential nutrients that can promote heart health and possibly reduce the risk of some chronic diseases, from cardiovascular disease and cancer to skin damage from sunlight.

Citrus fruits are valuable to human health, we should consume them daily to overcome and prevent micronutrient deficiencies. They are beneficial to the health of people suffering from obesity and diet-related chronic diseases as they contain no fat, sodium, or cholesterol. Because of the numerous nutritional and health benefits associated with the consumption of citrus fruits, they can be considered as part of a balanced diet.

This book, *Phytochemicals in Citrus*, contains 16 chapters providing a knowledge base on the chemical composition, bioactive components, biochemical properties, food use, and health benefits of citrus fruits. The information in this book will help readers to better understand the health benefits of citrus fruits, citrus products, and their dietary applications.

We thank all of the contributing authors for their cooperation in preparing this book, which we hope will serve as an excellent reference for those interested in citrus fruits, as health-promoting foods, and the science and technology of their bioactive components.

Xingqian Ye, PhD
Editor
Professor
College of Biosystems and Food Science
Zhejiang University
Zhejiang, China

Acknowledgments

This book is a product of extensive multidisciplinary collaborative efforts between scientists, engineers, academicians, and industry personnel. Therefore, the editor wishes to express his sincere gratitude and appreciation to all colleagues, graduate students, and reviewers for their assistance and support. Especially, the editor wishes to express his profound appreciation to Prof. Xueping Chen (Prof. Emeritus, Zhejiang University, China) for his advice on career development. The editor would also like to thank Ms. Qianying (Sophia) Ye (Purdue University), Prof. Yukio Kakuda and Dr. Joe Martin (University of Guelph, Canada), and Dr. Sophia Jun Xue (Guelph Food Research Center, Canada) for their review comments. The editor is also thankful to Dr. John Shi (CRC Series Editor and Senior Scientist at Agriculture and Agri-Food Canada) and Dr. Stephen Zollo (CRC/Taylor & Francis) for their valuable scientific advice, support, and encouragement during the editing process.

The editor also deeply thanks financial support from the China Nature Sciences Foundation (Project C200501).

Editor

Xingqian Ye, PhD, is a professor in the Department of Food Science and Nutrition, and the deputy dean of the College of Biosystems and Food Science at Zhejiang University in China. Prof. Ye earned a PhD degree in agriculture products storage and processing from Zhejiang Agricultural University. He has also conducted collaborative research in Bulgaria, Canada, the United States, and Greece. His research focuses on fruit and vegetable processing technology, nutraceuticals, and functional foods from fruits and vegetables. Recently, he has studied the processing and comprehensive utilization of Chinese bayberry, mandarin, and other local fruits and vegetables, especially the identification and separation of their phytochemicals. Prof. Ye has also worked on the development of mixing fruit, vegetables, nuts, and cereals to enhance their antioxidant capacity after processing. He has also studied the stability of flavonoids after ultrasonic treatment.

Dr. Ye has published more than 200 research papers in refereed scientific journals, along with 10 book chapters and has been invited to a number of presentations. He holds 35 Chinese patents and has edited *Chinese Dates: A Traditional Functional Food* (CRC Press, 2016), as well as 4 other books and a textbook on fruit and vegetable processing in Chinese (from the second edition to fourth edition). He has received several scholarly awards including the second place Award (three times) for Science and Technology from the Zhejiang Provincial Government. Prof. Ye was the guest editor of *LWT—Food Science and Technology* for the special issue on *Food Innovation in China* in 2014. He is an associate editor of *Food Quality and Safety* (Oxford University Press) and editorial board member of the *Journal of Food Engineering*, the *Journal of Chinese Institute of Food Science and Technology*, and four other Chinese journals.

Contributors

Elizabeth A. Baldwin
USDA–ARS
Ft. Pierce, Florida

Maria Susai Boobalan
Department of Chemistry
College of Natural and Computational Sciences
Haramaya University
Dire Dawa, Ethiopia

Nir Carmi
Institute of Plant Sciences
Agricultural Research Organization
Bet Dagan, Israel

Thais Borges Cesar
Faculdade de Ciencias Farmaceuticas UNESP
Araraquara, Brazil

Jianchu Chen
College of Biosystems and Food Science
Zhejiang University
Zhejiang, China

Jianle Chen
College of Biosystems and Food Science
Zhejiang University
Zhejiang, China

Shiguo Chen
College of Biosystems and Food Science
Zhejiang University
Zhejiang, China

Huan Cheng
College of Biosystems and Food Science
Zhejiang University
Zhejiang, China

Licinio Díaz
Plant Physiology Department
University of Murcia
Murcia, Spain

Tian Ding
College of Biosystems and Food Science
Zhejiang University
Zhejiang, China

Yoram Eyal
Institute of Postharvest and Food Sciences
Agricultural Research Organization
Bet Dagan, Israel

Shi Feng
Citrus Research and Education Center
University of Florida
Lake Alfred, Florida

Paula Souza Ferreira
Faculdade de Ciencias Farmaceuticas UNESP
Araraquara, Brazil

Danielle R. Gonçalves
Faculdade de Ciencias Farmaceuticas UNESP
Araraquara, Brazil

Sam-Long Hwang
Food Industry Research and Development Institute
Hsinchu, Taiwan

Gurpreet Kaur
Department of Phamaceutical Sciences and Drug Research
Punjabi University
Patiala, Punjab, India

Jasleen Kaur
Department of Phamaceutical Sciences and Drug Research
Punjabi University
Patiala, Punjab, India

Kee-Tae Kim
Bio/Molecular Informatics Center
Konkuk University
Seoul, Korea

Carlos A. Ledesma-Escobar
Department of Analytical Chemistry
University of Córdoba
Córdoba, Spain

Junhui Li
College of Biosystems and Food Science
Zhejiang University
Zhejiang, China

Shan Li
College of Biosystems and Food Science
Zhejiang University
Zhejiang, China

Donghong Liu
College of Biosystems and Food Science
Zhejiang University
Zhejiang, China

María D. Luque de Castro
Department of Analytical Chemistry
University of Córdoba
Córdoba, Spain

Yaqin Ma
Citrus Research Institute
Southwest University
Chongqing, China

Francisco Marin
Department of Production and Characterization of New Foods
University of Madrid
Madrid, Spain

Ana Ortuño
Plant Physiology Department
University of Murcia
Murcia, Spain

Hyun-Dong Paik
Department of Food Science and Biotechnology of Animal Resources
Konkuk University
Seoul, Korea

Jiang Ping
College of Biosystems and Food Science
Zhejiang University
Zhejiang, China

Ron Porat
Institute of Postharvest and Food Sciences
Agricultural Research Organization
Bet Dagan, Israel

Eran Raveh
Institute of Plant Sciences
Agricultural Research Organization
Bet Dagan, Israel

Laura Reuss
Citrus Research and Education Center
University of Florida
Lake Alfred, Florida

José Antonio del Río
Plant Physiology Department
University of Murcia
Murcia, Spain

Meenakshisundaram Selvamuthukumaran
School of Food Science and Postharvest Technology
Haramaya University
Dire Dawa, Ethiopia

John Shi
Guelph Research and Development Center
Agriculture and Agri-Food Canada
Guelph, Ontario, Canada

Ping-Hsiao Shih
Department of Pediatrics
Shuang Ho Hospital
New Taipei City, Taiwan

Joon Hyuk Suh
Citrus Research and Education Center
University of Florida
Lake Alfred, Florida

Yujing Sun
Department of Food Science and Technology
Zhejiang University of Technology
Hangzhou, China

Jinhu Tian
College of Biosystems and Food Science
Zhejiang University
Zhejiang, China

Ma. Ángeles Valdivia-López
Food and Biotechnology Department
National Autonomous University of Mexico
Mexico City, Mexico

Yu Wang
Food Science and Human Nutrition
University of Florida
Lake Alfred, Florida

Guihua Xu
Department of Food Science
Henan Institute of Science and Technology
Henan, China

Jianguo Xu
Zhejiang Citrus Research Institute
Zhejiang, China

Lufeng Yan
College of Biosystems and Food Science
Zhejiang University
Zhejiang, China

Qianying (Sophia) Ye
Department of Food Science
Purdue University
West Lafayette, Indiana

Xingqian Ye
College of Biosystems and Food Science
Zhejiang University
Zhejiang, China

Gow-Chin Yen
Department of Food Science and Biotechnology
National Chung Hsing University
Taichung, Taiwan

1 Citrus Fruits, Varieties, Chemical Properties, and Products in the Processing Industry

Xingqian Ye, Jianle Chen, Jianguo Xu, and Jianchu Chen

CONTENTS

1.1 CITRUS FRUITS AND THEIR BIOLOGICAL STRUCTURE

Citrus fruits, compared with other fruits, have specific characteristics. They differ botanically from berries and other fruits, and have a high nutritional value (Nagy and Shaw 1977; Ye et al. 2005).

1.1.1 The External Form of Citrus Fruits

Citrus fruit is usually globose, and its major morphological features are described in Figure 1.1 (Ye et al. 2005). The end with the pedicel is considered the base, and the opposite end is the top. A horizontal line is known as the equator. The part between the equator and the base is called the lower shoulder, and the part between equator and the top is called the upper shoulder. The part near the base is the pedicel side, and the part near the top end is the column end, where the navel is usually located. Different species vary greatly in navel size and shape. The fine points of the fruit surface are oil brittle points, with various essential oils contained in the interior (Figure 1.1).

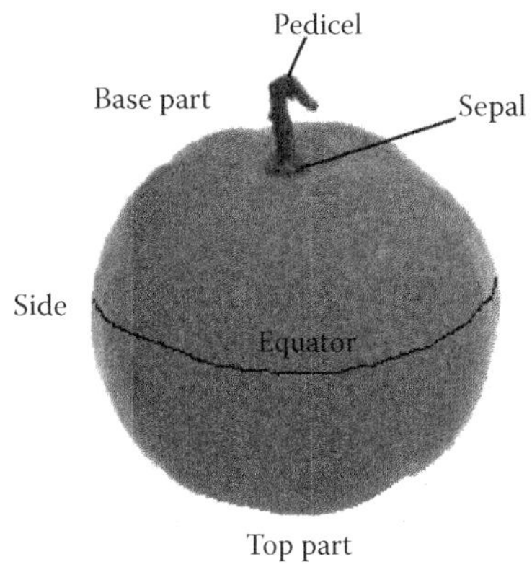

FIGURE 1.1 Citrus fruits and their nomenclature.

1.1.2 Size and Shape

Different species of citrus vary greatly in size. Generally speaking, the horizontal diameters of some common species are as follows: orange (5.7–9.5 cm), mandarin or tangerine (5.7–7.5 cm), lemon (4.4–6.4 cm), lime (3.5–5 cm), grapefruit (9.5–14.5 cm), pomelo (8–20 cm), and kumquat (1.5–3 cm). There are also great size differences between different cultivars in the same species (Ye et al. 2005; Deng et al. 2008; Citrus Pages 2014).

1.1.3 The Internal Structure of Citrus Fruits

Compared with other fruits, the internal structure of citrus fruits is more complex. From outside to inside, are peel (flavedo, oil gland, albedo, and vascular bundle),

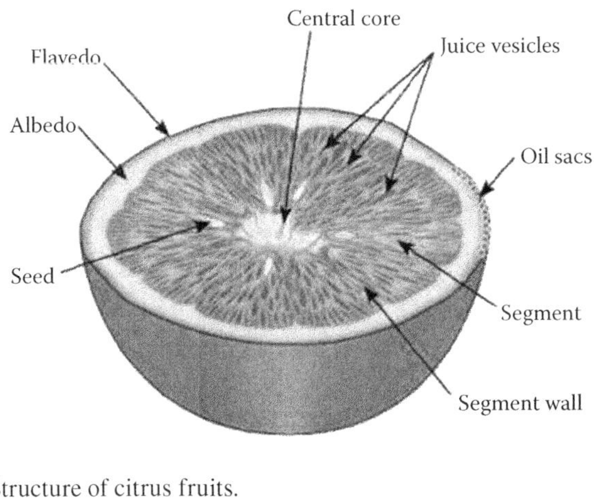

FIGURE 1.2 Structure of citrus fruits.

segments (segment membrane, juice vesicles or sac and seed), and the central core or pith (vascular bundle and parenchyma) (Figure 1.2) (Ye and Liu 2005; Kimball 1999).

1.1.3.1 Peel

Citrus peel can be divided into two layers, the epicarp (flavedo and albedo) and endocarp.

1.1.3.2 Epicarp

The epicarp is actually epidermis consisting of a wax plate of upper epicuticular wax, cuticle, and polygonal epidermic cells, and an outer wall that is horny. The exterior of epidermal cells is covered with a layer of wax that has a protective effect. The epicarp also has stomata consisting of a pair of guard cells that serves as one of the channels of fruit respiration. Generally, a citrus fruit has little or no porosity in the base, since the waxy layer is thicker at this point, and the stomata of citrus are often filled at maturity.

1.1.3.3 Mesocarp

The mesocarp includes two layers, the albedo and flavedo. The albedo contains oil sacs and pigments, while the flavedo consists of vascular tissue. Albedo cells are green, yellow, or orange, with oil sacs. The lower part of the epidermal layer consists of layers of small elongated cells, followed by layers of larger suborbicular cells, distributed before oil sacs, and finally a mix with albedo on top of the oil sacs. These cells are rich in plastids. When the fruit is ripe, when the weather turns cold, chloroplasts turn into chromoplasts. This causes the fruits to turn yellow or orange, and it can also be achieved with artificial de-greening treatments. As for the late ripening fruit, if the fruit were left on trees until the next spring, chromoplasts would turn into chloroplasts due to climate warming. Since mature citrus flavedo is rich in chromoplasts, the pigments therein are carotenoids and flavonoids, which are all beneficial to human health. In recent years, many papers have been published focused on the extraction of pigments from albedo in order to enhance or improve the color quality of citrus juice and other foods and for nutraceuticals.

1.1.3.4 Oil Sacs

Oil sacs, also called oil glands, consist of a cavity containing several aromatic oils, surrounded by a circle of cellular degradation. The outermost skin layer is the parenchyma of the albedo. These surrounding cells are the source of the oil, consisting of thin walls that are easily broken when mature. In a citrus fruit, the fruit stem has more oil glands than the top part, and the number of oil glands and the properties of the oils therein vary greatly between different *Citrus* species. Manual extraction of essential oils is efficient with lemon, orange, lime, tangerines, and others, but most mandarin oil is a low quality essential oil.

1.1.3.5 Flevedo

The flevedo is composed of a layer of loose, tubular cells with many branches. These cells are interwoven into a continuous network structure with large gaps. Cytoderm is rich in pectin, and an obvious thickening effect can be observed under the electron microscope since it is the site that has the highest pectin content within the citrus fruits. Hence, in general the raw material of artificial citrus pectin is extracted from the flavedo. The flavedo also contains flavonoids, some of which impart a bitter taste. Early on, people found that in grapefruit, oranges, lemons, and mandarin, flavonoid content in flavedo was higher than in juice, segment membrane, or albedo. The flavedo is also the main location of another bitter taste and health factor, limonoids. The thickness of flavedo varies in different *Citrus* species. Pomelos and grapefruits have the thickest, followed by oranges, natsumikans, and lemons, while mandarins and tangerines have the thinnest. In fruit juice production, the thickness of the flavedo is a major determinant of the appropriate juicer arrangement. If pressing is excessive, more flavedo is pressed into juice, making juice bitter with a content of flavonoids and limonoids substances that is too high. This is one of the causes of poor quality fruit juice by the deep-pressed method. When producing peeled mandarin segments in syrup production, the thickness of the flavedo is directly related to the ease of peeling. As for orange cake and fruit candy production, it is more related to the thickness and tenderness of the flavedo.

1.1.3.6 Citrus Segment

The interior part of the flavedo is the segment, which is thought to have developed from the endocarp. Generally, mandarin and tangerine have about 10 segments, kumquat has 3–6, while polemo has 13–25 and lemon has 7–10. The citrus segment is composed of a membrane, vesicle (sac), and seeds. The seeds of satsuma, a mandarin, degenerate because pollination does not occur. Each segment in the fruits is arranged in a ring with a central core or pith.

The orange vessels are a layer of reticulated vascular tissue that surround the segment and contain flavonoid glycosides, pectin, cellulose, and semicellulose. They may be used as a drug, but also as the material for pectin extraction. Their degree of adhesion to the segment affects the ease of peeling and processing.

There is a layer of waxy cuticle on the segment membrane consisting of six to eight layers with a relatively loose, full layer of two to three mesophyll cells between segments. The outmost layer is a mixture of mesophyll cells and flavedo. The structural components of segment membranes are very closely related with the separation of citrus segments. Generally, mandarins or tangerines are easier to peel and

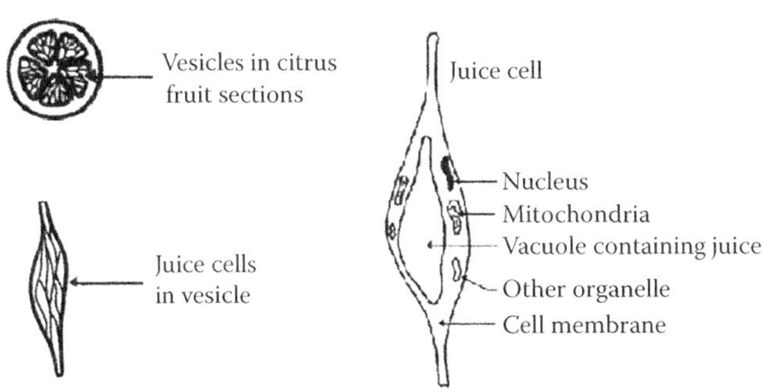

FIGURE 1.3 Citrus juice cell and vesicle (sac) cited. (From Kimball, D. A., *Citrus Processing—A Complete Guide*, 2nd ed., Aspen Publishers, Gaithersburg, MD, 1999.)

separate segments, and grapefruits are more difficult. Segment membranes also contains bitter substances, and one must be careful not to put too much pressure on the segment membrane during processing, especially in grapefruit, since they are thought to contain the highest content of limonin.

The sac is also called the juice vesicle (Figure 1.3). It is the edible part of citrus fruit located on the interior of the segment, and it is derived from independent meristems that form the sac and sac stem. The sac and sac stem surface contain wax (Kimball 1999). This is a major factor in maintaining the integrity of canned orange segments in syrup; however, in the manufacture of sac juice or drink, this can obstruct the separation of sac into a single. The sac surface also contains some pectin, and therefore treatment with calcium salt increases its hardness. The shape of the sac is related to the appearance of processed products, and canned mandarin segments in light syrup require slender sacs and short stems, firmly bonded with each other. The interior of sacs consists of thin-walled cells, characterized by large vacuoles, with lipid droplets, mitochondria, leucosomes and chromoplasts. These thin-walled cells are easily broken, thereby creating juice. Juice is mainly vacuolar sap, containing a wealth of nutrients, especially vitamins and minerals (Tisserat et al. 1990).

1.1.3.7 Citrus Seeds

Citrus seeds are often located in the interior of segments, developed from a fertilized ovule, but a fertilized ovule can also form seedless fruit. Citrus seeds contain a variety of fat, protein, and bitter substances that have a great impact on the processing and production of segments. Usually, seedless is preferred for fresh consumption. Seeds are also undesirable during juice processing, and pressure or fine filtering of crushed seed should be avoided; otherwise, it would easily make the juice bitter. On the other hand, from a comprehensive point of view, we can use seed oils and bitter substances in seeds for industrial and agricultural production services.

1.1.3.8 Central Core or Pith

The central core is also called the pith and is composed of several vascular bundles surrounded by a loose spongy group. The vasculature extends from the seed to the pedicel end. In citrus processing, the pith only affects the whole press of orange juice, and the vascular bundle and spongy tissue can increase the particle content of the fruit juice. In addition, spongy tissue is similar to flavedo in that excessive pressing will increase the glycoside extract and a mixture of other undesirable substances.

1.2 CITRUS FRUIT PRODUCTION

Citrus fruits are one of the most important agricultural products in the world, produced in more than 90 countries and regions in the world. World citrus production and consumption have experienced a strong growth period since the middle of the 1980s. The yield, total level of consumption, and consumption per capita of mandarins, tangerines, sweet oranges, pomelos, grapefruits, lemons, and limes have all experienced a rapid growth.

During the year from 1999 to 2000, annual production reached 90 million metric tons—more than grapes and bananas—the highest production among all the fruits.

International trade year volume was more than $65 billion, following wheat ($160 million) and maize ($100 billion), and a further increase in production and consumption is expected (FAO: http://www.fao.org/statistics/en/). Citrus production and export are important sources of income in many countries, while citrus are also a cheap source of vitamins, which is beneficial to nutrition, health, and food safety. With transportation and packaging costs reduced and the quality improved, the growth of citrus processing products is very fast, and is showing a strong momentum of development (Liu 2012).

TABLE 1.1
Fresh Citrus Fruit, Production, Supply, and Distribution in Selected Countries (1000 Metric Tons)

Sweet Orange	2011/2012	2012/2013	2013/2014	2014/2015	July 2016
Fresh production	53,830	49,871	52,144	48,772	45,763
Fresh domestic consumption	30,837	28,948	30,693	28,388	28,191
Processing	22,729	20,662	20,907	20,123	17,293
Export	3,932	3,889	4,002	4,051	4,309
Import	3,668	3,628	3,458	3,790	4,030
Tangerines/Mandarins					
Fresh production	23,906	24,679	26,695	28,557	28,963
Fresh domestic consumption	22,322	23,247	25,068	26,826	27,285
Processing	1,411	1,384	1,425	1,571	1,491
Export	2,389	2,165	2,483	2,337	2,322
Import	2,216	2,117	2,281	2,177	2,135
Grapefruit					
Fresh production	5,544	5,842	6,067	6,057	6,277
Fresh domestic consumption	4,602	4,818	5,087	5,260	5,540
Processing	873	954	907	732	674
Export	800	813	825	736	758
Import	731	743	752	671	695
Lemon/Lime					
Fresh production	6,524	6,512	6,216	7,301	7,001
Fresh domestic consumption	4,478	4,582	4,605	4,849	4,748
Processing	1,854	1,904	1,543	2,328	2,097
Export	1,689	1,552	1,590	1,725	1,791
Import	1,497	1,526	1,522	1,601	1,635

Source: USDA: http://www.fas.usda.gov/commodities/citrus-fruit.

TABLE 1.2
Potential Yield of Products from Citrus Fruits

Products	Orange/%	Grapefruit/%	Satsuma Mandarin/%
Juice	48%	48%	52%
Segment in light syrup			65%–70%
Dry peel			
Dry pulp w/o molasses, 10% H_2O	9.80	10.62	
Dry pulp, w/o molasses	7.11	7.25	
Molasses, 72 ^{0}Brix	3.43	3.89	
Ethanol from molasses (L, 100%)	0.49	–	
Cold pressed oil and limonene	0.74	0.26	0.76
Pulp wash solids	0.49–0.74	0.26–0.52	
Frozen pulp	4.9	7.77	
Oil-phase essence	0.0123	0.0077	
Aqueous essence at 15% alcohol	0.049	0.052	
Pectin (150 grade)	3.18	2.59	2.00
Seed oil	0.05	0.23	
Hesperidin	0.49	–	0.45
Naringin	–	0.52	
Limonoid		0.30	0.30

According to USDA statistics, the main producing countries in the world, during the years 2015 and 2016, had a total citrus production of 88 million tons, among which, oranges have 45.76 million tons, mandarins and tangerines have 28.96 million tons, grapefruits and pomelos have 6. 27 million tons, and lemons and limes have 7.00 million tons (Table 1.1). Analysis of nearly 5 years of production data has indicated that orange products are on the decline, but that mandarins or tangerines have increased the most, and that grapefruits, pomelos, lemons, and limes have had a slight rise (Table 1.2).

The leading producers of citrus are China, Brazil, the United States, Mexico, Turkey, Egypt, and so on. China has the largest production. In July 2016, it reached 31.3 million tons, accounting for 35.57% of the world's output; followed by Brazil with a production of 14.35 million tons, accounting for 16.31% of the world's output (Table 1.3).

Among *Citrus* species produced worldwide, the orange accounts for 52%, mandarin and tangerine for 33%, and lemon and pomelo for about 7% of total production. In terms of production of the total output of citrus, nearly 60% is for the fresh market consumption and about 40% is used for processing. But in Brazil, recent data show that the most abundant use is for processing of orange juice. Mandarins and tangerines are mostly used for fresh food, the main production country being China. In terms of raw materials of citrus fruit production, 80% are oranges, 8% are grapefruits, and 5% are lemons, mandarins, and tangerines (Table 1.4).

TABLE 1.3
Fresh Citrus Production in Selected Countries (1000 Metric Tons)

	Oranges	Tangerines/ Mandarins	Grapefruits	Lemons/ Limes	Total	%
Brazil	14,350				14,350	16.31%
China	7,000	20,000	4,300		31,300	35.57%
European Union	6,055	3,035	97	1,260	10,447	11.87%
United States	5,371	876	735	847	7,829	8.90%
Mexico	3,535		430	2,270	6,235	7.08%
Egypt	2,750				2,750	3.12%
Turkey	1,700	1,040	200	670	3,610	4.10%
South Africa	1,560		330	345	2,235	2.54%
Morocco	925	1,065			1,990	2.26%
Argentina	800	350		1500	2,650	3.01%
Vietnam	520				520	0.59%
Australia	455				455	0.52%
Costa Rica	315				315	0.36%
Guatemala	155				155	0.18%
Israel	105	240	185	60	590	0.67%
Japan		1,115			1,115	1.27%
South Korea		640			640	0.73%
Thailand		375			375	0.43%
Other	167	227	0	49	443	0.50%
Total	45,763	28,963	6,277	7,001	88,004	100.00%
	52.00%	32.91%	7.13%	7.96%	100%	

Source: USDA: http://www.fas.usda.gov/commodities/citrus-fruit.

TABLE 1.4
Citrus Production and Processing (1000 Metric Tons, %)

		2011/2012	2012/2013	2013/2014	2014/2015	July 2016
Orange	Fresh production	53,830	49,871	52,144	48,772	45,763
	Processing	22,729	20,662	20,907	20,123	17,293
	% of processing	42.22	41.43	40.09	41.26	37.79
Grapefruit	Fresh production	5,544	5,842	6,067	6,057	6,277
	Processing	873	954	907	732	674
	% of processing	15.75	16.33	14.95	12.09	10.74
Lemon and lime	Fresh production	6,524	6,512	6,216	7,301	7,001
	Processing	1,854	1,904	1,543	2,328	2,097
	% of processing	28.42	29.24	24.82	31.89	29.95
Mandarin and tangerine	Fresh production	23,906	24,679	26,695	28,557	28,963
	Processing	1,411	1,384	1,425	1,571	1,491
	% of processing	5.90	5.61	5.34	5.50	5.15

1.3 MAIN *CITRUS* SPECIES AND PRODUCTION

Citrus fruits belong to *Rutaceae aurantioideae*. Since 1753, when Linneaus designated *Citrus* L. many scholars have performed a series of morphological, cytological and molecular level studies. Notably, in the work of the American taxonomist W. T. Swingle, and the Japanese scholar T. Tanaka. The former divided *Citrus* L. into 16 species, and the latter divided *Citrus* L. into 159 species. Swingle's classification is simple and clear, but many of the horticultural species cannot be determined (Swingle and Reece 1967), while Tanaka's classification is conducive to the work of the horticultural industry, but is too complex for most of the food science industry and consumers (Tanaka 1977). Since they easily hybridize between species, and due to apomixis and clonal variations, the number of species of citrus is very difficult to accurately determine. According to the views of the pomologists and food scientists, this chapter summarizes the general classification and relationship of citrus fruit from citrus fruit characteristics in the perspective of economic value. It references China's latest research from Kimball (1999) and Deng (2008), but recent advances in molecular biology are not included.

- *Citrus sinensis* (L.) Osbeck—Sweet orange
- *Citrus reticulata* Blanco—Mandarin, tangerine
- *Citrus aurantium* L.—Sour or bitter orange
- *Citrus paradisi* Macfadyen—Grapefruit
- *Citrus grandis* (L.) Osbeck or *Citrus maxima* (Burm.) Merrill—Pomelo
- *Citrus limon* L. Burm. f.—Lemon
- *Citrus aurantiifolia* (Christm.)—Lime
- *Citrus limonia*—Rangpur
- *Citrus medica* L.—Citron

Main varietal group

- *Citrus reticulata* Blanco—Mandarin, tangerine

1.3.1 MANDARINS

Macroacrumen mandarins: Satsuma mandarin (*C. unshiu* Marc); Ougan (*C. suavissima* Hort. ex Tanaka); Jiaogan (*C. tankan* Hort.) (Tankan); King mandarin (*C. nobilis* Loureiro); Mediterranean mandarin (*C. deliciosu* Tenore).

Microacrumen tangerines: Ponkan (*C. reticulata* Blanco); Manju (*C. tardiferax* Hort. ex Tanaka); Hongju (red tangerine) (*C. tangerina* Hort. ex Tanaka); *Citrus paradisi* Macfadyen—Grapefruit (*Citrus paradisi* Macfadyen) (white or common, blood or pink, acidless); Pomelo [*Citrus grandis* (L.) Osbeck or *Citrus maxima* (Burm.) Merill] (white or common, blood or pink, acidless).

Dr. Mabberley, president of the IATP since 2005, deserves a special mention. He has presented the most interesting new views of citrus and the relationships between "the true citrus types." In his paper on native Australian citrus types, he introduces a new species and assigns the Swingle genera Microcitrus and Eremocitrus in the genus *Citrus*. In a classification for edible citrus, he states that there are only three *Citrus* species, citron, pomelo, and mandarin, which are then involved in several hybrids. In his more recent study on citrus linnaeus, he broadens the scope to include all the most common *Citrus* species. According to Mabberley, the following scheme provides a workable system of citrus for botanists and fruit growers alike. It is also helpful for understanding the origin of different kinds of citrus (Citrus Pages 2014).

1. *Citrus medica*, citron, is involved in: *Citrus limon* (citron × sour orange), lemon and similar hybrids like Palestine sweet lime and Volkamer lemon; *Citrus jambhiri* (citron × mandarin), rough lemon and similar hybrids like Rangpur lime, Mandarin lime, and other types like "Otaheite"; *Citrus aurantiifolia* (citron × lemon × Ichang papeda) lime (Mexican lime); and *Citrus bergamia* (citron × sour orange) bergamot, also considered a citron sour orange cross
2. *Citrus maxima*, pomelo, is involved in: *Citrus aurantium* (pomelo × mandarin) which includes three pomelo hybrids; *Citrus aurantium* (pomelo × mandarin) sour orange (the sour orange has inherited more features of the pomelo than the mandarin); *Citrus sinensis* (pomelo × mandarin) sweet orange (the sweet orange has inherited more features of the mandarin than the pomelo); this group also includes all the crosses of oranges, mandarins, and grapefruits such as tangors, ortaniques, tangelos, and their backcrosses like Page and Nova; and *Citrus paradisi* (pomelo × orange), the grapefruit

1.3.2 Mandarins and Tangerines

Mandarins (*Citrus reticulata* Blanco) are smaller and oblate rather than spherical like the common oranges (which are a mandarin hybrid). The taste is considered less sour than orange, as well as sweeter and stronger. A ripe mandarin is firm to slightly soft, heavy for its size, and pebbly-skinned. The peel is very thin, with a low bitter, white albedo, and they are usually easier to peel and split into segments. Generally mandarin and tangerine are considered part of the same group. The main species in the group are Satsumas (*Citrus unshiu* Marc.), Ponkan (*Citrus poonensis* Tanaka), and Clementines (*Citrus clementina* Tanaka). Other economically important species include Daisy, Pixie, and Changsha, among others.

Mandarins have been grown in China and Japan on a large scale since the sixteenth century. The biggest mandarin producers today are China, the European Union (mainly Spain), Japan, Morocco, and Turkey. Compared with the 2011 output, output in 2016 increased by 45.89%, 37.95%, 25%, 20.69%, and 18.86%, in Morocco, the United States, China, Argentina, and Turkey, respectively. The countries with the fastest increase in production were Morocco and the United States. The biggest mandarin exporters are China, Turkey, Morocco, the European Union (mainly Spain), and South Africa (Table 1.5).

TABLE 1.5
Fresh Fruit Production of Tangerines/Mandarins in Selected Countries (1000 Metric Tons)

	2011/2012	2012/2013	2013/2014	2014/2015	2015/2016	%
China	16,000	17,000	17,850	19,400	20,000	25.00%
European Union	3,099	2,927	3,213	3,474	3,035	–2.07%
Japan	1,001	846	1,124	1,070	1,115	11.39%
Morocco	730	662	1,160	1,003	1,065	45.89%
Turkey	875	876	880	960	1,040	18.86%
United States	635	660	700	803	876	37.95%
South Korea	586	667	672	697	640	9.22%
Thailand	360	375	375	375	375	4.17%
Argentina	290	300	370	350	350	20.69%
Israel	166	178	139	205	240	44.58%
Other	164	188	212	220	227	38.41%
Total	23,906	24,679	26,695	28,557	28,963	21.15%

Source: USDA: http://www.fas.usda.gov/commodities/citrus-fruit.

Mandarin types are grown in temperate Mediterranean-like areas, hot and humid tropical countries, and almost desert-type dry and arid environments. Additionally, some mandarin species tolerate cold and frost better than other citrus types. Certain species, for example, Satsumas, grown in China, can tolerate frost for several weeks.

Mandarins do not keep well on the tree after maturation. Due to their thin and often loose rind, they also tolerate storage and transportation less successfully than other citrus types. Mandarins are mostly eaten fresh, but a small amount of mandarin juice is produced. Canned segments in light syrup are the largest processing industry in China. They are used in fruit salads and other desserts, or as filling and decoration in cakes. Mandarin peel oil has an important position in the food industry as flavoring ingredients. It is used in sweets, gelatins, ice cream, chewing gum, pastry, and confectioneries. It is also used in soft drinks, mixers, essences, and flavorings, as well as in mandarin liqueurs and other alcohol products. Dried mandarin peels are important in traditional medicines, and serve as good material for the manufacturing of flavorings.

1.3.3 Oranges

The orange is a hybrid, between pomelos (*Citrus maxima*) and mandarins (*Citrus reticulata*). It has genes that are about 25% pomelo and about 75% mandarin. There are two kinds of oranges: one is sweet oranges (*Citrus sinensis*), and the other sour or bitter oranges (*Citrus aurantium*). The highest production is for sweet oranges, while only a very small amount of cultivation area is used for sour oranges, mainly for producing marmalades and other products. The orange fruits are round, flat round or oval, yellow to orange red, with a hard pericarp or being slightly easy to peel. Orange fruits have 9–12 segments, full or half-full cores, yellow, orange, purple or red flesh, a sweet or slightly sour taste, fewer seeds or no seeds. The flowering period is between March and May, the fruiting period from October to December, and the growing season of late maturing species extends from February to April of the following year.

The orange has a sweet and sour flavor, and is commonly peeled and eaten fresh or often used for juice production. The main production countries are Brazil, United States, Mexico, the European Union, and other countries and regions (USDA). The thick bitter rind is usually discarded, but can be processed into animal feed by desiccation, using pressure and heat. It is also used in certain recipes as a food flavoring or garnish. The outermost layer of the rind can be thinly grated with a zester to produce orange zest. The zest is popular in cooking because it contains oils and has a strong flavor similar to that of the orange pulp. The white part of the rind, including the pith, is a source of pectin and has nearly the same amount of vitamin C and other nutrients as the flesh. Over the past 5 years, the biggest producers of sweet oranges are Brazil, China, European Union, the United States, Mexico, Egypt, Turkey, and South Africa. However, during this time period, the total production volume of orange decreased by 15%, mainly in Brazil and the United States, with increases to 41% in Argentina, followed by Egypt, Australia, and other countries (Table 1.6). Oranges from Brazil and the United States are commonly used for orange juice production, the ratio of which accounts for more than 80%.

TABLE 1.6
Fresh Fruit Orange Production in Selected Countries (1000 Metric Tons)

	2011/2012	2012/2013	2013/2014	2014/2015	2015/2015	%
Brazil	20,482	16,361	17,870	16,716	14,350	−29.94%
China	6,900	7,000	7,600	6,900	7,000	1.45%
European Union	6,023	5,890	6,550	5,954	6,055	0.53%
United States	8,166	7,501	6,140	5,778	5,371	−34.23%
Mexico	3,666	4,400	4,533	4,158	3,535	−3.57%
Egypt	2,350	2,450	2,570	2,630	2,750	17.02%
Turkey	1,650	1,600	1,700	1,650	1,700	3.03%
South Africa	1,466	1,659	1,723	1,645	1,560	6.41%
Morocco	850	784	1,001	868	925	8.82%
Argentina	565	550	800	800	800	41.59%
Vietnam	530	520	520	520	520	−1.89%
Australia	390	435	430	430	455	16.67%
Costa Rica	370	325	315	315	315	−14.86%
Guatemala	150	155	155	155	155	3.33%
Israel	116	73	69	86	105	−9.48%
Other	156	168	168	167	167	7.05%
Total	53,830	49,871	52,144	48,772	45,763	−14.99%

Source: USDA: http://www.fas.usda.gov/commodities/citrus-fruit.

Main sweet orange cultivars can be divided into the following categories:

1. *Early sweet and common oranges*: Ambersweet, Hamlin, Rotuma Island, Berna, Jaffa (Florida Jaffa), Salustiana, Cadenera, Jincheng, Shamouti (Palestine Jaffa), Castellana, Marss, Trovita, Comuna, Parson Brown, Pineapple
2. *Valencia oranges*: Valencia, Delta Valencia, Midknight Valencia, and so on
3. *Navel oranges*
   a. Early navel oranges: Atwood, Navelina, Fisher, Skagg's bonanza, Thomson Zimmerman
   b. Mid-season navels: Cara Cara, Fukumoto, New Hall, Spring, Washington navel
   c. Late navel oranges: Autumn Gold, Lane Late, Ricalate, Barnfield, Navelate, Rohde Navel, and so on
4. *Blood oranges*: Rhode Red Valencia, Washington Sanguine, Ruby, Full blood oranges

1.3.4 Grapefruits

The grapefruit (*Citrus × paradisi*) is a subtropical citrus tree known for its sour to semisweet fruit. Grapefruits are a hybrid originating in the Barbados as an accidental cross between two introduced species, sweet oranges (*C. sinesis*) and pomelos, or shaddocks (*C. maxima*), both of which were introduced from Asia in the seventeenth century. When found, it was named the "forbidden fruit" (Morton 1987) and it has been misidentified with pomelos (Li et al. 2010). The fruit flesh is segmented and acidic, varying in color depending on the cultivars, which include white, pink and red pulps of varying sweetness (generally, the redder species are sweeter).

In many parts of the world, grapefruits are the customary breakfast fruits. Most grapefruits are chilled, cut in half, loosened from the peel and skin membranes with a special curved grapefruit knife and served fresh with perhaps a touch of sugar or honey. Grapefruit juice has increased in popularity, especially after its promotion as a diet drink started. Many weight loss diets include grapefruit juice. The pulp left over after commercial juice extraction is an important source of grapefruit oil, which is used as a flavoring in many soft drinks. The inner peel is a source of pectin and citric acid. Both are used by the food industry in the preservation of other fruits and in making jams and marmalades. Naringins, also extracted from grapefruit peels, gives tonic water its distinctive bitter flavor (Table 1.7).

1.3.5 Lemons and Limes

The lemon (*Citrus × limon*) is a species of small evergreen tree native to Asia. A study of the genetic origin of the lemons reported it to be a hybrid between bitter oranges (sour orange) and citrons. The tree's ellipsoidal yellow fruits are used for culinary and nonculinary purposes throughout the world, primarily for its juice, which has both culinary and cleaning uses (Citrus Pages 2014). The pulp and rind (zest) are also used in cooking and baking. The juice of the lemons is about 5% to 6% citric acid, which gives a sour taste. The distinctive sour taste of lemon juice makes it a key

TABLE 1.7
Fresh Grapefruit Production in Selected Countries (1000 Metric Tons)

	2011/2012	2012/2013	2013/2014	2014/2015	2015/2015	%
China	3200	3370	3717	3900	4300	34.38%
United States	1047	1092	950	807	735	–29.80%
Mexico	415	425	424	430	430	3.61%
South Africa	305	437	413	387	330	8.20%
Turkey	230	200	235	238	200	–13.04%
Israel	245	208	236	186	185	–24.49%
European Union	102	110	92	109	97	–4.90%
Other	0	0	0	0	0	0.00%
Total	5544	5842	6067	6057	6277	13.22%

Source: USDA: http://www.fas.usda.gov/commodities/citrus-fruit.

TABLE 1.8
Fresh Lemon and Lime Production in Selected Countries (1000 Metric Tons)

	2011/2012	2012/2013	2013/2014	2014/2015	2015/2015	%
Mexico	2055	2120	2187	2260	2270	10.46%
Argentina	1300	1350	780	1450	1500	15.38%
European Union	1264	1179	1308	1598	1260	–0.32%
United States	771	827	748	820	847	9.86%
Turkey	750	680	760	725	670	–10.67%
South Africa	260	245	312	339	345	32.69%
Israel	53	51	64	65	60	13.21%
Other	71	60	57	44	49	–30.99%
Total	6524	6512	6216	7301	7001	7.31%

Source: USDA: http://www.fas.usda.gov/commodities/citrus-fruit.

ingredient in drinks and foods such as lemonade and lemon meringue pie. The pulp, left over after commercial juice extraction, is an important source of citrus oil, pectin, and citric acid. These are used by the food, cosmetics, and pharmaceutical industries. Mexico, Argentina, and the European Union are the most important production countries of lemons and limes. The production has risen slightly in the past 5 years (Table 1.8).

1.4 SOME BIOACTIVE COMPONENTS IN CITRUS FRUITS

Citrus varieties such as oranges, grapefruits, mandarins, limes, and lemons, possess unique sensory properties and high nutritional value, and are among the most widely produced fruits in the world. Citrus originated in China where it is a highly profitable agricultural product. Citrus has a large cultivation area and long cultivation history.

TABLE 1.9
Major Events of Citrus Processing and Comprehensive Utilization

1950s	Successful development and commercialization of FMC citrus juice extractor and Brown juice extractor, enabling a large increase in the production of citrus fruits
1960s	Popularized frozen concentrated fruit juice as a food product
1970s	Studies on the comprehensive utilization of citrus; citrus essential oil recycling; application of citrus peel processing feed; selection of specific cultivars for canning in China
1977	*Citrus Science and Technology* published in the United States, which summarized global advances in research
1980s	Brazil becomes the largest producer of orange and orange juice; China promotes the spread of cultivars for canning, start of the promotion of low-temperature sterilization
1990s	China becomes the largest producer of canned citrus segments in light syrup, using a large number of low-temperature sterilization methods, and establishes a citrus juice production line
2000s	NFC chilly juice becomes popular; citrus polymethoxylated flavones, beta cryptoxanthin, zeaxanthin, coumarin, limonin are used as functional health factors; limonene application is used in the electronics industry as a cleaning agent

Among the 18 provinces and their cities in the southern part of the Yangtze River, the cultivation area has reached 1.5 million hm^2. Citrus production is abundant, which provides an incentive for the development and utilization of citrus resources. This has great potential, as citrus fruits are rich in nutrition, with not only high levels of pectin, oligosaccharide, organic acid, a variety of amino acids and other nutrients, but also a variety of nonnutritive physiologically active components such as limonin, alkaloid, volatile oils, phenolics, and other elements. These promote human health and have been associated with the prevention and treatment of some diseases. These trace elements give citrus fruits their unique flavor and nutritional value (Table 1.9).

Recently, phytochemicals in citrus fruits received attention in the literature, including phenolics (Abeysinghe et al. 2007; Kelebek et al. 2007), limonoids (Lam et al. 1994; Yu et al. 2005; Sun et al. 2005), coumarins (Ito et al. 2005; Widmer and Haun 2005), and alkaloids. Research has not been restricted to the edible parts of citrus fruits, but also byproducts created during processing. Phytochemicals in peels and seeds have attracted a wide attention from researchers. In addition, the scope of study has been extended to the immature fruits of citrus. The study found that the amount of these compounds in citrus peels and seeds are higher than that in the edible parts. For example, the content of flavonoids and other substances in immature citrus fruits is significantly higher than that in mature fruits. Hence, extracting these substances from citrus peels, seeds, and immature fruits might not only increase the value of citrus waste, but it will also provide a useful source of health-promoting food additives.

1.4.1 Flavonoid Compounds

1.4.1.1 Classification and Structural Characteristics of Citrus Flavonoids

Flavonoids are a class of phenolic compounds with the structure of C6–C3–C6, and are derivatives of chromone or chromane. Flavonoids are the most important active substances in citrus, and are found in fruit, bark, leaves and flowers of all species of sweet oranges, sour oranges, lemons, mandarins, and grapefruits. Based on the presence of the double bond between the C2 and C3 atoms in the carbon ring of its molecular structure, the presence of the hydroxyl group, the different connection sites of B ring in the C ring, and whether C ring is open, natural flavonoids can be divided into six categories. These are the flavones, flavanones, flavonols, isoflavones, anthocyanidins, and flavanols (or catechins) (Peterson and Dwyer 1998) (Figure 1.4). Thus far, more than 60 kinds of flavonoids have been identified in citrus, and they all belong to the first five categories listed above. Anthocyanidins, served as pigments of flowers and fruits, are very important in other plants, but are not as important in citrus—only as a substance existing in blood orange. Moreover, flavanones are found almost exclusively in citrus. Flavonoids are in the form of glycosides or aglycones, in the vast majority of varieties of citrus fruits. Those in the form of aglycones, can be divided into three categories: flavanone aglycones, flavone aglycones, and flavonol aglycones; those in the form of glycosides can be divided into two categories: Neohesperidosides and rutinosides (Macheix et al. 1990) (Figure 1.5).

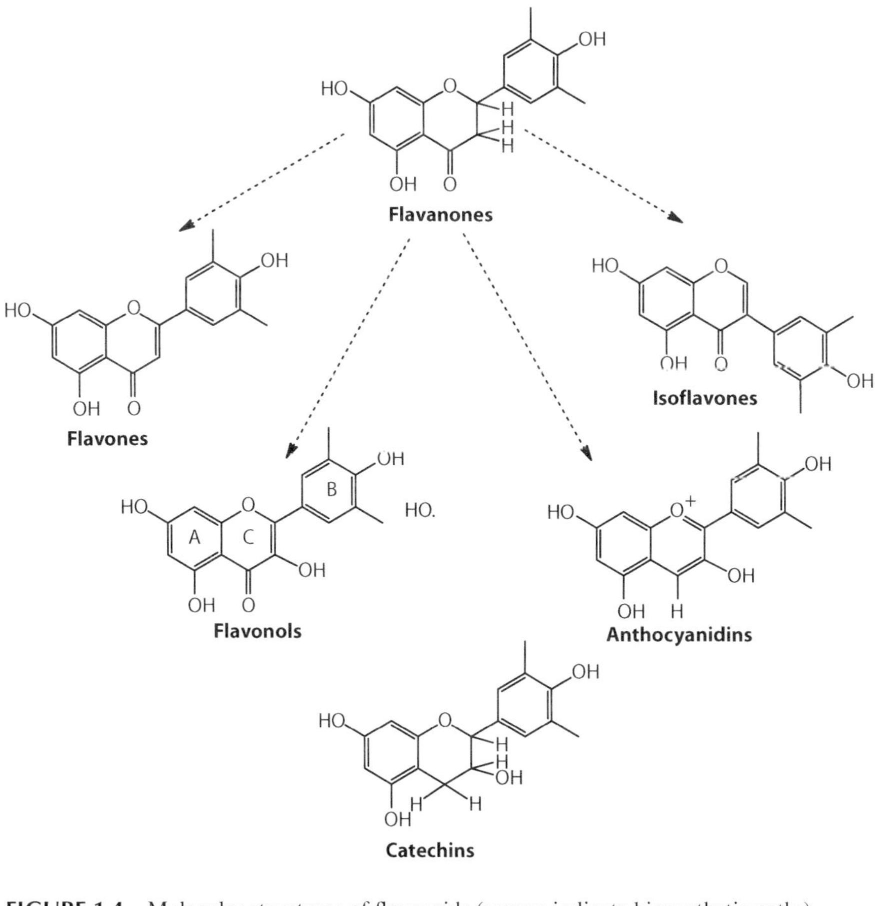

FIGURE 1.4 Molecular structures of flavonoids (arrows indicate biosynthetic paths).

Compound	Chemical name	Structural formula	Molecular formula
Flavanone aglycone forms:			
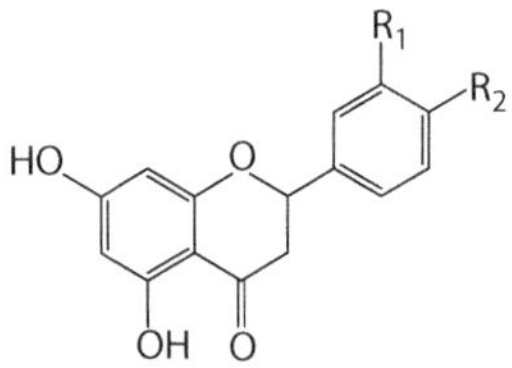			
	Naringenin	R_1=H; R_2=OH	$C_{15}H_{12}O_5$
	Hesperetin	R_1=OH; R_2=OCH_3	$C_{16}H_{14}O_6$
	Isosakuranetin	R_1=H; R_2=OCH_3	$C_{16}H_{14}O_5$
	Eriodictyol	R_1=OH; R_2=OH	$C_{15}H_{12}O_6$
Flavone aglycone forms:			
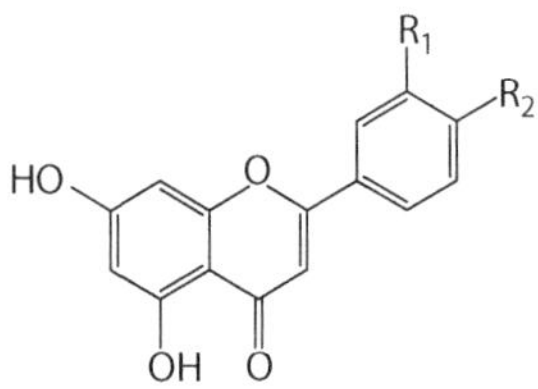			
	Apigenin	R_1=H; R_2=OH	$C_{15}H_{10}O_5$
	Luteolin	R_1=OH; R_2=OH	$C_{15}H_{10}O_6$
	Diosmetin	R_1=OH; R_2=OCH_3	$C_{16}H_{12}O_6$

FIGURE 1.5 Structural characteristics of main citrus flavonoids in the aglycone and glycoside forms. *(Continued)*

1.4.1.2 Flavanones

Among all species of citrus fruits, flavanone is the most abundant flavonoid compound, and naringenin and hesperetin are the most abundant flavanones. They are commonly found in the form of glycosides, rarely in the form of aglycones, and they can be found in various parts of citrus fruits. Based on the difference of glycoside, they can then be divided into two categories: neohesperidosides and rutinosides. Flavanone neohesperidosides are mainly containing naringin, neohesperidin, and neoeriocitrin, and so on, which have a bitter taste. Flavanone rutinosides are mainly hesperidin, narirutin, and didymin, and so on, with no bitter taste (Figure 1.5). In citrus fruits, flavanones usually exist in the form of diglycosides, which are associated with the typical flavor of citrus fruits. Peterson made a comprehensive overview and summary of the content, distribution and composition of flavanones in mandarins, sweet oranges, sour oranges, lemons, and grapefruits (Peterson et al. 2006a,b). Naringin and neohesperidin can be hydrolyzed to dihydrochalcone, which are 300

Compound	Chemical name	Structural formula	Molecular formula
Flavonol aglycone forms:			
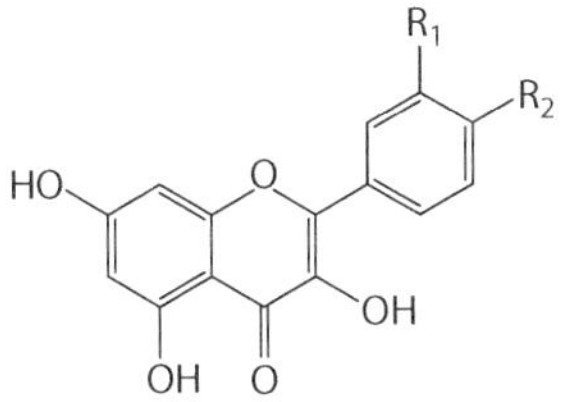			
	Quercetin	R_1=OH; R_2=OH	$C_{15}H_{10}O_7$
	Kaempferol	R_1=H; R_2=OH	$C_{15}H_{10}O_6$
Flavanone neohesperidoside forms:			
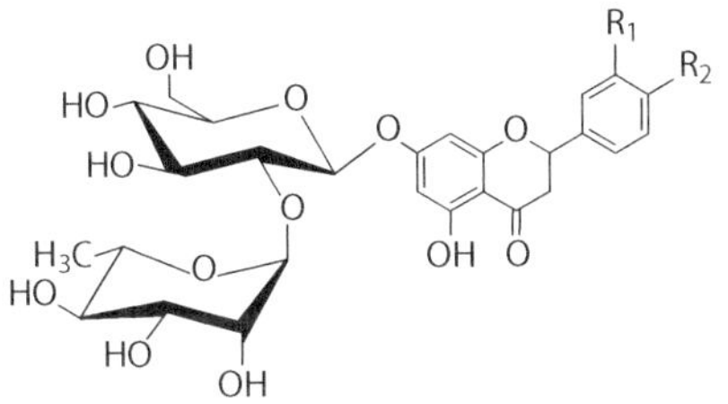			
	Naringin	R_1=H; R_2=OH	$C_{27}H_{32}O_{14}$
	Neohesperidin	R_1=OH; R_2=OCH_3	$C_{28}H_{34}O_{15}$
	Poncirin	R_1=H; R_2=OCH_3	$C_{28}H_{34}O_{14}$
	Neoeriocitrin	R_1=OH; R_2=OH	$C_{27}H_{32}O_{15}$
Flavanone rutinoside forms:			
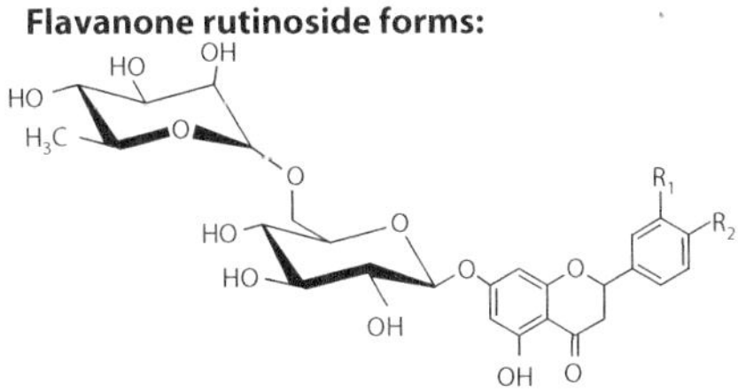			
	Narirutin	R_1=H; R_2=OH	$C_{27}H_{32}O_{14}$
	Hesperidin	R_1=OH; R_2=OCH_3	$C_{28}H_{34}O_{15}$
	Didymin	R_1=H; R_2=OCH_3	$C_{28}H_{34}O_{14}$
	Eriocitrin	R_1=OH; R_2=OH	$C_{27}H_{32}O_{15}$

FIGURE 1.5 (CONTINUED) Structural characteristics of main citrus flavonoids in the aglycone and glycoside forms.

and 1000 times sweeter than sucrose, respectively (Del Río et al. 1997; Horowitz and Gentili 1969).

1.4.1.3 Polymethoxylated Flavonoids

Like other flavonoids, the parent nucleus of polymethoxylated flavones (PMFs) is a 2-benzene chromone with a skeleton of C6–C3–C6, but the benzene rings, especially the A ring, are replaced by a lot of methoxy, (sometimes hydroxy) groups, having a weaker polarity compared to flavanone glycosides. Polymethoxylated flavones are widely found in various species of citrus fruits, and almost only found in the citrus genus (Ye et al. 2005; Tang et al. 2006). In addition, each species is unique, although the content is small. It has many special effects and can be used as a commercial maker for citrus juice adulteration (Marini and Balestrieri 1995; Ooghe and Detavernier 1997) and in identification of citrus classification. Flavonoids are usually detected by HPLC-MS. Mouly et al. (1998) divided these compounds into two groups based on the different elution time: flavanone glycoside is the first eluted compound and polymethoxylated flavones are the last, which are less polar (PMFs). More than 20 kinds of flavonoids have been isolated and identified from citrus plants, among which the most common ones are nobiletin, tangeretin, and sinensetin; their chemical structures are shown in Figure 1.6.

1.4.1.4 Distribution and Determination of Flavonoids in the *Citrus* Genus

Flavonoids are widely found in citrus plants. Present in the largest amounts in citrus fruits we find neohesperidin and naringin in flavanone neohesperidoside forms, narirutin and hesperidin in flavanone rutinoside forms, and naringin and

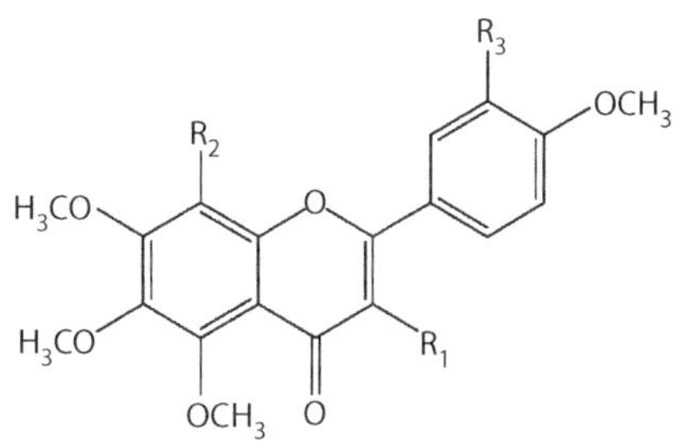

Polymethoxylated flavonoids	R1	R2	R3
Scutellarein	H	H	H
Sinensetin	H	H	OCH_3
Tangeretin	H	OCH_3	H
Hexamethoxyflavone	OCH_3	H	OCH_3
Nobiletin	H	OCH_3	OCH_3
Heptamethoxyflavone	OCH_3	OCH_3	OCH_3

FIGURE 1.6 Structural characteristics of common citrus polymethoxylated flavones.

polymethoxylated flavones in flavone aglycone forms—Nobiletin, Tangeretin, and Sinensetin. Flavanones with 7-*O*-glycosyl are the most abundant flavonoid compounds in all citrus fruits (Benavente-García et al. 1995). For example, lemon peels are rich in glycosidic flavonoids (Park et al. 1983). The distribution of flavonoids varies in the different varieties of citrus fruits. Naringin, neohesperidin and poncirin in flavanone neohesperidoside forms are mainly found in bergamot, grapefruit, and bitter orange juice, while hesperidin, narirutin, and didymin, in flavanone rutinoside forms, are mainly found in bergamot, oranges, tangerines, and lemon juice (Horowitz 1986). Bitter orange is an important source of neohesperidin and naringin. These compounds may be used to produce sweeteners. Lemons are rich in neoeriocitrin and hesperidin. The predominant flavonoid compounds in pummelo are naringin, hesperidin, neohesperidin, narirutin, and other flavanone glycosides, among which, naringin accounts for more than 80% of the total content (Russell et al. 1987).

The composition of flavonoids differs across different fruit tissues, and as such flavanone glycosides have a different composition whether in the seeds and peels, or in the juice. For example, in lemon peels and seeds, as well as *Citrus reticulata* seeds, there exists naringin; however, it is not present in their juice (Ooghe and Detavernier 1997). On the other hand, the compositions of flavanone compounds are different in citrus seeds and peels. For example, lemon seeds mainly contain neoeriocitrin and hesperidin, while peels are rich in neoeriocitrin, naringin, and neohesperidin (Tripoli et al. 2007). Also flavanone glycoside contents are quite different. In lemon peels, neoeriocitrin and naringin have similar content, but in lemon seeds, the content of neoeriocitrin is 40 times higher than that of naringin (Bocco et al. 1998).

Kelebek et al. (2008) determined, using HPLC-DAD, the content of five flavanones in two blood orange varieties (Moro and Sanguinello) and found that they mainly contained narirutin, naringin, hesperidin, neohesperidin, and didymin, and that Moro juice had a higher content of these five compounds than Sanguinello juice. Wang et al. (2007) determined the content of three flavanones (naringin, hesperidin, and neohesperidin), two flavone aglycones (diosmetin and luteolin), two flavonol aglycones (quercetin and kaempferol), one flavonol glycosides (rutin), and one polymethoxylated flavone (sinensetin) in the edible parts of eight citrus fruits from Taiwan using the method of RP-HPLC, and documented the differences in flavonoid composition and content between the peels and edible parts.

He et al. (1997) used HPLC-MS to isolate and identify eight flavonoids in *Citrus aurantium* extract, including isonaringin, naringin, hesperidin, neohesperidin, naringenin, nobiletin, and naringenin. Ortuno et al. (1997) used the RP-HPLC method to compare these flavonoids and found that citrus (including *Citrus aurantium*) had the highest content of hesperidin, naringin, and neohesperidin. The most important flavonoids components are naringin, rhoifolin, poncirin, and neohesperidin in the epicarp of the immature fruit of *Citrus grandis* (exocarpium citri grandis [Hua Ju Hong] or Rutaceae citrus *Citrus grandis* "Tomentosa") (Su and Lin 2001; Lei et al. 2000). Weber et al. (2006) used a combination of HPLC-MS (LC-MS) and HPLC-NMR-PDA (MS liquid NMR) methods to isolate and identify polymethoxylated flavones in molecular distillation cold pressed orange peel oil, and they detected sinensetin, nobiletin, tangeretin, quercetogetin, heptamethoxyflavone, and others. Wang and Chen (1995) used HPLC to determine the naringin content in *Citrus aurantium* L. and also compared the naringin content in different parts of the fruits and noted the changes of the naringin contents during different harvest seasons.

1.4.2 Phenolic Acids

1.4.2.1 Types and Structural Characteristics of Phenolic Acids

Phenolic acids are a class of substances characterized by a benzene ring with a hydroxyl group; they account for about one-third of plant-derived phenolic compounds in foods. According to their structure, they can be divided into hydroxybenzoic and hydroxycinnamic acids. Common hydroxylated cinnamic acid derivatives include coumaric acid, caffeic acid, and ferulic acid, which are usually esterified with quinic acid or glucose. The hydroxylated benzoic acid derivatives of the most common phenolic acids are hydroxybenzoic acid, vanillic acid, protocatechuic, and so on, mainly in the glycoside form (Herrmann 1989). The structures and classification of common phenolic acids are shown in Figure 1.7. Citrus phenolic acids mainly exist in free and bound forms. Bound phenolic acids are typically bonded with an ester bond or an ether bond with a number of compounds (Xu et al. 2007). Phenolic acids, as a potential protective factor against cancer and heart disease, have received much attention in recent years partly because of their potential antioxidant and anti-inflammatory activity, their effects on chronic diseases, and their widespread

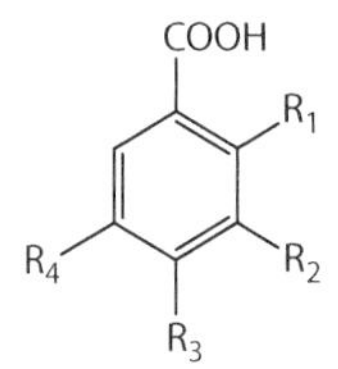

	R_1	R_2	R_3	R_4
Benzoric acid	H	H	H	H
Salicylic acid	OH	H	H	H
Gentisic acid	OH	H	H	OH
Vanillic acid	H	OH	OMe	H
Gallic acid	H	OH	OH	OH
Syringic acid	H	OMe	OH	OMe

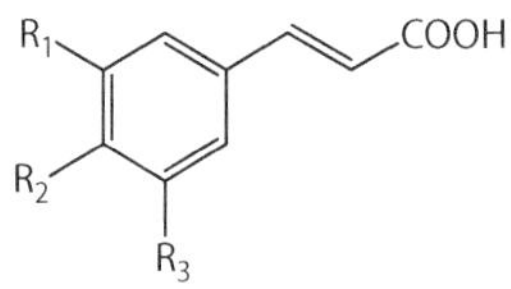

	R_1	R_2	R_3
Cinnamic acid	H	H	H
Coumaric acid	H	OH	H
Caffeic acid	OH	OH	H
Ferulic acid	OMe	OH	H
Sinapic acid	OMe	OH	OMe

FIGURE 1.7 Molecular structures of phenolic acids.

existence in plant-derived foods (Morton et al. 2000; Mattila and Kumpulainen 2002; Lodovici et al. 2001).

1.4.2.2 Distribution and Determination of Phenolic Acids in Citrus Fruits

Kelebek et al. (2008) used HPLC-DAD to determine the content of two kinds of benzoic phenolic acids (gallic acid and protocatechuic acid) and five cinnamic acids (caffeic acid, chlorogenic acid, *p*-coumaric acid, ferulic acid, and sinapic acid) in two blood oranges, Moro and Sanguinello. The results showed that Moro juice has a total of 9.25 mg/L of benzoic phenolic acids and 74.35 mg/L of cinnamic phenolic acids, which is more than in Sanguinello juice. Gorinstein et al. (2004) conducted a study on the composition and content of phenolic acids in the pulp and peel of Jaffa White and its variant, Jaffa Sweeties. Experimental results showed that the content of gallic acid, protocatechuic acid, 4-hydroxybenzoic acid, vanillic acid, *p*-coumaric acid, salicylic acid, ferulic acid, anisic acid, and erucic acid in the peel of Jaffa White and its variant is significantly higher than that in the pulp. The total content of these four hydroxylated cinnamic acids (caffeic acid, *p*-coumaric acid, ferulic acid, and erucic acid) is higher in Jaffa White pulp (362 nmol/g) and peel (1513 nmol/g), compared to the content in its variant (272 and 1277 nmol/g, respectively; Figure 1.9). Bocco et al. (1998) determined the content of esterified phenolic acids (four kinds of cinnamic acid: caffeic acid, *p*-coumaric acid, ferulic acid, and erucic acid) in citrus fruit peel and seeds of *Citrus reticulata*, *Citrus sinensis* (L.) Osbeck, lemon, *Citrus maxima*, and *Citrus aurantium* L. originated from South Africa and Italy. They found a significantly higher content of phenolic acids in the seeds than in the peel. Rapisarda et al. (2003) conducted a study on the change, of four cinnamic phenolic acids (caffeic acid, *p*-coumaric acid, ferulic acid, and erucic acid) present in the citrus juices of blood oranges, tangerines and their hybrids; with different maturity they found that the trend in change has a significant relationship with citrus varieties. Peleg et al. (1991) also reported the composition and content of cinnamic phenolic acids of oranges and grapefruits at different stages of maturity. They found that with the increase in maturity, the content of erucic acid in orange juices increased, while the content of other three phenolic acids dropped.

1.4.3 Limonoid Substances

1.4.3.1 Types and Structural Characteristics of Limonoids

Limonoids are also called limonins, and they are a class of highly oxidized secondary metabolites of tetracyclic triterpenoids found in plants, mainly present in the rutaceae and meliaceae families. Citrus limonin compound is one of the main bitter flavor components (Cai and Hashinaga 1996). So far, about 300 kinds of lemon bitter analogues have been isolated. About 38 kinds of aglycone limonoids and 20 kinds of limonoid glycosides have been isolated from citrus plants (Sawabe et al. 1999; Saipetch et al. 2004).

In citrus plants, limonoid compounds exist in two forms: one form is limonoids aglycone, the structural features of which are that, on D ring C-17 is connected with a pyran ring, C-3, C-4, C-7, C-16, and C-17 are connected with the oxygen-containing functional groups. Except for deoxy limonin, epoxy structures exist on the site of both C-14 and C-15 (Liu et al. 2007). Aglycone limonoid compounds can be divided into two parts: limonene A ring lactone and limonene D ring lactone. Complete fruits contain large amounts of limonene A ring lactone (LARL) (without bitterness), which under acidic conditions are soon transformed into limonene D ring lactone (with bitterness). This reaction is accelerated by the presence of limonene D ring lactone hydrolase, which is the reason why citrus juice produces a bitter aftertaste (Hasegawa et al. 1991). The main citrus aglycones have limonin, nomilin, and knock Obama ketone (structures shown in Figure 1.8).

The other form is limonoid glycoside without bitterness (Ohta et al. 1992) and are constituted of a pentose in form of glycosidic bond synthetic by limonoids aglycone molecule D ring after ring opening at the position C-17, which experienced a series of very complex biochemical processes in the plant body. At present, only 19 kinds of glycosides have been identified in the *Citrus* genus. The common types of limonoid glycosides are: limonin 17-β-glucopyranoiside, nomilin acid 17-β-D-glucopyranoside, ichchangin 17-β-D-glucopyranoiside, isolimonic acid 17-β-D-glucopyranoside, deactylnomilic acid 17-β-D-glucopyranosied, obacunone 17-β-D-glucopynoside, and so on. Figure 1.9 shows the structure of limonin glucoside and nomilin glucoside (Bennett et al. 1989).

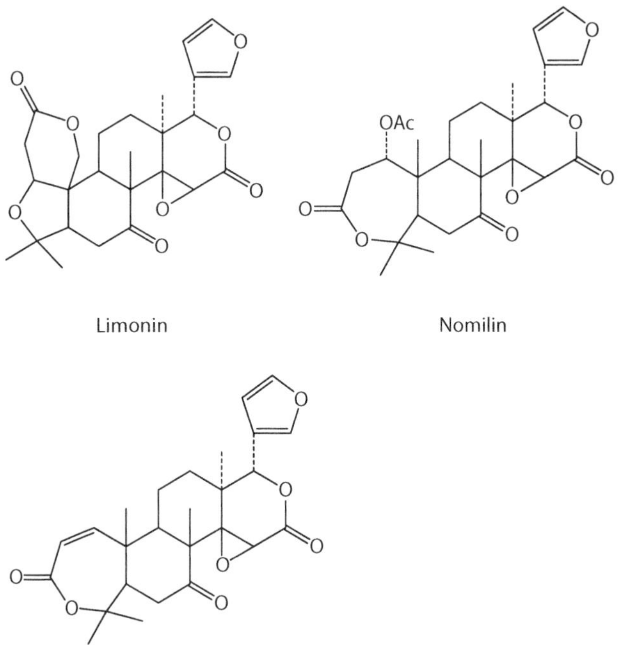

FIGURE 1.8 Structures of the main aglycone of limonoids.

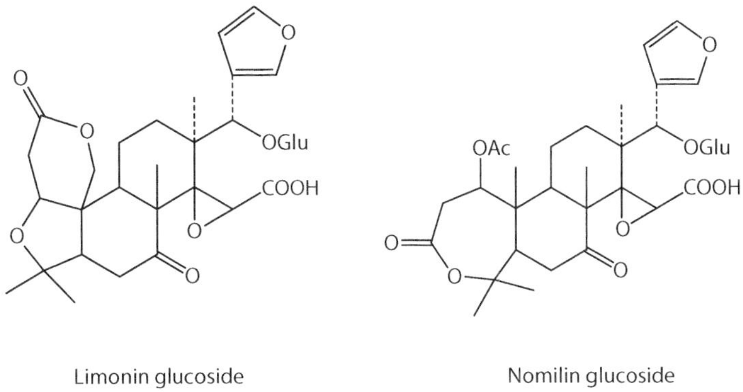

FIGURE 1.9 Structures of limonin and nomilin glucosides.

1.4.3.2 Distribution and Determination of Limonoids Glycoside Compounds in *Citrus* Genus Plants

Limonoids are widely distributed in a variety of species in the *Citrus* genus, and the total fruit content and composition of limonoids differs between varieties. Ohta and Hasegawa (1995) conducted a study on the composition of limonoids on 16 species of mature pomelo fruits and found that pummelo juice contains 18 ppm limonin and 29 ppm limonin glycoside (LG). Seeds contained limonene, nomilin, obama knock one, and a trace amount of deacetylation nomilin and its glycosides. In seeds, the content of limonoid aglycone is between 773 and 9900 ppm, and its glycoside has a content of 130 to 1912 ppm. Chen et al. (2006) used HPLC method to determine limonin and naringin in citrus juice from three different places, and compared the content differences.

Different parts of citrus fruits have different limonoids content. Seeds have the highest, followed by peels, while pulp has the lowest (Luo and Pan 2008); and with the content of limonene, nomilin, obama knock ketone from high to low (Hashinaga et al. 1983). But there are exceptions, such as citrons where nomilin content is greatest in all parts except the seeds, followed by limonin, which suggests citron bitterness is mainly due to nomilin content (Cai et al. 1992, 1993). In addition, grapefruit seeds are a major source of limonoid substances, in a proportion of about 1.5% of the fresh weight of seeds (Brano et al. 2000). Sun et al. (2004) determined the content of limonin and nomilin in different tissues of ripe Zhoushan Gaoxie pomelo fruits. In the epicarp, no limonin and nomilin were detected, while the coat had a higher content of limonin and nomilin than the peel. Zeng et al. (2003) determined the content of limonoids and limonin from different parts (peel, seed, juice) of different varieties of citrus, and found that limonoid content in the seed was higher than in peels, but lower in the juice. Limonoids content in seeds is highest in Guanximiyou (208.4 mg/kg), followed by Dengken Grapfruit (182.4 mg/kg), Nanchung sweet oranges (165.2 mg/kg), Jin oranges (34.2 mg/kg), and finally, Da-Huang-Bao red oranges (26.2 mg/kg). Moreover, limonins content in citrus fruits is also related to fruit maturity. For Huyou, pomelo, Wenzhoumigan, and Penggan, during growth and maturation, the contents of Limonin and Nomilin increase from May to September, and diminish afterwards until the end of October when the contents stabilize at a lower level (Sun et al. 2005).

1.4.4 Alkaloids

1.4.4.1 Structural Characteristics of the Main Alkaloids in Citrus

Alkaloids (alkaloids) generally refer to alkaline nitrogenous compounds found in plants, most of which are heterocyclic with optical rotation and obvious physiological effect. Alkaloids in citrus include octopamine, synephrine, tyramine, *N*-methyltyramine, hordenine, and other phenethylamine alkaloids. Among these, synephrine is the most important (Fumiyo et al. 1992), and its structure is as shown in Figure 1.10. It is reported that synephrine can affect the body's metabolism, help obese people lose weight by a mechanism in which stimulated lipolysis leads to a metabolic rate and fat oxidation acceleration (Carpéné et al. 1999; Tang et al. 2006). Synephrine and *N*-methyltyramine are two identified essential ingredients in *Citrus aurantium*, which strengthen the heart, increase cardiac output, contract blood vessels and enhance total peripheral vascular resistance to raise left ventricular pressure and arterial blood pressure, and can be used for the treatment of bronchial asthma and anesthesia hypotension, orthostatic hypotension, shock, and collapse after surgery.

1.4.4.2 Distribution and Determination of Alkaloids in Citrus

Synephrine alkaloids are widely distributed in many plants belonging to the Rutaceae citrus family, with various contents in each variety, which decrease with fruit maturity. In China, citrus plant alkaloid research is mainly aimed at young citrus fruits and traditional Chinese herbal medicine *Citrus aurantium* and its alkaloid

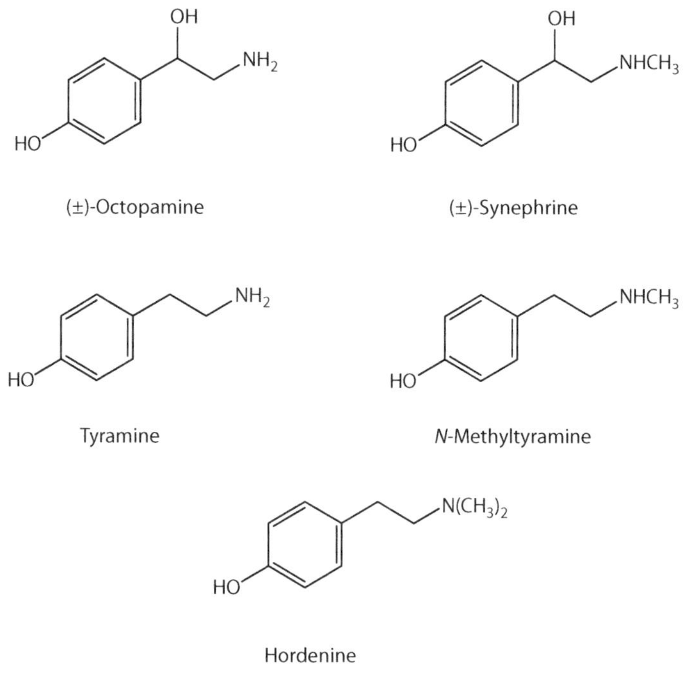

FIGURE 1.10 Chemical structures of phenethylamine alkaloids isolated from citrus.

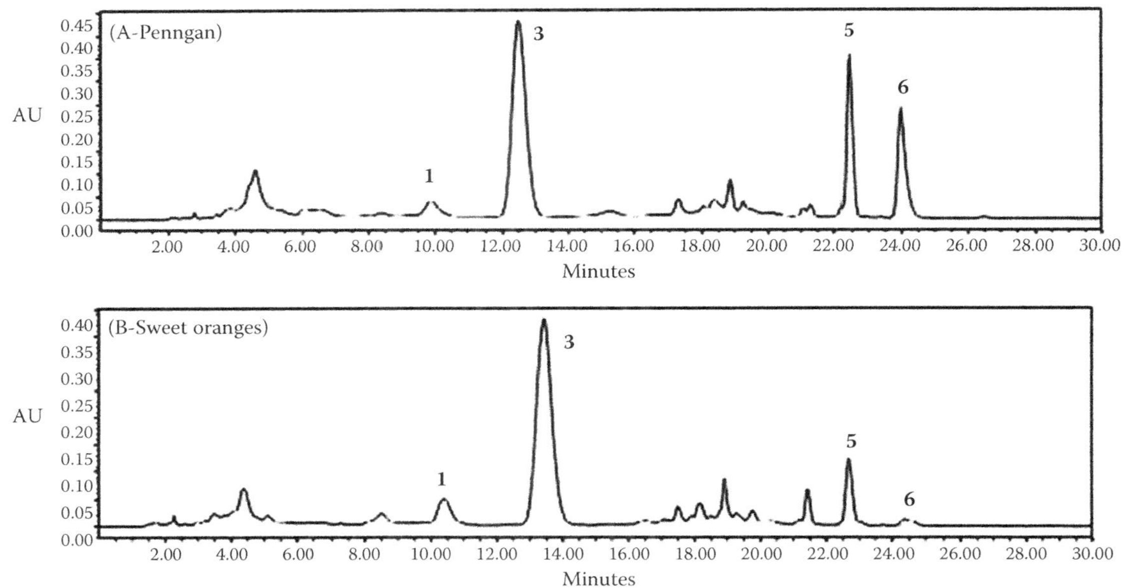

FIGURE 1.11 HPLC profile of the PGs and PMFs of the sample at 330 nm (*1*: narirutin, *2*: naringin, *3*: hesperidin, *4*: neohesperidin, *5*: nobiletin, *6*: tangeretin). (*Continued*)

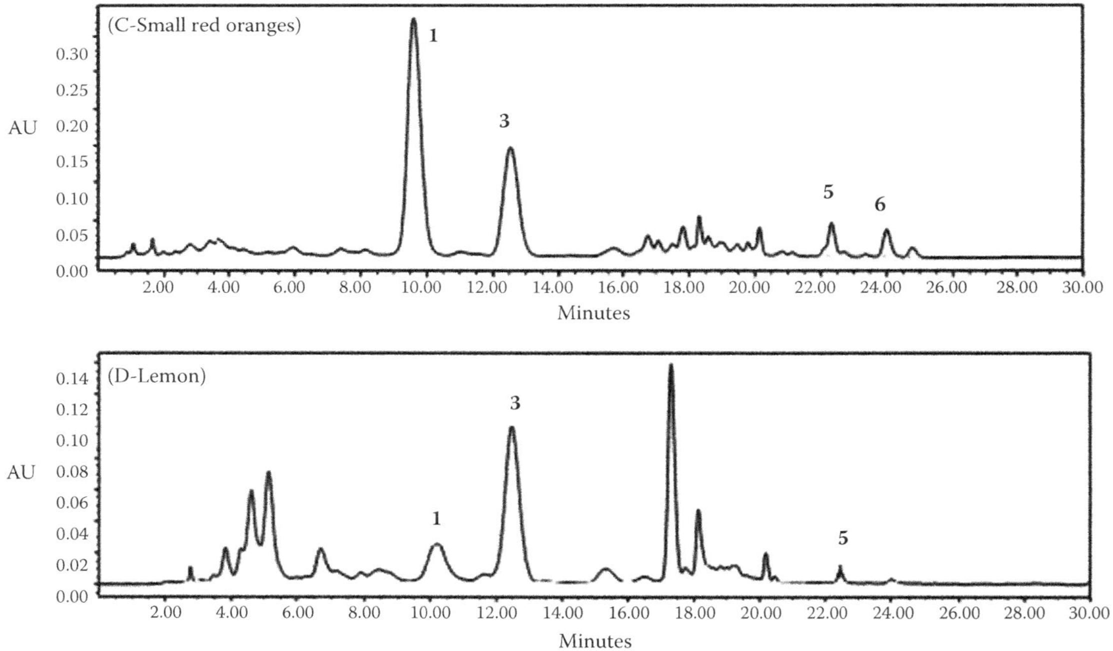

FIGURE 1.11 (CONTINUED) HPLC profile of the PGs and PMFs of the sample at 330 nm (*1*: narirutin, *2*: naringin, *3*: hesperidin, *4*: neohesperidin, *5*: nobiletin, *6*: tangeretin). (*Continued*)

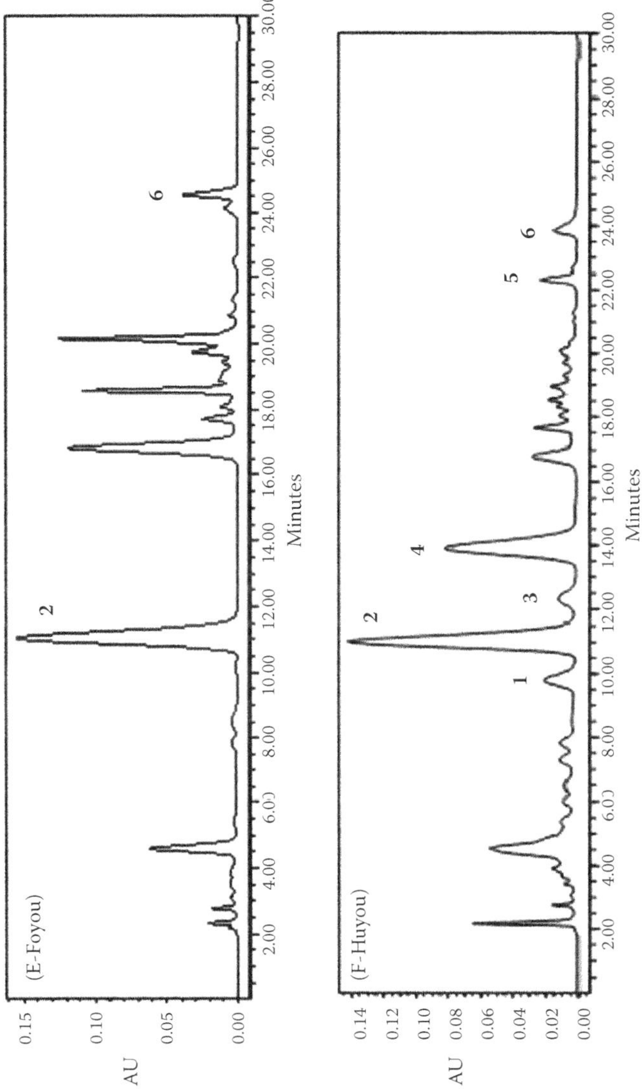

FIGURE 1.11 (CONTINUED) HPLC profile of the PGs and PMFs of the sample at 330 nm (*1*: narirutin, *2*: naringin, *3*: hesperidin, *4*: neohesperidin, *5*: nobiletin, *6*: tangeretin). *(Continued)*

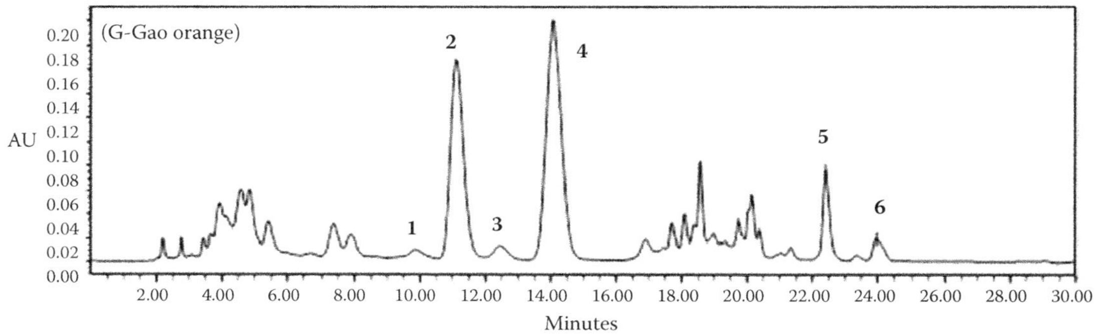

FIGURE 1.11 (CONTINUED) HPLC profile of the PGs and PMFs of the sample at 330 nm (*1*: narirutin, *2*: naringin, *3*: hesperidin, *4*: neohesperidin, *5*: nobiletin, *6*: tangeretin). (From Ye, X. Q., and D. H. Liu, *Citrus Processing and Utilization*, Beijing, Light Industry Publishing House, 2005.)

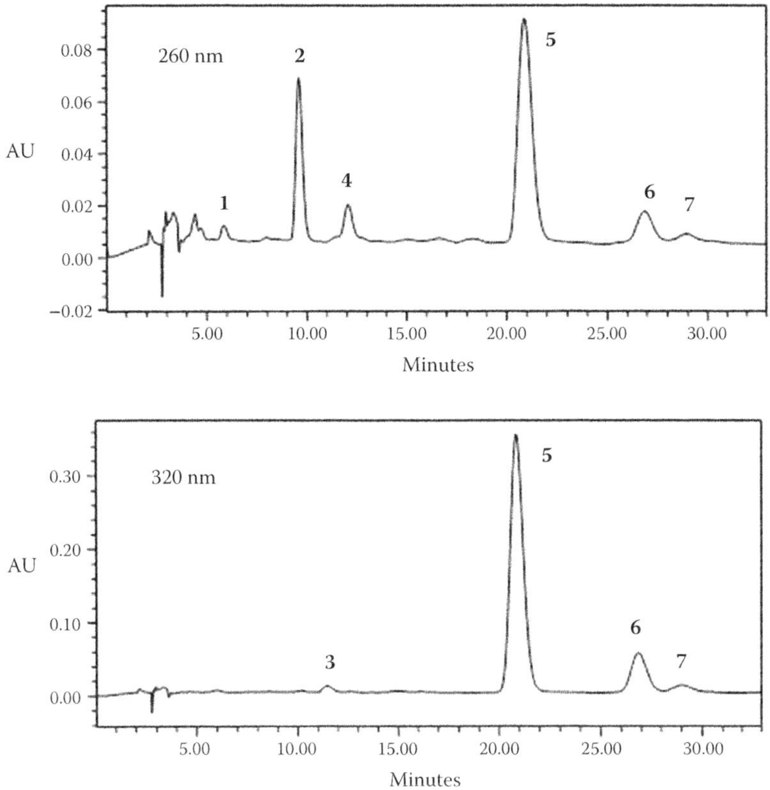

FIGURE 1.12 HPLC graphs of the phenolic acids of young lemon fruits at 260 nm and 320 nm (*1*: protochatechuic; *2*: *p*-hydroxybenzoic; *3*: caffeic; *4*: vanillic; *5*: *p*-coumaric; *6*: ferulic; *7*: sinapic). (From Ye, X. Q., and D. H. Liu, *Citrus Processing and Utilization*, Beijing, Light Industry Publishing House, 2005.)

types, content and their physiological functions. Pellati and Benvenuti (2002) used RP-HPLC to determine the content of *N*-methyltyramine, synephrine, and tyramine in fresh *Citrus aurantium*, *Citrus aurantium* dry goods and a commercial *Citrus aurantium* extract, and at the same time, performed quantitative separation of DL-synephrine in all *Citrus aurantium* samples using chiral columns with beta cyclodextrin as the stationary phase. The researchers found that D-synephrine content in dry *Citrus aurantium* samples was significantly higher than that of L-synephrine. Li et al. (2004) determined the synephrine and *N*-methyltyramine content of Qingpi and Sihua Qingpi from three and four places of origin respectively, and their different growth periods. Some chemical compositions by HPLC profiles of the PGs and PMFs of the orange samples, and some phenolic acids from lemon young fruits are shown in Figures 1.11 and 1.12, respectively (Ye and Lui 2005).

Chen (2005) summed up the synephrine and *N*-methyltyramine content of *Citrus aurantium* from 21 different places of origin. As shown in the results by species, the content of synephrine in *Citrus aurantium* from highest to lowest was: *Citrus reticulate* > sweet orange > sour orange > *Citrus sinensis* > citron (Chen 2005). Different storage periods also had an impact on synephrine content. Zhou and Gui (1997) conducted a study of bran fried *Citrus aurantium* quality during different storage periods, and found that along with the extended storage, bran fried *Citrus aurantium* synephrine levels decreased significantly. Synephrine decreased by 56.7% when comparing a 4-year sample with one that had not been stored at all.

1.4.5 Research Progress of Antioxidant Ability of Some Bioactive Components on Citrus Fruits

1.4.5.1 Antioxidant Ability of Flavonoids in Citrus Fruits

The antioxidant property of flavonoids (such as naringin, and so on) is mainly caused by the removal of peroxide, hydroxyl radical scavenging, and the antilipid peroxidation (LPO) activity (Mohamad-Reza et al. 2003). Wilmsen et al. (2005) used the DPPH radical scavenging test and reported the antioxidant activity of hesperidin, compared with that of Trolox, and found they are at the same level of antioxidant activity. Yu et al. (2005) used the β-carotene bleaching method, DPPH free radical scavenging experiment, superoxide free radical scavenging experiment, and the LDL oxidation experiment to evaluate the antioxidant activity of limonoids, flavonoids (flavanone aglycone, flavanone glycoside, flavonols) and coumarin. They found that the majority of flavonoids had strong antioxidant activities, while limonoids and coumarin, which have less hydroxyl groups, had relatively weak antioxidant activities. Majo et al. (2005) studied the antioxidant activity of nine flavanones (aglycone, and glucoside, and so on), and found that antioxidant ability of flavanone was related to the sites, number and *O*-methylated degree of free hydroxyl groups in the flavanone molecule.

Based on the hydroxyl sites and the total number of hydroxyl group antioxidant mechanisms, flavanone aglycones' *O*-methylation has nearly no impact on antioxidant ability; however, as for flavanone glycosides, *O*-methylation can decrease antioxidant ability significantly. Kim and Lee (2004) also found that the number of free hydroxyl group and the antioxidant ability of flavonoids was significantly positively correlated. The antioxidant ability of polyphenol is generally stronger than that of monophenol, but glycosylation reduces the flavanones antioxidant ability. In recent years, there have been many papers reporting flavanones and their glycosides antioxidant ability, but antioxidant capacity of polymethoxylated flavonoid *in vitro* has rarely been reported.

1.4.5.2 Antioxidant Ability of Phenolic Acids in Citrus Fruits

Literature on the evaluation of the antioxidant ability of phenolic acids stems mostly from *in vitro* studies. Kikuzaki et al. (2002) used multiple antioxidant methods and determined the antioxidant ability of 30 kinds of phenolic acids. The DPPH free radical scavenging ability was, from highest to lowest: caffeic acid > sinapic acid > ferulic acid > ferulic acid ester > coumaric acid. The results also showed that under the system of linoleic acid, esterification enhances the antioxidant ability of ferulic acid. According to reports, cinnamic acid can effectively prevent the oxidation of low-density lipoprotein (LDL), and the antioxidant ability is, from strongest to lowest: caffeic acid > ferulic acid > coumaric acid (Castelluccio 1996). Gulcin (2006) conducted a comprehensive assessment of the *in vitro* antioxidant capacity of caffeic acid. The methods used were: ABTS radical scavenging experiment, DPPH radical scavenging test, reducing power test, ferric thiocynanate method (FTC), superoxide anion radical scavenging experiments, and experimental metal chelate; the results showed that caffeic acid has a good antioxidant activity *in vitro.*

1.4.5.3 Antioxidant Ability of Limonin in Citrus Fruits

Opinions on the antioxidant ability of limonoids substances are also different. Sun et al. (2005) conducted a study using β-carotene bleaching assay and showed that the antioxidant ability of limonin and nomilin varied across different tissues and citrus varieties, and that the antioxidant ability of limonin and nomilin was about 2.9 to 8.3 times of that of ascorbic acid. Yu et al. (2005) evaluated the antioxidant ability of 11 active substances in citrus and found that, at 10 μM concentration, the inhibition rate of limonin glycoside to β-carotene-linoleic acid system free radical is less than 7%, while the inhibition rate of flavonoids could be as high as 51.3%. In the DPPH system, the free radical scavenging rate of limonene and limonene glycoside was only 0.5% and 0.25% respectively, which was significantly lower than rutin, baicalin, and other flavonoids. The peroxide radical scavenging rate of limonene, limonene glycosides, and coumarin was 2.5–10%, while flavonoids had the highest peroxide radical scavenging rate at 64.1%. Breksa and Manners (2006) used four *in vitro* antioxidation systems, ORAC (the oxygen radical absorbance capacity), TEAC (Trolox equivalent antioxidant capacity), DPPH and β-carotene-linoleic acid bleaching to study the antioxidant activity of limonin and nomolin, and claimed that pure limonin and nomilin had no antioxidant capacity, because they did not have the basic molecular structure to exhibit an antioxidant ability.

1.5 CITRUS FRUIT PROCESSING AND UTILIZATION

1.5.1 Citrus Product Development

Due to its complex structure, citrus is one of the most difficult fruits to process. Before the breakthrough technology of citrus juice in the 1950s, there was not much processing of citrus juice. However, after the 1950s, there was a rapid increase in the production of fruit juice, such that citrus juice became the first major juice considered to be a daily fruit juice drink. The United States and Brazil are the main countries processing citrus juice. Brazil began mass production since the 1960s, and has since maintained its position as the top producer in the world.

Large quantities of mandarins and tangerines are processed in Asia and the Mediterranean region. From the perspective of processing characteristics, mandarins are suitable for processing into segments. Spain is an important manufacturer of canned products in Europe. Japan used to be the main producer and exporter of canned citrus, and the major country for importations; however, since the middle of the 1980s, it has gradually been replaced by China. Since the 1960s, canned citrus segments in light syrup are still one of the main agricultural exports from China.

From a technical point of view, the citrus processing industrialization process has been helped by some significant advances in technology. For example, as for orange juice processing, special juice extractors were invented by FMC Corporation in 1946.

Through decades of efforts, new technologies have been invented and modified to ensure the quality of citrus juice, such as finding of the basic cause of orange juice bitterness and the method to eliminate it (prevention). Also, inventing the highly efficient concentrating machine to make frozen concentrated orange juice (FCOJ), which has become one of the main commodities in world trade.

As a large-scale industrial process, the peeling, segmenting, de-segmenting of membrane using chemical methods during canned citrus processing can be a mechanized operation. At the end of the 1980s, a large number of applications of low temperature sterilization technologies were applied and greatly improved the quality of segments, and especially the preservation of texture. A large number of applications of commercial sterilization technologies in the canning industry, used as an alternative to traditional heat insulation, have promoted the improvement of the quality of canned mandarin oranges (Shan 2016).

The comprehensive utilization of citrus was traditionally mostly for pectin production and essential oil production. In the 1960s, commercial production of beverages appeared as well as other fruit peel products with essential oils as traditional products. For example, limonene, gradually showing its advantages, has now become a stable natural organic compound.

There have been a large number of handbooks on the chemistry and utilization of citrus. In 1977, *Citrus Science and Technology*, was published, which includes relevant documents and reports. *Citrus Processing: Quality Control and Technology,* was published in 1991 (Kimball 1991); *Citrus Processing—A Complete Guide*, and the *Handbook of Citrus By-Products and Processing Technology*, were published in 1999; *Citrus Limonoids: Functional Chemicals in Agriculture and Foods* was published in 2000 (Berhow et al. 2000). The publication of these books has promoted the comprehensive approach to citrus processing (Braddock 1999; Ye et al. 2005).

1.5.2 The Current Situation of the Citrus Processing Industry

According to the U.S. Department of Agriculture's incomplete statistics, the world's citrus production was 88 million tons in 2016, among which, oranges account for 46 million tons, tangerines for 29 million tons, grapefruits and pomelos for 6277 thousand tons, lemons and limes for 7 million tons; the processed proportion for each of these are 37.8%, 10.7%, 29.9%, 5.2%, respectively. From the perspective of species, oranges are the most processed, followed by lemons and limes, grapefruits, and mandarins, while tangerines are the least processed. As for oranges, Brazil, the United States, the European Union, Mexico, and China have the highest processing activities, with rates of 65.4%, 64.2%, 22.23%, 4.6%, and 85.7% of total harvest yields, respectively. Grapefruit is mainly grown for fresh consumption, and the main processing countries are the United States, South Africa, Israel, Mexico, and the European Union. In 2016, annual processing reached only (in thousands of tons) 324, 137, 110, 86, and 17, respectively, with processing rates of 44.1%, 41.5%, 59.5%, 20.2%, and 17.5%, respectively. The main countries processing lemons are Argentina, Mexico, the United States, the European Union, and South Africa, and the 2016 annual processed amounts reached (in thousand of tons) 1,200, 350, 245, 181, and 76, respectively, with processing rates of 80%, 15.4%, 28.9%, 14.6%, and

22.9%, respectively. Mandarins and tangerines are mainly processed in China, the European Union, the United States, Argentina, and South Korea, and the 2016 annual production reached (in thousands of tons) 660, 328, 135, 110, and 95, respectively, with processing rate of 3,3%, 10.8%, 15.4%, 31.4%, and 13.3%, respectively. As for lemons, the main processing countries are Spain, the United States, Italy, and Turkey (USDA 2016).

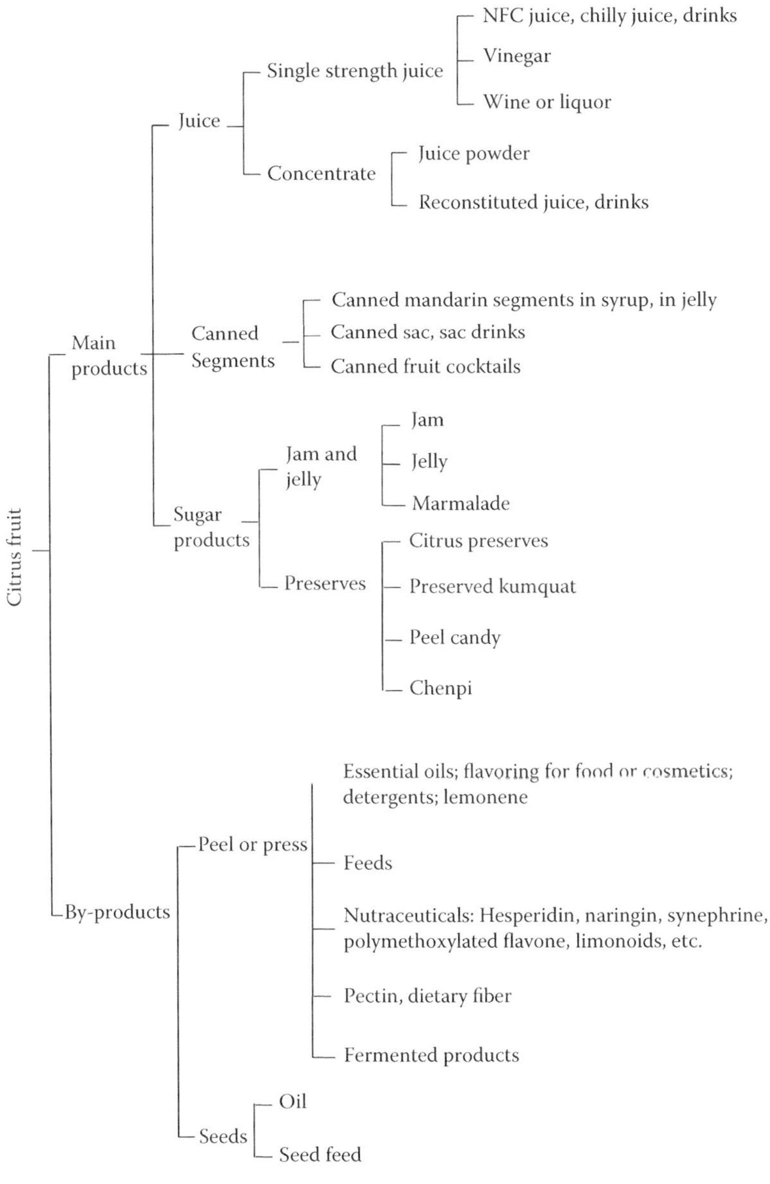

FIGURE 1.13 Citrus fruit utilization and their products.

1.5.3 Citrus By-Product Utilization

Juice and their products used to be processed from orange and canned segments with light syrup from mandarin and tangerine. The pulp, peels from industry, and even the young dropped fruits from orchards are used as raw materials for the extraction of nutraceuticals. There are more than a thousand types of processed citrus by-products using citrus as the basic raw material. The main commercial products are shown in Table 1.2 and Figure 1.13.

ACKNOWLEDGMENT

Financial support from the China Nature Sciences Foundation (Project C200501) is acknowledged.

REFERENCES

Abeysinghe, D. C., X. Li, C. Sun, W. Zhang, C. Zhou, and K. Chen. 2007. Bioactive compounds and antioxidant capacities in different edible tissues of citrus fruit of four species. *Food Chemistry* 104(4):1338–1344.

Benavente-García, O., J. Castillo, F. Sabater, and J. A. Del Rio. 1995. Characterisation of S-adenosyl-L-methionine: Eriodictyol 40-O-methyltransferase from *Citrus aurantium*. Developmental changes in the levels of 40-O-methoxyflavonoids and S-adenosyl-L-methionine: Eriodictyol 40-O-methyltransferase activity. *Plant Physiology and Biochemistry* 33:227–263.

Bennett, R. D., S. Hasegawa, and Z. Herman. 1989. Glucosides of acidic limonoids in *Citrus*. *Phytochemistry* 28(10):2777–2781.

Berhow, M. A., S. Hasegawa, and G. D. Manners. (Eds.). 2000. *Citrus Limonoids: Functional Chemicals in Agriculture and Food*. American Chemical Society.

Bocco, A., M. E. Cuvelier, H. Richard, and C. Berset. 1998. Antioxidant activity and phenolic composition of citrus peel and seed extracts. *Journal of Agricultural and Food Chemistry* 46(6):2123–2129.

Braddock, R. J. 1999. *Handbook of Citrus By-products Processing Technology*. John Wiley.

Brano, T. 2000. Grapefruit: The last decade acquisitions. *Fitolerapia* 71:29–37.

Breksa, A. P., and G. D. Manners. 2006. Evaluation of the antioxidant capacity of limonin, nomilin, and limonin glucoside. *Journal of Agricultural and Food Chemistry* 54(11):3827–3831.

Cai, H., and F. Hashinaga. 1996. The prospect and current studies on the limonoids in citrus. *Acta botanica sinica* 38(4):328–336.

Cai, H., G. P. Kharel, and F. Hashinaga. 1992. Studies on the limonoids in citron. *Journal of the Japanese Society for Horticultural Science* 61:628–629.

Cai, H., G. P. Kharel, and F. Hashinaga. 1993. Changes in the limonoid content during development of Citrus fruit (*Citrus medica L.*). *Journal of the Japanese Society for Horticultural Science* 62(1):215–220.

Carpéné, C., J. Galitzky, E. Fontana et al. 1999. Selective activation of β3-adrenoceptors by octopamine: Comparative studies in mammalian fat cells. *Naunyn-Schmiedeberg's Archives of Pharmacology* 359(4):310–321.

Castelluccio, C., G. P. Bolwell, C. Gerrish, and C. Rice-Evans. 1996. Differential distribution of ferulic acid to the major plasma constituents in relation to its potential as an antioxidant. *Biochemical Journal* 316(2):691–694.

Chen, Z. F. 2005. Study on medicinal *Citrus* species and their active constituents. *Chinese Wild Plant Resources* 24(4):38–40.

Chen, J., Y. X. Gao, W. L. Wu, and S. Z. Li. 2006. Determination of bitter components in citrus juice by high performance liquid chromatography. *Chinese Journal of Chromatography* 24(2):157–160.

Citrus Pages. 2014. http://users.kymp.net/citruspages/home.html (http://citruspages.free.fr/)

Del Río, J. A., M. D. Fuster, F. Sabater, I. Porras, A. García-Lidón, and A. Ortuño. 1997. Selection of citrus varieties highly productive for the neohesperidin dihydrochalcone precursor. *Food Chemistry* 59(3):433–437.

Deng, X. X. 2008. *Citrus Cultivars in China*. Beijing: China Agriculture Publishing House.

FAO. http://www.fao.org/statistics/en/

Fumiyo, K., X. D. LI, and T. Kiyoko. 1992. Determination of synephrine and N-methyltyramine in Citrus aurantium by high-peformance liquid chromatography with electrochemical detector. *Chemical and Pharmaceutical Bulletin* 40(12):3284–3286.

Gorinstein, S., Z. Zachwieja, E. Katrich, E. Pawelzik, R. Haruenkit, S. Trakhtenberg, and O. Martin-Belloso. 2004. Comparison of the contents of the main antioxidant compounds and the antioxidant activity of white grapefruit and his new hybrid. *Lebensmittel-Wissenschaft und-Technologie* 37(3):337–343.

Gulcin, I. 2006. Antioxidant activity of caffeic acid (3,4-dihydroxycinnamic acid). *Toxicology* 217(2–3):213–220.

Hasegawa, S., P. Ou, C. H. Fong, Z. Jr., Herman, C. W. Coggins, and D. R. Atkin. 1991. Changes in the limonoate A-ring lactone and limonin 17-β-D-glucopyranoside content of navel orange during fruit growth and maturation. *Journal of Agricultural and Food Chemistry* 39(2):262–265.

Hashinaga, F., and S. Itoo. 1983. Seasonal changes of limonoids in the component parts of several citrus fruits. *Proceedings of the International Society of Citriculture* 2:901–905.

He, X. G., L. Z. Lian, L. Z. Lin, and M. W. Bernart. 1997. High-performance liquid chromatography–electrospray mass spectrometry in phytochemical analysis of sour orange (*Citrus aurantium* L.). *Journal of Chromatography A* 791(1–2):127–134.

Herrmann, K. 1989. Occurrence and content of hydroxycinnamic and hydroxybenzoic acid compounds in foods. *Critical Reviews in Food Science and Nutrition* 28(4):315–347.

Horowitz, R. M. 1986. Taste effects of flavonoids. *Progress in Clinical and Biological Research* 213:163–175.

Horowitz, R. M., and B. Gentili. 1969. Taste and structure in phenolic glucosides. *Journal of Agricultural and Food Chemistry* 17:696–700.

Ito, C., M. Itoigawa, M. Ju Ichi, N. Sakamoto, H. Tokuda, H. Nishino, and H. Furukawa. 2005. Antitumor-promoting activity of coumarins from citrus plants. *Planta Medica* 71(1):84–87.

Kelebek, H., A. Canbas, and S. Selli. 2008. Determination of phenolic composition and antioxidant capacity of blood orange juices obtained from cvs. Moro and Sanguinello (*Citrus sinensis* [L.] Osbeck) grown in Turkey. *Food Chemistry* 107:1710–1716.

Kikuzaki, H., M. Hisamoto, K. Hirose, K. Akiyama, and H. Taniguchi. 2002. Antioxidant properties of ferulic acid and its related compounds. *Journal of Agricultural and Food Chemistry* 50(7):2161–2168.

Kim, D., and C. Y. Lee. 2004. Comprehensive study on vitamin C equivalent antioxidant capacity (VCEAC) of various polyphenolics in scavenging a free radical and its structural relationships. *Critical Reviews in Food Science and Nutrition* 44:253–273.

Kimball, D. (Ed.). 1991. *Citrus Processing: Quality control and technology*. Springer Science & Business Media.

Kimball, D. A. 1999. *Citrus Processing—A Complete Guide*, 2nd ed. Gaithersburg, MD: Aspen Publishers.

Lam, L. K., T. Zhang, and J. Hasegawa. 1994. S. Citrus limonoid reduction of chemically induced tumorigenesis. *Food Technology* 48(11):104–108.

Lei, H. M., W. J. Sun, and W. H. Lin. 2000. Chemical composition of exocarpium citri gradis. *Northwest Pharmaceutical Journal* 15(5):203–204.

Li, X. D., Z. J. Ma, S. Lin, X. Z. Gu, and S. J. Mao. 2004. Determination of synephrine and *N*-methyltyramine in *pericarpium citri reticulatae viride* by HPLC. *China Journal of Chinese Materia Medica* 29(6):537–539

Li, X., R. Xie, Z. Lu, and Z. Zhou. 2010. The origin of cultivated citrus as inferred from internal transcribed spacer and chloroplast DNA sequence and amplified fragment length polymorphism fingerprints. *Journal of the American Society for Horticultural Science* 135(4):341–350.

Liu, L., X. Y. Qi, and X. Y. Dong. 2007. New advances in study of limonoids in citrus. *Academic Periodical of Farm Products Processing* 7:37–41.

Liu, Y. Q., H. Emily, and S. A. Tanumihardjo. 2012. History, global distribution, and nutritional importance of citrus fruits. *Comprehensive Reviews in Food Science and Food Safety* 11, 530–545.

Lodovici, M., F. Guglielmi, M. Meoni, and P. Dolara. 2001. Effect of natural phenolic acids on DNA oxidation in vitro. *Food Chemistry and Toxicology* 39(12):1205–1210.

Luo, S. Z., and L. H. Pan. 2008. Advances in study on and use of limonoids. *The Beverage Industry* 11(1):4–7.

Macheix, J. J., A. Fleuriet, and J. Billot. 1990. The main phenolics of fruits. In: *Fruit Phenolics*. Boca Raton, FL: CRC Press.

Majo, D. D., M. Giammanco, M. L. Guardia, E. Tripoli, S. Giammanco, and E. Finotti. 2005. Flavanones in citrus fruit: Structure-antioxidant activity relationships. *Food Research International* 38(10):1161–1166.

Marini, D., and F. Balestrieri. 1995. Multivariate analysis of flavanone glycosides in citrus juices. *Italian Journal of Food Science* 7(3):255–264.

Mattila, P., and J. Kumpulainen. 2002. Determination of free and total phenolic acids in plant-derived foods by HPLC with diode-array detection. *Journal of Agricultural and Food Chemistry* 50(13):3660–3667.

Mohamad-Reza, S., and Nasrin, S. 2003. Effects of flavonoids on the susceptibility of low-density lipoprotein to oxidative modification. *Prostaglandins, Leukotrienes Essential Fatty Acids* 69(1):73–77.

Morton, J. F. 1987. Grapefruit. *Fruits of Warm Climates*. Miami, FL: Julia F. Morton, pp. 152–158.

Morton, L. W., R. A. A. Caccetta, I. B. Puddey, and K. D. Croft. 2000. Chemistry and biological effects of dietary phenolic compounds: Relevance to cardiovascular disease. *Clinical and Experimental Pharmacology and Physiology* 27(3):152–159.

Mouly, P., E. M. Gaydou, and A. Auffray. 1998. Simultaneous separation of flavanone glycosides and polymethoxylated flavones in citrus juices using liquid chromatography. *Journal of Chromatography A* 800(2):171–179.

Nagy, S., P. Shaw, and M. K. Veldhuis. 1977. *Citrus Science and Technology*, Vol. 1 and Vol. 2. The AVI Publishing Company, Inc., Westport, CT.

Ohta, H., and S. Hasegawa. 1995. Limonoids in pummelos (*Citrus grandis* L. Osbeck). *Journal of Food Science* 60(6):1284–1285.

Ohta, H., M. Berhow, R. D. Bennett, and S. Hasegawa. 1992. Limonoids in seeds of citrus hanaju. *Phytochemistry* 31(11):3905–3907.

Ooghe, W. C., C. M. Detavernier. 1997. Detection of the addition of *Citrus reticulata* and hybrids to *Citrus sinensis* by flavonoids. *Journal of Agricultural and Food Chemistry* 45(5):1633–1637.

Ortuño, A., I. Reynaldo, M. D. Fuster, J. Botía, D. G. Puig, F. Sabater, A. G. Lidón et al. 1997. Citrus cultivars with high flavonoid contents in the fruits. *Scientia Horticulturae* 68(1–4):231–236.

Park, G. L., S. M. Avery, J. L. Byers, and D. B. Nelson. 1983. Identification of bioflavonoids from citrus. *Food Technology* 37(12):98–105.

Peleg, H., M. Naim, R. L. Rouseff, and U. Zehavi. 1991. Distribution of bound and free phenolic acidin oranges (*Citrus sinensis*) and grapefruits (*Citrus paradisi*). *Journal of the Science of Food and Agriculture* 57(3):417–426.

Pellati, F., and S. Benvenuti. 2002. Determination of adrenergic agoinists from extracts and herbal products of *Citrus aurantium* L. var. *amara* by LC. *Journal of Pharmaceutical and Biomedical Analysis* 29:1113–1119.

Peterson, J., and J. Dwyer. 1998. Flavonoids: Dietary occurrence and biochemical activity. *Nutrition Research* 18(12):1995–2018.

Peterson, J. J., G. R. Beecher, S. A. Bhagwat, J. T. Dwyer, S. E. Gebhardt, D. B. Haytowitz, and J. M. Holden. 2006a. Flavanones in grapefruit, lemons, and limes: A compilation and review of the data from the analytical literature. *Journal of Food Composition and Analysis* 19:S74–S80.

Peterson, J. J., J. T. Dwyer, G. R. Beecher, S. A. Bhagwat, S. E. Gebhardt, D. B. Haytowitz, and J. M. Holden. 2006b. Flavanones in oranges, tangerines (mandarins), tangors, and tangelos: A compilation and review of the data from the analytical literature. *Journal of Food Composition and Analysis* 19:S66–S73.

Rapisarda, P., P. Pannuzzo, G. Romano, and G. Russo. 2003. Juice components of a new pigmented citrus hybrid *Citrus sinensis* (L.) Osbeck × *Citrus clementina* Hort. ex Tan. *Journal of Agricultural and Food Chemistry* 51(6):1611–1616.

Russell, L. R., F. M. Shirley, and O. Y. Charles. 1987. Quantitative survey of naringin, hesperidin and nechesperidin in citrus. *Journal of Agricultural and Food Chemistry* 35:1027–1030.

Saipetch, K. S. 2004. Separation and quantitation of limonoids and flavonoids in juice and by-products of sweet orange (*Citrus sinensis*). (Doctoral dissertation, Michigan State University).

Sawabe, A., M. Morita, T. Kisu, H. Kishine, Y. Ohtsubo, T. Minematsu, Y. Mastsubara et al. 1999. Isolation and characterization of new limonoid glycosides from citrus unshiu peels. *Carbohychase Research* 315(1):142–147.

Shan, Y. 2016. *Canned Citrus Processing: Techniques, Equipment, and Food Safety*. Beijing: China Science Press.

Su, W. W., and H. D. Lin. 2001. Research advances in exocarpium citri grandis. *Chinese Traditional Patent Drug* 8:608–610.

Sun, C. D., K. S. Chen, Q. J. Chen, Y. P. Ren, and S. L. Zhang. 2004. Extraction, identification and determination of natural limonin and nomolin from citrus fruits. *Journal of Chinese Institute of Food Science and Technology* 4(1):6–11.

Sun, C. D., K. S. Chen, Y. Chen, and Q. J. Chen. 2005. Contents and antioxidant capacity of limoin and nomilin different tissues of citrus fruit of four cultivars during fruit growth and maturation. *Food Chemistry* 93(4):599–605.

Swingle, W. T., and P. C. Reece. 1967. The botany of citrus and its wild relative. *The Citrus Industry* 190–430.

Tanaka, T. 1977. Fundamental discussion of citrus classification. *Studia Citrogia* 14:1–6.

Tang, F., L. Tao, X. Luo, L. Ding, M. Guo, L. Nie, and S. Yao. 2006. Determination of octopamine, synephrine and tyramine in citrus herbs by ionic liquid improved 'green' chromatography. *Journal of Chromatography A* 1125(2):182–188.

Tisserat, B., D. Daniel Jones, and P. D. Galletta. 1990. Juice vesicle populations in citrus fruit. *Botanical Gazette* 151(1):64–72.

Tripoli, E., M. Guardia, S. Giammanco, D. D. Majo, and M. Giammanco. 2007. Citrus flavonoids: Molecular structure, biological activity and nutritional properties: A review. *Food Chemistry* 104(2):466–479.

USDA, United States Department of Agriculture, Foreign Agricultural Service. The Citrus: World Markets and Trade, January 25, 2017. FAS Online at: https://apps.fas.usda.gov/psdonline/circulars/citrus.pdf.

Wang, S. C., and Z. M. Chen. 1995. Determination of naringin in sour orange (*C. aurantum*) by HPLC. *Journal of Hengyang Medical College* 23(1):4–7.

Wang, Y. C., Y. C. Chuang, and Y. H. Ku. 2007. Quantitation of bioactive compounds in citrus fruits cultivated in Taiwan. *Food Chemistry* 102(4):1162–1171.

Weber, B., B. Hartmann, D. Stockigt, K. Schreiber, M. Roloff, H. J. Bertram, and C. O. Schmidt. 2006. Liquid chromatography/mass spectrometry and liquid chromatography/nuclear magnetic resonance as complementary analytical techniques for unambiguous identification of polymethoxylated flavones in residues from molecular distillation of orange peel oils (*Citrus sinensis*). *Journal of Agricultural and Food Chemistry* 54(2):274–278.

Widmer, W., and C. Haun. 2005. Variation in furanocoumarin content and new furanocoumarin dimers in commercial grapefruit (*Citrus paradisi* Macf.) juices. *Journal of Food Science* 70(4):205–209.

Wilmsen, P. K., D. S. Spada, and M. Salvador. 2005. Antioxidant activity of the flavonoid hesperidin in chemical and biological systems. *Journal of Agricultural and Food Chemistry* 53(12):4757–4761.

Xu, G. H., X. Q. Ye, J. C. Chen, and D. H. Liu. 2007. Effect of heat treatment on the phenolic compounds and antioxidant capacity of citrus peel extract. *Journal of Agricultural and Food Chemistry* 55(2):330–335.

Ye, X. Q., and D. H. Liu. 2005. *Citrus Processing and Utilization*. Beijing: Light Industry Publishing House.

Yu, J., L. Wang, R. L. Walzem, E. G. Miller, L. M. Pike, and B. S. Patil. 2005. Antioxidant activity of citrus limonoids, flavonoids, and coumarins. *Journal of Agricultural and Food Chemistry* 53(6):2009–2014.

Zeng F. K., L. S. Zou, and B. L. Jiao. 2003. Study on the content and distribution of limonoides in citrus fruits. *Journal of Chinese Institute of Food Science and Technology* 3(4):79–81.

Zhou J. G., and D. H. Gui. 1997. Studies on the quality of baked *Fructus aurantii* immaturus with bran of different storage periods. *China Journal of Chinese Materia Medica* 22(2):8–12.

2 Bioactive Components in Citrus Fruits and Their Health Benefits

Meenakshisundaram Selvamuthukumaran, Maria Susai Boobalan, and John Shi

CONTENTS

2.1 INTRODUCTION

Citrus fruits have been collected and used by man for centuries for medicinal, herbal and agricultural purposes. Citrus fruits belong to the family of Rutaceae, and are one of the main fruit tree crops grown throughout the world (Waterman and Grundon 1983). All citrus fruits share in common their sweet and sour flavor; they are acidic, exotic fruits with juicy, bitter fruit segments inside. They possess refreshing juice and are available almost all year round. *Citrus* species are small to medium-size shrubs or trees that are cultivated throughout the tropics and subtropics. They are native to parts of India, China, and northern Australia. India ranks sixth in the production of citrus fruit cultivation in the world. Other major citrus-producing countries are Spain, United States, Japan, South Africa, Israel, Brazil, Turkey, and Cuba (Harley et al. 2006). The citrus fruit is well known for therapeutic properties such as antitumor, anti-inflammatory and anticancer, due to phytovitamins and nutrients present in them. Citrus fruits or their products are part of the daily human intake in one form or the other all over the world.

Citrus fruits, as such, have long been valued for their wholesome nutritious and antioxidant properties. It is scientifically established that citrus fruits, especially lemons and oranges, by virtue of their richness in vitamins and minerals, have many proven health benefits. Moreover, we are now beginning to appreciate the other biologically active, nonnutrient compounds found in citrus fruits such as phytochemical antioxidants, soluble as well as insoluble dietary fiber—helpful in reducing the risk of cancers, many chronic diseases like arthritis, obesity, and coronary heart diseases.

Fresh citrus fruits act as a rich source of dietary fiber and hence are recognized as an important component of the healthy human diet. In India, the species *Citrus reticulate* L., *Citrus medica* L., and *Citrus limon* L. (commonly known as orange, citron, and lemon, respectively) are widely grown for edible consumption.

2.2 ORANGE—*CITRUS SINENSIS*

The orange is a citrus fruit (family Rutaceae) originating in Southeastern Asia. There are many different varieties of oranges; Valencia, Persian, Naval, Jaffa and Blood. Oranges are classified into two categories, bitter and sweet, with the sweet being the most popularly consumed. The fruit was first cultivated in the Middle East in the ninth century and was introduced to Europe by the fifteenth century by the Moors, Portuguese, and Italians.

The main production regions of oranges are North and South America (led by Brazil, Mexico, and Argentina), the Mediterranean basin (led by Spain, Italy, Egypt, and Turkey), and the South and East Asian regions (led by China, India, and Japan).

Orange is an evergreen flowering tree. The height of an orange tree is generally 9–10 m (although very old specimens have reached 15 m). The orange fruit is a hesperidium. It is a type of berry that ranges widely in size, color, shape, and juice quality. Fruits are globose to ovoid in shape. Wild orange fruit has a smooth skin, and the petiole wings are entire. The petioles of sour orange leaves are much larger than those of the sweet orange. The word "orange" is derived from the Sanskrit term "narang." Mainly, there are 11 individual pieces present in a typical fruit. Oranges are round citrus fruits with finely textured skin that is, of course, orange in color just like their pulpy flesh. Oranges usually range from approximately 2 to 3 inches in diameter.

2.2.1 Nutritional and Functional Value

Citrus fruits have long been valued for their wholesome nutritious and antioxidant properties. It is a scientifically established fact that citrus fruits, especially oranges by virtue of their abundance in vitamins, antioxidants, and minerals can benefit in many ways (Table 2.1). Moreover, it is now an acknowledged fact that the other biologically active, nonnutrient compounds in citrus fruits such as phytochemical antioxidants, soluble and insoluble dietary fiber help in cutting down cancer risk, chronic diseases like arthritis, obesity, and coronary heart diseases.

The fruit is low in calories, contains no saturated fats or cholesterol, but is rich in dietary fiber pectin. Pectin, by its virtue as a bulk laxative, helps protect the mucous membrane of the colon by decreasing its exposure time to toxic substances, as well as by binding to cancer-causing chemicals in the colon. Pectin has also been shown to reduce blood cholesterol levels by decreasing its reabsorption in the colon by binding to bile acids.

TABLE 2.1
Nutritional Value of Oranges

Proximate	Nutrient Value	Percentage of RDA (U.S.)
Energy	47 kcal	2.5%
Carbohydrates	11.75 g	9%
Protein	0.94 g	1.5%
Total fat	0.12 g	0.5%
Cholesterol	0 mg	0%
Dietary fiber	2.40 g	6%
	Vitamins	
Folates	30 μg	7.5%
Niacin	0.282 mg	2%
Pantothenic acid	0.250 mg	5%
Pyridoxine	0.060 mg	4.5%
Riboflavin	0.040 mg	3%
Thiamin	0.100 mg	8%
Vitamin C	53.2 mg	90%
Vitamin A	225 IU	7.5%
Vitamin E	0.18 mg	1%
Vitamin K	0 μg	0%
	Electrolytes	
Sodium	0 mg	0%
Potassium	169 mg	3.5%
	Minerals	
Calcium	40 mg	4%
Copper	39 μg	4%
Iron	0.10 mg	1%
Magnesium	10 mg	2.5%
Manganese	0.024 mg	1%
Zinc	0.08 mg	1%

Oranges, like other citrus fruits, are an excellent source of vitamin C (provide 53.2 mg per 100 g, about 90% of RDA), which is a powerful natural antioxidant. Consumption of foods rich in vitamin C helps the body develop resistance against infectious agents and scavenges harmful, proinflammatory free radicals from the blood.

Oranges also contain very high levels of vitamin A, which is required for maintaining healthy mucus membranes and skin, and essential for proper vision. It is also a very good source of B-complex vitamins such as thiamin, pyridoxine, and folates. These vitamins are essential in the sense that the human body requires them from external sources to replenish itself.

Orange fruits also contain a very high amount of minerals like potassium and calcium. Potassium is an important component of cell and body fluids that helps control heart rate and blood pressure through countering the pressing effects of sodium.

2.2.2 Phytoconstituents of Oranges

Orange fruit contains 1.5% essential oil. The main phytoconstituents present in orange fruit are D-limonene (90%), citral, citronellal, nootkaton, sinesal, n-nonanal, n-decanal, n-dodecanal, linalylacetate, geranyl acetate, citronelyl acetate, and anthranil acid methyl ester (Table 2.2 and Figure 2.1). Lipophilic flavonoids and furanocumarines are reported in pressed oils. There is some evidence that active ingredients in the orange stimulate the secretion of gastric juice. Orange also contains several

TABLE 2.2
Phytoconstituents of Orange

Phytoconstituents	Plant Parts
Flavone glycosides: Neohesperidin, naringin, hesperidin, narirutin	Fruit peel
Triterpene: Limonene, citrol	
Pigment: Anthocyanin, β-cryptoxanthin, cryptoxanthin, zeaxanthin, rutin, eriocitrin, and homocysteine	
Polymethoxylated flavones: Tangeretin and nobiletin	
Flavonoids: Citacridone, citbrasine, and noradrenaline	
Terpenoids: Linalool, β-elemene	Leaves
Triterpenes: Limonene	Flowers
Vitamins: B_1, B_2, B_3, B_5, B_6, and vitamin C	Fruits
Minerals: Calcium, iron, magnesium, zinc, phosphorus, and potassium	

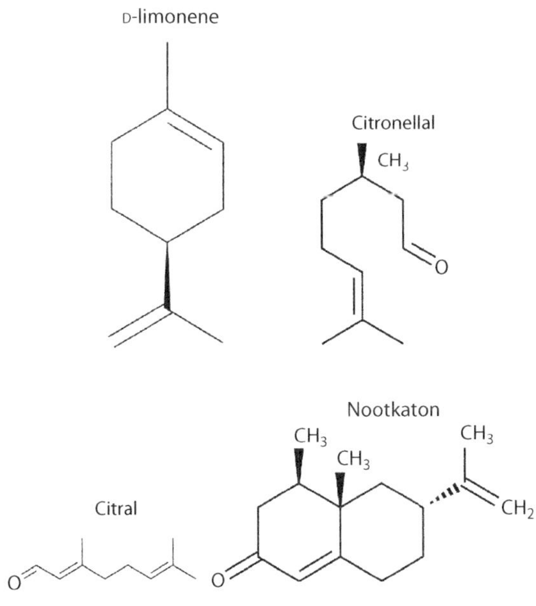

FIGURE 2.1 Chemical structures of orange bioactive constituents. *(Continued)*

Neohesperidin

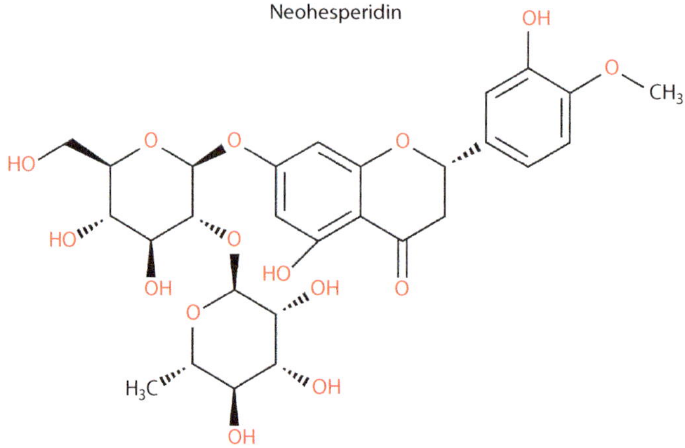

Naringin

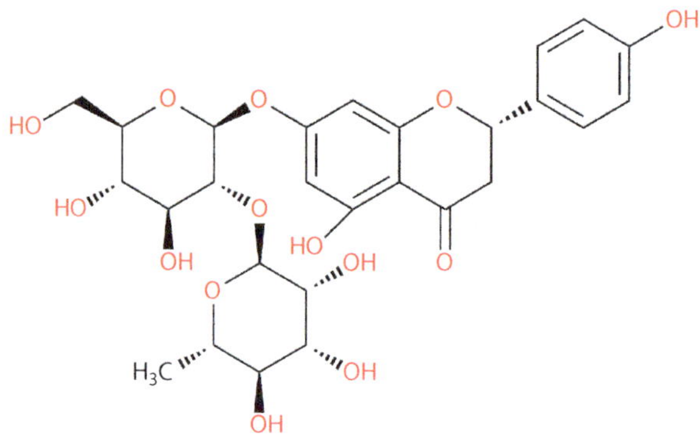

FIGURE 2.1 (CONTINUED) Chemical structures of orange bioactive constituents.

(Continued)

bitter flavone glycosides like neohesperidin and naringin, whose sugar component is neohesperidose, and rutin whose sugar component is rutinose. Both sugars are disaccharides of glucose and rhamnose (6-desoxymannose) (Ihrig 1995).

2.2.3 Bioactive Properties

2.2.3.1 Antioxidant Property

Oranges are a rich source of vitamin C, flavonoids, phenolic compounds, and pectins. The main flavonoids found in *Citrus* species are hesperidine, narirutin, naringin, and eriocitrin (Guarnieri et al. 2007; Ghasemi et al. 2009). Just one orange provides around 116% of the daily requirement for vitamin C, the primary water-soluble antioxidant, which prevents free radical generation in the body and damage to the tissues in the aqueous environment both inside and outside cells. Drinking of orange juice without salt and sugar is associated with reduced severity of inflammatory

Narirutin

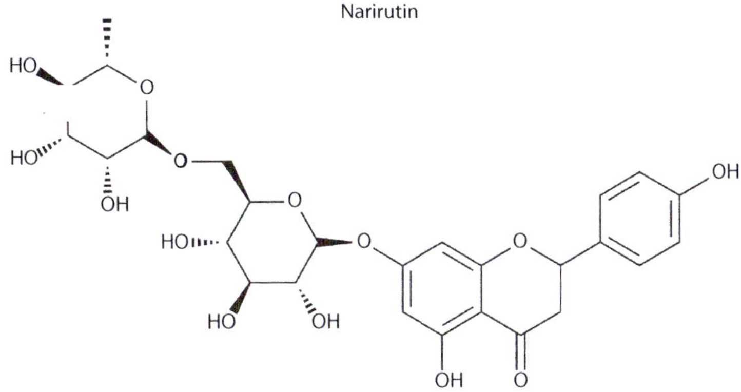

Hesperidine

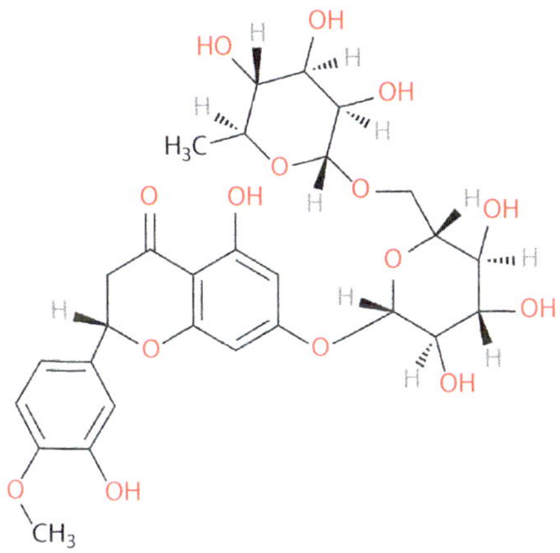

FIGURE 2.1 (CONTINUED) Chemical structures of orange bioactive constituents.

(Continued)

conditions like asthma, osteoarthritis and rheumatoid arthritis. Vitamin C is also necessary for the proper functioning of immune system, and thus is good for preventing colds, coughs, and recurrent ear infections.

2.2.3.2 Protection against Cardiovascular Disease

According to a recent World Health Organization report, citrus fruits offer protection against cardiovascular diseases by reducing levels of homocysteine. Orange fruit contains vitamin C, carotenoids, and flavonoids, which are cardioprotective. The cholesterol lowering effect of the orange is produced by Limonene. Polymethoxylated flavones (PMFs) are present in citrus fruit peel, and can lower cholesterol more effectively than some prescription drugs, without showing any side effect. Although a variety of citrus fruits contain PMFs, the most common PMFs are tangeretin and nobiletin, which are found in the peels of oranges. PMFs work like statin drugs that inhibit the synthesis of cholesterol and triglycerides inside the liver. Therefore, grating a tablespoon or so of the peel of orange each day and using it to flavor tea, salads,

Eriocitrin

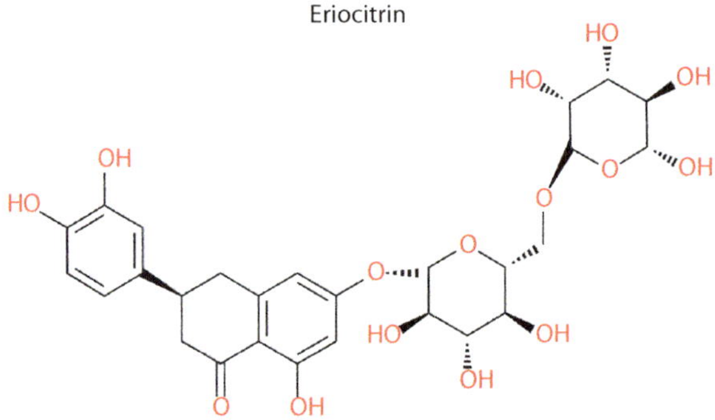

Tangeretin

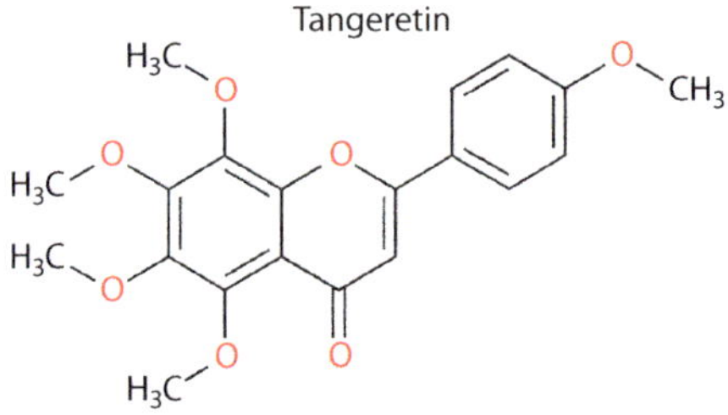

Nobiletin

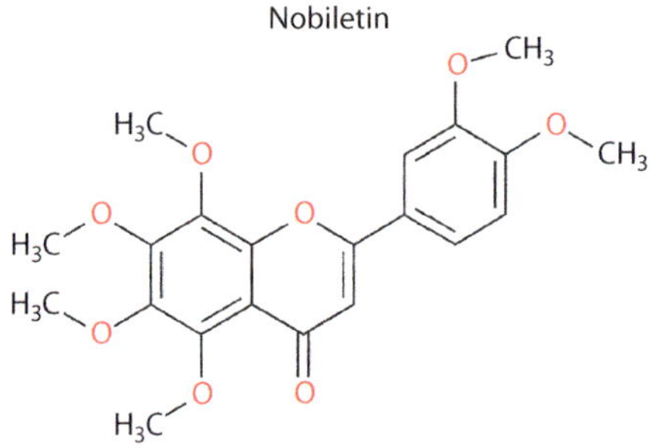

FIGURE 2.1 (CONTINUED) Chemical structures of orange bioactive constituents.

(Continued)

yogurt, soups, snacks, or rice may be a practical way of achieving some cholesterol-lowering benefits (Kurowska and Manthey 2004).

2.2.3.3 Anticarcinogenic Property

Limonene, one of the main constituents of orange, reduces the risk of mouth, skin, lung, breast, stomach, and colon cancer. Another constituent of orange is hesperidin, and its flavone analogue, diosmin, has also exhibited anticarcinogenic activities in various *in vivo* studies. Anticarcinogenic activity mainly depends on antioxidant properties of the molecules, as well as their ability to modulate the activity of detoxifying hepatic enzymes. The polymethoxylated flavones have shown strong antiproliferative action against cancer cells and antigen activated T-lymphocytes.

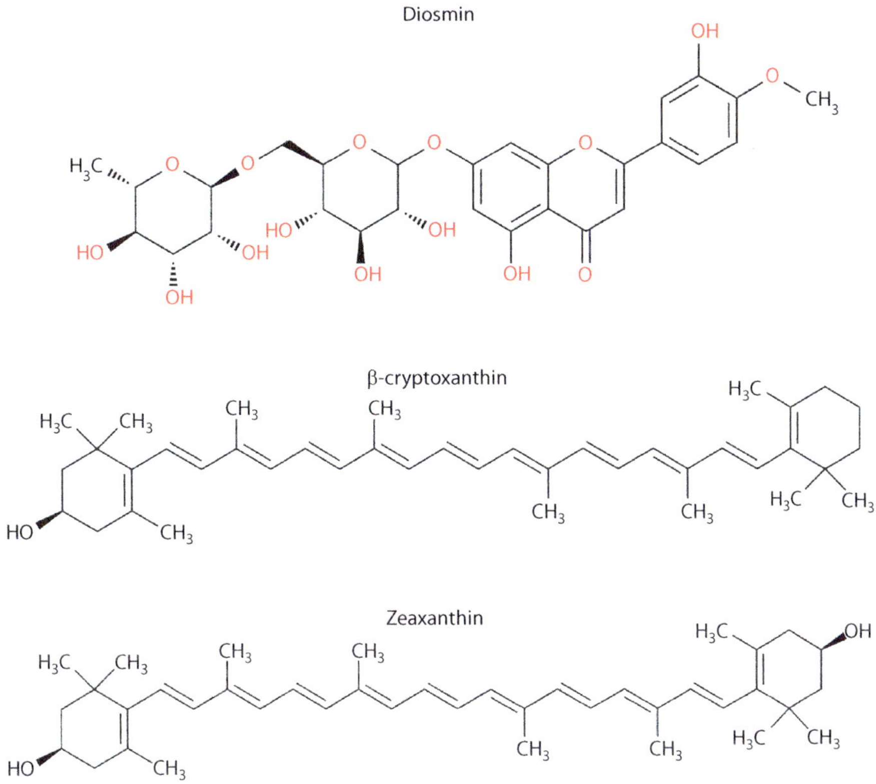

FIGURE 2.1 (CONTINUED) Chemical structures of orange bioactive constituents.
(Continued)

β-cryptoxanthin (an orange-red carotenoid) is present in highest amounts in oranges, and it may significantly lower one's risk of developing lung cancer (Tanaka et al. 1997).

2.2.3.4 Reduced Risk of Kidney Stones

A study published in the *British Journal of Nutrition* found that when women drank 1/2 liter of orange juice daily, their urinary pH value and citric acid excretion increased thereby significantly diminishing the risk of forming calcium oxalate stones (Honow et al. 2003).

Alkalizing beverages are highly effective in preventing the recurrence of calcium oxalate (Ox), uric acid, and cystinelithiasis. Hanow et al. (2003) evaluated the influence of grapefruit and apple juice consumption on the excretion of urinary variables and the risk of crystallization in comparison with orange juice. All investigations were carried out on nine healthy female subjects without any history of stone formation, aged 26–35 years. Each juice was tested over a 5-day study. During the study, the subjects received a standardized diet. Fluid intake of 2.75 L was composed of 2.25 L neutral

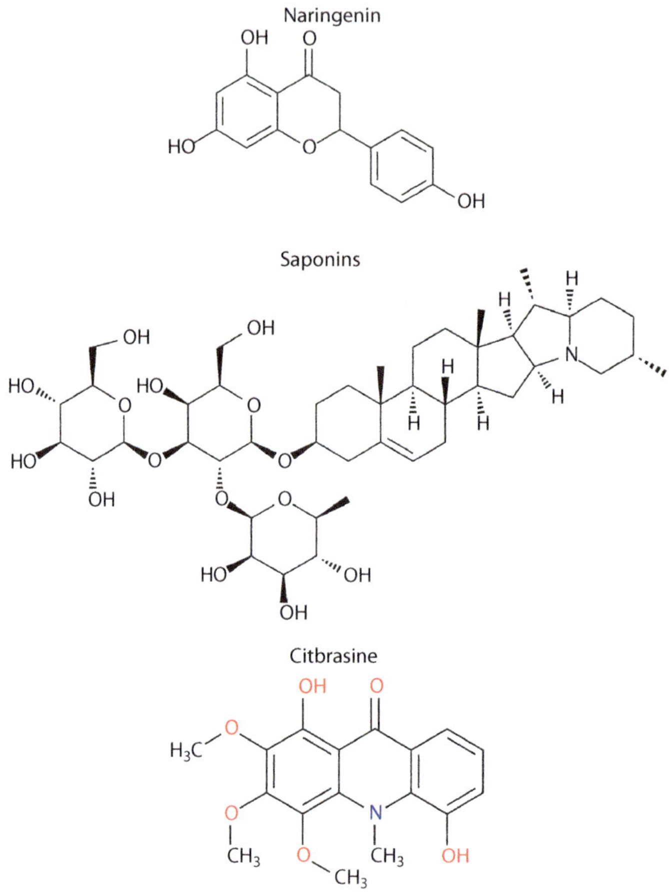

FIGURE 2.1 (CONTINUED) Chemical structures of orange bioactive constituents.
(Continued)

mineral water, 0.4 L coffee and 0.1 L milk. On the fourth and fifth days, 0.5 L mineral water was partly substituted by 0.5 or 1.0 L of the juice under investigation, respectively. The influence on urinary variables was evaluated in 24-hour urine samples. In addition, the BONN risk index of CaOx, relative supersaturation (RS) CaOx crystallization was determined. Due to an increased pH value and an increased citric acid excretion after consumption of each juice, the RSCaOx decreased significantly for grapefruit juice. The BONN risk index yielded a distinct decrease in the crystallization risk. They showed that both grapefruit and apple juice reduce the risk of CaOx stone formation at a magnitude comparable with the effects obtained from orange juice.

2.2.3.5 Antiulcer Property

Intake of orange juice on a regular basis reduced the infection incidence with *Helicobacter pylori*, thus preventing development of ulcers (Simon et al. 2003). They examined the relation between serum ascorbic acid and *Helicobacter pylori* serology

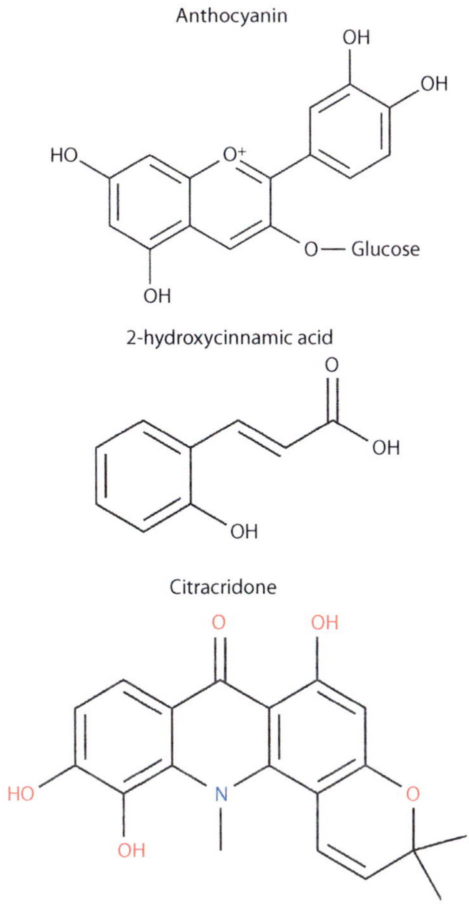

FIGURE 2.1 (CONTINUED) Chemical structures of orange bioactive constituents.

from a probability sample of U.S. adults. Among the U.S. adult population selected, 32% had a positive serology for *H. pylori*, and of these 54% were positive for the CagA antigen. Among this population, a 0.50 mg/dL increase in serum ascorbic acid level was associated with a decreased seroprevalence of *H. pylori*. They concluded that regular consumption of orange juice reduces ulcers.

2.2.3.6 Antianxiety Effect

Aromatherapy is the use of essential oils as an alternative treatment for medical purposes. Despite the lack of sufficient scientific proof, it is considered a holistic complementary therapy employed to enhance comfort and decrease distress. Citrus fragrances have been used by aromatherapists particularly for the treatment of anxiety symptoms. Based on this claim, Faturi et al. (2010) investigated the effects of *Citrus sinensis* essential oil on Wistar male rats. The rats were evaluated in the elevated plus-maze followed by the light/dark paradigm. The animals were exposed to the orange aroma (100, 200, or 400 μL) for 5 min while in a Plexiglas chamber and were then immediately submitted to the behavioral tests. At all doses, *C. sinensis* oil demonstrated anxiolytic activity in at least one of the tests and, at the highest dose, it presented significant effects in both animal models, as indicated by increased exploration of the open arms of the elevated plus-maze (time: $p = 0.004$; entries: $p = 0.044$) and of the lit chamber of the light/dark paradigm (time: $p = 0.030$). In order to discard the possibility that this outcome was due to nonspecific effects of any odor exposure, the behavioral response to *Melaleuca alternifolia* essential oil was also evaluated, using the same animal models, but no anxiolytic effects were observed. These results suggest an acute anxiolytic activity of orange essence, giving some scientific support to its use as a tranquilizer by aromatherapists.

2.2.3.7 Antityphoid Activity

Typhoid fever, caused by *Salmonella typhi*, is a major public health problem, particularly in developing countries. Kumar et al. (2010) studied the constituents of orange fruit that are responsible for the antityphoid activity and their study showed that the flavonoid compounds, such as citracridone, citbrasine, and saponins, are responsible for an antityphoid activity.

2.2.3.8 Antimicrobial Activity

Oranges are eaten to allay fever, the roasted pulp is prepared as a poultice for skin diseases, and the fresh peel is rubbed on acne. A decoction of the dried leaves and flowers is taken in Italy and France as an antispasmodic, cardioprotective, and antiemetic agent. A decoction of husked orange seeds is prescribed for urinary ailments in China. Orange peel oil produces a lethal effect on fleas, fire ants, and houseflies due to its 90%–95% limonene content. Orange peel is medicinally used against fungi (Strange et al. 1993). In 1993, Strange et al. found that wounding citrus fruits induces the production of antimicrobial compounds. They isolated new compounds from the peel of injured grape and orange fruits and identified the extracts of these fruits' wound material, which produced positive compounds, that is, 3-(4-hydroxy, 3-(3-methyl-2-butenyl)-phenyl)-2-(E)-propenol, that had an antimicrobial activity against *P. digitatum* and *C. cucumerinum*.

Citrus reticulata essential oil is an effective inhibitor of the biodegrading and storage-contaminating fungus *Aspergillus niger*. Major antifungal constituents of orange are limonene (84.2%), linalol (4.4%), and myrcene (4.1%) (Sharma and Tripathi 2008).

2.2.3.9 Larvicidal Activity

The saponins present in the peel possess larvicidal properties. Wiesman and Chapagain (2005) studied the effect of aqueous extracts of the fruit pulp, seed kernel, roots, bark, and leaves of *Balanites aegyptiaca* Del. (Zygophyllacea) on the larvae of the *Culex pipiens* mosquito. Early fourth instars larvae of *C. pipiens* mosquitoes were exposed, for up to three days, to a dilution of 0%, 0.1%, 0.25%, 0.5%, 1.0%, and 2.0% aqueous extracts of fruit pulp, seed kernel, roots, bark, and leaves. They found that the tested extracts produced larval mortality because of the presence of saponin, and noticed that the larval mortality was greatest with the aqueous root extract. The lowest concentration of root extract (0.1%) showed 100% larval mortality after three days, whereas a 0.5% concentration of aqueous bark extract was needed for 100% larval mortality. Aqueous extracts of leaf, fruit pulp, and seed kernel produced less larval mortality compared to the root and/or bark extracts. They suggested that all parts of *B. aegyptiaca* had larvicidal properties that could be developed and used as natural insecticides for mosquito control.

2.2.3.10 Antidiabetic Activity

The antidiabetic activity of orange is due to bioflavonoids such as hesperidin and naringin present in citrus fruit peels. These peels play an antidiabetic role in C57BL/Ks J-db/db mice via the regulation of glucose regulatory enzymes. These enzymes decrease the activity of glucose-6-phosphatase and hosphoenol pyruvate. The antidiabetic potential of orange peel and juice appear to be mediated via antiperoxidation, inhibition of α-amylase enzyme activity that is responsible for the conversion of complex carbohydrates to glucose, increased hepatic glycogen content, stimulation of insulin secretion, and repair of secretory defects of pancreatic β-cells (Parmar and Kar 1995, 2008).

2.2.3.11 Anti-Inflammatory, Healing, and Antiarthritic Activity

The anti-inflammatory activity of *Citrus reticulata* is due to the presence of polymethoxy flavones. The polymethoxy flavone content, especially nobiletin, appears to be responsible for the anti-inflammatory activities of certain citrus peel extracts (Lin et al. 2003). Wounds are generally defined as physical injuries that result in an opening or breaking of the skin. The healing property of an orange depends on a wide variety of phytonutrients such as citrus flavones (hesperidin and naringenin), anthocyanins, hydroxycinnamic acids, and a variety of polyphenols. The most important flavone in orange is hesperidin, which has been shown to reduce high blood pressure and cholesterol in animal studies. Importantly, most of this phytonutrient is found in the peel and inner white pulp of the orange, rather than in its liquid orange center. This beneficial compound is too often removed during processing of oranges into juice (Sandhya et al. 2011).

Carotenoids, zeaxanthin, and β-cryptoxanthin are phytonutrients that remarkably reduce the risk of rheumatoid arthritis. Those persons consuming high amounts of zeaxanthin and cryptoxanthin had 52% less likelihood of developing rheumatoid arthritis. *Citrus sinensis* peel extracts contain bioflavonoids, including polymethoxylated flavones (PMFs), which have anti-inflammatory, antioxidant, and hypolipidemic effects (Oben et al. 2009).

2.2.3.12 Orange as a Folklore Medicine and Its Traditional Uses

From the time immemorial, the entire orange plant including fruits, leaves, flowers, peels, and juice has been used as traditional medicine. Orange is a good source of vitamins (B_1, B_2, B_3, B_5, A, B_6, C), flavonoids, terpenes, potassium, and calcium. Oranges are effective in the management of many diseases such as arthritis, asthma, Alzheimer's disease, Parkinson's disease, macular degeneration, Diabetes mellitus, gallstones, multiple sclerosis, cholera, gingivitis, optimal lung function, cataracts, ulcerative colitis, and Crohn's disease.

Traditionally, orange juice is used as a general tonic. It helps to eliminate toxins from the body, helps to maintain hydration, which is useful in cases of anxiety disorder and stress. It is also used as a Mexican traditional medicine for the treatment of tuberculosis and stomach upsets. It improves appetite and prevents constipation. The humble orange has a long history in Chinese medicine as a cooling agent for coughs, colds, and respiratory disorder. It is a traditional Chinese symbol of good luck and prosperity. It is used in the treatment of obesity. Orange symbolizes innocence and fertility. In France, it is used for the treatment of angina, hypertension, constipation, diarrhea, menstrual disorder, and palpitations.

2.2.4 Processed Orange Products

2.2.4.1 Orange Marmalade

The preservation of fruit by marmalade making is a familiar process carried out on a small scale by housewives in many parts of the world. This sugar preserve is defined as a semisolid or gel-like product prepared from fruit ingredients together with one or more sweetening ingredients and may contain suitable food acids and pectins; the ingredients are concentrated by cooking to such a point that the total soluble solids (TSS) of the finished marmalade is not below 65% (Table 2.3 and Figure 2.2). Gel formation is dependent on the presence in the fruit of carbohydrate pectin, which at a pH of 3.2–3.4 and in the presence of a high concentration of sugar, has the property of forming a viscous semisolid. During marmalade boiling, all microorganisms are destroyed within the product, and it is filled hot into clean receptacles which are subsequently sealed, and then inverted so that the hot marmalade contacts the lid surface, preventing spoilage by microorganisms during storage.

2.2.4.2 Orange Fruit Juice Powder

2.2.4.2.1 Extraction of Fruit Juice

Sweet mature fruits are preferred for the production of juice powder. Fruits are washed in cold tap water and drained. Fruits are manually cut up in to two halves and the seeds are removed. The juice that is localized in the sacs is extracted with the help of a manually operated household hand juicer. The juice obtained had a deep

TABLE 2.3
Marmalade Recipe

Ingredient	Quantity
Fruit	55 lb
Water	–
Sugar	55 lb
Pectin (150 grade)	225 g
Citric acid (50% sol.)	300 cc

Select unblemished fruits

↓

Wash the fruits with water and drain

↓

Halve the fruits and squeeze out the juice

↓

Preserve the peel in clean containers

↓

Filter the juice twice, once through a thin and once through a thick cloth

↓

Weigh the juice to calculate the amount of sugar and heat to slowly bring to a boil with the pot covered

↓

Add several large pieces of peel, from which the albedo (white part) has been removed

↓

Soak the peel in the juice for 15 minutes

↓

Remove the peel

↓

Weigh the sugar so that it equals the weight of the juice

↓

Preserve 1% of the sugar to mix it with the pectin at a later stage

↓

Add the sugar to the boiling juice, dissolving it rapidly without leaving crystals on the walls of the pot

↓

Weigh the pectin so that it accounts for 0.5% of the total expected weight of the juice-sugar mixture, and blend with the 1% of sugar that was preserved

↓

Cut the peel in thin strips (3–4 mm wide and 3 cm long), after having removed the albedo (white part)

↓

Add the peel to the boiling mixture and cook for 5 minutes with the pot covered

↓

Add the pectin to the juice and dissolve properly

↓

Boil on a high flame until 64°–65° Brix or a temperature of 104°C is reached

↓

Pour the mixture in the jars, making sure that they are filled to the brim
Close the jars and turn them upside down to cool

↓

Clean the jars and seal the lids with adhesive tape

↓

Label and store

FIGURE 2.2 Flowchart for preparing marmalade from orange fruit.

FIGURE 2.3 Orange fruit juice powder.

orange color. The orange juice obtained is vacuum-filtered through an 11 μm nylon mesh, transferred into a vessel, and stored in a refrigerator for 1 hour. The top layer of clear liquid is then carefully removed.

2.2.4.2.2 *Spray Drying of Fruit Juice Powder*

Pilot plant spray dryer (S.M. Scientech, India) with a cocurrent air flow is used for spray drying. The blower speed was set at 2400 rpm for drying. Distilled water is pumped into the dryer at a set flow rate at 10 rpm (10 rpm = 30 mL/min) to achieve the inlet/outlet temperatures of 200°C and 120°C, respectively. The dryer is run at this condition for about 10 min before the feed is introduced. The feed is prepared by blending orange juice and maltodextrin at a ratio of 80:20. The slurry is fed to the spray dryer at above set conditions, drying is performed, and the powder (Figure 2.3) is collected in a preweighed, insulated glass bottle connected at the end of a cyclone collector.

2.3 CITRON—*CITRUS MEDICA*

Citrus medica, commonly known as a citron in English and bijapura in Ayurvedic literature, is a shrub or small tree (Figure 2.4). Its leaflets are 3–6 inches long, elliptic-ovate or ovate-lanceolate with sort, wingless or nearly wingless petioles; flowers are 5–10 in a raceme, small or middle-sized; petals are generally more or less pink; and fruit is globose ovoid or oblong often mamillate at the apex. This plant is apparently found wild in Kumaon, Pachamarhi, Sikkim, the Khasia Hills, the Garo Hills, Chittagong, the upper Yunzalin Valley, the Western Ghats, and the Satpura Range in Central India.

FIGURE 2.4 Variety of *Citrus medica* with oblong shape.

It is native to India, Nepal, the Philippines, Bangladesh, Indonesia, Malaysia, and other Southeast Asian countries. It grows in any kind of soil whether sandy or loamy. This fruit is like a cross between sweet lime and lemon. It is almost as big as a sweet lime with a rough, dark green outer layer. It can be cut like a sweet lime and the juice is as easy to extract as for the lemon. The juice is less acerbic than that of the lemon and has a pleasant odor and flavor. Adding water, sugar, and a pinch of salt to the extracted juice can be very refreshing and replenishes electrolytes in the body. Citron is used around the world in a wide variety of foods and beverages and often served as garnish with fish or meat, and with tea, both hot and iced. Additionally, a wide range of sweets are prepared using citron flavoring, such as candies, jams, sherbet, cookies, cakes, puddings, pies, tarts, and confectionery. Citron tree extracts are widely used in cosmetics and beauty products in several countries.

2.3.1 Nutritional and Functional Value

The fruit contains a high amount of fiber (Table 2.4), which helps to keep the digestive system healthy. In addition it can help prevent heart disease, diabetes, weight gain, and some cancers. It also contains enormous amount of minerals, viz. calcium, phosphorus and iron. The calcium function in the body plays a key role in cell signaling, blood clotting, muscle contraction and nerve function; the main function of phosphorus is in bones and teeth formation, as well as for producing protein for growth and the maintenance and repair of cells and tissues; iron is an essential element for blood production and is needed for a number of highly complex processes, continuously taking place on a molecular level, and indispensable to human life, for example, transportation of oxygen around the body. Citron also contains a significant amount of vitamin C, which is required for growth and repair of tissues in all parts of the body. It is also used to make skin, tendons, ligaments, and blood vessels. It heals wounds and forms scar tissues.

TABLE 2.4
Nutritive Value of *Citrus medica*

Moisture	86.2 g
Protein	0.073 g
Fat	0.05 g
Fiber	1.13 g
Ash	0.44 g
Calcium	36.9 mg
Phosphorus	16.8 mg
Iron	0.59 mg
Carotene	0.009 mg
Thiamine	0.056 mg
Riboflavin	0.027 mg
Niacin	0.126 mg
Ascorbic acid	369 mg

Note: Values per 100 g of edible portion.

2.3.2 Traditional Uses

Various parts of bijapura are widely used in the Indian traditional system of medicine. The ripe fruit is a potent antiscorbutic, stomachic, cardiac tonic, stimulant, sedative, analgesic, and is used in dyspepsia, bilious vomiting, cold, fever, palpitation, sore throat, cough, asthma, thirst, hiccough, and earache; the root is analgesic, antispasmodic and used in diarrhea, piles, and constipation; the seeds are anthelmintic, stomachic, sedative, cardiac tonic, and useful in palpitation; the flowers and buds are astringent and used in blood disorders; and the peels are anthelmintic (Ayurvedic Pharmacopoeia of India 2001). Leaves are useful to induce sleep. Fruit extracts have also shown good antioxidant activity (Jayaprakash and Patil 2007). In ancient literature, citron was mentioned as an antidote for every kind of poison (Álvarez-Arias and Ramón-Laca 2005). Both the leaves and juice of the citron are used by the people of South-Eastern Nigeria for febrile illness. In South India, the juice is highly recommended for high blood pressure since it can lower it. It also contains potassium, which also improves heart health. It is a well-known fact that juice consumption with warm water also helps purify the blood and liver.

2.3.3 Phytochemical Properties

Tables 2.5, 2.6, and 2.7 show the phytochemical properties of the fruit decoction, peels (rind of fruit) and leaves of *Citrus medica*. The fruit decoction have alkaloids, flavonoids, phenols, carbohydrates, and mucilage; the peels have alkaloids, flavonoids, steroids, phenols, and carbohydrates; the leaves contain alkaloids, flavanoids, steroids, and glycosides (Archana et al. 2010; Kabra et al. 2012; Bairagi et al. 2011). The flavonoids reported from the fruits are hesperidin:3,5,6-trihydroxyl-4,7-dimethoxy flavone, 3,5,6-trihydroxy-3′,4′,7-trimethoxy flavones (Figure 2.5). The peel is reported to contain coumarins, limettin, scoparone, scopoletin, and umbelliferone, while the seeds

TABLE 2.5
Phytochemical Properties of Fruit Decoction, Peels (Rind of Fruit), and Leaves of *Citrus medica*

Phytochemical	Fruit Decoction	Peels	Leaves
Alkaloids	+	+	+
Flavonoids	+	+	+
Tannins	–	–	–
Terpenoids	–	–	–
Steroids	–	+	+
Glycosides	–	–	+
Phenols	+	+	–
Carbohydrates	+	+	–
Mucilage	+	–	–

Note: (+) = Present, (–) = Absent.

TABLE 2.6
Major Constituents of Leaf Essential Oil from *Citrus medica*

Name of Component	Content (%)
Erucylamide	27.3
Limonene	19.21
Citral	13.17
Mehp	7.96
2,6-Octadien-1-ol, 3,7-dimethyl-, acetate, (Z)-	6.14
6-Octenal, 3,7-dimethyl-	4.89
1,2-Cyclohexanediol, 1-methyl-4-(1-methylethenyl)-	3.48
Methoprene	3.31

TABLE 2.7
Major Constituents of Peel Essential Oil from *Citrus medica*

Name of Component	Content (%)
Isolimonene	39.87
Citral	22.15
Limonene	22.40
β-Myrcene	2.12
Neryl acetate	2.71
Neryl alcohol	2.41

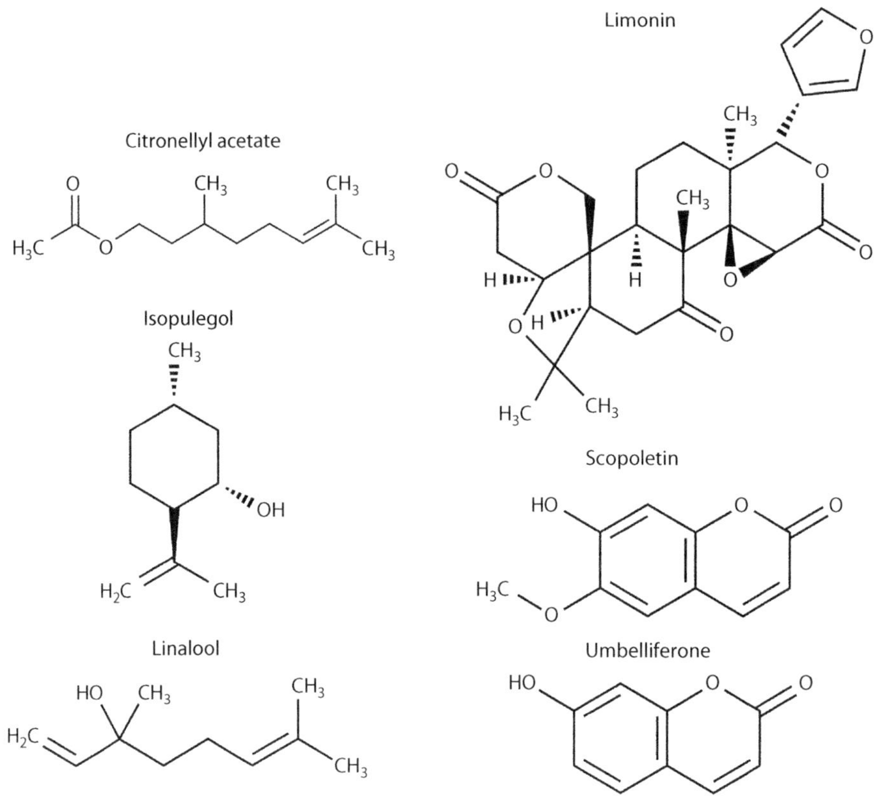

FIGURE 2.5 Chemical structures of citron bioactive constituents. *(Continued)*

contain limonin, limonol, and nomilinic acid (Govindachari 2000; Bhatia 2001). Singh et al. (1999) reported that the major constituents in leaf oil are citronellal, citronellol, limonene, citronellyl acetate, isopulegol, and linalool. Many constituents have been identified in leaf and peel essential oil (19 and 43, respectively). Tables 2.6 and 2.7 show the major essential oils of peels and leaf of *Citrus medica*. The leaf and peel oils are a complex mixture of numerous compounds, many present in trace amounts. It is worth mentioning here that there is a great variation in the chemical composition of the leaf and peel essential oil of *Citrus medica*. Erucylamide and isolimonene are the main and most important components in leaf and peel oil of the Bangladesh variety, but it is totally absent in all other reported oils. This indicates that erucylamide and isolimonene are the first reported component in *Citrus medica* leaf and peel oils. *Citrus medica*, growing widely in Bangladesh, may be utilized as a source to isolate natural erucylamide and isolimonene (Bhuiyan and Begum 2009). Its high concentration of erucylamide and isolimonene in leaf and peel oil makes this fruit potentially useful in medicine because these oils exhibit fungitoxicity (Essien et al. 2008).

Limonol

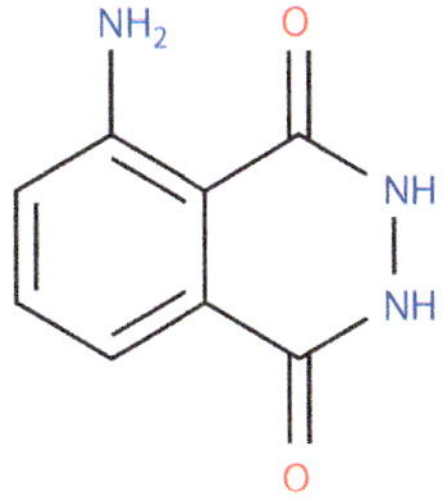

Scoparone

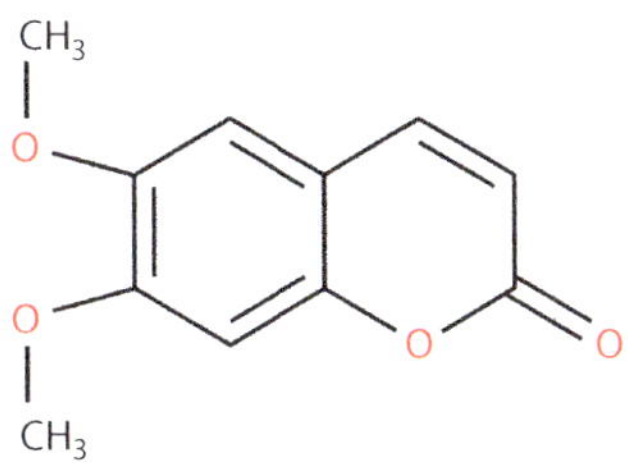

Erucylamide

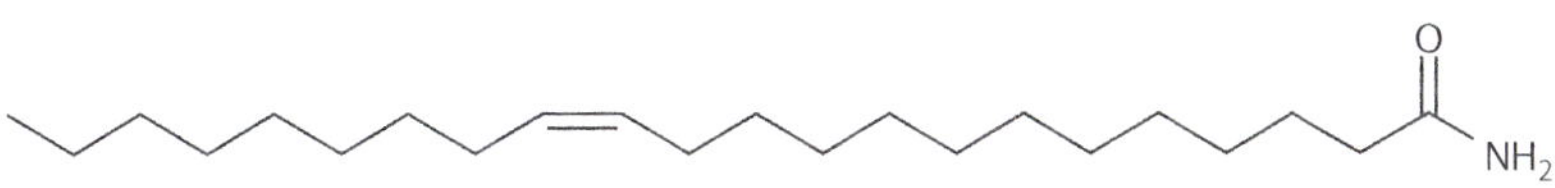

FIGURE 2.5 (CONTINUED) Chemical structures of citron bioactive constituents.

(Continued)

2.3.3.1 Analgesic Activity

Using the hot plate method, all the three doses (1, 2, and 4 mL/kg) of fruit decoction of *Citrus medica* were found to be effective. However, only two doses (2 and 4 mL/kg) of decoction were found to be effective in the tail immersion method; a 1 mL/kg dose was found to be ineffective in this analgesic model for evaluating centrally acting drugs. The decoction of *C. medica* in a dose of 4 mL/kg inhibited the pain, produced by hot plate and tail immersion methods, with a potency comparable to a diclofenac sodium injection. Hence, it can be concluded that this study validates the traditional use of a *C. medica* decoction as an analgesic. Its analgesic activity could be due to flavonoids and phenolic compounds, as the analgesic effect of these compounds have been well-documented (Archana et al. 2010).

2.3.3.2 Hypoglycemic and Anticholinesterase Activity

Oxidative damage, caused by the action of free radicals, may initiate and promote the progression of a number of chronic diseases, including diabetes and Alzheimer's disease. The *n*-hexane extract of Diamante citron (*Citrus medica* L. cv.

Nomilinic acid

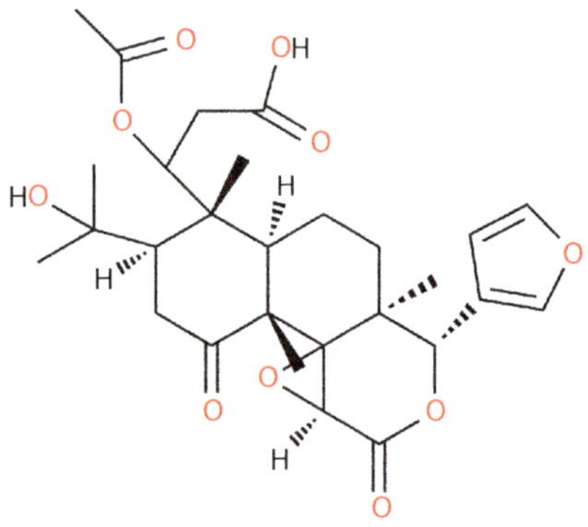

Limettin

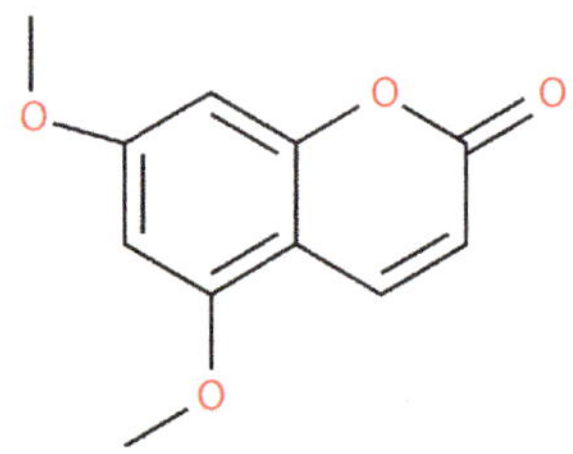

FIGURE 2.5 (CONTINUED) Chemical structures of citron bioactive constituents.

(Continued)

Diamante) peel, which is characterized by the presence of monoterpenes and sesquiterpenes, showed significant antioxidant activity, demonstrated using different assays (DPPH test, β-carotene bleaching test, and bovine brain peroxidation assay). Diamante citron peel extract showed a hypoglycemic activity and an anticholinesterase effect. Thus, these *in vitro* activities of Diamante citron suggest that it can be used in the treatment of diabetes and Alzheimer's disease (Filomena et al. 2007).

2.3.3.3 Anticancer Activity

Some fruits and vegetables are considered important anticancer foods because of their abundant antioxidants such as phenols, vitamin C, vitamin E, β-carotene, and lipotene. Citrus is the most interesting among these fruits. The vital capacity test and Ames test were used to evaluate the anticancer effect of *Citrus medica*, with special emphasis on the application of *Salmonella typhimurium* to identify antimutagenesis and anticancer levels of chemicals. In this research, both half-ripe and ripe fruit juice displayed an anticancer and antimutagenesis effect, but half-ripe fruit juice was the more effective (Maliheh et al. 2009).

2.3.3.4 Antidiabetic, Hypocholesterolemic, and Hypolipidemic Activity

The petroleum ether extract of *C. medica* seeds (200 and 400 mg/kg) induced significant reduction ($p < 0.05$) of fasting blood glucose, serum cholesterol, serum

3,5,6-Trihydroxy-7,4'-dimethoxyflavone

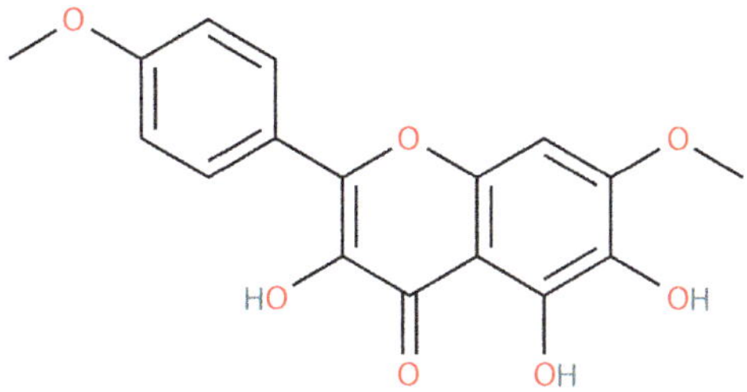

α-Terpinene

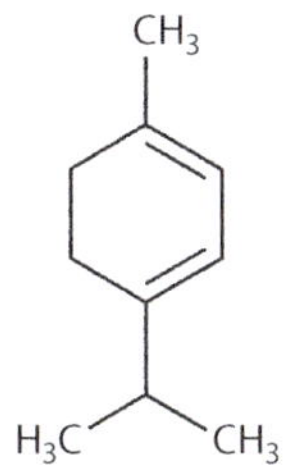

Isolimonene

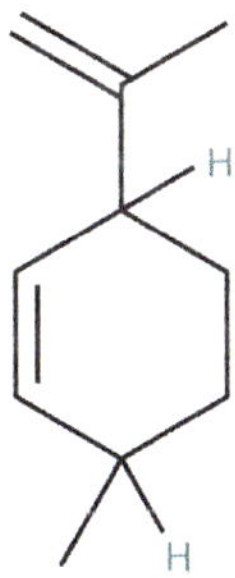

FIGURE 2.5 (CONTINUED) Chemical structures of citron bioactive constituents.

(Continued)

triglycerides, LDL, and VLDL in a dose dependent manner after 15 days of drug administration. Although 200 mg/kg/day of seed extract for 15 days did not produce any change in HDL level, a 400 mg/kg/day dose did significantly increase HDL level in diabetic rats. So it was concluded that *C. medica* seed extract had a significant antidiabetic, hypocholesterolemic, and hypolipidemic activity (Archana et al. 2011).

Citrus medica L. cv Diamante peel extract was also able to reduce plasma glucose concentration and lower the levels of plasma cholesterol and triglycerides, producing *in vivo* metabolic effects in mice. *Citrus medica* could be used as new potential

Trans α-bergamotene

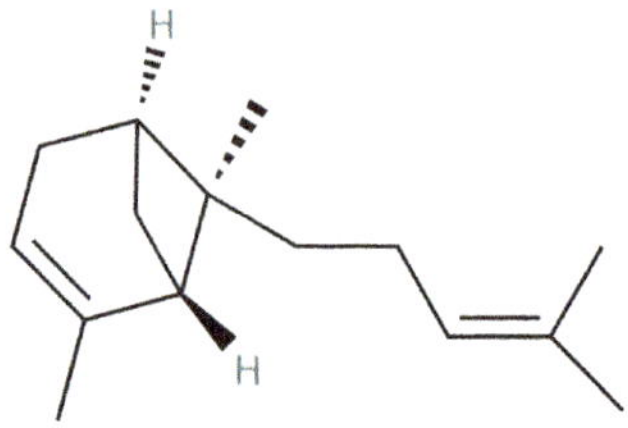

β-Bisabolene

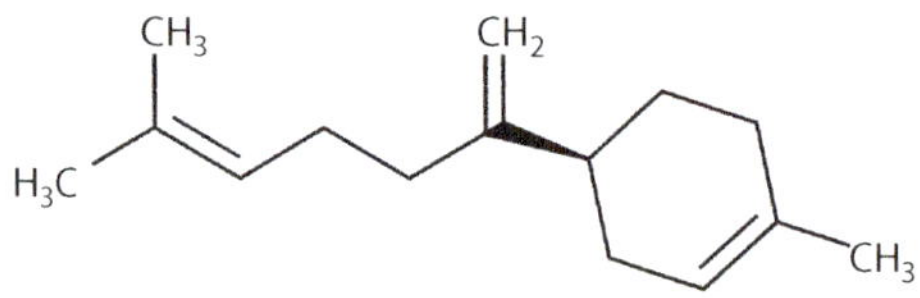

3',5,6-Trihydroxy-3,4',7-trimethoxyflavone

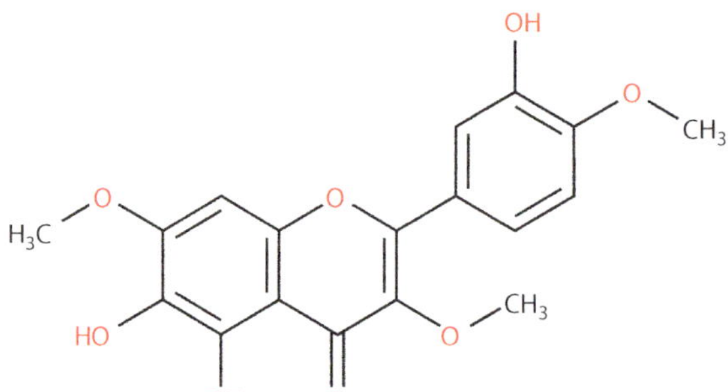

FIGURE 2.5 (CONTINUED) Chemical structures of citron bioactive constituents.

source, with functional properties, for food or nutraceutical products (Menichini et al. 2011).

2.3.3.5 Insulin Secretagogue Activity

In vivo, the safety, hypoglycemic, and antidiabetic activity of *Citrus medica* L. var. *Sarcodactylis* (Finger citron) was tested in Sprague-Dawley–SPF rats and Wistar DIO rats, respectively. Its insulin secretagogue effect was confirmed through kinetic analysis of the hypoglycemic patterns of the intraperitoneal glucose tolerance and the insulin–glucose tolerance tests. Therefore, Finger citron fruits that concomitantly possess insulin secretagogue and slimming effects would be very beneficial to type 2 diabetes mellitus patients (Peng et al. 2009).

2.3.3.6 Anthelmintic Activity

In an *in vitro* study, petroleum ether extracts of *Citrus medica* leaves have shown to possess dose-dependent anthelmintic activity when compared to Piperazine citrate. *Citrus medica* has been confirmed as an anthelmintic against the Indian adult earthworm (*Pheretima posthumad*). The mechanism of the anthelminthic activity of *Citrus medica* cannot be explained; however, it may be due to its effect on glucose uptake inhibition in the parasites and depletion of its glycogen synthesis. *Citrus medica* may also activate a nicotinic cholinergic receptor in the worms resulting in either persistent depolarization or hyperpolarization. There is a need for further studies to identify the active constituent responsible for this anthelmintic activity. Alcoholic extracts of the rind of *Citrus medica* has also shown moderate *in vitro* anthelmintic activity against human *Ascaris lumbricoides* (Raj 1975).

2.3.3.7 Antimicrobial Activity

Plant-based antimicrobials have enormous therapeutic potential as they can serve the purpose with less side effects often associated with synthetic antimicrobials. Continued further exploration of plant-derived antimicrobials is needed today. *In vitro* antibacterial activity of an ethanolic extract of *Citrus medica* peels was performed using the agar cup method. It was found that the peel extract was effective against *Staphylococcus aureus*, *Proteus vulgaris*, *Klebsiella pneumonia*, *Escherichia coli*, *Bacillus subtilis*, and *Pseudomonas aeruginosa*. Further research is necessary to determine the identity of the antibacterial compounds from the peels of *Citrus medica*, and also to determine their full spectrum of efficacy. *Citrus medica* peel extract possesses a broad spectrum of activity against a panel of bacteria responsible for most common bacterial diseases. This extract opens the possibility of finding new clinically effective antibacterial compounds (Kabra et al. 2012).

2.3.3.8 Antiulcer Activity

Aqueous extracts of the fruits of *C. medica* caused statistically significant decreases in the ulcer scores, percentage of ulcers, and ulcer index against ethanol-induced ulcers in rats. The antiulcer effect of *C. medica* could be due to the presence of flavonoids as one of its constituents, as polyphenolic compounds are known to exhibit gastroprotective effects by virtue of their antioxidant properties. These observations were further substantiated by the histopathological findings where decreased mucosal ulceration, decreased inflammatory infiltration in mucosa, and a decrease in edema in sub mucosa were observed in the extract-pretreated groups compared to the untreated group. It was concluded that the *C. medica* fruit extract possesses antiulcer activity and also validated the traditional use of the aqueous extract of *C. medica* as an antiulcer remedy (Nagaraju et al. 2012).

2.3.3.9 Estrogenic Activity

Oral administration of the petroleum ether extract of *Citrus medica* leaves at 400 mg/kg of bodyweight increased uterus weight and showed estrogen-like activity in ovariectomized rats. It exhibited a significant estrogenic activity ($p < 0.05$) at a dose of 400 mg/kg of bodyweight. Petroleum ether extract of leaves could be useful as a safe natural source for estrogenic activity for postmenopausal women (El-Alfy et al. 2012). Estrogenic and antiestrogenic activities of petroleum ether extract of *Citrus medica* seeds were also studied in albino rats. Its result strongly indicates the potent estrogenic nature of the petroleum ether extract of *Citrus medica* seeds, which may be used as an antifertility agent (Sharangouda and Patil 2007).

2.3.4 Processed Citron Products

2.3.4.1 Citron Pickles

Good quality citrons are selected, washed, and cut into small pieces and the juice from at least five citron fruits is extracted with domestic squeezer and kept aside. The seasoning powder is prepared by dry roasting and grinding the ingredients (dry red chilies, mustard seeds, and asafetida) on a low flame after addition of salt.

In a sterilized glass jar, add the extracted citron lime juice and mix in salt along with the cut pieces of citron. Mix them well and let rest for 5 days in a sterile and airtight container. After 5 days, add the prepared seasoning powder according to taste. Let it remain for another day. Add edible oil and mustard seeds. Mix it well and use.

2.3.4.2 Citron Syrup

Synthetic or fruit-based syrups are very popular as summer drinks in different parts of India. They are generally prepared from extracts of natural material or by using artificial flavors and colors. These are added to heavy sugar syrups of 70–75° Brix. The syrup is prepared by heating sugar in water to which a little acid is added to invert the sugar. The sharbhat is finally packed in sterilized glass bottles and stored at room temperature. The added heavy sugar syrup will act as a preservative against several microorganisms. The product has to be diluted at the ratio of 1:3 when served. The greater advantages of this product is that it can be utilized during peak season period and it quenches the thirst of the consumer during the summer season period.

2.3.4.3 Citron Candy

The most important part of the Citron is the peel or rind, which is a fairly important product in international trade. The fruits are halved, depulped, and immersed in seawater or ordinary salt water to ferment for about 40 days. The brine is changed every 2 weeks, the fruits rinsed and put in denser brine in wooden barrels for storage and for export. After partial desalting and boiling to soften the peel, it is candied in a strong sucrose/glucose solution. The candied peel is sundried or put up in jars for future use. Candying is done mainly in England, France, and the United States. The candied peel is widely employed in the food industry, especially as an ingredient in fruitcake, plum pudding, buns, sweet rolls, and candy. In India, during the British period, the peel-candy was sold in confectionery shops.

2.3.4.4 Citron Oil

Citron oil is the second most widely used essential oil after sweet orange oil. It is obtained from the outer peel and is used in all kinds of beverages, soft drinks as well as in food products like cake, pastries, pies, candies, confectionery, aerated water, and so on. It is also used in perfumery, toilet waters, eau de colognes, soaps, and so on.

2.4 LEMON—*CITRUS LIMON*

Citrus limon originated in Southeast Asia, probably in India or Southern China (Figure 2.6). The plant is cultivated in the Mediterranean and subtropical climates worldwide. The lemon tree is an evergreen, growing up to 6 m in height. Its toothed leaves are light green and the fruit is small, oval shape with green to yellow color. Unlike other citrus varieties, the lemon tree bears fruit continuously. The fruit is a modified berry with a tough, leathery rind. The peel contains many volatile oil glands in pits. The interior flesh is composed of segments called carpels, which are made up of numerous juice-filled vesicles. Matured lemons turn yellow from green, measure about 5–8 cm in diameter, and weigh about 50–80 g in weight.

2.4.1 Nutritive Value of Lemons

Lemons are packed with numerous health benefiting nutrients (Table 2.8). They contain very little fat and protein and consist mainly of carbohydrates (9.22%) and water (89.15%). The carbohydrates in lemons are primarily composed of fibers and simple sugars such as glucose, fructose, and sucrose. The main fiber in lemon is pectin. Soluble fibers like pectin can lower blood sugar levels by slowing down the digestion of sugar and starch. A medium-sized lemon only contains about 29 calories and are an excellent source of vitamin C (ascorbic acid) that provides about 88% of daily recommended intake. Ascorbic acid is a powerful water-soluble natural antioxidant. This vitamin is helpful in preventing scurvy. Besides, consumption of foods rich in vitamin C helps the human body to develop resistance against infectious agents and scavenges harmful, proinflammatory free radicals from the blood. The fruit is also a good source of B-complex vitamins such as pantothenic acid, pyridoxine, and folates. These are essential in the sense that the body requires them from external sources to replenish itself. Additionally, they are also composed of minute levels of vitamin A, required for maintaining healthy mucus membranes and skin, and essential for maintaining proper vision. Further, they carry a healthy amount of minerals

FIGURE 2.6 Lemon—*Citrus limon.*

TABLE 2.8
Nutritive Value of Lemon Fruit

Name of Component	Content
Water (%)	89.15
Protein (g)	1.12
Carbohydrate (g)	9.22
Fiber (g)	2.70
Fat (g)	0.33
Calories (kcal)	29
Vitamin A (IU)	22.0
Thiamine (mg)	0.03
Niacin (mg)	0.11
Riboflavin (mg)	0.03
Pyridoxine (mg)	0.09
Pantothenic acid (mg)	0.18
Folates (μg)	11.6
Vitamin C (mg)	53.6
Vitamin E (mg)	0.17
Sodium (mg)	2.09
Potassium (mg)	139
Calcium (mg)	27.1
Iron (mg)	0.59
Magnesium (mg)	8.12
Zinc (mg)	0.05

like iron, potassium, and calcium. Potassium is an important component of cells and body fluids, and helps control the heart rate and blood pressure.

2.4.2 Bioactive Properties of Lemons

Pharmacologically, the lemon is primarily important for its vitamin C and potassium content. Epidemiological studies associate the intake of citrus fruit with a reduction in the risk of various diseases. Citrus fruits, as such, have long been valued for their wholesome nutritious and antioxidant properties. It is a scientifically established fact that citrus fruits, especially lemons and oranges, by virtue of their richness in vitamins and minerals, have many proven health benefits. Moreover, it is now beginning to be appreciated that the other biologically active, nonnutrient compounds found in citrus fruits such as phytochemical antioxidants (Figure 2.7), and soluble as well as insoluble dietary fiber are helpful in reducing the risk for cancer and many chronic diseases such as arthritis, obesity, and coronary heart disease.

2.4.2.1 Antioxidant Effects

German studies in the late 1980s, have related this effect to the peel (Carper 1988). Pectin fiber and lemon oil also possess antioxidant properties. Several groups of

Nomilin

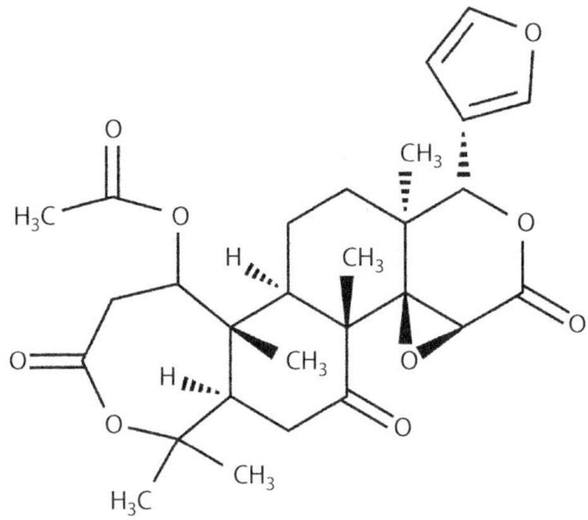

Vitamin C

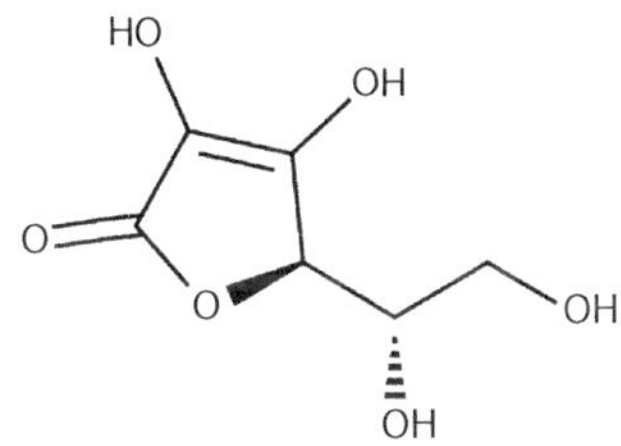

FIGURE 2.7 Chemical structures of lemon bioactive constituents. *(Continued)*

researches, having identified eriocitrin, hesperidin, and coumarins as antioxidants, pursued experiments in diabetic rats, and venous endothelial cells, as well as in activated Epstein–Barr virus models. Their studies identified antioxidant mechanisms and radical scavenging effects that resulted in the inhibition of radical formation (Miyake et al. 1998, 2007; Manners 2007).

2.4.2.2 Anticancer Properties

In an experiment with the flavonoid eriocitrin and its metabolites, and with coumarins extracted from lemon fruit, apoptosis has been demonstrated in acute myelomonocytic leukemia cells (Miyake et al. 2007; Ogata et al. 2000). A meta-analysis of epidemiological studies associates the consumption of citrus fruit with a larger protective effect against oral cancer. The structure of various chemical constituents of lemon and their relationship to cancer prevention has been investigated (Manners 2007; Pavia et al. 2006; Benavente-Garcia et al. 2007).

Jacob et al. (2000) reported the potential of citrus limonoids as anticancer agents in mice, where it was found that five limonoids aglycones (limonin, nomilin, obacunone, isoobacunoic acid, ichangin) induced significant amounts of glutathione-S-transferase (GST) in the liver and intestinal mucosa. GST is a major detoxifying enzyme system, which catalyzes the conjugation of glutathione with

Folic acid

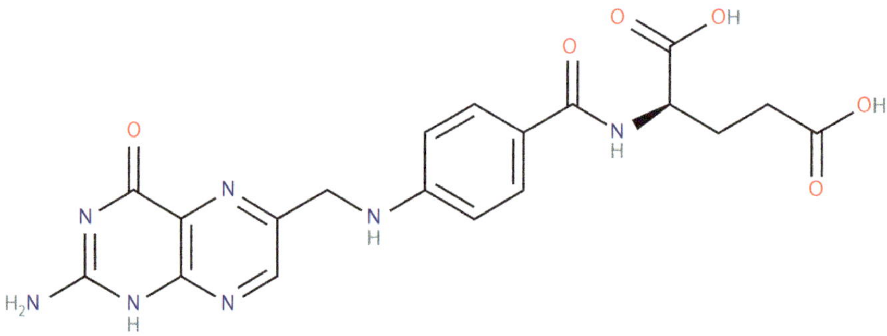

Pectin (polygalacturonic acid)

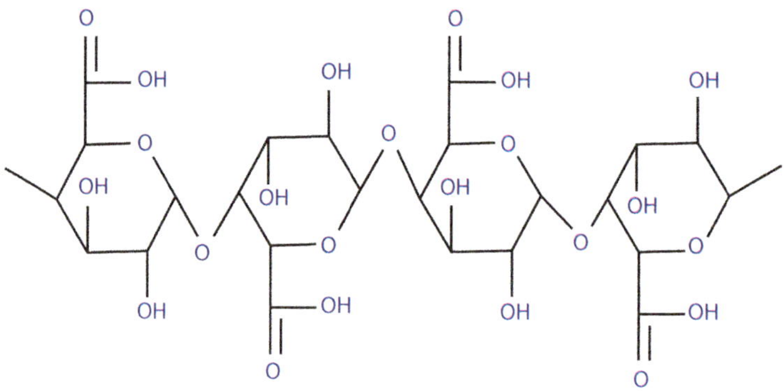

Coumarins

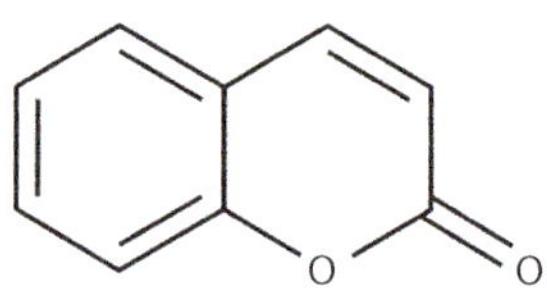

FIGURE 2.7 (CONTINUED) Chemical structures of lemon bioactive constituents.

(Continued)

many potentially carcinogenic compounds that are highly electrophilic in nature. A study of the inhibitory effects of two limonoid aglycones (limonin and nomilin) on the formation of benzo[a]pyrene induced neoplasia in the forestomach of ICR/Ha mice showed that the incidence of tumors could be reduced by more than 50% at a 10 mg/dose.

Jansen (2002), reported on the anticancer and health protective properties of citrus fruit components. Bioactive components present in citrus fruits that are implicated

Obacunone

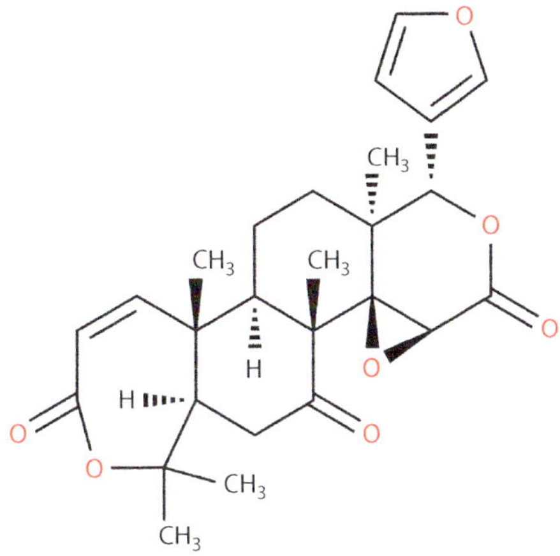

Ichangin

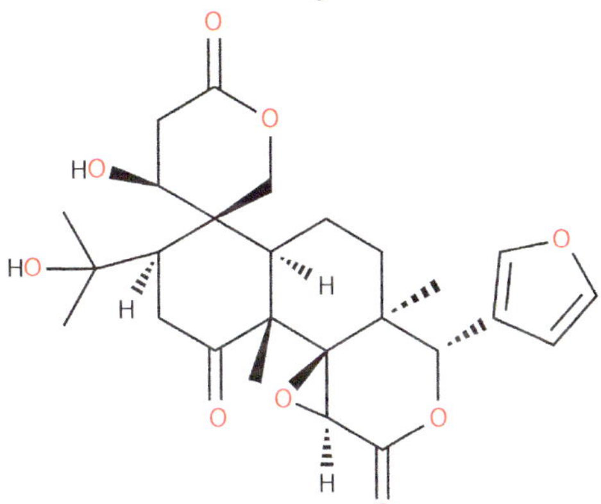

Isoobacunoic acid

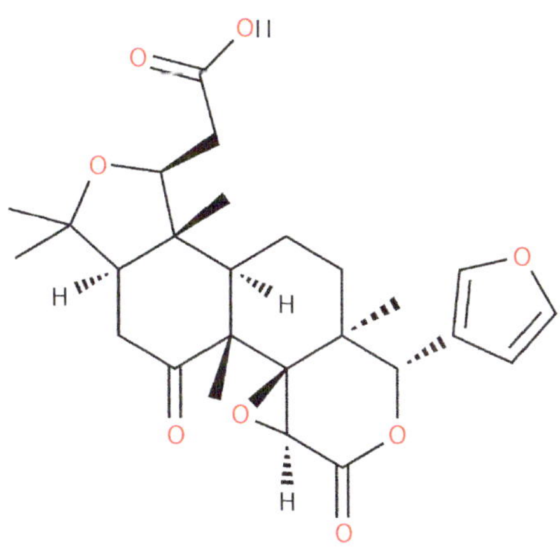

FIGURE 2.7 (CONTINUED) Chemical structures of lemon bioactive constituents.

in degenerative disease prevention include vitamin C, β-carotene, flavonoids, limonoids, folic acid, and dietary fibers. Vitamin C, flavonoids, and β-carotene are potential antioxidants protecting against oxidation of biomolecules such as DNA, protein and lipid membranes, thereby reducing the risk of cancer, cataract, and cardiovascular disease. Limonoids may protect against a variety of cancers by inducing GST activity to neutralize carcinogenic free radicals. Folic acid plays an important role in amino acid metabolism and hence, it is a critical factor for growth. Lemons may help reduce the risk of many types of cancers, including breast cancer. This is mainly due to plant compounds like hesperidin and D-limonene.

2.4.2.3 Nephrolithiasis

Lemon juice has been shown to increase citrate levels in patients with hypocitraturic calcium nephrolithiasis, in a small, long-term trial (mean duration, 44.4 months); 120 mL diluted lemon juice containing 5.9 g of dietary citrate consumed daily resulted in a clinically important reduction in stone formation (Kang et al. 2007; Seltzer et al. 1996).

2.4.2.4 Hypolipidemic Activity

Khan et al. (2010) reported on an evaluation of the hypolipidemic effect of citrus lemon. They compared effects of citrus lemon on cholesterol, triglycerides, LDL, and HDL at the dose of 1 mL/kg citrus lemon for 30 and 45 days respectively in animals fed a high cholesterol diet. After 30 days animals showed highly a significant decrease in cholesterol (150.8 mg/dL) in comparison to control animals (345.3 mg/dL). After 45 days, the decrease in cholesterol continued and a highly significant decrease was observed in comparison to the control animals. Similarly, a highly significant decrease in LDL concentration was observed after 30 days (122.6 mg/dL) in comparison to control animals (273 mg/dL). The decrease in the levels of cholesterol and LDL persisted even after 45 days.

HDL concentration was increased significantly after 30 days (7.27 mg/dL) compared to controls (3.07 mg/dL). However, increase in HDL level became insignificant after 45 days (2.15 mg/dL) compared to controls (2.1 mg/dL). Decrease in triglycerides was insignificant after 30 days compared to control animals; however, a highly significant decrease was observed in triglyceride levels after 45 days (26.69 mg/dL) in comparison to controls (40.3 mg/dL).

2.4.2.5 Antimicrobial Activity

Kumar et al. (2011) reported an antimicrobial activity from solvents extracted from citrus fruit peels viz. *Citrus sinensis* and *Citrus limon*. They reported on the antibacterial activity of different citrus fruit peel solvent extracts (ethyl acetate, acetone, ethanol, petroleum ether, and water) against five pathogenic bacteria's viz. *Staphylococcus aureus, Bacillus subtilis, Escherichia coli, Klebsiella pneumonia,* and *Salmonella typhi*. They found that the highest antibacterial potentiality was exhibited by the acetone peel extract of *Citrus sinensis,* followed by the ethyl acetate peel extract of *Citrus limon*. The peel extract of *Citrus sinensis* and *Citrus limon* can be considered to be as equally potent as antibiotics such as metacillin and penicillin. Lemon juice and lemon oil have been evaluated for antimicrobial action. The oil shows some bacteriostatic and antiviral action, which may be due to the presence of citral and linalool. Lemon has been shown to inhibit the growth of *Aspergillus* mold and has been used to disinfect drinking water or also used to inactivate the rabies virus (Fisher and Phillips 2006; Manners 2007; Ballot et al. 1987; Alderman and Marth 1976; D'Aquino and Teves 1994).

2.4.2.6 Prevention of Anemia

Anemia is often caused by iron deficiency, and it is most common in premenopausal women. Lemons contain small amounts of iron, but are a great source of vitamin C and citric acid, which can increase the absorption of iron from other foods, helping indirectly to prevent anemia (Ballot et al. 1987).

2.4.2.7 Prevention of Kidney Stones

The citric acid in lemons can increase urine volume, which may help prevent the formation of kidney stones. Some studies have shown that lemon juice and lemonade can be effective in this regard (Penniston et al. 2007; Kang et al. 2007).

2.4.2.8 Cardiovascular Health

Cardiovascular disease, including heart attacks and strokes, is the world's most common cause of death. Intake of fruits high in vitamin C is linked to reduced cardiovascular disease. Low levels of vitamin C in the blood are also linked to increase risk of stroke, especially among those who are overweight or have high blood pressure. Intake of isolated fibers from citrus fruits has been shown to decrease blood cholesterol levels, and the essential oils in lemons can protect LDL cholesterol particles from becoming oxidized (Wisker et al. 1994). Recent studies on rats show that the plant compounds hesperidin and diosmin may have beneficial effects on some key risk factors for heart disease (Srinivasan and Pari 2013; Kim et al. 2003).

2.4.3 Other Uses

Lemons were the primary commercial source of citric acid before the development of fermentation-based processes. The juice of the lemon may be used for cleaning. A halved lemon dipped in salt or baking powder is used to brighten copper cookware. The acid dissolves the tarnish and the abrasives assist the cleaning. As a sanitary kitchen deodorizer the juice can deodorize, remove grease, bleach stains, and disinfect; when mixed with baking soda, it removes stains from plastic food storage containers. The oil of the lemon's peel also has various uses. It is used as a wood cleaner and polish, where it's solvent property is employed to dissolve old wax, fingerprints, and grime. Lemon oil and orange oil are also used as a nontoxic insecticide treatment. Lemon oil may be used in aromatherapy. Lemon oil aroma does not influence the human immune system, but may enhance mood. The low pH of juice makes it antibacterial, and in India, the lemon is used in the Indian traditional medicines Siddha and Ayurveda.

2.4.4 Processed Products from Lemons

2.4.4.1 Squash

This is a type of fruit beverage containing at least 25% fruit juice and 40%–50% total soluble solids, commercially. It also contains about 1% acid and 350 ppm sulfur dioxide or 600 ppm sodium benzoate. It is diluted before serving. The squash is prepared by blending extracted lemon juice with acidified sugar syrup, followed by the addition of preservatives, before being bottled and stored at room temperature (Table 2.9 and Figure 2.8).

2.4.4.2 Ready-to-Serve

This is a type of fruit beverage (Table 2.9 and Figure 2.9), which contains at least 10% fruit juice and 10% total soluble solids besides about 0.3% acid. It is not diluted before serving, hence it is known as ready-to-serve (RTS).

2.4.4.3 Cordials

It is a sparkling, clear, and sweetened fruit juice from which pulp and other insoluble substances have been completely removed. It contains at least 25% juice and 30% TSS (Table 2.9 and Figure 2.10). It also contains about 1.5% acid and 350 ppm sulfur dioxide. This is very suitable for blending with wines. Both lime and lemon are suitable for making cordial.

2.4.4.4 Pickles

The preservation of food in common salt or in vinegar is known as pickling. It is one of the most ancient methods of preserving fruits and vegetables. Pickles are good appetizers and add to the palatability of a meal. They stimulate the flow of gastric juice and thus help in digestion. Lemon pickles are prepared with salt, vinegar, oil, or with a mixture of salt, oil, spices, and vinegar (Table 2.9 and Figure 2.11). It is very popular in India and is consumed along with curd rice in most of the South.

2.5 SUMMARY

Citrus fruits are the main fruit trees grown throughout the world and are well appreciated for their refreshing juice and many health benefits. Numerous therapeutic properties, such as anticanceral, antiviral, antitumoral, anti-inflammatory, have been attributed to citrus fruits. Citrus have effects on capillary fragility as well as an ability to inhibit platelet aggregation. These numerous health benefits are linked to the high amounts of photochemical and bioactive compounds such as flavonoids, carotenoids, vitamins, and minerals contained in citrus fruits. These phytonutrients may act as antioxidants, stimulate the immune systems, induce protective enzymes in the liver, or block damage to genetic material. Phytonutrients and vitamins may be responsible for their antioxidant, anticancer, and anti-inflammatory properties. The *Citrus* species has numerous applications in herbal medicine, and the products derived from their fruits bring numerous health benefits to consumers around the world.

TABLE 2.9
Lemon Processed Products Recipe

Lemon Squash		Lemon RTS		Lemon Cordials		Lemon Pickles	
Ingredient	**Quantity**	**Ingredient**	**Quantity**	**Ingredient**	**Quantity**	**Ingredient**	**Quantity**
Juice	100 lb	Juice	100 ml	Clarified juice	100 lb	Fruit	1 kg
Sugar	168 lb	Sugar	100 g	Sugar	129 lb	Salt	200 g
Citric acid	1 lb	Citric acid	–	Water	170 lb	Red chili powder	5 g
Water	127 lb	Water	900 ml	Color	Optional	Cinnamon powder	10 g
Lemon essence	2 lb	Lemon essence	–	Lemon essence	–	Cumin powder	10 g
Potassium metabisulfite	350 ppm	Potassium metabisulfite	70 ppm	Potassium metabisulfite	350 ppm	Cardamom powder	10 g
–	–	–	–	–	–	Black pepper powder	10 g
–	–	–	–	–	–	Clove powder	5 g

Fruits
↓
Washing
↓
Trimming
↓
Cutting or grating
↓
Juice extraction
↓
Straining
↓
Juice measuring
↓
Preparation of syrup
sugar + water + acid, heating just
to dissolve)
↓
Straining
↓
Mixing with juice
↓
Addition of preservative
(0.6 g KMS or 1.0 g sodium
benzoate/liter squash)
↓
Bottling
↓
Capping
↓
Storage

FIGURE 2.8 Flowchart for processing lemon squash.

Fruits

(Pulp/juice)

↓

Mix with strained syrup solution
(sugar + water + acid, heated just
to dissolve) according to recipe

↓

Homogenization

↓

Bottling

↓

Crown corking

↓

Pasteurization
(at about 90°C) for 25 minutes

↓

Cooling

↓

Storage

FIGURE 2.9 Flowchart for processing RTS beverages.

Fruits
↓
Washing
↓
Cut into halves
↓
Juice extraction
↓
Straining
↓
Addition of preservative
(KMS 1 g/liter juice)
↓
Storing in glass container for 10–15 days
for clarification
(suspended material settles down)
↓
Syphoning off the
supernatant clear juice
↓
Straining and measuring
↓
Preparation of syrup
↓
Straining

Mixing of juice with syrup
↓
Addition of preservative
(KMS 0.6 g/liter product)
↓
Bottling
↓
Capping
↓
Store in cool and dry place

FIGURE 2.10 Flowchart for processing lemon cordials.

Select lemon

↓

Wash

↓

Cut into four pieces

↓

Squeeze out juice from quarter of the fruits

↓

Mix spices and salt with juice

↓

Mix with lime pieces and fill in jars

↓

Cover with lid and keep in sun for 4–6 days (shake jars at least twice a day)

↓

Store at ambient temperature (in cool and dry place)

FIGURE 2.11 Flowchart for processing lemon pickles.

REFERENCES

Alderman, G. G., and E. H. Marth. 1976. Inhibition of growth and aflatoxin production of *Aspergillusparasiticus* by citrus oils. *Zeitschrift für Lebensmittel-Untersuchung und Forschung* 160:353–358.

Álvarez-Arias, B., and L. Ramón-Laca. 2005. Pharmacological properties of citrus and their ancient and medieval uses in the Mediterranean region. *Journal of Ethnopharmacology* 97:89–95.

Archana, N., V. Juyal, and A. B. Melkani. 2010. Analgesic activity of fruit decoction of *Citrus medica* Linn. *Journal of Pharmacy Research* 3:2119–2121.

Archana, N., A. Joshi, V. Juyal, and T. Kumar. 2011. Antidiabetic and hypolipidemic activity of *Citrus medica* L. seed extract in streptozotocin induced diabetic rats. *Pharmacognosy Journal* 3:80–84.

Ayurvedic Pharmacopoeia of India. 2001. Part I, Vol. III, New Delhi: Gout of India. pp. 27–28.

Bairagi, G. B., A. O. Kabra, and R. J. Mandade. 2011. Anthelmintic activity of *Citrus medica* L. leaves in Indian adult earthworm. *International Journal of PharmTech Research* 3:664–667.

Ballot, D., R. D. Baynes, T. H. Bothwell, M. Gillooly, J. Macfarlane, A. P. Macphail, G. Lyons et al. 1987. The effects of fruit juices and fruits on the absorption of iron from a rice meal. *The British Journal of Nutrition* 57:331–343.

Benavente-García, O., J. Castillo, M. Alcaraz, V. Vicente, J. A. Del-Río, and A. Ortuño. 2007. Beneficial action of citrus flavonoids on multiple cancer-related biological pathways. *Current Cancer Drug Targets* 7:795–809.

Bhatia, S. C. 2001. *Perfumes and Cosmetics*. Vol. II. New Delhi: CBS Publishers and Distributors. pp. 627.

Bhuiyan, M. N. I., and J. Begum. 2009. Constituents of peel and leaf essential oils of *Citrus medica* L. *Journal of Scientific Research* 1:387–392.

Carper, J. 1988. *The Food Pharmacy*. New York: Bantam Publishing.

D'Aquino, M., and S. A. Teves. 1994. Lemon juice as a natural biocide for disinfecting drinking water. *Bulletin of the Pan American Health Organization* 28:324–330.

El-Alfy, T. S., M. H. Hetta, N. Z. Yassin, R. F. Rahman, K. Abdel, and M. Esraa. 2012. Estrogenic activity of *Citrus medica* L. leaves growing in Egypt. *Journal of Applied Pharmaceutical Science* 2:180–185.

Essien, E. P., J. P. Essien, B. N. Ita, and G. A. Ebong. 2008. Physicochemical properties and fungitoxicity of the essential oil of *Citrus medica* L. against groundnut storage fungi. *Turkish Journal of Botany* 32:161–164.

Faturi, C. B., J. R. Leite, P. B. Alves, A. C. Canton, and F. Teixeira-Silva. 2010. Anxiolytic effect of sweet orange aroma in Wistar rat. *Progress in Neuro-Psychopharmacology and Biological Psychiatry* 34:605–609.

Filomena, C., A. S. Giancarlo, T. Rosa, R. L. Monica, and M. Francesco. 2007. In vitro activities of *Citrus medica* L. cv. Diamante (*Diamante citron*) relevant to treatment of diabetes and Alzheimer's disease. *Phytotherapy Research* 21:427.

Fisher, K., and C. A. Phillips. 2006. The effect of lemon, orange and bergamot essential oils and their components on the survival of *Campylobacter jejuni*, *Escherichia coli* O157, *Listeria monocytogenes*, *Bacillus cereus* and *Staphylococcus aureus* in vitro and in food systems. *Journal of Applied Microbiology* 101:1232–1240.

Ghasemi, K., Y. Ghasemi, and M. A. Ebrahimzadeh. 2009. Antioxidant activity, phenol and flavonoid contents of 13 citrus species peels and tissues. *Pakistan Journal of Pharmaceutical Science* 22:277–281.

Govindachari, T. R. 2000. Antifungal activity of some tetranor triterpenoids. *Fitoterapia* 71:317.

Guarnieri, S., P. Riso, and M. Porrini. 2007. Orange juice vs vitamin C: Effect on hydrogen peroxide-induced DNA damage in mononuclear blood cells. *British Journal of Nutrition* 97:639–643.

Harley, I., R. S. Manner, V. Buker, D. W. Easton-Smith, and R. E. Craig. 2006. Species Profiles for Pacific Island Agro Forestry. www.agroforestry.net.

Honow, R., N. Laube, A. Schneider, T. Kessler, and A. Hesse. 2003. Influence of grapefruit, orange-, and apple-juice consumption on urinary variables and risk of crystallization. *British Journal of Nutrition* 90:295–300.

Ihrig, M. 1995. Qualitätskontrolle von süßem Orangenschalenöl. *Pharmazeutische Zeitung* 140:2350–2353.

Jacob, R., S. Hasegawa, and G. Manners. 2000. The potential of *Citrus limonoids* as anticancer agents. *Perishables Handling Quarterly* 102:6–8.

Jansen, S. 2002. Anticancer and health protective properties of citrus fruits components. *Asia Pacific Journal of Clinical Nutrition* 11:79–84.

Jayaprakash, G. K., and B. S. Patil. 2007. In vitro evaluation of the antioxidant activities in fruit extracts from citron and blood orange. *Food Chemistry* 101:410–418.

Kabra, A. O., G. B. Bairagi, A. S. Mahamuni, and R. S. Wanare. 2012. In vitro antimicrobial activity and phytochemical analysis of the peels of *Citrus medica*. *International Journal of Research in Pharmaceutical and Biomedical Sciences* 3:34.

Kang, D. E., R. L. Sur, G. E. Haleblian, N. J. Fitzsimons, K. M. Borawski, and G. M. Preminger. 2007. Long-term lemonade based dietary manipulation in patients with hypocitraturic nephrolithiasis. *Journal of Urology* 177:1358–1362.

Khan, Y., R. A. Khan, A. Syeda, and S. Afshan. 2010. Evaluation of hypolipidemic effect of *Citrus lemon*. *Journal of Basic and Applied Sciences* 6:39–43.

Kim, H. K., T. S. Jeong, M. K. Lee, Y. B. Park, and M. S. Choi. 2003. Lipid-lowering efficacy of hesperetin metabolites in high cholesterol fed rats. *Clinical Chimica Acta* 327:129–37.

Kumar, A., M. Narayani, A Subanthini, and M. Jayakumar. 2011. Antimicrobial activity and phytochemical analysis of citrus fruits peel utilization of fruit waste. *International Journal of Engineering Science and Technology* 3:5414–5421.

Kumar, V., R. Nandini, S. Shashidhara, and S. Anitha. 2010. Anti typhoid activity of aqueous extract of fruit peel citrus sinensis. *International Journal of Pharmaceutical Research and Development* 2:217–219.

Kurowska, E. M., and J. A. Manthey. 2004. Hypolipidemic effects and absorption of citrus polymethoxylated flavones in hamsters with diet-induced hypercholesterolemia. *Journal of Agricultural and Food Chemistry* 19:2879–2886.

Lin, N., T. Sato, Y. Takayama, Y. Mimaki, Y. Sashida, M. Yano, and A. Ito. 2003. Novel anti-inflammatory actions of nobiletin, a citrus polymethoxy flavonoid, on human synovial fibroblasts and mouse macrophages. *Biochemical Pharmacology* 65:2065–2071.

Maliheh, E. 2009. Antimutagenicity and anticancer effects of *Citrus medica* fruit juice. *Acta Medica Iranica* 47:374–77.

Manners, G. D. 2007. Citrus limonoids: Analysis, bioactivity, and biomedical prospects. *Journal of Agriculture and Food Chemistry* 55:8285–8294.

Menichini, F., R. Tundis, M. R. Loizzo, M. Bonesi, B. Liu, P. Jones, S. J. Persaud et al. 2011. *C. medica* cv Diamante peel chemical composition and influence on glucose homeostasis and metabolic parameters. *Food Chemistry* 124:1083–1089.

Miyake, Y., K. Yamamoto, N. Tsujihara, and T. Osawa. 1998. Protective effects of lemon flavonoids on oxidative stress in diabetic rats. *Lipids* 33:689–695.

Miyake, Y., M. Mochizuki, M. Okada, M. Hiramitsu, Y. Morimitsu, and T. Osawa. 2007. Isolation of antioxidative phenolic glucosides from lemon juice and their suppressive effect on the expression of blood adhesion molecules. *Bioscience, Biotechnology, and Biochemistry* 71:1911–1919.

Nagaraju, B., S. C. Anand, N. Ahmed, J. N. Narendra Sharath-Chandra, F. Ahmed, and G. V. Padmavathi. 2012. Antiulcer activity of aqueous extract of *Citrus medica* Linn. Fruit against ethanol-induced ulcer in rats. *Advances in Biological Research* 6:24–29.

Oben, J., E. Enonchong, S. Kothari, W. Chambliss, R. Garrison, and D. Dolnick. 2009. Phellodendron and citrus extracts benefit joint health in osteoarthritis patients: A pilot, double-blind, placebo controlled study. *BioMed Central* 1:8–38.

Ogata, S., Y. Miyake, K. Yamamoto, K. Okumura, and H. Taguchi. 2000. Apoptosis induced by the flavonoid from lemon fruit (*Citrus limon* Burm. f.) and its metabolites in HL-60 cells. *Bioscience, Biotechnology, and Biochemistry* 64:1075–1078.

Parmar, H. S., and A. Kar. 1995. Possible amelioration of atherogenic diet induced dyslipidemia, hypothyroidism and hyperglycemia by the peel extracts of *Mangiferaindica*, *Cucumismelo* and *Citrullus vulgaris* fruits in rats and liver antioxidative enzyme activity in rats. *Bioscience, Biotechnology, and Biochemistry* 59:595–601.

Parmar, H. S., and A. Kar. 2008. Medicinal values of fruit peels from *Citrus sinensis*, *Punicagranatum* and *Musa paradisiaca* with respect to alterations in tissue lipid peroxidation and serum concentration of glucose, insulin and thyroid hormones. *Journal of Medicinal Food* 11:376–381.

Pavia, M., C. Pileggi, C. G. Nobile, and I. F. Angelillo. 2006. Association between fruit and vegetable consumption and oral cancer: A meta-analysis of observational studies. *American Journal of Clinical Nutrion* 83:1126–1134.

Peng C. H., Y. B. Ker, C. F. Weng, C. C. Peng, C. N. Huang, L. Y. Lin, and R. Y. Peng. 2009. Insulin secretagogue bioactivity of finger citron fruit (*Citrus medica* L. var. *Sarcodactylis* Hort, Rutaceae). *Journal of Agricultural and Food Chemistry* 57:8812–8819.

Penniston, K. L., T. H. Steele, and S. Y. Nakada. 2007. Lemonade therapy increases urinary citrate and urine volumes in patients with recurrent calcium oxalate stone formation. *Urology* 70:856–60.

Raj, R. K. 1975. Screening of indigenous plants for anthelmintic action against human *Ascaris lumbricoides*: Part—II. *Indian Journal of Physiology and Pharmacology* 19(1).

Sandhya, S., P. SaiKumar, K. R. Vinod, D. Banji, and K. Kumar. 2011. Plants as potent anti diabetic and wound healing agents. *Hygeia Journal for Drugs and Medicines* 3:11–19.

Seltzer, M. A., R. K. Low, M. McDonald, G. S. Shami, and M. L. Stoller. 1996. Dietary manipulation with lemonade to treat hypocitraturic calcium nephrolithiasis. *Journal of Urology* 156:907–909.

Sharangouda, and S. B. Patil. 2007. Estrogenic activity of petroleum ether extract of *Citrus medica* L. on immature albino rats. *International Journal of Green Pharmacy* 2:91–94.

Sharma, N., and A. Tripathi. 2008. Effects of *Citrus sinensis* (L.) Osbeck epicarp essential oil on growth and morphogenesis of *Aspergillus niger* (L.). *Microbiological Research* 163:337–344.

Simon, J. A., E. S. Hudes, and G. I. Perez-Perez. 2003. Relation of serum ascorbic acid to *Helicobacter pylori* serology in U.S. adults: The third national health and nutrition examination survey. *Journal of the American College of Nutrition* 22:283–289.

Singh, G., I. P. S. Kapoor, O. P. Singh, P. A. Leclercq, and N. Klinkby. 1999. Chemical constituents of peel and leaf essential oils of *Citrus medica* L. *Journal of Essential Oil-Bearing Plants* 2:119.

Srinivasan, S., and L. Pari. 2013. Antihyperlipidemic effect of diosmin: A citrus flavonoid on lipid metabolism in experimental diabetic rats. *Journal of Functional Foods* 5:484–492.

Strange, R. R., S. L. Miland, J. W. Eckert, and J. J. Sims. 1993. An antifungal compound produced by grapefruit and valencia orange after wounding of the peel. *Journal of Natural Products* 56:1627–1629.

Tanaka, Y., H. Makita, K. Kawabata, H. Mori, M. Kakumoto, K. Satoh, A. Hara et al. 1997. Chemoprevention of azoxymethane-induced rat colon carcinogenesis by the naturally occurring flavonoids, diosmin and hesperidin. *Carcinogenesis* 18:957–965.

Waterman, P. G., and M. F. Grundon. 1983. *Chemistry and Chemical Taxonomy of the Rutales*. New York: Academic Press.

Wiesman, Z., and B. P. Chapagain. 2005. Larvicidal of aqueous extracts of *Balanites aegyptiaca* (desert date) against the larvae of *Culexpipiens* mosquitoes. *African Journal of Biotechnology* 4:1351–1354.

Wisker E., M. Daniel, and W. Feldheim. 1994. Effects of a fiber concentrate from citrus fruits in humans. *Nutrition Research* 14:361–372.

3 Citrus Bioactives as Prospective Therapeutic Agents for the Modulation of Colon Cancer

Jasleen Kaur and Gurpreet Kaur

CONTENTS

3.1 INTRODUCTION

Recently, there has been a growing realization that increased consumption of fruits and vegetables is associated with innumerable health benefits. The health-promoting action of fruits and vegetables is due to a number of nutritive (vitamins and minerals) and non-nutritive (polyphenols, terpenes, alkaloids, steroids, and fiber) components. Both these components are synthesized by plants and undergo numerous bioactivities (Bourgaud et al. 2001). Polyphenols are the largest group of phytochemicals and have at least one aromatic ring with reactive hydroxyl groups. Depending upon the structural characteristics, these are classified as flavonoids, nonflavonoids (phenolic acids—hydroxybenzoic acid and hydroxycinnamic acid), stilbenes, and lignans. Polyphenols possess a wide spectrum of biological activities such as anti-inflammatory, anticarcinogenic, neuroprotective, antiallergic, oestrogenic, antithrombotic, hepatoprotective, antibiotic, antiviral, antiulcer, antilipidaemic, and vasorelaxing (Heilmann 2009). They are recognized as promising cancer preventive agents because of their capacity to regulate the altered cellular homeostasis in cancerous cells. Their main chemopreventive mechanisms encompass induction of apoptosis, interference in cell cycle stages, prevention of DNA damage, angiogenesis, and modulation of cell-signaling cascades.

The two major families that have outstanding annual fruit production and are important sources of polyphenols include Rutaceae (orange, grapefruit, lemon, lime, mandarin, and pomelo) and Rosaceae (apple, pear, peach, cherry, apricot, raspberry, and strawberry). The single genus *Citrus* belonging to family Rutaceae contains many power-packed fruits that are economical and highly consumed throughout the world. In addition to nutrition, they are good sources of various bioactive components that play an important role in preventing chronic diseases.

3.2 *CITRUS*—A COMPLETE FRUIT

Citrus is one widely cropped genus of the Rutaceae family. Citrus fruits are the most economical fruits in the world. The family Rutaceae comprises six genera: *Citrus*, *Poncirus*, *Fortunella*, *Microcitrus*, *Eremocitrus*, and *Clymania*. However, only three genera—*Citrus*, *Poncirus*, and *Fortunella* possess commercial importance. The genus *Citrus* is divided into two subgenera: *Eucitrus* (edible fruit with pulp vesicles not containing drops of acrid or bitter oil) and *Papeda* (inedible fruit with acrid oil droplets in juice vesicles). Out of total citrus production, 65% is used for fresh consumption while the remaining 35% is used for industrial processing (Moore 2001).

All citrus fruits are fleshy fruits botanically classified as modified berries and known as hesperidium. Citrus fruits comprise a fleshy portion having fluid-filled vesicles protected by tough aromatic rind. They are consumed as fresh fruit or juice because of their nutritional value, distinctive (refreshing) aroma, and flavor (mix of sweetness and tartness). All citrus fruits are acidic in nature and exhibit marked detoxifying properties. Most of the protective action of citrus fruits has been attributed to vitamin C. However, experimental work and epidemiological studies have highlighted that the health benefits are associated with consumption of whole citrus fruits rather than vitamin C alone. Citrus fruits, in addition to vitamin C, are rich in a number of phytochemicals such as flavonoids, limonoids, carotenoids, folate, and dietary fiber. These bioactives play an important role in prevention of a variety of cancers and degenerative diseases (Pao and Fellers 2003).

A number of by-products and waste products are generated by the citrus fruit industries after the extraction of juice. These include citrus pulp, seed meals, molasses, and peels. The major by-product after processing of citrus fruit is citrus peel, which contributes 50%–65% of whole fruit weight, and is employed in the preparation of infusions, tinctures, candies, and wine. Citrus peels have been used for hundreds of years in traditional Chinese and Japanese medicine systems. The peel comprises two parts: epicarp (flavedo, the outer colored peripheral part) and mesocarp (albedo/pith, the white spongy inner portion). The flavedo portion mostly consists of cellulosic material and contains essential oils, pigments (carotenoids, chlorophyll, and flavonoids), phenolic acids, paraffin waxes, steroids, and triterpinoids. The albedo is the inner part of the peel and is a rich source of dietary fiber such as pectin and cellulose, which constitute approximately one-third of the dry weight of albedo, while the remainder is composed of flavonoids, polysaccharides, and monosaccharides such as glucose and fructose.

The other waste product of juice industries is seeds, which have been found to be a rich source of triterpene derivatives, chemically known as limonoids. Although their intense bitterness is responsible for the delayed bitterness in citrus juice, their outstanding biological actions have proven to be of utmost importance in medicine. Table 3.1 depicts the bioactive compounds present in citrus fruits.

3.3 COLON CANCER

The colonic mucosa is constantly exposed to external stimuli that make it vulnerable to a variety of chronic ailments. Disturbances in colonic homeostasis due to various impulses result in diverse arrays of intestinal diseases such as amoebiasis, diarrhea, constipation, Crohn's disease, ulcerative colitis, irritable bowel syndrome, and colon cancer. Some of these pathological conditions are reasonably innocuous, but the majority of them are devitalizing and life threatening. For example, colon cancer is the third leading cause of cancer-related death in women and the fourth in men, with 693,600 deaths reported worldwide in 2012. The incidence of colon cancer is increasing dramatically, presumably due to erratic dietary habits, smoking and high alcohol intake, and sedentary lifestyles. The age of a person is another factor that may increase the risk of colon cancer. About 70% of colon cancer cases are reported in patients older than 65 years, and 40% in those older than 75 years (Benson 2007).

The manifestation of colon cancer may take place over many years as a consequence of strong interaction between genetic and environmental factors. The genetic factors responsible for the development of colon cancer may be inherited or noninherited mutations. Hereditary forms comprise less than 5% of all colon cancer cases; the sporadic (nonhereditary) form arises in approximately 70%–80% of patients. The two common forms of hereditary colon cancer are familial adenomatous polyposis (FAP) and hereditary nonpolyposis colon cancer (HNPCC) (Lynch syndrome). The FAP syndrome originates from inherited mutations in tumor suppressor genes like adenomatous polyposis coli (APC), while Lynch syndrome results from inherited mutations in any of five mismatched repair (MMR) genes. The sporadic/nonhereditary mutations normally occur in the same genes, along with Kristen rays (K-rays), p53, and p16, as a result of exposure to environmental factors (Giardiello, Brensinger, and Petersen 1997; Aaltonen et al. 1998; Ahsan et al. 1998; Wijnen, Vasen, and Khan 1998).

The nonhereditary environmental factors encompass low intake of fiber, fruits and vegetables, and vitamins and minerals, as well as high intake of fatty foods, alcohol, and red meat and lack of physical activity. Dietary components contribute a major role as nonhereditary environmental factors. These factors alter the sensitivity of the colon to genetic damage (Huxley et al. 2009; Wei et al. 2009; Basterfield and Mathers 2010; Chan and Giovannucci 2010; Thompson et al. 2011). Similarly, inflammatory

TABLE 3.1
Important Chemical Constituents of Citrus Fruits

Chemical Constituents	Basic Chemical Structure	Examples
Flavonoids 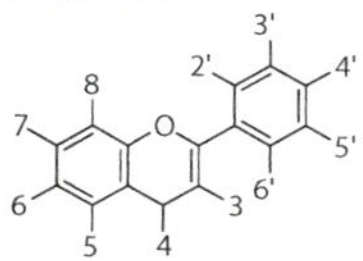	Flavanones 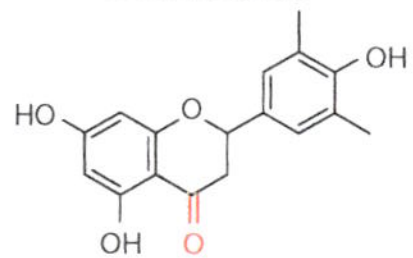	Hesperetin, naringenin, eriodictyol and isosakuranetin
	Flavones 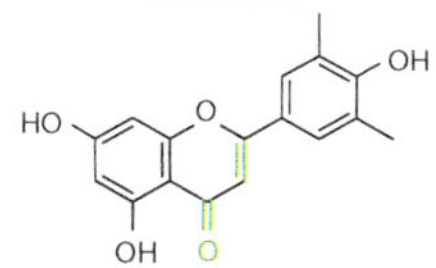	Apigenin, luteolin, chrysoeriol, diosmetin
	Polymethoxyflavones (PMFs) 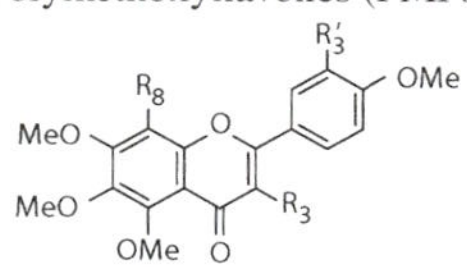	Nobiletin, tangeretin, hydroxylated-PMFs (H-PMFs)
	Flavonols 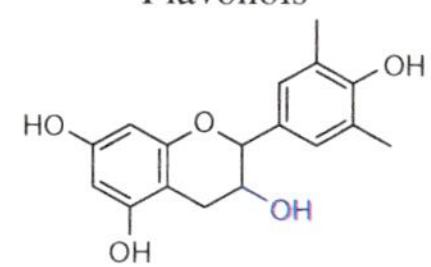	Kaempferol, quercetin, myrcetin, isorhamnetin
Carotenoids	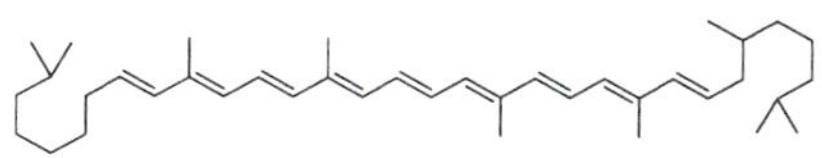	Lutein, β-cryptoxanthin, β-carotene, lycopene, zeaxanthin

(Continued)

TABLE 3.1 (CONTINUED)
Important Chemical Constituents of Citrus Fruits

Chemical Constituents	Basic Chemical Structure		Examples
Limonoids	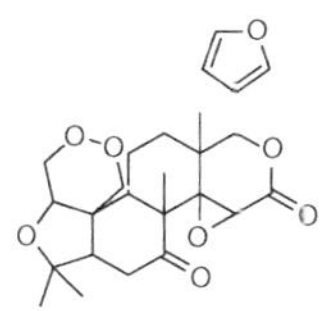	Limonin Calamine Ichangensin 7-α-acetate limonoids	Limonin, nomilin, obacunone, deacetylnomilinic acid, isolimonic acid, ichanexic acid
Coumarins 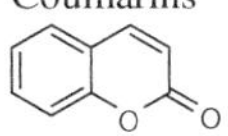	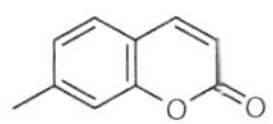 Simple coumarins 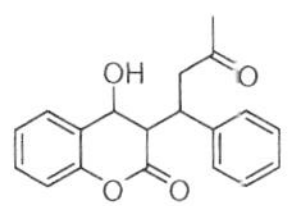 Pyranocoumarins	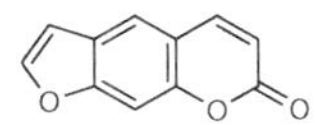 Furanocoumarins 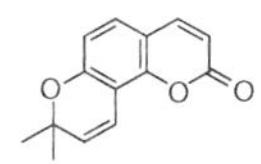 Pyrone substituted coumarins	Auraptene, collinin, umbelliferone, esculin, esculetin, 4-methylesculetin, limettin, isopimpinellin

Note: The colored text indicates the characteristic functional groups and examples of respective categories of flavonoids.

conditions and previous history of polyps in patients also increase their susceptibility to developing colon cancer. Further, the alteration in digestion and its related biochemical, physiological, and behavioral processes due to an irregular lifestyle can also trigger the initiation of colon cancer (Derry et al. 2013).

Genetic mutations and environmental factors produce molecular abnormalities that disrupt the balance between proliferation, differentiation, and apoptosis processes (Nambiar, Gupta, and Misra 2010). This results in abnormal cellular proliferation and inflammation, leading to aberrant crypt foci (ACF). These ACFs are the putative preneoplastic lesions of colonic neoplasia and appear in the early stages of colon cancer. ACFs subsequently transform to polyps, adenomas (ADs), or adenomatous polyps and finally grow to a stage of adenocarcinoma (ADC). ADC accounts for greater than 90% of all large bowel cancers and involves superficial mucosal epithelial cells, which usually invade locally through the bowel wall and then metastasize to distant organs (Benson 2007). Detailed etiology, imbalance in various processes, and crucial molecular targets responsible for progression of colon cancer are represented in Figure 3.1.

Several histological (morphological), biochemical, and proliferation biomarkers can be employed to elucidate colon cancer (Tanaka, Kohno, and Mori 2001). The number/incidence or multiplicity of ACF, AD, and ADC is the various biomarkers indicating histological changes. In colon cancer, the occurrence of ACF points out several biological aberrations, including gene mutations, amplification, and increased cell proliferation activity (Bird 1995). 5′-Bromodeoxyuridine (BrdU)-labeling index, proliferating cell nuclear antigen (PCNA)-labeling index, and silver-stained nucleolar region (AgNOR) number are indicators of proliferation (Tanaka 1997). A decrease in these biomarkers indicates a decline in proliferation activity. Ornithine decarboxylase (ODC) and polyamines are involved in normal cellular proliferation and play a pivotal role in carcinogenesis including colon tumorigenesis (Lan, Trempus, and Gilmour 2000). Literature reports suggest that an imbalance between the phase I carcinogen-activating enzymes and the phase II detoxifying enzymes is critical in determining an individual's risk for cancer (Wilkinson and Clapper 1997). Any deficiency in phase II detoxifying enzyme activity, especially glutathione-*S*-transferase (GST) and quinone reductase (QR) might increase the risk for development of colon cancer (Szarka et al. 1995). Therefore, ODC, polyamines, GST, and QR might constitute effective biochemical biomarkers for chemopreventive studies. In addition, inflammatory mediators such as cyclooxygenase (COX), lipooxygenase (LOX), phospholipase A_2 (PLA_2), inducible nitric oxide synthase (iNOS), and nuclear factor-kappa B (NF-κB) serve as inflammation biomarkers (Tanaka et al. 2001).

3.4 THE ROLE OF NATURAL PHYTOCHEMICALS IN COLON CANCER TREATMENT

Currently, pharmacotherapy and surgical intervention remain the first choice for the treatment of colon diseases. However, a number of drawbacks are associated with these therapies, which include hair loss, mouth sores, loss of appetite, nausea and vomiting, diarrhea, increased chance of infections, bleeding, and fatigue (Benson 2007).

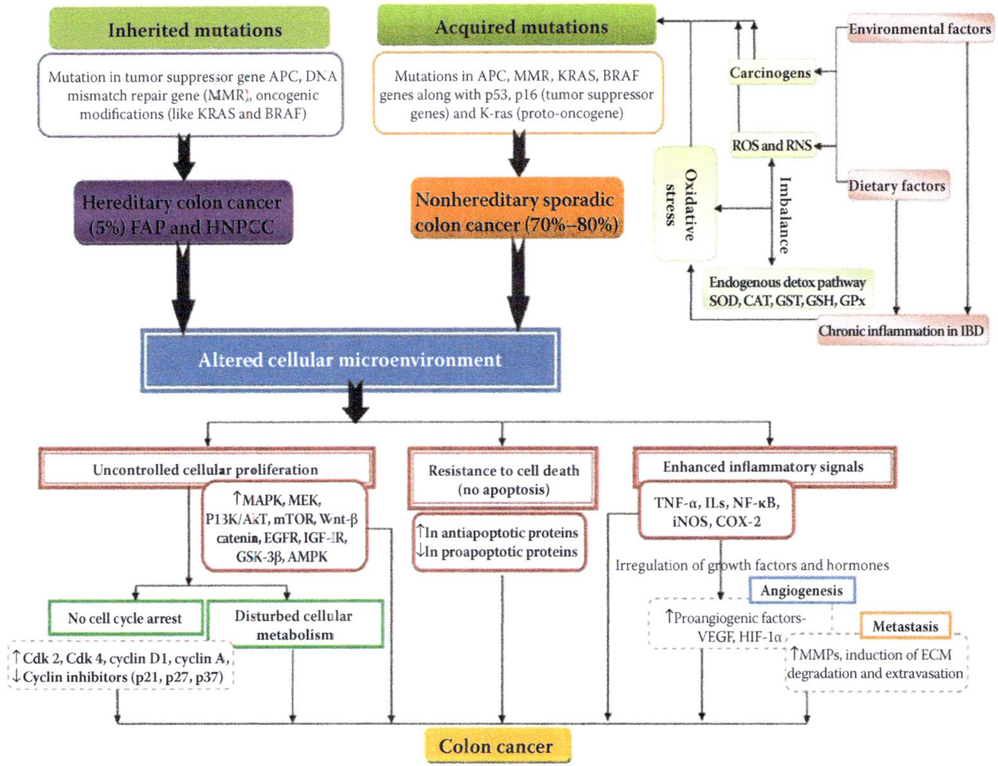

FIGURE 3.1 Brief description of etiology, imbalance in various processes, and crucial molecular targets involved in the progression of colon cancer. (Reprinted from *Journal of Functional Foods*, 13, Jasleen Kaur and Gurpreet Kaur, An insight into the role of citrus bioactives in modulation of colon cancer, 239–261, Copyright 2015, with permission from Elsevier.)

Therefore, to prevent these side effects, improve the quality of life, and reduce the cost involved in treatment, a promising option for the management of colon cancer is the administration of one or more naturally occurring compounds or their derivatives that have pleiotropic biological and nontoxic effects on normal human cells (Martin and Appel 2010; Hur et al. 2012). These active principles may not entirely prevent the occurrence of colon cancer, but can slow or reverse its progression. These plant phytochemicals are known to possess enormous beneficial effects like free radical scavenging, mitigation of oxidative stress-induced tissue damage (Urquiaga and Leighton 2000; Martin and Appel 2010), downregulation of tumor necrosis factor (TNF-α) and other proinflammatory biomarkers (Hur et al. 2012; Gupta et al. 2014), and modulation of intestinal barrier defects (Shigeshiro, Tanabe, and Suzuki 2013). In addition to these effects, numerous biological processes are also affected by these phytochemicals. These include altered gene expression, intracellular signaling pathways, P-glycoprotein activation, and enzyme activities associated with carcinogen activation. They also increase the apoptotic action, blood vessel dilation, detoxification, and chelation of transition metals and prevent reactive oxygen species (ROS) formation (Duthie, Duthie, and Kyle 2000; Andrade et al. 2006; Martin and Appel 2010).

The presently available anticancer therapy is based on modulating one target and is often associated with several side effects. However, plant-derived phytochemicals such as polyphenols, alkaloids, saponins, coumarins, limonoids, volatile oils, and carotenoids have the potential to modulate multiple targets. These phytochemicals have been found to exert anti-inflammatory activity via their action on molecular targets like COX, LOX, PLA_2, iNOS, NF-κB, peroxisome proliferator-activated receptors (PPARs), and nonsteroidal anti-inflammatory drug-activated genes (NAG-1) (Yoon and Baek 2005). Some of the altered signal transduction pathways in colon cancer, for example, mitogen-activated protein kinases (MAPK), phosphatidylinositol-3-kinase (PI3K/AKT/mTOR), glycogen synthase kinase-3β (GSK-3β), leptin, epidermal growth factor receptor (EGFR), extracellular signal regulated kinase (ERK½), transformation growth factor (TGF-β), Wnt/β-catenin, and activator protein (AP-1) are also known to be modulated by polyphenols. The polyphenols are also found to demonstrate cytotoxic action by increasing the expressions of proapoptotic proteins Bax and Bak, and decreasing the antiapoptotic proteins Bcl-2 and Bcl-xl levels. In addition, they are also involved in obstructing the cell cycle of mutated cells by accumulating the cells in the different mitotic stages like G_0/G_1, S, G_2, and M phases. Reports also suggest that polyphenols offer themselves as third-generation P-glycoprotein (P-gp) inhibitors as they help in overcoming resistance against drugs. They have also been found to prevent angiogenesis by acting on two angiogenic factors: vascular endothelial growth factors (VEGFs) and matrix metalloproteinases (MMP-2).

Scientists are exploring the role of these phytochemicals for targeting important regulators of colon cancer. An in-depth understanding of the role of these dietary components in modulating gut health and gut-associated disease is the current area of research. The chemopreventive profile and multiple downstream molecular targets moderated by citrus bioactives as proposed by the researchers are depicted in Figure 3.2.

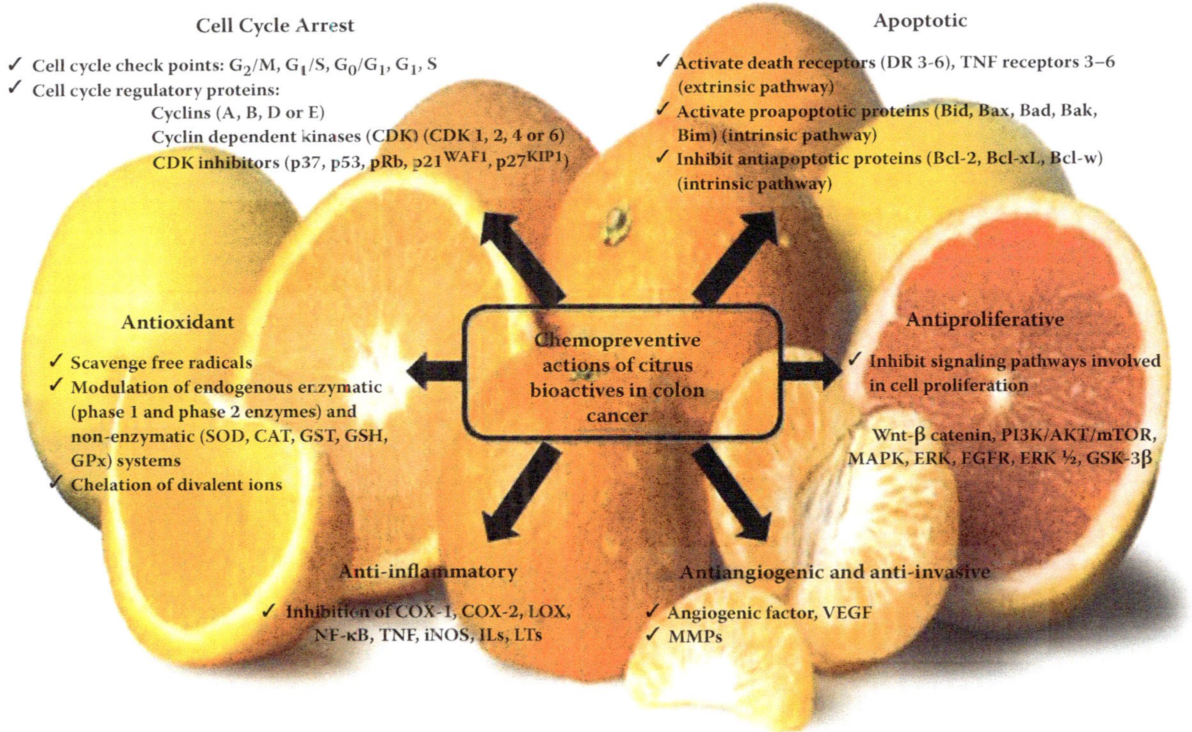

FIGURE 3.2 Chemopreventive actions and molecular targets moderated by citrus bioactives for the treatment of colon cancer. (Reprinted from *Journal of Functional Foods*, 13, Jasleen Kaur and Gurpreet Kaur, An insight into the role of citrus bioactives in modulation of colon cancer, 239–261, Copyright 2015, with permission from Elsevier.)

The medicinal importance of the citrus bioactives present in different consumed and nonconsumed portions of citrus fruits and their experimental studies in modulation of colon cancer are described in the next section.

3.4.1 Flavonoids

Flavonoids comprise a major group of naturally occurring polyphenolic compounds present especially in the peel of citrus fruits. The various flavonoid compounds vary in their basic benzo-γ-pyrone structure with respect to –OH, $-OCH_3$ groups, and *O*-glycosidic substitutions. The major citrus flavonoids include flavanones, flavones, flavonols, and anthocyanins (only in blood oranges). Flavanones and polymethoxyflavones are known to be exclusively present in high quantities in citrus peel. Flavones and flavonols are present in relatively lower concentrations in citrus fruits. These are generally present in their *O*/*C*-glycosidic forms. Flavonoids occur predominantly in nature as their glycosides (mono-, di-, and triglycosidic forms) with either rutinose or neohesperidose as sugar moieties and, less commonly, in aglycone form (Khan and Dangles 2014; Kaur and Kaur 2015).

3.4.1.1 Citrus Flavanones

Citrus flavanones are found in high concentration in white spongy inner portion of citrus peel as compared to the fleshy part of citrus fruits (Nogata et al. 2006). The flavanones usually exist as diglycosidic forms—7-*O*-rutinoside (rut) and 7-*O*-neohesperidoside (nh)—and confer the typical taste to citrus fruits. The most abundant flavanones in citrus fruits are hesperetin, naringenin, eriodictyol, and isosakuranetin. Hesperetin (4′-methoxy-5,7,3-trihydroxyflavanone), a bioactive flavanone, is present in abundance in grapes, oranges, and lemon juice. It has been employed for ages in the Chinese traditional medicinal system owing to its antioxidant and anticarcinogenic activities. Hesperetin possesses multiple biological and pharmacological activities such as specific alteration of signal transduction pathways, anti-inflammatory, antihypertensive, and antiatherogenic properties, enzyme inhibition, antiproliferative capability, and capacity to scavenge free radicals, inhibit low-density lipoprotein oxidation *in vitro*, and HMG-CoA reductase activity *in vivo* (Bok et al. 1999; Shin et al. 1999; Garg et al. 2001; Lin, Sato, and Takayama 2003; Cai et al. 2004). The glycosidic form of hesperetin occurs as both hesperidin and neohesperidin. Hesperidin is in a predominantly glycosylated form as compared to neohesperidin. It is found in abundance in lemons, limes, sweet oranges, tangerines, and tangor species of citrus fruits (Cano, Medina, and Bermejo 2008), whereas neohesperidin is mainly found in sour oranges and tangelos, along with lemon peel (Peterson et al. 2006a,b; Gonzalez-Molina et al. 2010). The major glycosidic form (i.e., hesperidin) has been studied for its chemopreventive actions for colon, skin, and bladder cancers. Studies carried out by Aranganathan et al. (2008) revealed that oral supplementation of hesperetin (10, 20, and 30 mg/kg) decreased the activity of undesirable fecal and colonic bacterial enzymes and thus the exposure of the intestinal lumen to harmful components. When given orally (20 mg/kg), hesperetin reduced mucinase activity, thus maintaining the barrier properties of GIT; as a result, a significant decrease (75%) in incidence of ACF was observed (Aranganathan et al. 2008).

Similar antiproliferative action was earlier reported by Miyagi et al. (2000), where a reduction in incidence of tumor and ACF formation was observed after administration of hesperidin-rich orange juice.

In 2009, Aranganathan and Nalini (2009) observed that oral administration of hesperetin (20 mg/kg) in male Wistar rats decreased the levels of reactive oxygen species (ROS) by reducing the lipid peroxidation in colonic mucosa and increasing the levels of antioxidant enzymes superoxide dismutase (SOD), catalase (CAT) and glutathione peroxidase (GPx), glutathione reductase (GR), and reduced glutathione (GSH). This diminished the damaging effect of ROS on lipids, proteins, and DNA present in colonic mucosa. In further studies, Aranganathan et al. (2009) observed that hesperetin (20 mg/kg) lowered the number of ACF/colon from 90 ± 8.4 to 40.2 ± 3.7 in DMH-treated male albino Wistar rats. The main mechanism attributed to this action was the modulation of phase 1 and phase 2 enzymes that regulates the detoxification and excretion of carcinogens. Further, Aranganathan and Nalini (2012) explored the antiproliferative action of hesperetin in male albino Wistar rats. A reduction in proliferation markers like PCNA labeling index (80.30%), AgNORs/nucleus (39.5%), and morphological markers such as ACF (90%) was also observed after administration of dietary hesperetin.

Sivagami et al. (2012) studied the apoptotic activity of hesperetin and its analog on HT-29 cells. The investigation suggested that HA demonstrated greater antiproliferative action than hesperetin as was evident from the IC_{50} values (HA: 32.15 μM; hesperetin: 70.25 μM) at the end of 24 h. The IC_{50} values further reduced to lower concentration after 48 h (HA: 27.28 μM; hesperetin: 41.79 μM) and 72 h (HA: 21.35 μM; hesperetin: 30.95 μM) treatments. This increased cytotoxic activity was ascribed to the higher number of methoxyl groups and absence of hydroxyl groups in the hesperetin analog. Both hesperetin and its analog were also found to maintain an oxidant/antioxidant balance as indicated by the significant increase in thiobarbituric acid (TBARS) and protein carbonyl content (PCC) levels and depletion of enzymic antioxidants (i.e., SOD, CAT, and GPx).

Studies carried out by Tanaka et al. (1997) revealed that dietary administration of hesperidin and diosmin resulted in chemopreventive effects in male F44 rats. It was observed that administration of diosmin (900 ppm) and hesperidin (100 ppm) decreased the incidence of ACF/colon (75%) and the incidence and multiplicity of neoplasm (44%). This reduction was suggested to be due to decrease in ODC activity in the colonic mucosa. The resulting inhibition of cell proliferation was also supported by the significant reduction of the BrdU index and AgNORs number in crypt cells.

Saiprasad et al. (2013) reported that hesperidin (25 mg/kg body weight) reduced the number of ACF/colon to 15.6 ± 1.54 (treated in the initiation phase) and 16.2 ± 1.6 (treated in the postinitiation phase) from 38.2 ± 3.4 (azoxymethane [AOM]-treated animals). The mechanism behind this suppression was found to be a decrease in oxidative stress and enhanced antioxidant status of colonic mucosa (enzymic: SOD, CAT, GPx, and GR; nonenzymic: GSH, vitamins C and E). The authors also revealed an increase in the thickness of mucosa and high infiltration of inflammatory cells into mucosa. A prominent downregulation of inflammatory markers such as NF-κB, iNOS, and COX-2 further suggested the anti-inflammatory action of hesperidin.

Investigations carried out by Tanaka et al. (2000) accentuated the chemopreventive action of mandarin juice (MJ) (rich in hesperidin and β-cryptoxanthin) prepared from satsuma mandarin (*Citrus unshui*). Three batches of mandarin juice with different ratios of hesperidin and β-cryptoxanthin (i.e., MJ [0.8:79], MJ2 [1.7:84], and MJ5 [3.9:100]) were evaluated for their anticancer action. MJ5 was found to show maximum activity. A reduction in frequency of ACF (15%) and multiplicity (80%) was observed in male F344 rats after treatment with MJ5. Further, a reduction in the PCNA-positive index (28.0 ± 6.0% and 41.7 ± 6.0% in AD and ADC, respectively) and the cyclin D1-positive index (13.8 ± 1.9% and 22.0 ± 3.3% in AD and ADC, respectively) was also found to be associated with MJ5. The apoptotic index of colonic AD did not significantly differ among the groups. However, the apoptotic index of colonic ADC was found to show a significant increase in animals administered MJ, MJ2, or MJ5 as compared with those given tap water ($p > 0.01$ or $p > 0.02$) (Tanaka et al. 2000). The downregulation of protein expression of proCASP3 and upregulation of CASP3 also manifested the apoptotic action of hesperidin in human colon cancer cells (SNU-C4) (Park et al. 2008).

Naringenin, 5,7,4′-trimethoxyflavanone, is another major citrus flavanone, present in significant amounts in white grapefruits. Naringenin possesses a bitter taste due to its glucose moiety. It is predominantly present in nature as its glycosidic forms, naringin and narirutin. The former constitutes a major flavonoid in sour oranges and lemon peel while minor amounts are detected in grapefruits (Igual et al. 2013). Narirutin is present in tangor, sweet oranges, tangerines, and tangelos (Peterson et al. 2006a).

Naringenin has been found to depict antioxidative and antiproliferative actions. Frydoonfar, McGrath, and Spigelman (2003) examined the antiproliferative action of naringenin in HT-29 colon cancer cell lines. A significant antiproliferative activity of naringenin was reported in concentration ranges of 0.75 to 2.85 mmol. Nevertheless, the concentration range of 0.02–0.09 mmol caused an increased proliferation of cell lines. Naringenin was also found to demonstrate cytotoxic activity against *Caco*-2 cells as demonstrated by EC_{50} values (557 ± 48 μM) after a treatment period of 48 h (MTT assay). A dose-dependent apoptotic action was observed at low concentration (250 and 500 μM) as indicated by the appearance of hypodiploid state in sub-G_1 cells. However, high concentration (1000 μM) of naringenin exhibited necrotic action. These results correspond to high susceptibility of *Caco*-2 cells to naringenin (Kanno et al. 2005). Previous studies have shown that the varied effects of flavonoids are attributable to the presence of substituted functional groups (Constantinou, Kamath, and Murley 1998; Kris-Etherton et al. 2002). Furthermore, it has also been suggested that the –OH group attached to a 5- or 7-carbon of basic flavonoid structure determines the antioxidant and antiapoptotic activities (Lee et al. 2005, 2007). Thus, to demonstrate whether the substitution at C7 is critical to the apoptosis-inducing activity of naringenin, Lee et al. (2008) synthesized eight derivatives by substitution of the –OH group by specific benzyl or amino acid moieties (Table 3.2). The loss in cell viability was more profound for KUF-1 and KUF-2 (IC_{50}: 10 and 15 μM) as compared to naringenin (>150 μM), while no reduction in the cell viability was observed with other derivatives. It was observed that KUF-1 and KUF-2 demonstrated a dramatic 10- to 15-fold greater growth inhibitory activity as

TABLE 3.2
Naringenin Derivatives Synthesized by Substitution of –OH Group at C7 Position

Structure of Naringenin

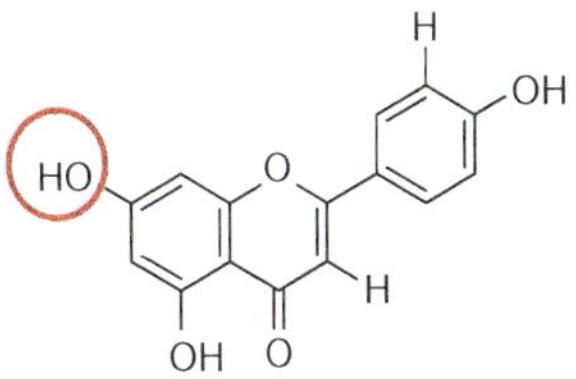

Derivatives of Naringenin

Substitution of –OH Group at C7 Position with Benzyl or Amino Acid Moieties[a]	
KUF-1	Benzyl
KUF-2	m-Metoxybenzyl
KUF-3	p-Fluorobenzyl
KUF-4	m-Iodobenzyl
KUF-5	2-Naphtylmethyl
KUF-6	Benzoxycarbonylmethyl
KUF-7	MeO-L-Leu-D-Pro-carbonylmethyl
KUF-8	MeOGly-D-Pro carbonylmethyl

Substitution of OH Group at C7 Position with Bulky Groups[b]	
N1	Thiophenecarboxylate
N2	Methylbenzoate
N3	Isobutyrate
N4	Allyloxy
N5	Phenyl carbonate

Note: Gly: glycine; Leu: leucine; Pro: proline.

[a] Lee, E. R. et al., 2008. *Journal of Cellular Biochemistry* 104:259–273.

[b] Yoon, H. et al., 2013. *Bioorganic & Medicinal Chemistry Letters* 23:232–238.

compared to naringenin in MTT assay. The percentage of apoptotic cells was also increased markedly by KUF-1 and KUF-2 in RKO cell cultures as evidenced by the apoptosis associated loss in mitochondrial inner transmembrane potential. The induction of peroxisome proliferator activated receptor (PPAR) cleavage and activation of caspase-3 and caspase-8 enhancement in intracellular ROS production was found to be the downstream events involved in apoptosis induced by the derivatives. Moreover, the upregulation of ERK½ phosphorylation was observed as the mechanism behind the apoptotic cell death in RKO cells.

Similarly, studies carried out by Yoon et al. (2013) highlighted the importance of substitution of –OH with bulky groups at 7-C position of naringenin (Table 3.2) in governing the anticancer activity. The IC_{50} values of the substituted derivatives (N1–N5) were found to be very low (1.20, 6.03, 15.87, 1.91, and 20.01 μM) as compared to unsubstituted naringenin (36.75 μM). Naringenin has already been known to arrest G_1 stage of cell cycle in HCT116 colon cancer cells (Lee, Bode, and Dong 2011). However, cyclin-dependent kinase (CDK)-2 binding assay and *in silico* docking studies revealed that substituted derivatives of naringenin were much more able to accumulate the HCT116 cells in the G_1 stage of the cell cycle. This accumulation of cells was ascribed to the hydrophobic interaction and H-bonding of substituted naringenin with the CDK-2 enzyme (control the cell cycle progression) as compared to unsubstituted one (docking studies). This interaction decreased the CDK-2 enzymatic activity sufficiently to induce the G_1 phase cycle arrest. The highest inhibition activity was exhibited by N4 followed by N1. The order of inhibitory activity was found to be N4 > N1 > N6 > N2 > N3 > N5 > naringenin. Further, the scores obtained in *in silico* docking studies explained the differences in hydrophobic interactions and hydrogen bonding of naringenin and their derivatives with amino acid residues of CDK-2 enzymes. These derivatives were also evaluated for their antioxidant action. The order for radical scavenging effects was found to be vitamin C (89.1%) > naringenin (59.3%) > N4 (20.7%) > N5 (15.7%) > N3 (15.4%) > N2 (14.2%) > N1 (11.2%), as depicted in Figure 3.3 (Yoon et al. 2013).

Naringin, the glycosylated form, has been found to be effective only in reducing the number of preneoplastic lesions (oral dose of 200 mg/kg) in the middle and proximal colon region and has a limited effect on advanced tumor lesions present in the distal colon region. Histopathological studies indicated an increase in the expression of mucins and densities of goblet cells, and a reduction was observed in crypt dysplasia. A significant lowering in the number of AgNORs/nucleus and mitosis and an increase in homogenous antioxidant mineral distribution (Cu, Mg, Se, and Zn) were also noticed. The reduced levels of 8-Oxo-2′-deoxyguanosine (8-OHdG) in genomic DNA of colonic tissue (treated with 100 and 200 mg/kg naringin) indicated lower oxidative damage to genomic DNA in a dose-dependent manner (Sequetto et al. 2014).

Eriodictyol exists in nature as eriocitrin (eriodictyol-7-*O*-rut) and neoeriocitrin (eriodictyol-7-*O*-nh). Eriodictyol is mainly obtained from lemon juice and seeds, whereas lemon peel and sour oranges are rich in neoeriocitrin. Eriodictyol has been found to possess a range of biological and pharmacological properties including

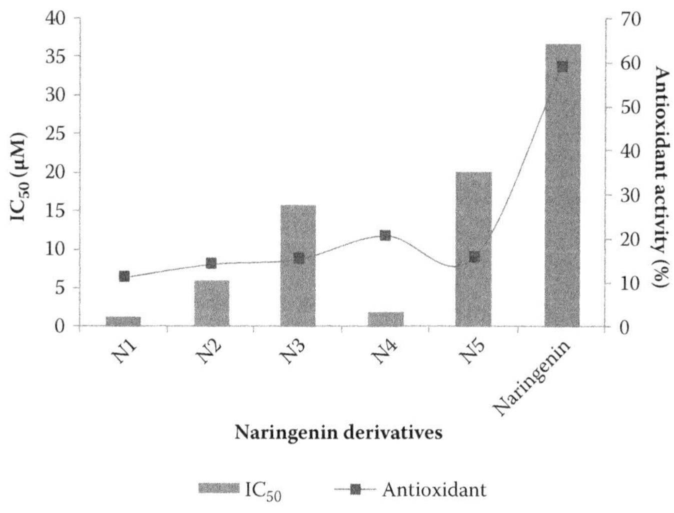

FIGURE 3.3 Comparison of IC_{50} value and antioxidant activity of naringenin derivatives.

antioxidant, anti-inflammatory, antineoplastic, antimalarial, and antinociception activities (Ramon et al. 1999; Chatzopoulou et al. 2010; Liu et al. 2011; Rossato et al. 2011), as well as to be active against several tumor cell lines, and shows a weak proliferative activity against normal cell lines. Isosakuranetin, another flavanone, is found in *Citrus* species as aglycone as well as in a glycosidic form (didymin/neoponcirin [rut] and poncirin [nh], respectively). Didymin constitutes a major component of sweet orange, while grapefruits and orange juice contain poncirin (Peterson et al. 2006a,b). The glycosidic form didymin has been tested in human gastric cancer cells (SGC-7901) and poncirin in neuroblastoma and human lung cancer cell lines (Hung et al. 2010; Singhal et al. 2012; Zhu et al. 2013). The research studies associated with these two flavanones and their glycosidic forms for their anticancer action in the colon have not been investigated in detail until now.

3.4.1.2 Flavones

Flavones are specifically distributed in flavedo portions of citrus peel and are present in very low concentrations in the fleshy portion of citrus fruits (Arcas et al. 2000; Kim et al. 2011). The prominent examples include apigenin, luteolin, diosmin, and chrysoeriol, found largely as 7-*O*-rut and 7-*O*-nh glycosides. The number of di-*C*-glycosides (apigenin 6,8-di-*C*-glucoside, diosmentin 6,8-di-*C*-glucoside, chrysoeriol 6,8-di-*C*-glucoside) is higher than mono-*C*-glycosides in citrus juices (Gattuso et al. 2007).

Apigenin, 5,7,4′-trihydroxyflavone (aglycone) is a common dietary flavone and occurs typically in glycosidic forms in fruits and vegetables. The various glycosidic forms of apigenin are apiin, apigenin 7-*O*-glucoside, apigenin 8-*C*-glucoside (vitexin), apigenin 6-*C*-glucoside (isovitexin), and apigenin 7-*O*-nh (rhoifolin) (Gattuso et al. 2007; Gonzalez-Molina et al. 2010; Lefort and Blay 2011). Apigenin demonstrates less cytotoxic effect on normal cells as compared to cancerous cells and is relatively nontoxic and nonmutagenic. It has been found to possess distinct clinical activities such as anti-inflammatory, antiplatelet, and antitumor properties. Apigenin suppresses both the initiation and promotion of carcinogenesis, inhibits the mutagenicity of the carcinogens 2-aminoanthracene and benzo[a]pyrene. It has shown apoptotic action in different types of cancer cells that include monocytic and lymphocytic leukemia and cervical, lung, breast, and colon cancer. It induces mitochondrial death cascade by inhibition of VEGF and hypoxia inducible factor (HIF-1) expressions via pI3K/AKT/p70S6K1 and HDM2/p53 pathways (Choudhury et al. 2013).

Luteolin (3′,4′,5,7 tetrahydroxyflavone) is present in citrus fruits as luteolin 7-*O*-rut, and lucenin-2 (luteolin 6,8-di-*C*-glucoside) (Gattuso et al. 2007). It has been recognized as a strong antioxidant, a radical scavenger, an inhibitor of protein kinase C, and an antiallergic, antimitotic, apoptotic, antimutagenic, and antitumorigenic agent. The distinctive chemopreventive actions associated with apigenin and luteolin, which led to its exploration in the treatment of colon cancer, are described in Table 3.3.

Diosmetin (3′,5,7-trihydroxy-4′-methoxyflavone) and its glycosidic form diosmin (3′,5,7-trihydroxy-4′-methoxyflavone-7-rhamnoglucoside) are one of the main flavones present in citrus fruits (Yin, Cheng, and Lou 2004; Yoo et al. 2007).

TABLE 3.3
Anticancer Effects and Mechanisms Associated with Apigenin and Luteolin

Anticancer Effects	Apigenin	Luteolin
Antiproliferative	Inhibition of mTOR C2 (mammalian target of rapamycin complex-2) via downregulation of RICTOR expression (Guo et al. 2011)	↓ The expressions of β-catenin, GSK-3β and cyclin D (key components of Wnt signaling pathway) (Pandurangan and Ganapasam 2011; Pandurangan et al. 2013a)
	↓ In mTOR and cyclin D1 expression (Turktekin et al. 2011)	Inhibit the activation of IGF-1R signaling pathway and inactivate the Akt and ERK1/2 expressions in HT-29 cells (Lim et al. 2012)
	↑ In CD26 (cell surface protein), ecto-ADA (adenosine deaminase) binding (Lefort and Blay 2011)	↓ The ERK phosphorylation in KRAS mutated HCT15 cells while decreasing the Akt phosphorylation in BRAF mutated CO115 cells (Xavier et al. 2009)
	Phosphorylation and upregulation of Fas associated protein with death domain (FADD) and expression of ataxia-telangiectasia mutated (ATM) kinases pathway (Zhong et al. 2010)	Modulate the expressions of GST-α, μ, and Nrf2 (Pandurangan et al. 2013b)
Apoptotic expressions	Apigenin combined with 5-fluorouracil (5-FU) ↑ the apoptotic percentage from 24.92% to 29.13% (Turktekin et al. 2011)	↑ Bax and caspase-3 while ↓ the Bcl-2 expression (Pandurangan and Ganapasam 2013a)
	Apigenin antagonized the ABT-263 induced upregulation of Mcl-2 and upregulated the Bim and Bax expressions in human colon cancer cells (Shao et al. 2013)	4.5 fold ↑ in caspase-3 activity (luteolin + BITC) by downregulating $p21^{waf1/cip1}$ mRNA expressions (Sakai et al. 2012)
Anti-inflammatory	Attenuate inflammatory mediators such as COX-2 and iNOS (Vanamala et al. 2006; Leonardi et al. 2010)	↓ The expressions of iNOS and COX-2 (Pandurangan et al. 2014a)
Antiangiogenic and anti-invasive	↓ In HIF-1α and VEGF expression in both normoxic and hypoxic conditions (Fang et al. 2007)	↓ The expressions of MMP-9 and MMP-2 (Pandurangan et al. 2014b)
	Upregulation of TAGLN (transgelin) and a ↓ in MMP-9 expression in a dose-dependent manner (Chunhua et al. 2013)	
	Luteolin	
Cell cycle arrest	Arrest G_2/M phase (Pandurangan et al. 2013a)	
Antioxidant	↑ GSH, PSH, and total thiols, ↓ GSSG (Pandurangan and Ganapasam 2013b)	
	↑ The levels of CAT, SOD, GPx, GR in colonic tissues and GSH, vitamin A, C, E in colonic tissue and plasma while ↓ the plasma and colonic levels of LPO and OH. (hydroxyl radical) (Pandurangan and Ganapasam 2008)	

(*Continued*)

TABLE 3.3 (CONTINUED)
Anticancer Effects and Mechanisms Associated with Apigenin and Luteolin

Anticancer Effects	Luteolin
Overcome the resistant nature of drugs	Luteolin in combination with BITC overcame its resistance in p-53 positive HCT-116 cells (Sakai et al. 2012)
	Restore the sensitivity of oxaliplatin resistant HCT116 and SW62 cells and exhibit a synergistic action with it (Chian et al. 2014)
Effect on drug's metabolism	↓ The phase 1 enzymes in colon and liver while increasing the levels of phase 2 enzymes (Pandurangan et al. 2013b)
	Inhibit the phase 1 enzyme metabolism and phase 3 transport, which increase the intracellular accumulation or bioavailability of radioactively labeled B(a)P (Bothe et al. 2010)
Biomarkers (biochemical, histological, and proliferation)	↓ Number of AgNORs and PCNA labeling index
	↓ The polyamine levels (putrescine, spermidine, and spermine) (Pandurangan and Ganapasam 2011)
	↓ Lipid peroxidation end products—that is, protein carbonyl (PC), malonaldehyde, and conjugated dienes (CD)
	↓ Glycoconjugates levels (hexose, hexosamine, mucoprotein, sialic acid, and fucose in colonic tissue) and levels of hexose, hexosamine, and sialic acid in serum
	↓ The incidences of mucin depleted foci (MDF) significantly (Pandurangan et al. 2012)
	↓ ALP and LDH (membrane damage markers); ↑ ROS production (Pandurangan and Ganapasam 2013c)
	↓ Cellular glycoproteins like fucose and hexose (undergo alterations during carcinogenesis) while restoring the levels of sialic acid (Vaiyapuri and Namasivayam 2009)
Miscellaneous	↓ The activities of Na^+K^+-ATPase, Mg^+-ATPase while ↑ the activities of Ca^+-ATPase
	↓ The levels of lysosomal enzymes (Pandurangan et al. 2013c)
	Completely protect the DNA from oxidative damage in short period of preincubation (Ramos, Pereira-Wilson, and Collins 2010)

Diosmin possesses anti-inflammatory, free-radical scavenging, and antimutagenic properties. It is employed as a vascular-protecting agent to treat chronic venous insufficiency, hemorrhoids, lymphedema, and varicose veins (Le Marchand et al. 2000; Cesarone et al. 2006; Camarda et al. 2007). Diosmetin has been found to attenuate the dextran sodium sulfate (DSS)-induced colitis in BALB/c mice by increasing the prostaglandins (PGs) production, decreasing the neutrophil infiltration, and inhibiting the upregulation of 1L-1β in colon mucosa (Villegas et al. 2003). A study conducted by Yoo et al. (2007) described the inhibitory activity of 10 citrus flavonoids on P-gp mediated drug efflux action where diosmin was reported to markedly increase the accumulation of rhodamine-123 and digoxin (established substrate for P-gp-pump) in *Caco*-2 cells. Further, diosmin produced a concentration-dependent increase in accumulation of rhodamine-123 within the range of 10–100 μM. This concentration range was also found to be free of cytotoxicity. The findings suggested that diosmin (50 μM) increased the transport of digoxin from apical (A) to basal (B), but decreased the basal to apical transport. The apparent permeability coefficients $P_{app\ A\text{–}B}$ (A: cell monolayer surface area; B: basal cell layer) and $P_{app\ B\text{–}A}$ for diosmin (50 μM) were found to be 7.6 ± 0.06 and 17.1 ± 0.10 with a transport ratio of 2.3 as compared to $P_{app\ A\text{–}B}$: 9.5 ± 0.06 and $P_{app\ B\text{–}A}$: 10.4 ± 0.42 with a transport ratio of 1.1 for control verapamil. Thus, diosmin can be used as a potent P-gp substrate and increases the absorption of drugs. A comparative study on cytotoxicity revealed that diosmetin (IC_{50}: 82.9 μM) showed better cytotoxic action than luteolin (IC_{50}: 96.9 μM), glycosidic forms: diosmetin 7-(6″-*O*-*p*-hydroxyphenylacetyl)-*O*-β-D-glucopyranoside and diosmetin 7-*O*-β-D-glucopyranoside (both having IC_{50}– > 200 μM) in Colon205 (human colon cancer cells). This study emphasized that the aglycone forms are much more potent than the glycosidic forms (Xie et al. 2009).

Chrysoeriol (4′,5,7-trihydroxy-3′-methoxyflavone) exists in various citrus fruit juices such as *Citrus limon*, *Citrus bergamia*, *Citrus sinensis*, and *Citrus reticulate*. It has recently been proven to possess anticancer activity against breast cancer cells (Takemura et al. 2010; Tan et al. 2013; Amrutha et al. 2014), human prostate cancer (Kim et al. 2013), and multiple myeloma cells (Yang et al. 2010). However, its role in colon cancer treatment has not been explored yet.

3.4.1.3 Flavonols

Flavonols are found to be present in minor quantities in citrus fruits. The major flavonols present are quercetin, kaempferol, rutin, myricetin, and isorhamnetin. Their glycosidic forms exist as flavonol-*O*-diglycosides (quercetin-3-*O*-rut, isorhamnetin-3-*O*-rut, and kaempferol-3-*O*-rut), flavonol-*O*-triglycosides (quercetin-3-*O*-rut-7-*O*-glucoside, kaempferol-3-*O*-rutinoside-7-*O*-glucoside, isorhamnetin-3-*O*-rut-7-*O*-glucoside, tamarixetin-3-*O*-rut-7-*O*-glucoside), and dihydroflavonol-*O*-diglycosides (dihydroquercetin-7-*O*-rut, dihydrokaempferol-7-*O*-rut, and dihydroisorhamnetin-7-*O*-rut). The biological activities of flavones, such as antioxidant, antithrombotic, anti-inflammatory, antiproliferative, and antiangiogenic properties, are well documented (Abad-Garcia et al. 2012). Table 3.4 highlights the anticancer reports of flavonols in colon cancer.

3.4.1.4 Polymethoxyflavones

Polymethoxyflavones (PMFs) constitute a separate subclass of flavonoids that bear two or more methoxy groups on a basic benzo-γ-pyrone skeleton. The most commonly studied PMFs are permethoxylated PMFs (PM-PMFs) such as tangeretin, nobiletin, and 3,5,6,7,8,3′,4′-hexamethoxyflavone (HMF). These PMFs exist in high concentrations in the peel of various *Citrus* species. Recently, a new class of PMFs has been isolated in aged orange peel extracts known as hydroxylated PMFs (H-PMFs). During the long-term storage of citrus fruits, the PM-PMFs undergo autohydrolysis and give rise to H-PMFs that include 5-hydroxy-6,7,8,3′,4′-pentamethoxyflavone (5HPMF), 5-hydroxy-3,6,7,8,3′,4′-hexamethoxyflavone (5HHMF), and 5-hydroxy-6,7,8,4′-tetramethoxyflavone (5HTMF) (Qiu et al. 2010). They are present in high concentration in citrus juice and have been reported to possess higher biological

TABLE 3.4
Anticancer Reports of Flavonols in Colon Cancer

Chemopreventive Mechanism of Action	Flavonols	Molecular Targets	Reference
Antiproliferative action by modulating signal transduction pathways	Quercetin	AMPK-p38 MAPK (AMPK-5′ adenosine monophosphate-activated protein kinase), AMPK (AMP-activated protein kinase), AMPK-COX-2, Wnt/β-catenin pathways, MAPK pathway	Kim et al. 2010, 2012a, 2014; Lee et al. 2009; Shan, Wang, and Li 2009; Dihaal et al. 2008
	Kaempferol	Regulate MAPK pathway via RSK inhibition and selectively inhibit the PI3k-Akt-mTOR pathway	Tsimplouli et al. 2010
	Isorhamnetin	↓ The phosphorylation levels of Akt, phosph-p70S6 kinase and phosph-4E-BP1 protein and selectively inhibit the PI3k-Akt-mTOR pathway, inhibit oncogenic Src activity and β-catenin nuclear translocation	Li et al. 2014; Saud et al. 2013
Apoptotic action	Quercetin	Induce hypoxia in HCT116 cells, ↑ the expressions of sestrin-2 (downstream effector of p53), ROS generation and ↓ the survivin mRNA and protein expressions (antiapoptotic survival gene), sulfated derivative of quercetin (quercetin-5′,8 disulfonate) ↑ the ROS dependent apoptotic action	Kim et al. 2012a, 2014; Del Follo-Martinez et al. 2013; Zhang et al. 2012
	Kaempferol	Kaempferol in combination with tumor necrosis factor related apoptosis inducing ligand (TRAIL) overcomes the TRAIL's resistance and induces apoptosis by upregulating the proapoptotic receptors (DR5 and DR4; death receptors), time, and dose-dependent cleavage of PARP	Yoshida et al. 2008; Li et al. 2009a
Arrest of cell cycle phases	Quercetin	Accumulate the cells in G_2/M, G_1, and S phase	Shan et al. 2009; Kim et al. 2010; Zhang et al. 2012

(Continued)

TABLE 3.4 (CONTINUED)
Anticancer Reports of Flavonols in Colon Cancer

Chemopreventive Mechanism of Action	Flavonols	Molecular Targets	Reference
	Kaempferol	Tac (semisynthetic kaempferol glucoside) delays the transition from G_2 to M phase, G_1 to S phase	Tsimplouli et al. 2010
	Isorhamnetin	Accumulate the cells in G_2/M phase	Li et al. 2014; Jaramillo et al. 2010
Anti-inflammatory action	Quercetin	↓ The proliferation index by inhibiting the expressions of inflammatory mediators like COX-1 and COX-2	Warren et al. 2009; Lee et al. 2009; Turner et al. 2007
		↓ The macrophage infilteration in intestinal villi leading to decrease in intestinal polyp multiplicity	Murphy et al. 2011
	Kaempfero	Alter the lipid peroxidation, ↑ the level of colonic mucosal TBARS, antioxidant enzymes CAT, SOD and rejuvenate GPx and GSH enzymes	Nirmala and Ramnathan 2011
	Rutin	Monoglucosyl-rutin (100 or 500 ppm) significantly ↓ the number of ACF, BrdU labeling indices and anti-PCNA antibody	Matsunaga et al. 2000
Synergistic cytotoxic action	Quercetin	In combination with resveratrol (1:1), it decreases the ROS formation by 2.25-fold, ↑ its antioxidant capacity by threefold and apoptosis by inducing cleavage of caspase-3 and PARP	Del Follo-Martinez et al. 2013
		In combination with rutin, it ↓ the DNA damage by scavenging ROS and reduces the oxidative stress (induced by high doses of PhIP(2-amino-1-methyl-6-enylimidazo[4,5-b]pyridine) and IQ (2-amino-3-methylimidazo[4,5-f]quinolone)	Kurzawa-Zegota et al. 2012
	Kaempferol	In combination with 5-FU, it ↑ the p53 dependent apoptotic action	Xavier et al. 2011
	Rutin	In combination with quercetin, it ↑ the apoptotic index, cyclin D1 expressions, and ↓ the focal area of dysplasia (FAD)	Yang et al. 2000

potencies compared to PMFs (Lai et al. 2007; Li et al. 2007c; Pan et al. 2007). Although methoxylation leads to low solubility, it results in high oral bioavailability due to a substantial increase in metabolic stability and membrane transport in liver and intestine (Van de Waterbeemd, Lennernas, and Artursson 2003; Li et al. 2009b). A wide range of biological activity has been found to be associated with PMFs, including anti-inflammatory (Ho et al. 2012), anticarcinogenic (Lai et al. 2007; Tang et al. 2007), and antiatherosclerotic (Saito, Abe, and Sekiya 2007) properties. H-PMFs are found to scavenge free radicals, and methoxylated PMFs are known to inhibit various free radical generating enzymes like NADPH oxidase and iNOS (Murakami et al. 2000a,b; Choi et al. 2007).

Investigations have demonstrated that PMFs inhibit the initiation and progression of cancer due to their multifaceted activities such as anti-inflammatory, antiproliferative, apoptosis, selective cytotoxicity, blockade of metastatic cascade, and reduction in lymphocyte infiltration (Li et al. 2009b; Walle 2007). The *in vitro* and *in vivo* experiments to explore the anticancer action of PMFs in pathogenesis of colon cancer have drawn much attention in the last few years.

Miyamoto et al. (2008) studied the chemoprotective effect of nobiletin in ICR (Institute of Cancer Research) mice by evaluating its action on serum adipocytokine levels. Adipocytokines such as adiponectin, IL-6, and leptin are adipocyte secreted proteins. Leptin synthesis is known to be highly influenced by various factors such as insulin, TNF-α, PGs, glucocorticoids, and reproductive hormones. They play a central role in pathogenesis of metabolic syndromes such as obesity and insulin resistance and they also influence the proliferation of malignant cells. Obesity has been considered as one of the risk factors involved in the progression of colon cancer. In this study, the authors examined the serum leptin levels before and after administration of nobiletin. It was observed that treatment with AOM followed by 1% DSS increased the serum leptin levels to six times the usual level without any alteration in triglycerides, adiponectin, and IL-6. After administration of nobiletin (100 ppm), a 75% reduction in serum leptin levels was observed. A decrease in leptin secretion and leptin-dependent proliferation of HT-29 colon cancer cells via inactivation of MAPK/extracellular signaling protein kinase was proposed. However, Miyamoto et al. (2010) observed a complete opposite action of nobiletin on serum leptin levels in obese *db/db* mice. These genetically modified C57BL/Ksj-*db/db* male mice are used to validate the relationship between obesity and colon cancer. These obese animals usually suffer from hyperlipidemia, hyperinsulinemia, hyperleptinemia, hyperglycemia, and hypercholesterolemia and are reported to be highly susceptible to colon carcinogenesis. It was found that nobiletin treatment reduced the ACF incidence by 68%, but did not lower the serum leptin levels. This study showed that a weak leptin decreasing activity of nobiletin, however, produced a dramatic decrease in insulin growth factor (IGF-1). The authors attributed this suppression in proliferation of preneoplastic lesions in this study to the decreased activity of the insulin signaling pathway only.

The inhibitory effects of hydroxylated PMFs (H-PMFs) and permethoxylated PMFs (PM-PMFs) were compared in two colon cancer cell lines: HCT116 cells and HT-29 cells. The H-PMFs comprised 5-hydroxy-6,7,8,3′,4′-pentamethoxyflavone (5HPMF), 5-hydroxy-3,6,7,8,3′,4′-hexamethoxyflavone (5HHMF), and 5-hydroxy-6,7,8,4′-tetramethoxyflavone (5HTMF). The PM-PMFs used were nobiletin, tangeretin, and

3,5,6,7,8,3′,4′-heptamethoxyflavone (HMF). The H-PMFs displayed strong inhibitory effect on colon cancer cells as compared to PM-PMFs. The cytotoxic effect (IC_{50} value) of both H-PMFs and PM-PMFs in both HCT116 and HT-29 cells is depicted in Figure 3.4. It was observed that in HCT116 cells, 5HHMF accumulated the cell population in both G_0/G_1 and G_0/M phases, while 5HHMF and 5HTMF increased it in the G_2/M phase only. Further, an increase of 3.7-, 3.4-, and 5.4-fold in the G_2/M cell population was observed with 5HPMF (36 μM), 5HHMF (18 μM), and 5HTMF (6 μM) in HT-29 cells. The apoptotic cell population of HCT116 cells increased by 2.2-fold with 5HPMF (8 μM) and 5HHMF (4 μM) and 4.4-fold with 5HTMF (3 μM). In HT-29 cells, the order of apoptosis-inducing potency was found to be 5HTMF > 5HHMF > 5HPMF. However, PM-PMFs were not found to increase the apoptotic action. H-PMFs also inhibited the AOM-induced phosphorylation of EGFR, ERK½, and inactivated the Wnt-β catenin pathway in a dose-dependent manner. The increased bioavailability and bioactivities of both PMFs were attributed to the higher number of methoxy groups. The hydrophobic nature of these groups enhances their cellular uptake and binding with the plasma membrane (Qiu et al. 2010).

Lai et al. (2011) demonstrated the *in vivo* chemoprotective efficacy of hydroxylated PMFs (H-PMFs) against colon carcinogenesis for the first time. They reported a reduction of 48% and 80% in the occurrence of microadenoma when ICR mice were fed diets containing 0.01% and 0.05% H-PMFs, respectively. The number of large ACFs was found to decrease from 23 ± 3 to 13 ± 4 (0.01% H-PMFs) and 23 ± 3 to 10 ± 2 (0.05% H-PMFs), confirming their antiproliferative action. The proapoptotic effects were confirmed by 3.6- and 5.2-fold increases (0.01% and 0.05% H-PMFs, respectively) in the activation of caspase-3. A decrease in cytosolic and nuclear accumulation of β-catenin, expressions of cyclin D1, phosphorylation of Akt

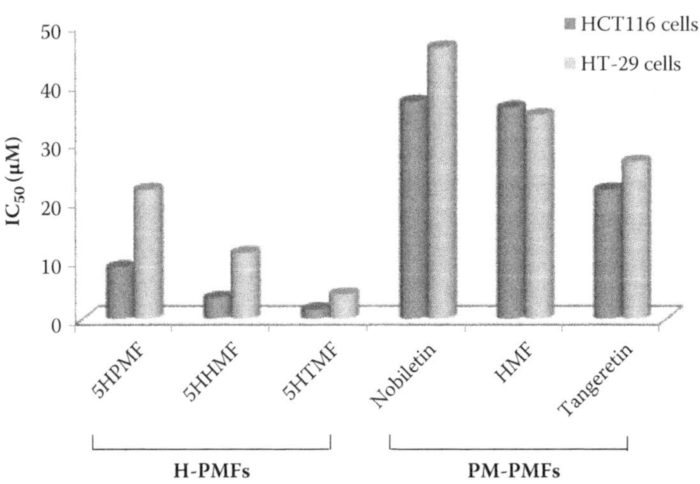

FIGURE 3.4 Comparison of IC_{50} values of H-PMFs and PM-PMFs in HCT116 and HT-29 cells.

and GSK-3β and inflammatory mediators, OD, and phosphorylation of STAT-3 in a dose-dependent manner by 0.01% and 0.05% H-PMFs were hypothesized to be the underlying molecular mechanism of chemoprevention. H-PMFs also inhibited the AOM-induced phosphorylation of EGFR and ERK½ and suppressed the RAS activation in a dose-dependent manner. Further, a reduction in the expression of inflammatory enzymes (iNOS and COX2) and protein expression of MMP-9, VEGF, and cyclin D1 upon long-term dietary H-PMF feeding was also observed.

Nobiletin has also been found to be very effective in reducing the number of ACFs in PhIP-treated F344 male rats. PhIP is a potent carcinogenic agent present in red meat. Nobiletin significantly decreased the number of ACF (ACF less than or equal to three crypts in the transverse colon and ACF with greater than or equal to four crypts per focus in the whole colon) as compared to the control (Tang et al. 2011).

A study carried out by Qiu et al. (2011a) reported that 5HTMF and 5HPMF showed high antiproliferative activity in HCT116 $p53^{+/+}$ and HCT116 $p53^{-/-}$ cell populations, respectively. The H-PMFs-5HTMF and 5HPMF produced an early apoptotic action in HCT116 $Bax^{+/-}$ cells; however, 5HHMF was found to be less sensitive to HCT116 $Bax^{-/-}$ cells. The different human colon cancer cell lines responded differently such that the HCT116 $p53^{+/+}$ cells accumulated in the G_0/G_1 and G_2/M phases, but HCT116 $p53^{-/-}$ accumulated in the G_2/M phase with no increase, but rather a decrease in the G_0/G_1 phase. These results further strengthened that all three H-PMFs induced G_0/G_1 arrest through p53 dependent mechanisms and G_2/M phase arrest through p53 independent mechanisms. Qiu et al. (2011b) also investigated the effect of 5HHMF on colony formation of different human colon cancer cells SW620, HCT116, and HT29. A 91% inhibition in growth was achieved in all cancer cells; however, SW620 colon cancer cells were found to be the most sensitive (IC_{50}: 3 μM) as compared to HCT116 (IC_{50}: 4 μM) and HT29 (IC_{50}: 5 μM) cells. The inhibitory actions of 5HHMF were also documented in oncogenic signaling pathways like Wnt/β-catenin, EGFR/K-Ras/Akt, and NF-κB, which are frequently involved in colon carcinogenesis. The investigation also revealed that inhibition of growth was not mediated via ROS formation, but rather through apoptosis and inhibition of angiogenesis, as well as blocking of Wnt/β-catenin and EGFR/K-Ras/Akt signaling pathways. The dose-dependent alterations in β-catenin signaling proteins and dose-dependent decrease in membrane and downstream signaling proteins of the EGFR/K-Ras pathway are described in Figure 3.5.

Lai et al. (2013a) studied the anticancer efficacy of gold lotion (GL) after its oral administration in male ICR mice. "Gold lotion" (a citrus peel extract) was originally used as a cosmetic product to protect the skin from UV radiations. It has been found to be rich in flavonoids (450 ppm or 0.1 mg/mL)—naringin and hesperidin and PMFs (106 ppm/0.1 mg/mL)—sinesetin, nobiletin, tangeretin, 5HTMF, 5HPMF, and 5HHMF. The researchers observed a decrease in the number of large sized ACF, downregulation of gene and protein expressions of iNOS and COX-2, OD, VEGF, and MMP-9 in colonic tissues of mice, suggesting GL as a novel product capable of preventing inflammation-associated colon carcinogenesis. The studies showed that it is free of any toxicity after its application on skin as well as after oral ingestion.

Cell lines	Effect of 5HHMF at different concentration
HT-29	1,3,4,5 μM produce a 15, 22, 70, 91% decrease, respectively, in number of colonies of cells
HCT116	1,2,3,4 μM produce a 30, 35, 70, 90% decrease, respectively, in number of colonies of cells
SW620	1,2,2.5,3 μM produce 91% inhibition at 3 μM concentration

SW620 human colon cancer cells were found to be the most sensitive cell lines

Angiogenesis inhibition

5HHMF decrease the formation of HUVECS tubes in a dose-dependent manner

No association with ROS formation

Confirmed by increase in ROS by N-acetyl-cysteine (NAC-antioxidant) in HT-29, HCT116, and SW620 colon cancer cells cell lines

Colon carcinogenesis

Wnt/β-catenin signaling

Dose-dependent alterations in β-catenin signaling proteins

Dose-dependent alterations in Wnt/β-catenin signaling proteins

	3 μM	6 μM	9 μM
Nuclear levels of β-catenin	No change	50% reduction	70% reduction
E-cadherin levels	Slight increase	Moderate increase	110% increase
β-catenin/ E-cadherin ratio	10% decrease	40% decrease	90% decrease

EGFR/K-Ras signaling

Dose-dependent decrease in membrane proteins and downstream signaling proteins

Dose-dependent alterations in EGFR/K-Ras signaling proteins

	3 μM	6 μM	9 μM
Membrane associated K-Ras	30	60	90
Cytosolic fraction of K-Ras	No significant changes		
Akt phosphorylation levels	40	50	90
Nuclear translocation of NF-κB/p65	20	50	90

FIGURE 3.5 Cytotoxic actions of 5HHMF in HCT116 cells.

Lee et al. (2013a) explored the role of PMFs in altering the expression of GADD45α (growth arrest and DNA damage-inducible protein α), which plays an important role in suppressing the tumor by inducing apoptosis. The aim was to establish a relation between antiproliferative actions of PMFs via induction of GADD45α expression and to elucidate the mechanism behind the expression of this gene. The order of potency in terms of growth inhibition of HCT116 cells was found to be HTMF > nobiletin > tangeretin. The PMFs have been found to upregulate the expressions of GADD45α and CHEK1 (a cell cycle regulator) by altering the acetylation levels of histone H3K14ac. Further, tangeretin treatment was found to increase the levels of H3K27a around intron 1, whereas nobiletin and tangeretin decrease the H3K4me1 levels. The study suggested that the GADD45α gene can act as a potential target in the treatment of colon cancer.

Lai et al. (2013b) investigated the effect of demethylation and acetylation at the C5 position on a tangeretin structure for its anticancer action in HCT116 cells and HT-29 cells. The derivatives, including 5-demethyltangeretin and 5-acetyloxytangeretin, showed better cytotoxic action and cell cycle arrest at the G_2/M phase with 5-acetyloxytangeretin possessing higher activity. 5-Demethyltangeretin and 5-acetyloxytangeretin also exhibited an apoptotic action. The higher activity of 5-acetyloxytangeretin than 5-demethyltangeretin in HT-29 cells was attributed to their antiproliferative and apoptotic effects.

3.4.2 Volatile Oils

Essential oils (EOs) are mixtures of low molecular weight (less than 500 Da) aromatic and aliphatic volatile compounds having pleasant aromas and distinctive tastes. They are synthesized by plants as volatile metabolites and stored in special secretory structures such as oil ducts, resin ducts, glands, or trichomes (glandular hairs) of different plant parts (flowers, buds, seeds, leaves, twig bark, herbs, wood, fruits, and roots) (Baser and Demirci 2007). These are extracted from plants by steam distillation and maceration techniques. Chemically, the major constituents of EOs encompass terpenoids and phenylpropanoids with various other aromatic and aliphatic compounds such as terpenes, alcohols, acetones, phenols, acids, aldehydes, and esters. The composition of EOs is mainly governed by the cultivation, extraction, and separation methods, which further decide their biological activities. Being complex mixtures of chemical compounds, they can be classified into three main groups: monoterpenes, sesquiterpenes, and oxygenates (Bakkali et al. 2008; Carson and Hammer 2011). Natural pigments (mainly carotenoids and chlorophylls) are also present in the EOs. Citrus essential oils (CEOs) contain 85%–99% of volatile and 1%–15% of nonvolatile components (Smith et al. 2001). The 16 species of the genus *Citrus* contribute a large sector for the production of CEOs.

EOs are mainly extracted from the pericarp of citrus peel by cold pressing. This involves scraping or breaking the oil cells near the fruit's surface by pressing and grinding the fruit peels or seeds with the use of granite milestones or stainless steel presses. Water is used to drag the oil in the form of an emulsion, which is then centrifuged to obtain the cold pressed oil. This process produces heat through friction; however, the temperature is maintained below 120°F (49°C) or even below for any oil to be considered cold pressed. Thus, cold pressed oils retain all of their flavor, aroma, and nutritional value and prevent oxidation of oils. Nowadays, folding of essential oils (distilled and concentrated from its already highly concentrated form) is done to protect the oils from oxidation and to increase their solubility in water and organoleptic properties (Di Giacomo and Di Giacomo 2002; Moyler 2002). CEOs are mainly folded (concentrated) using high vacuum fractional distillation. The main folding degrees commercially available are twofold to fivefold for lime and mandarin oils, 2-fold to 10-fold for lemon and grapefruit oils, and 2-fold to 20-fold for orange oil (Nguyen et al. 2009; Lopez-Munoz and Balderas-Lopez 2014).

CEOs have been widely employed for providing flavor and aroma in cosmetics and food industries. They have also been found to possess widespread application in traditional systems of medicine due to a variety of biological activities like antioxidant, antimicrobial, antiviral, antimutagenic, anticancer, and anti-inflammatory properties. All these bioactivities have compelled researchers to explore their potential in the treatment of various types of cancers (Fisher and Phillips 2008). The major mechanisms that may be responsible for the anticancer action of volatile oils include induction of phase II enzymes, activation of ERK and caspase-dependent mitochondrial death pathways, cell cycle arrest, alteration in signal transduction, apoptosis, and inhibition of metastasis through suppression of VEGF expressions (Lu et al. 2004; Chen et al. 2006; Ji et al. 2006). The incidence of different types of malignancies like glioma, gastric, liver, pulmonary, breast, and leukemia is also reported to decrease after treatment with essential oils (Kaefer and Milner 2008; Hamid, Aiyelaagbe, and Usman 2011; Raut and Karuppayil 2014; Patel and Gogna 2015).

Patil et al. (2009) explored the antiproliferative action of volatile oil extracted from the fruits of *Citrus aurantifolia* (lime) on human colon carcinoma cells (SW480). The lime volatile oil was found to be rich in two major compounds: D-limonene (30.13%) and D-dihydrocarvone (30.47%). A dose-dependent inhibitory action was obtained at dose levels of 6.25, 12.5, 25.0, 50.0, 100.0, and 200.0 μg/mL. The percentage of inhibition of growth of these cells was 75% (100 μg/mL) as compared to 22.62% by positive control camptothecin (25 mg/mL) in SW480 cells (MTT assay). Moreover, lime volatile oil was found to be safe as manifested by no significant inhibition of growth of noncancerous cells (N1H3T3; Swiss mouse embryo fibroblast). Inhibitions of 55.17% and 44.76% was achieved by volatile oil and camptothecin, respectively, at a dose of 25 mg/mL. The release of lactate dehydrogenase (LDH) from the cells indicated damage to the cell membrane. This suggested the occurrence of necrosis and apoptosis in cultured human colon cancer cells. The authors proposed that high lipophilicity of compounds present in volatile oils caused the membrane disintegration and release of LDH from SW480 cells. Further, apoptosis-mediated cell death was also supported by increased DNA fragmentation, elevation of caspase 3 content (1.8- to 2-fold) and expression of Bax/Bcl 2 genes (2- to 4.3-fold) after 24 and 48 h, respectively.

Jayaprakash et al. (2013) carried out the chemoprofiling of *Citrus limettiodes* (Palestine sweet lime; PSL) and examined the volatile oil obtained from PSL for antiproliferative activities. The PSL volatile oil was found to be rich in 13 monoterpenes, nine sesquiterpenes, one acyclic ester, and one straight chain hydrocarbon. The four main compounds were D-limonene (54.6%), triacontane (7.99%), sabinene

(6.4%), and myrcene (5.2%). The PSL was found to depict a dose-dependent antiproliferative action with 58.4% and 77.3% inhibition at 100 and 200 ppm, respectively, at the end of 24 h. Furthermore, the cytotoxic actions on SW480 cells were also manifested by their accumulation in early and late phases of apoptosis with 100 ppm of volatile oil. A maximum of 13.9- and 17.3-fold increases in Bax/Bcl-2 expressions was achieved with 100 ppm of volatile oil and 50 ppm of D-limonene (positive control), respectively. The PSL volatile oil was also found to suppress the inflammatory markers—that is, COX-2 by 65% (50 ppm) and 80% (100 ppm). These results were consistent with suppression of inflammatory markers by D-limonene in human colon cancer SW480 cells. The authors warranted the need of *in vivo* studies for possible applications of PSL volatile oil for preventing fatal diseases.

In another study, Murthy, Jayaprakasha, and Patil (2012) investigated the limonene-rich hydrodistilled volatile oils from the fruits of *Citrus sinensis* (blood oranges) for their antiproliferative actions against SW480 and HT-29 cells. The blood orange volatile oil (BOVO) and D-limonene (used as reference standard) were formulated into an emulsion. Inhibition of growth of SW480 and HT-29 cells by 74.2% and 53.4%, respectively, was obtained at a dose of 100 ppm after treatment for 72 h. A 1.7-fold increase in proapoptotic markers was also observed in SW480 cells at 100 ppm BOVO emulsion, while in HT-29 cells, Bax/Bcl-2 expressions increased to 11.3-fold after administration of 100 ppm BOVO emulsion. These results revealed the antiproliferative and apoptotic action of BOVO emulsion. Further, a dose-dependent decrease in the expressions of VEGF and MMP-9 in both SW480 and HT-29 colon cancer cells and decreased human umbilical vein endothelial cell (HUVEC) formation suggested its antiangiogenesis action.

Some selected molecules (such as limonene) from the citrus essential oils have already been granted as generally regarded as safe (GRAS) status by the U.S. Food and Drug Administration (FDA). However, only restricted numbers of studies have delved into the role of these volatile oils on different crucial molecular targets of colon cancer.

3.4.3 Carotenoids

Citrus fruits are one of the greatest reservoirs of carotenoids. Chemically, carotenoids are tetraterpenes having 3–13 conjugated double bonds along the 40-carbon polyene chain. Carotenoids with an oxygen atom in their structure are called as xanthophylls (lutein, zeaxanthin), whereas nonoxygenated are known as carotenes (α-carotene, β-carotene, and lycopene). Carotenoids are the pigments that provide color to fruits and vegetables and they are involved in a plethora of physiological functions in plants and bioactivities in humans. The most beneficial effects of carotenoids are their antioxidant and provitamin A activity (Tanaka, Shnimizu, and Moriwaki 2012). In addition, they have also been found to play an important role in intercellular communication, immune system activity, and reducing the incidence of cancer, cardiovascular diseases, macular degeneration and cataracts. Carotenoids such as α-carotene, β-carotene, β-cryptoxanthin, lycopene, lutein, zeaxanthin, violaxanthin, neoxanthain, canthaxanthin, astaxanthin, fucoxanthin, and siphonaxanthin have displayed anticancer activity in different cancer cells, such as colon, liver, breast, prostate, cervix, and leukemia (Nishino et al. 2000; Ajila and Brar 2012; Tanaka et al. 2012; Haddad et al. 2013; Rokkaku et al. 2013; Manabe, Hirata, and Sugawara 2014).

The main mechanisms of cancer chemoprevention by carotenoids involve alterations in cellular pathways leading to cell growth or cell death. These include immune modulation, hormone and growth factor signaling, regulatory mechanisms of cell cycle progression, cell differentiation, and apoptosis (Sporn and Suh 2002; Tanaka et al. 2012). Narisawa et al. (1996) reported protective anticancer action of α-carotenes, lycopene, and lutein on preneoplastic colorectal ADC lesions. The same authors in another study found that oral administration of β-cryptoxanthin (25 ppm) along with diet lowered the incidence of colon carcinogenesis to 68% as compared to control (96%) (Narisawa et al. 1999). Xanthophylls such as astaxanthin and canthaxanthin have also shown an inhibitory effect on cellular proliferation in rats with AOM-induced colon cancer (Tanaka et al. 1995). Tanaka et al. (2000) reported chemopreventive actions of satsuma mandarin. The pulp, rich in β-cryptoxanthin and hesperidin, showed a reduction in occurrence of ACF (15%), frequency (78%), and multiplicity (80%) of colonic ADC, respectively.

Studies have revealed suppression in colitis and obesity-associated colon carcinogenesis after oral administration of segment membrane of *Citrus unshui* (rich in β-cryptoxanthin and fiber) in male *db/db* mice (Suzuki et al. 2007; Tanaka et al. 2008). β-Cryptoxanthin has been found to repair DNA oxidation in human colon carcinoma *Caco*-2 cells (Lorenzo et al. 2009) and induce apoptosis in colo 320 and colo 205 human colon cancer cells (Ugocsai et al. 2005). β-Cryptoxanthin and phytosterols (β-sitosterol, campesterol, and stigmasterol) alone or in combination have been reported to possess apoptosis-mediated antiproliferative action in human colon ADC cells (*Caco*-2 cells). The resulting effect was not synergistic or additive, but rather indicated the nonantagonistic nature of both phytoconstituents. *In vitro* studies showed a decrease in cell viability with all treatments due to accumulation of significantly higher proportion of cells in the sub-G_1 phase. This was accompanied by dephosphorylation of BAD, mitochondrial depolarization, and caspase 3-dependent PARP cleavage, with intracellular Ca^{2+} influx and increase of RONS levels as initial triggers (Cilla et al. 2015).

The inherent antioxidant actions along with their tendency to modulate various molecular targets illuminate the excellent therapeutic action of carotenoids. However, the number of investigations affirming the use of carotenoids present mainly in *Citrus* species (such as lutein, zeaxanthin, astaxanthin, and canthaxanthin) is scarce. Thus, a deeper exploration of these carotenoids is warranted before they can be exploited clinically.

3.4.4 Coumarins

The term "coumarin" is derived from the French word "coumarou," the vernacular name of the tonka bean (*Dipteryx odoranta*). Chemically, the basic nucleus of coumarin compounds is benzo-α-pyrone, in which the benzene ring is joined to a pyrone ring. Based upon the nature of the heterocyclic ring joined to the benzene ring, coumarins can be categorized as (a) simple coumarins, (b) furanocoumarins,

(c) pyranocoumarins, and (d) pyrone-substituted coumarins (Table 3.1). The fruits belonging to the families Rutaceae and Umbelliferae are the richest source of coumarins (Keating and O'Kennedy 1997; Lacy and O'Kennedy 2004). The coumarins and the furanocoumain (psoralens) compounds are nonvolatile in nature, found only in cold-pressed citrus peel oils and not in distilled oils. These comprise a large class of compounds, particularly present in high amounts in essential oils. The highest amount of coumarin is present in cold-pressed lime oils (7%, w/w) (Stanley and Jurd 1971). Coumarins possess a versatile backbone on which substitutions can be carried out, thus positioning them as a novel therapeutic agent for the future. Although their physiological role in plants is not well defined, they have been found to demonstrate multiple pharmacological activities such as antioxidant, anti-inflammatory, antiviral, bacteriostatic, antithrombotic, antinociceptive, antituberculer, antidepressant, anticholinesterase, and antitumor (in malignant melanoma, renal cell, carcinoma, prostate cancer, and leukemia) properties (Rajabi, Feiz, and Luque 2015). Studies have also highlighted the effectiveness of coumarins in treating chemotherapy- and radiotherapy-related side effects (Grotz et al. 2001). The molecular processes that are modulated by coumarins include cell cycle arrest, angiogenesis, mitosis, and inhibition of telomerase, kinase, and heat shock proteins (HSP 90) (Thakur, Singla, and Jaitak 2015). The common coumarins belonging to different subtypes are listed in Table 3.1.

Prenyloxycoumarins (simple coumarins with prenyl substitution at C6 position)—for example, auraptene (AUR) and collinin—significantly attenuate the expression of inflammatory mediators, iNOS and COX-2, and TNF-α. Kohno et al. (2006) explored the chemopreventive action of AUR and collinin in inflammatory bowel disease (IBD)-associated colon cancer. The histopathological studies confirmed that in AOM/DSS-induced colon cancer, the incidence and multiplicity of AD, ADC, and colonic inflammation were significantly lowered after treatment with AUR (0.01%) and collinin (0.05%). This reduction was also indicated by a decrease in PCNA-labeling index, expressions of nitrotyrosine, iNOS, COX-2 in colonic epithelium, and an enhanced apoptotic index. These results emphasized that both AUR and collinin were effective in inhibiting the colitis/IBD-related colon cancer.

Kaneko, Tahara, and Takabayashi (2007) conducted a study to determine the suppressive action of esculetin and esculin in DMH-induced oxidative damage and ACF formation in colons of male Fischer 344 rats. Previous reports have demonstrated that animals treated with oxidative carcinogens possess higher or increased levels of 8-oxo-7,8-dihydroguanine (8-oxoG) or 8-oxo-2′-deoxyguanosine (8-oxodG) (Nakae et al. 1997). A major oxidative DNA adduct, 8-oxoG enhances the possibility of mutagenesis. The oral administration of esculetin (0.01%) and esculin (0.02% and 0.05%) in drinking water was found to decrease the content of 8-oxodG and TBARS in colon mucosa. This excellent action of esculin was attributed to the change induced in the metabolic pathway of DMH (Kaneko et al. 2007).

Citrus AUR has also been evaluated for its suppressive actions on preneoplastic lesions in genetically altered female c57bl/ksj-*db/db* mice and its wild type. A decrease in the number of ACF, β-catenin accumulated crypts (BCAC) and PCNA labeling index and an increase in the apoptotic index propounded that AUR was able to suppress the early phase of colon carcinogenesis in both phenotypes.

Although, AUR did not produce any marked alterations in serum total cholesterol, glucose, and leptin profiles in wild type mice, a decrease in triglyceride levels was observed in *db/db* mice. This reduced incidence of ACF and BCAC in *db/db* mice was attributed to a decrease in serum triglyceride levels (Hayashi et al. 2007). Such an association has been reported in animals as well as in humans (Yamada et al. 1998; Niho et al. 2003a,b, 2005).

Citrus unshiu segment membrane (CUSM) has been evaluated for its preventive actions in AOM-induced colon cancer in obese/diabetic *db/db* mice. CUSM is a waste product produced after squeezing the juice; however, it is rich in fiber and biologically active compounds hesperidin and AUR. A study conducted by Tanaka et al. (2008) suggested that CUSM inhibited the development of ACF by 53%, 54%, and 59% at a concentration of 0.02%, 0.1%, and 0.5%, respectively, and BCAC by 36%, 53%, and 74% at a concentration of 0.02%, 0.1%, and 0.5% in male *db/db* mice. This inhibition was attributed to triglyceride lowering action without any alteration in other biochemical parameters.

The ethanolic extracts of peel and juice of *Citrus* species rich in PMF and coumarin were investigated for their anticancer activity in human colon cancer HT-29 cells using hesperidin as a reference compound. The highest anticancer activity was obtained with HMF (IC_{50}: 28 μM) followed by 5,7,8,3′,4′-pentamethoxyflavone (IC_{50}: 43 μM), AUR (IC_{50}: 47 μM), tangeretin (IC_{50}: 48 μM) (Hirata et al. 2009).

Tanaka et al. (2010) evaluated two novel prodrugs formed from the inclusion complexes of 3-(4′-geranyloxy-3′-methoxyphenyl)-2-transpropenoic acid (GOFA) and AUR with β-cyclodextrin (β-CD) for their chemopreventive activity in colitis-associated colon cancer. Reductions of 63% ($p < 0.05$) and 83% ($p < 0.001$) in the multiplicity of colonic ADC were noticed with 100 and 500 ppm of GOFA/β-CD, respectively, at the end of 18 weeks. Furthermore, 100 and 500 ppm of AUR/β-CD suppressed the development of colonic ADCs. The effects of GOFA/β-CD and AUR/β-CD on cell proliferation and apoptosis of colonic adenocarcinoma were examined using a PCNA labeling index and a terminal deoxynucleotidyl transferase (TdT)-mediated deoxyuridine triphosphate (dUTP)–biotin nick end labeling (TUNEL) method, respectively, and on apoptosis-inhibiting activity by positive rate of survivin. The incidence and multiplicity of the colonic AD and ADC and mean labeling indices of PCNA, TUNEL-positive rates, and positive rates of survivin of treated groups were significantly lower than in the untreated groups. Table 3.5 summarizes the effect of GOFA/β-CD and AUR/β-CD on colonic AD, ADC, inflammation, development of mucosal ulcer, high-grade dysplasia, PCNA labeling index, TUNEL-positive rates, and positive rate of survivin. The mean scores of the immunohistochemical expression of the proinflammatory cytokines (NF-κB, Nrf2, TNF-α, Stat3, IL-6, and IL-1β) in colonic ADCs of treated groups were also found to be significantly less in comparison to untreated groups. The study suggested that these novel prodrugs were able to inhibit colitis-related colon carcinogenesis due to the modulation of inflammation, proliferation, and expression of proinflammatory cytokines.

Park et al. (2011) elucidated the underlying molecular mechanism of esculetin-induced inhibition of cancer cell growth in HCT116 cells. A decrease in cell proliferation was observed to occur via arrest of the G_1 phase of cell cycle through the downregulation of cyclin D1/CDK4 and cyclin E/CDK2, which are the regulating

TABLE 3.5
Effect of GOFA/β-CD and AUR/β-CD on Colonic AD, ADC, Inflammation, Development of Mucosal Ulcer, High-Grade Dysplasia, PCNA Labeling Index, TUNEL-Positive Rates and Positive Rate of Survivin

Development of Colonic AD and ADC

	Incidence (%)		Multiplicity (No. of Tumors/Colon)	
Treatment	AD	ADC	AD	ADC
AOM/1.5% DSS	61	64	1.39 ± 1.50	1.96 ± 2.24
AOM/1.5% DSS/100 ppm GOFA/β-CD	40	24	0.72 ± 1.06	0.52 ± 1.16
AOM/1.5% DSS/500 ppm GOFA/β-CD	25	13	0.33 ± 0.64	0.25 ± 0.74
AOM/1.5% DSS/100 ppm AUR/β-CD	46	46	0.96 ±1.27	1.21 ± 1.61
AOM/1.5% DSS/500 ppm AUR/β-CD	50	25	1.00 ±1.32	0.42 ± 0.83

Colonic Inflammation and Development of Mucosal Ulcer and High-Grade Dysplasia

Treatment	Inflammation Score (Incidence, %)	No. of Colonic Mucosal Ulcer/Colon (Incidence, %)	No. of High-Grade Dysplasia/Colon (Incidence, %)
AOM/1.5% DSS	2.79 ± 0.96	1.29 ± 1.36 (75)	2.21 ± 1.83 (82)
AOM/1.5% DSS/100 ppm GOFA/β-CD	1.52 ± 1.05	0.36 ± 0.64 (28)	0.64 ± 1.19 (32)
AOM/1.5% DSS/500 ppm GOFA/β-CD	0.75 ± 0.90	0.33 ± 0.56 (29)	0.50 ±1.14 (25)
AOM/1.5% DSS/100 ppm AUR/β-CD	1.71 ± 0.69	0.42 ± 0.58 (38)	1.25 ± 1.67 (50)
AOM/1.5% DSS/500 ppm AUR/β-CD	1.17 ± 0.87	0.33 ± 0.70 (21)	0.75 ± 1.45 (33)

Labeling Indices of PCNA, TUNEL-Positive Rates and Positive Rates of Survivin

Treatment	Antiproliferative Activity by PCNA Labeling Index	Apoptotic Activity TUNEL-Positive Rates	Apoptotic Inhibiting Activity by Positive Rate of Survivin
AOM/1.5% DSS	84.5 ± 9.4	7.67 ± 1.28	58.1 ± 12.6
AOM/1.5% DSS/100 ppm GOFA/β-CD	56.3 ± 11.2	12.00 ± 4.56	38.0 ± 8.5
AOM/1.5% DSS/500 ppm GOFA/β-CD	57.3 ± 9.6	13.70 ± 4.04	23.0 ± 7.6
AOM/1.5% DSS/100 ppm AUR/β-CD	71.5 ± 9.4	10.55 ± 3.62	53.4 ± 11.3
AOM/1.5% DSS/500 ppm AUR/β-CD	61.7 ± 9.4	13.17 ± 2.79	41.3 ± 6.6

factors governing the G_1/S phase cell cycle progression. This G_1 phase arrest was also found to be associated with the upregulation of p27KIP expressions (CDK inhibitor), blockade of the ERK½ function (pathway involved in proliferation), and Ras (inactivate ERK½ pathway). A series of novel polysulfide derivatives of coumarin were synthesized by Saidu et al. (2012) and their cytotoxic activity was studied in HCT116 cells. A reduction of 30%, 18%, or 22% in cell viability of HCT116 cells was observed by novel coumarin polysulfide compounds SV25 (disulfides), SV28 (trisulfides), and SV29 (tetrasulfides) (Figure 3.6) at concentration of 25 μM after 24 h. The study revealed that the coumarin tetrasulfides accumulated 63% of treated cells in the G_2-phase manner. The SV29-mediated ROS production also increased the overall apoptotic action. The proposed mechanism for the enhanced G_2-phase arrest of the cell cycle was the inhibition of cdc25 phosphatase activity (one of the key enzymes responsible for G_2/M phase transition). The tetrasulfide derivatives were found to be more potent than the trisulfide and disulfide, suggesting the prominent role of sulfur in increasing the inhibitory activity.

In vitro anticancer studies in SW480 cells were carried out using coumarins (5-geranyloxy-7-methoxycoumarin, limettin, and isopimpinellin) isolated from the hexane extract of lime (*Citrus aurantifolia*). Among these coumarins, geranyloxy-7-methoxycoumarin demonstrated the highest antiproliferative activity. A 67% inhibition in cell proliferation was observed at 25 μM, while both limettin and isopimpinellin showed less than 40% inhibition after 72 h of incubation, emphasizing geranyloxy-7-methoxycoumarin as a potent inhibitor of SW480 cell proliferation. It was also observed that the percentage of cells in the sub-G_1/G_0 phase increased from 5.3% (in control cells) to 18% (50 μM) after 48 h of incubation. An inhibition of caspase 3 through activation of caspase 8; downregulation of Bcl2; increased expressions of p53, as well as its phosphorylated product; and phosphorylation of p38 MAPK were proposed to be the underlying mechanism of apoptosis in SW480 colon cancer cells (Patil et al. 2013).

Hamidinia, Ramezani, and Mojtahedi (2013) investigated the cytotoxic activity of umbelliprenin (prenyloxycoumarins) in invasive SW48 cells and noninvasive SW116

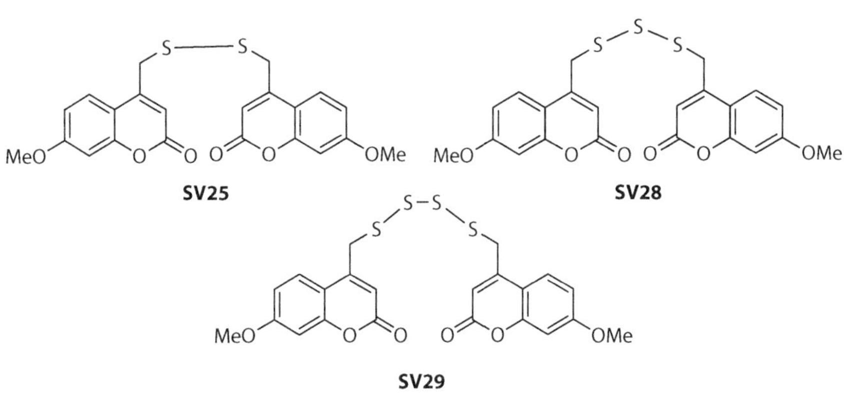

FIGURE 3.6 Polysulfide derivatives of coumarin SV25 (disulfides), SV28 (trisulfides), and SV29 (tetrasulfides).

cells. A significant cytotoxic activity was observed against invasive SW48 cells at all concentrations (12.5, 25, 50, 100, and 200 μM) except for 6.25 μM. The IC_{50} values of 117, 77, and 69 μM after 24, 48, and 72 h, respectively, suggested dose- and time-dependent cytotoxic actions. In the noninvasive SW116 cells, the cytotoxic action was observed only at higher concentrations of 100 and 200 μM (34% and 64% cell death, respectively); however, a significant proliferative effect was observed at lower concentrations, as was evident in microscopic images. The difference in the activity of umbelliprenin on invasive SW48 and noninvasive SW116 cells highlighted the importance of a personalized medicine approach in cancer treatment.

Esculetin (6,7-dihydroxycoumarin) has also been found to act as a potent antagonistic agent against β-catenin-mediated colon cancer. The computer docking studies showed a disruption in the formation of β-catenin T-cell factor complexes of the Wnt signaling pathway by direct binding of esculetin to the amino acid residues of β-catenin (Lee et al. 2013b).

3.4.5 Limonoids

Limonoids are intensely bitter principles present mainly in the Rutaceae, Meliaceae, and Cneoraceae families. They are highly oxygenated terpenoids with a prototypical structure of 4,4,8-trimethyl-17-furanylsteroid skeleton (C_{26} triterpenes) (Fraser, Mulholland, and Fraser 1997; Roy and Saraf 2006). Limonoids have been categorized into four groups—limonin, calamine, ichangensin, and 7-α-acetate limonoids—as depicted in Table 3.1 (Manners et al. 2003; Roy and Saraf 2006; Jayaprakasha et al. 2008). Limonoids exist in nature as aglycone and glycosidic forms. The aglycone form is chiefly present in seeds (70%, w/w) and peels (80%, w/w), while the glycosides are abundant in juice (61%, w/w) and pulp (76%, w/w). It has been suggested that these limonoids are translocated from the fruit tissue to seeds and are the main reason behind the bitterness of juice extracted from the seedless plant as compared to the fruits with seeds. Because the aglycone forms such as limonin, nomilin, obacunone, ichangin, and deacetyl nomilin are water insoluble, they are responsible for the bitter taste, whereas the glycosidic forms are tasteless in nature (Jacob, Hasegawa, and Manners 2000). The seeds of *Citrus sudachi* and *Citrus reticulate* and the peel of *Citrus unshiu* are rich sources of limonoids (Sawabe et al. 1999; Khalil, Maatooq, and El Sayad 2003). To date, 62 limonoids have been reported in *Citrus* species (Jayaprakasha et al. 2008; Kim et al. 2012b).

A wide array of biological activities is associated with limonoids, including antiviral, cardioprotective, immunomodulatory, radical scavenging, and mucous clearing actions (Rohr et al. 2002; Raphael and Kutan 2003; Battinelli et al. 2003; Poulose, Harris, and Patil 2005; Yu et al. 2005). Their potential cytotoxic actions have been explored in liver, breast, lung, stomach, brain, pancreas, and blood cancers (Borradaile, Carroll, and Kurowska 1999; Lam et al. 2000; Guthrie, Kurowska, and Carroll 2001; Tian et al. 2001; Poulose et al. 2005; Poulose, Harris, and Patil 2006; Jayaprakasha et al. 2008; El-Readi et al. 2010; Patil et al. 2010; Murthy, Jayaprakasha, and Patil 2011a). The anticancer action of citrus limonoids has been associated with their GST- and NADH:QR-inducing action. Various limonoids such as limonin, limonin glucoside (LG), deacetyl nomilinic acid glucoside (DNAG), defuran limonin, and limonin-7-methoxamine have been studied for their enzyme-inducing capacities/capabilities (Kelly, Jewell, and O'Brien 2003; Perez et al. 2009, 2010). Citrus limonoids have also been evaluated for their antioxidant actions. Out of limonoid glucosides limoin 17-D-glucopypranoside (LG), obacunone 17-D-glucopyranoside (OG), nomilinic acid 17-D-glucopyranoside (NAG), and deacetylnomilinic acid 17-D-glucopyranoside (DNAG), NAG exhibited the highest and LG the lowest radical scavenging action. However, these activities were comparable to those of vitamin C and SOD (Poulose et al. 2005).

In addition to this, isolimonic acid, ichanexic acid, limonexic acid (in combination with β-sitosterol glucoside), and obacunone exhibited G_2/M phase arrest (Jayaprakasha et al. 2008, 2009), while G_1 and G_0/G_1 phase arrests were observed with obacunone and methyl nomilinate, respectively (Murthy et al. 2011a; Kim et al. 2012b). Obacunone was found to be more effective than its glucoside as evidenced from its IC_{50} value of 97.0 μM as compared to its glucoside value (109.7 μM). This inhibitory action was found to occur due to decrease in Bcl2/Bax expressions, activation of caspase 3, and fragmentation of DNA (Murthy et al. 2011a). On the other hand, the glucosidic form of limonin shows antiproliferative action at a low concentration of 37.39 μM as compared to that of limonin (54.74 μM) (Murthy et al. 2011b). It also increased the doxorubicin-mediated cytotoxic activity by 2.98-fold by decreasing the P-gp activity (El-Readi et al. 2010). A 96% inhibition of cell proliferation has also been reported for limonin in combination with curcumin (Murthy, Jayaprakasha, and Patil 2013). All these studies suggested the suppressive actions of citrus limonoids in colon cancer. However, issues such as water insolubility, bitterness, and low yield upon extraction need to be addressed before these can be employed as novel chemopreventive agents to treat colon cancer.

3.5 SUMMARY

Citrus fruits are widely consumed fruits with versatile biological activities due to the presence of a diverse array of phytochemicals. However, not only the fruit, but also the peel, pulp, and seeds (waste products/by-products left over by the citrus juice industry) contain substantial amounts of active constituents that possess potential anticancer activity. Citrus bioactives inhibit the various signaling cascades (MAPK, PI3K/AKT/mTOR, GSK-3β, leptin, EGFR, ERK½, TGF-β, and Wnt/β-catenin), modulate the expressions of proapoptotic proteins (Bax and Bak) and anti-apoptotic proteins (Bcl-2, Bcl-xl) associated with extrinsic and intrinsic apoptotic pathways and arrest the G_0/G_1, S, G_2, and M phases of the cell cycle. They have also been reported to scavenge ROS, suppress the inflammatory mediators (COX, LOX, PLA_2, iNOS, and NF-κB), inhibit angiogenic factors (VEGF and MMP-2), and restore the levels of antioxidant enzymes (SOD, CAT, GPx, GR, and GSH). This chapter describes the various *in vitro* and *in vivo* studies of these bioactive constituents (flavonoids, volatile oils, coumarins, carotenoids, and limonoids) in modulation of colon cancer. These studies suggest the role of citrus bioactives in successfully mitigating the sporadic and inherited, as well as inflammation and obesity associated colon cancer.

Although, numerous studies available in the literature have elucidated the mechanism of action of citrus phytoconstituents, only limited research has been focused to determine their pharmacokinetic fate. Thus, in-depth investigations are required to determine the pharmacokinetic parameters and effective therapeutic and toxic concentration of citrus bioactives before these can be proposed as prospective therapeutic agents for colon cancer treatment.

ACKNOWLEDGMENT

The authors are grateful to the University Grants Commission (UGC, BSR Research Fellowship Scheme 2013–2014 granted to the Department of Pharmaceutical Sciences and Drug Research, Punjabi University, Patiala, Appointment Letter No. 13003) for financial support for successful completion of this work.

REFERENCES

Aaltonen, L. A., R. Salovaara, P. Kristo, F. Canzian, A. Hemminki, P. Peltomäki, and J. P. Mecklin. 1998. Incidence of hereditary nonpolyposis colorectal cancer and the feasibility of molecular screening for the disease. *New England Journal of Medicine* 338:1481–1487.

Abad-Garcia, B., S. Garmon-Lobato, L. A. Berrueta, B. Gallo, and F. Vicente. 2012. Online characterization of 58 phenolic compounds in *Citrus* fruit juices from Spanish cultivars by high-performance liquid chromatography with photodiode-array detection coupled to electrospray ionization triple quadrupole mass spectrometry. *Talanta* 99:213–224.

Ahsan, H., A. I. Neugut, G. C. Garbowski, J. S. Jacobson, K. A. Forde, M. R. Treat, and J. D. Waye. 1998. Family history of colorectal adenomatous polyps and increased risk for colorectal cancer. *Annals of Internal Medicine* 128:900–905.

Ajila, C. M., and S. K. Brar. 2012. Role of dietary antioxidants in cancer. In Shankar, S. and R. K. Srivastava (eds.), *Nutrition, diet and cancer* (pp. 377–412). Dordrecht: Springer.

Amrutha, K., P. Nanjan, S. K. Shaji, D. Sunilkumar, K. Subhalakshmi, L. Rajakrishna, and A. Banerji. 2014. Discovery of lesser known flavones as inhibitors of NF-κB signaling in MDA-MB-231 breast cancer cells—A SAR study. *Bioorganic & Medicinal Chemistry Letters* 24:4735–4742.

Andrade, R. G., Jr., J. S. Ginani, G. K. Lopes, F. Dutra, A. Alonso, and M. Hermes-Lima. 2006. Tannic acid inhibits in vitro iron-dependent free radical formation. *Biochimie* 88:87–96.

Aranganathan, S., and N. Nalini. 2009. Efficacy of the potential chemopreventive agent, hesperetin (citrus flavanone), on 1,2-dimethylhydrazine induced colon carcinogenesis. *Food and Chemical Toxicology* 47:2594–2600.

Aranganathan, S., and N. Nalini. 2012. Antiproliferative efficacy of hesperetin (citrus flavanoid) in 1,2-dimethylhydrazine-induced colon cancer. *Phytotherapy Research* 27:999–1005.

Aranganathan, S., J. P. Selvam, and N. Nalini. 2008. Effect of hesperetin, a citrus flavonoid, on bacterial enzymes and carcinogen-induced aberrant crypt foci in colon cancer rats: A dose-dependent study. *Journal of Pharmacy and Pharmacology* 60:1385–1392.

Aranganathan, S., J. P. Selvam, N. Sangeetha, and N. Nalini. 2009. Modulatory efficacy of hesperetin (citrus flavanone) on xenobiotic-metabolizing enzymes during 1,2-dimethylhydrazine-induced colon carcinogenesis. *Chemico-Biological Interactions* 180:254–261.

Arcas, M. C., J. M. Botía, A. M. Ortuno, and J. A. Del Río. 2000. UV irradiation alters the levels of flavonoids involved in the defence mechanism of *Citrus aurantium* fruits against *Penicillium digitatum*. *European Journal of Plant Pathology* 106:617–622.

Bakkali, F., S. Averbeck, D. Averbeck, and M. Idaomar. 2008. Biological effects of essential oils—A review. *Food and Chemical Toxicology* 46:446–475.

Baser, K. H. C., and F. Demirci. 2007. Chemistry of essential oils. In Berger, R. G. (ed.), *Flavours and fragrances: Chemistry, bioprocessing and sustainability*, pp. 43–86. Springer: Berlin.

Basterfield, L., and J. C. Mathers. 2010. Intestinal tumours, colonic butyrate and sleep in exercised Min mice. *British Journal of Nutrition* 104:355–363.

Battinelli, L., F. Mengoni, M. Lichtner, G. Mazzanti, A. Saija, C. M. Mastroianni, and V. Vullo. 2003. Effect of limonin and nomilin on HIV-1 replication on infected human mononuclear cells. *Planta Medica* 69:910–913.

Benson, A. B., III. 2007. Epidemiology, disease progression, and economic burden of colorectal cancer. *Journal of Managed Care Pharmacy* 13:5–18.

Bird, R. P. 1995. Role of aberrant crypt foci in understanding the pathogenesis of colon cancer. *Cancer Letters* 93:55–71.

Bok, S. H., S. H. Lee, Y. B. Park, K. H. Bae, K. H. Son, T. S. Jeong, and M. S. Choi. 1999. Plasma and hepatic cholesterol and hepatic activities of 3-hydroxy-3-methyl-glutaryl-CoA reductase and acyl CoA: Cholesterol transferases are lower in rats fed citrus peel extract or a mixture of citrus bioflavonoids. *Journal of Nutrition* 129:1182–1185.

Borradaile, N. M., K. K. Carroll, and E. M. Kurowska. 1999. Regulation of HepG2 cell apolipoprotein B metabolism by the citrus flavanones hesperetin and naringenin. *Lipids* 34:591–598.

Bothe, H., C. Gotz, N. Stobbe-Maicherski, E. Fritsche, J. Abel, and T. Haarmann-Stemmann. 2010. Luteolin enhances the bioavailability of benzo(a)pyrene in human colon carcinoma cells. *Archives of Biochemistry and Biophysics* 498:111–118.

Bourgaud, F., A. Gravot, S. Milesi, and E. Gontier. 2001. Production of plant secondary metabolites: A historical perspective. *Plant Science* 161:839–851.

Cai, Y. Z., Q. Luo, M. Sun, and H. Corke. 2004. Antioxidant activity and phenolic compounds of 112 traditional Chinese medicinal plants associated with anticancer. *Life Sciences* 74:2157–2184.

Camarda, L., V. Di Stefano, S. F. Del Bosco, and D. Schillaci. 2007. Antiproliferative activity of citrus juices and HPLC evaluation of their flavonoid composition. *Fitoterapia* 78:426–429.

Cano, A., A. Medina, and A. Bermejo. 2008. Bioactive compounds in different citrus varieties. Discrimination among cultivars. *Journal of Food Composition and Analysis* 21:377–381.

Carson, C. F., and K. A. Hammer. 2011. Chemistry and bioactivity of essential oils. In Thormar, H. (ed.), *Lipids and essential oils as antimicrobial agents*, pp. 203–238. New York: John Wiley & Sons.

Cesarone, M. R., G. Belcaro, L. Pellegrini, A. Ledda, G. Vinciguerra, A. Ricci, A. Di Renzo, I. Ruffini, G. Gizzi, E. Ippolito, F. Fano, M. Dugall, G. Acerbi, U. Cornelli, M. Hosoi, and M. Cacchio. 2006. Venoruton vs. Daflon: Evaluation of effects on quality of life in chronic venous insufficiency. *Angiology* 57:331–339.

Chan, A. T., and E. L. Giovannucci. 2010. Primary prevention of colorectal cancer. *Gastroenterology* 138:2029–2043.

Chatzopoulou, A., A. Karioti, C. Gousiadou, V. Lax Vivancos, P. Kyriazopoulos, S. Golegou, and H. Skaltsa. 2010. Depsides and other polar constituents from *Origanum dictamnus* L. and their *in vitro* antimicrobial activity in clinical strains. *Journal of Agricultural and Food Chemistry* 58:6064–6068.

Chen, J., M. Lu, Y. Jing, and J. Dong. 2006. The synthesis of L-carvone and limonene derivatives with increased antiproliferative effect and activation of ERK pathway in prostate cancer cells. *Bio-organic and Medicinal Chemistry* 14:6539–6547.

Chian, S., Y. Y. Li, X. J. Wang, and X. W. Tang. 2014. Luteolin sensitizes two oxaliplatin-resistant colorectal cancer cell lines to chemotherapeutic drugs via inhibition of the Nrf2 pathway. *Asian Pacific Journal of Cancer Prevention* 15:2911–2916.

Choi, S. Y., H. C. Ko, S. Y. Ko, J. H. Hwang, J. G. Park, S. H. Kang, S. H. Han, S. H. Yun, and S. J. Kim. 2007. Correlation between flavonoid content and the NO production inhibitory activity of peel extracts from various citrus fruits. *Biological and Pharmaceutical Bulletin* 30:772–778.

Choudhury, D., A. Ganguli, D. G. Dastidar, B. R. Acharya, A. Das, and G. Chakrabarti. 2013. Apigenin shows synergistic anticancer activity with curcumin by binding at different sites of tubulin. *Biochimie* 95:1297–1309.

Chunhua, L., L. Donglan, F. Xiuqiong, Z. Lihua, F. Qin, L. Yawei, Z. Liang, W. Ge, J. Linlin, Z. Ping, L. Kun, and S. Xuegang. 2013. Apigenin up-regulates transgelin and inhibits invasion and migration of colorectal cancer through decreased phosphorylation of AKT. *Journal of Nutritional Biochemistry* 24:1766–1775.

Cilla, A., A. Attanzio, R. Barberá, L. Tesoriere, and M. A. Livrea. 2015. Anti-proliferative effect of main dietary phytosterols and β-cryptoxanthin alone or combined in human colon cancer *Caco*-2 cells through cytosolic Ca^{+2} and oxidative stress-induced apoptosis. *Journal of Functional Foods* 12:282–293.

Constantinou, A. I., N. Kamath, and J. S. Murley. 1998. Genistein inactivates bcl-2, delays the G2/M phase of the cell cycle, and induces apoptosis of human breast adenocarcinoma MCF-7 cells. *European Journal of Cancer* 34:1927–1934.

Del Follo-Martinez, A., N. Banerjee, X. Li, S. Safe, and S. Mertens-Talcott. 2013. Resveratrol and quercetin in combination have anticancer activity in colon cancer cells and repress oncogenic microRNA-27a. *Nutrition and Cancer* 65:494–504.

Derry, M. M., K. Raina, C. Agarwal, and R. Agarwal. 2013. Identifying molecular targets of lifestyle modifications in colon cancer prevention. *Frontiers in Oncology* 3:1–20.

Di Giacomo, A., and G. Di Giacomo. 2002. Essential oil production. In Dugo, G., and A. Di Giacomo (eds.), *Citrus: The genus Citrus*, pp. 114–147. London: Taylor & Francis.

Dihaal, A. A., H. Van der Woude, P. J. Hendriksen, H. Charif, L. J. Dekker, L. IJsselstijn, V. C. de Boer, G. M. Alink, P. C. Burgers, I. M. Rietjens, R. A. Woutersen, and R. H. Stierum. 2008. Transcriptome and proteome profiling of colon mucosa from quercetin fed F344 rats point to tumor preventive mechanisms, increased mitochondrial fatty acid degradation and decreased glycolysis. *Proteomics* 8:45–61.

Duthie, G. G., S. J. Duthie, and J. A. Kyle. 2000. Plant polyphenols in cancer and heart disease: Implications as nutritional antioxidants. *Nutrition Research Reviews* 13:79–106.

El-Readi, M. Z., D. Hamdan, N. Farrag, A. El-Shazly, and M. Wink. 2010. Inhibition of P-glycoprotein activity by limonin and other secondary metabolites from *Citrus* species in human colon and leukaemia cell lines. *European Journal of Pharmacology* 626:139–145.

Fang, J., Q. Zhou, L. Z. Liu, C. Xia, X. Hu, X. Shi, and B. H. Jiang. 2007. Apigenin inhibits tumor angiogenesis through decreasing HIF-1*a* and VEGF expression. *Carcinogenesis* 28:858–864.

Fisher, K., and C. Phillips. 2008. Potential antimicrobial uses of essential oils in food: Is citrus the answer? *Trends in Food Science and Technology* 19:156–164.

Fraser, L. A., D. A. Mulholland, and D. D. Fraser. 1997. Classification of limonoids and protolimonoids using neural networks. *Phytochemical Analysis* 8:301–311.

Frydoonfar, H. R., D. R. McGrath, and A. D. Spigelman. 2003. The variable effect on proliferation of a colon cancer cell line by the citrus fruit flavonoid naringenin. *Colorectal Disease* 5:149–152.

Garg, A., S. Garg, L. J. Zaneveld, and A. K. Singla. 2001. Chemistry and pharmacology of the citrus bioflavonoid hesperidin. *Phytotherapy Research* 15:655–669.

Gattuso, G., D. Barreca, C. Gargiulli, U. Leuzzi, and C. Caristi. 2007. Flavonoid composition of citrus juices. *Molecules* 12:1641–1673.

Giardiello, F. M., J. D. Brensinger, and G. M. Petersen. 1997. The use and interpretation of commercial APC gene testing for familial adenomatous polyposis. *New England Journal of Medicine* 336:823–827.

Gonzalez-Molina, E., R. Dominguez-Perles, D. A. Moreno, and C. Garcia-Viguera. 2010. Natural bioactive compounds of *Citrus limon* for food and health. *Journal of Pharmaceutical and Biomedical Analysis* 51:327–345.

Grotz, K. A., P. Wustenberg, R. Kohnen, B. Al-Nawas, H. H. Henneicke-von Zepelin, A. Bockisch, and W. Wagner. 2001. Prophylaxis of radiogenic sialadenitis and mucositis by coumarin/troxerutine in patients with head and neck cancer—A prospective, randomized, placebo-controlled, double-blind study. *British Journal of Oral and Maxillofacial Surgery* 39:34–39.

Guo, Z., Y. Zhou, B. M. Evers, and Q. Wang. 2011. Apigenin inhibits mTORC2 through down-regulation of RICTOR expression in colorectal cancer cells. *Gastroenterology* 140:S-400–S-401.

Gupta, S. C., A. K. Tyagi, P. Deshmukh-Taskar, M. Hinojosa, S. Prasad, and B. B. Aggarwal. 2014. Downregulation of tumor necrosis factor and other proinflammatory biomarkers by polyphenols. *Archives of Biochemistry and Biophysics* 559:91–99.

Guthrie, N., E. M. Kurowska, and K. K. Carroll. 2001. Compositions and methods of treatment of neoplastic diseases and hypercholesterolemia with citrus liminoids and flavonoids and tocotrienols; patent number: US6251400; Publication date: 2001-06-26.

Haddad, N. F., A. J. Teodoro, F. L. de Oliveira, N. Soares, R. M. de Mattos, F. Hecht, R. S. Dezonne, L. Vairo, R. C. dos Santos Goldenberg, F. C. A. Gomes, and D. P. de Carvalho. 2013. Lycopene and beta-carotene induce growth inhibition and proapoptotic effects on ACTH-secreting pituitary adenoma cells. *PloS One* 8:62773.

Hamid, A. A., O. O. Aiyelaagbe, and L. A. Usman. 2011. Essential oil: Its medicinal and pharmacological uses. *International Journal of Current Research* 33:86–98.

Hamidinia, M., M. Ramezani, and Z. Mojtahedi. 2013. Cytotoxic/proliferative effects of umbelliprenin on colon cancer cell lines. *Annals of Colorectal Research* 1:101–105.

Hayashi, K., R. Suzuki, S. Miyamoto, Y. Shin-ichiroh, H. Kohno, S. Sugie, S. Takashima, and T. Tanaka. 2007. Citrus auraptene suppresses azoxymethane-induced colonic preneoplastic lesions in C57BL/KsJ-db/db mice. *Nutrition and Cancer* 58:75–84.

Heilmann, J. 2009. New medical applications of plant secondary metabolites. In Wink, M. (ed.), *Annual plant reviews, functions and biotechnology of plant secondary metabolites*, pp. 348–366. New York: John Wiley & Sons.

Hirata, T., M. Fujii, K. Akita, N. Yanaka, K. Ogawa, M. Kuroyanagi, and D. Hongo. 2009. Identification and physiological evaluation of the components from citrus fruits as potential drugs for anti-corpulence and anticancer. *Bioorganic and Medicinal Chemistry* 17:25–28.

Ho, C. T., M. H. Pan, C. S. Lai, and S. Li. 2012. Polymethoxyflavones as food factors for the management of inflammatory diseases. *Neuropathology* 20:337–341.

Hung, J. Y., Y. L. Hsu, Y. C. Ko, Y. M. Tsai, C. J. Yang, M. S. Huang, and P. L. Kuo. 2010. Didymin, a dietary flavonoid glycoside from citrus fruits, induces Fas-mediated apoptotic pathway in human non-small-cell lung cancer cells *in vitro* and *in vivo*. *Lung Cancer* 68:366–374.

Hur, S. J., S. H. Kang, H. S. Jung, S. C. Kim, H. S. Jeon, I. H. Kim, and J. D. Lee. 2012. Review of natural products actions on cytokines in inflammatory bowel disease. *Nutrition Research* 32:1–16.

Huxley, R. R., A. Ansary-Moghaddam, P. Clifton, S. Czernichow, C. L. Parr, and M. Woodward. 2009. The impact of dietary and life style risk factors on risk of colorectal cancer: A quantitative overview of the epi-demiological evidence. *International Journal of Cancer* 125:171–180.

Igual, M., E. Garcia-Martinez, M. M. Camacho, and N. Martinez-Navarrete. 2013. Jam processing and storage effects on β-carotene and flavonoids content in grapefruit. *Journal of Functional Foods* 5:736–744.

Jacob, R., S. Hasegawa, and G. Manners. 2000. The potential of citrus limonoids as anticancer agents. *Perishables Handling Quarterly* 102:6–8.

Jaramillo, S., S. Lopez, L. M. Varela, R. Rodriguez-Arcos, A. Jimenez, R. Abia, R. Guillen, and F. J. Muriana. 2010. The flavonol isorhamnetin exhibits cytotoxic effects on human colon cancer cells. *Journal of Agricultural and Food Chemistry* 58:10869–10875.

Jayaprakasha, G. K., Y. Jadegoud, G. A. Nagana-Gowda, and B. S. Patil. 2009. Bioactive compounds from sour orange inhibit colon cancer cell proliferation and induce cell cycle arrest. *Journal of Agricultural and Food Chemistry* 58:180–186.

Jayaprakasha, G. K., K. K. Mandadi, S. M. Poulose, Y. Jadegoud, G. A. Nagana Gowda, and B. S. Patil. 2008. Novel triterpenoid from *Citrus aurantium* L. possesses chemopreventive properties against human colon cancer cells. *Bioorganic and Medicinal Chemistry* 16:5939–5951.

Jayaprakasha, G. K., K. N. Murthy, R. M. Uckoo, and B. S. Patil. 2013. Chemical composition of volatile oil from *Citrus limettioides* and their inhibition of colon cancer cell proliferation. *Industrial Crops and Products* 45:200–207.

Ji, J., L. Zhang, Y. Y. Wu, X. Y. Zhu, S. Q. Lv, and X. Z. Sun. 2006. Induction of apoptosis by D-limonene is mediated by a caspase-dependent mitochondrial death pathway in human leukemia cells. *Leukemia & Lymphoma* 47:2617–2624.

Kaefer, C. M., and J. A. Milner. 2008. The role of herbs and spices in cancer prevention. *Journal of Nutritional Biochemistry* 19:347–361.

Kaneko, T., S. Tahara, and F. Takabayashi. 2007. Inhibitory effect of natural coumarin compounds, esculetin and esculin, on oxidative DNA damage and formation of aberrant crypt foci and tumors induced by 1,2-dimethylhydrazine in rat colons. *Biological and Pharmaceutical Bulletin* 30:2052–2057.

Kanno, S. I., A. Tomizawa, T. Hiura, Y. Osanai, A. Shouji, M. Ujibe, T. Ohtake, K. Kimura, and M. Ishikawa. 2005. Inhibitory effects of naringenin on tumor growth in human cancer cell lines and sarcoma S-180-implanted mice. *Biological and Pharmaceutical Bulletin* 28:527–530.

Kaur, J., and G. Kaur. 2015. An insight into the role of citrus bioactives in modulation of colon cancer. *Journal of Functional Foods* 13:239–261.

Keating, G. J., and R. O'Kennedy. 1997. The chemistry and occurrence of coumarins. In O'Kennedy, R., and R. D. Thornes (eds.), *Coumarins: Biology, applications and mode of action*, pp. 23–66. New York: John Wiley & Sons.

Kelly, C., C. Jewell, and N. M. O'Brien. 2003. The effect of dietary supplementation with the citrus limonoids, limonin and nomilin on xenobiotic-metabolizing enzymes in the liver and small intestine of the rat. *Nutrition Research* 23:681–690.

Khalil, A. T., G. T. Maatooq, and K. A. El Sayed. 2003. Limonoids from *Citrus reticulata*. *Journal of Biosciences* 58:165–170.

Khan, M. K., and O. Dangles. 2014. A comprehensive review on flavanones, the major citrus polyphenols. *Journal of Food Composition and Analysis* 33:85–104.

Kim, C., M. C. Kim, S. M. Kim, D. Nam, S. H. Choi, S. H. Kim, K. S. Ahn, E. H. Lee, S. H. Jung, and K. S. Ahn. 2013. *Chrysanthemum indicum* L. extract induces apoptosis through suppression of constitutive STAT3 activation in human prostate cancer DU145 cells. *Phytotherapy Research* 27:30–38.

Kim, G. T., S. H. Lee, J. I. Kim, and Y. M. Kim. 2014. Quercetin regulates the sestrin 2-AMPK-p38 MAPK signaling pathway and induces apoptosis by increasing the generation of intracellular ROS in a p53-independent manner. *International Journal of Molecular Medicine* 33:863–869.

Kim, H. G., G. S. Kim, J. H. Lee, S. Park, W. Y. Jeong, Y. H. Kim, J. H. Kim, S. T. Kim, Y. A. Young Ah Cho, W. S. Lee, S. J. Lee, J. S. Jin, and S. C. Shin. 2011. Determination of the change of flavonoid components as the defence materials of *Citrus unshiu* marc. fruit peel against *Penicillium digitatum* by liquid chromatography coupled with tandem mass spectrometry. *Food Chemistry* 128:49–54.

Kim, H. J., S. K. Kim, B. S. Kim, S. H. Lee, Y. S. Park, B. K. Park, S. J. Kim, J. Kim, C. Choi, J. S. Kim, S. D. Choe, J. W. Jung, K. H. Roh, K. S. Kang, and J. Y. Jung. 2010. Apoptotic effect of quercetin on HT-29 colon cancer cells via the AMPK signaling pathway. *Journal of Agricultural and Food Chemistry* 58:8643–8650.

Kim, H. S., T. Wannatung, S. Lee, W. K. Yang, S. H. Chung, J. S. Lim, W. Choe, I. Kang, S. S. Kim, and J. Ha. 2012a. Quercetin enhances hypoxia-mediated apoptosis via direct inhibition of AMPK activity in HCT116 colon cancer. *Apoptosis* 17:938–949.

Kim, J., G. K. Jayaprakasha, A. Vikram, and B. S. Patil. 2012b. Methyl nomilinate from citrus can modulate cell cycle regulators to induce cytotoxicity in human colon cancer (SW480) cells *in vitro*. *Toxicology In Vitro* 26:1216–1223.

Kohno, H., R. Suzuki, M. Curini, F. Epifano, F. Maltese, S. P. Gonzales, and T. Tanaka. 2006. Dietary administration with prenyloxycoumarins, auraptene and collinin, inhibits colitis-related colon carcinogenesis in mice. *International Journal of Cancer* 118:2936–2942.

Kris-Etherton, P. M., K. D. Hecker, A. Bonanome, S. M. Coval, A. E. Binkoski, K. F. Hilpert, and T. D. Etherton. 2002. Bioactive compounds in foods: Their role in the prevention of cardiovascular disease and cancer. *American Journal of Medicine* 113:71–88.

Kurzawa-Zegota, M., M. Najafzadeh, A. Baumgartner, and D. Anderson. 2012. The protective effect of the flavonoids on food-mutagen-induced DNA damage in peripheral blood lymphocytes from colon cancer patients. *Food and Chemical Toxicology* 50:124–129.

Lacy, A., and R. O'Kennedy. 2004. Studies on coumarins and coumarin-related compounds to determine their therapeutic role in the treatment of cancer. *Current Pharmaceutical Design* 10:3797–3811.

Lai, C. S., S. Li, C. Y. Chai, C. Y. Lo, C. T. Ho, Y. J. Wang, and M. H. Pan. 2007. Inhibitory effect of citrus 5-hydroxy-3,6,7,8,3′, 4′-hexamethoxyflavone on 12-O-tetradecanoylphorbol 13-acetate-induced skin inflammation and tumor promotion in mice. *Carcinogenesis* 28:2581–2588.

Lai, C. S., S. Li, C. B. Liu, Y. Miyauchi, M. Suzawa, C. T. Ho, and M. H. Pan. 2013a. Effective suppression of azoxymethane-induced aberrant crypt foci formation in mice with citrus peel flavonoids. *Molecular Nutrition and Food Research* 57:551–555.

Lai, C. S., M. L. Tsai, A. C. Cheng, S. Li, C. Y. Lo, Y. Wang, X. Hang, C. T. Ho, Y. J. Wang, and M. H. Pan. 2011. Chemoprevention of colonic tumorigenesis by dietary hydroxylated polymethoxyflavones in azoxymethane-treated mice. *Molecular Nutrition and Food Research* 55:278–290.

Lai, C. S., J. Y. Zeng, S. Li, Q. Huang, C. T. Ho, and M. H. Pan. 2013b. Inhibitory effects of 5-demethyltangeretin and 5-acetyloxy-6,7,8,4′-tetramethoxyflavone on human colon cancer cells. In Ho, C.-T., C. Mussinan, F. Shahidi, and E. T. Contis (eds.), *Nutrition, Functional and Sensory Properties of Foods*, pp. 281–290. Cambridge: Royal Society of Chemistry.

Lam, L., S. Hasegawa, C. Bergstrom, S. Lam, and P. Kenney. 2000. Limonin and nomilin inhibitory effects on chemical-induced tumorigenesis. In Berhow, M., S. Hasegawa, and G. M. Manners (eds.), *Citrus Limonoids: Functional Chemicals in Agriculture and Foods*, pp. 185–200. Washington, DC: ACS Symposium Series 758, American Chemical Society.

Lan, L., C. Trempus, and S. K. Gilmour. 2000. Inhibition of ornithine decarboxylase (ODC) decreases tumor vascularization and reverses spontaneous tumors in ODC/Ras transgenic mice. *Cancer Research* 60:5696–5703.

Lee, A. C. L., W. C. Hsiao, D. E. Wright, S. Y. Chong, S. K. Leow, C. T. Ho, C. F. Kao, and Y. C. Lo. 2013a. Induction of *GADD45*α expression contributes to the anti-proliferative effects of polymethoxyflavones on colorectal cancer cells. *Journal of Functional Foods* 5:616–624.

Lee, E. R., Y. J. Kang, H. J. Kim, H. Y. Choi, G. H. Kang, J. H. Kim, B. W. Kim, H. S. Jeong, Y. S. Park, and S. G. Cho. 2008. Regulation of apoptosis by modified naringenin derivatives in human colorectal carcinoma RKO cells. *Journal of Cellular Biochemistry* 104:259–273.

Lee, E. R., Y. J. Kang, J. H. Kim, H. T. Lee, and S. G. Cho. 2005. Modulation of apoptosis in HaCaT keratinocytes via differential regulation of ERK signaling pathway by flavonoids. *Journal of Biological Chemistry* 280:31498–31507.

Lee, E. R., J. H. Kim, Y. J. Kang, and S. G. Cho. 2007. The anti-apoptotic and anti-oxidant effect of eriodictyol on UV-induced apoptosis in keratinocytes. *Biological and Pharmaceutical Bulletin* 30:32–37.

Lee, K. W., A. M. Bode, and Z. Dong. 2011. Molecular targets of phytochemicals for cancer prevention. *Nature Reviews Cancer* 11:211–218.

Lee, S. Y., T. G. Lim, H. Chen, S. K. Jung, H. J. Lee, M. H. Lee, D. J. Kim, A. Shin, K. W. Lee, A. M. Bode, Y. J. Surh, and Z. Dong. 2013b. Esculetin suppresses proliferation of human colon cancer cells by directly targeting β-catenin. *Cancer Prevention Research* 6:56–64.

Lee, Y. K., S. Y. Park, Y. M. Kim, W. S. Lee, and O. J. Park. 2009. AMP kinase/cyclooxygenase-2 pathway regulates proliferation and apoptosis of cancer cells treated with quercetin. *Experimental and Molecular Medicine* 41:201–207.

Lefort, E. C., and J. Blay. 2011. The dietary flavonoid apigenin enhances the activities of the anti-metastatic protein CD26 on human colon carcinoma cells. *Clinical and Experimental Metastasis* 28:337–349.

Le Marchand, L., S. P. Murphy, J. H. Hankin, L. R. Wilkens, and L. N. Kolonel. 2000. Intake of flavonoids and lung cancer. *Journal of National Cancer Institute* 92:154–160.

Leonardi, T., J. Vanamala, S. S. Taddeo, L. A. Davidson, M. E. Murphy, B. S. Patil, N. Wang, R. J. Carrol, R. S. Chapkin, J. R. Lupton, and N. D. Turner. 2010. Apigenin and naringenin suppress colon carcinogenesis through the aberrant crypt stage in azoxymethane treated rats. *Experimental Biology and Medicine* 235:710–717.

Li, C., X. Yang, C. Chen, S. Cai, and J. Hu. 2014. Isorhamnetin suppresses colon cancer cell growth through the PI3K-Akt-mTOR pathway. *Molecular Medicine Reports* 9:935–940.

Li, S., M. H. Pan, C. Y. Lo, D. Tan, Y. Wang, F. Shahidi, and C. T. Ho. 2009b. Chemistry and health effects of polymethoxyflavones and hydroxylated polymethoxyflavones. *Journal of Functional Foods* 1:2–12.

Li, S., S. Sang, M. H. Pan, C. S. Lai, C. Y. Lo, C. S. Yang, and C. T. Ho. 2007c. Anti-inflammatory property of the urinary metabolites of nobiletin in mouse. *Bioorganic and Medicinal Chemistry Letters* 1:5177–5181.

Li, W., B. Du, T. Wang, S. Wang, and J. Zhang. 2009a. Kaempferol induces apoptosis in human HCT116 colon cancer cells via the ataxia-telangiectasia mutated-p53 pathway with the involvement of p53 upregulated modulator of apoptosis. *Chemico–Biological Interactions* 177:121–127.

Lim, D. Y., H. J. Cho, J. Kim, C. W. Nho, K. W. Lee, and J. H. Park. 2012. Luteolin decreases IGF-II production and downregulates insulin-like growth factor-I receptor signaling in HT-29 human colon cancer cells. *BMC Gastroenterology* 12:1–10.

Lin, N., T. Sato, and Y. Takayama. 2003. Novel anti-inflammatory actions of nobiletin, a citrus polymethoxy flavonoid, on human synovial fibroblasts and mouse macrophages. *Biochemical Pharmacology* 65:2065–2071.

Liu, C., A. Bayer, S. E. Cosgrove, R. S. Daum, S. K. Fridkin, R. J. Gorwitz, S. L. Kaplan, A. W. Karchmer, D. P. Levine, B. E. Murray, and M. J. Rybak. 2011. Clinical practice guidelines by the Infectious Diseases Society of America for the treatment of methicillin-resistant *Staphylococcus aureus* infections in adults and children. *Clinical Infectious Diseases* 52:18–55.

Lopez-Munoz, G. A., and J. A. Balderas-Lopez. 2014. Photothermal characterization of citrus essential oils and their derivatives. *Thermochimica Acta* 579:40–44.

Lorenzo, Y., A. Azqueta, L. Luna, F. Bonilla, G. Domínguez, and A. R. Collins. 2009. The carotenoid β-cryptoxanthin stimulates the repair of DNA oxidation damage in addition to acting as an antioxidant in human cells. *Carcinogenesis* 30:308–314.

Lu, X. G., L. B. Zhan, B. A. Feng, M. Y. Qu, L. H. Yu, and J. H. Xie. 2004. Inhibition of growth and metastasis of human gastric cancer implanted in nude mice by D-limonene. *World Journal of Gastroenterology* 10:2140–2144.

Manabe, Y., T. Hirata, and T. Sugawara. 2014. Suppressive effects of carotenoids on the antigen-induced degranulation in RBL-2H3 rat basophilic leukemia cells. *Journal of Oleo Science* 63:291–294.

Manners, G. D., R. A. Jacob, P. Andrew, III, T. K. Schoch, and S. Hasegawa, S. 2003. Bioavailability of citrus limonoids in humans. *Journal of Agricultural and Food Chemistry* 51:4156–4161.

Martin, K. R., and C. L. Appel. 2010. Polyphenols as dietary supplements: A double-edged sword. *Nutrition and Dietary Supplements* 2:1–12.

Matsunaga, K., N. Yoshimi, K. Shimoi, Y. Yamada, M. Katayama, K. Sakata, T. Kuno, K. Yoshida, Z. Qiao, N. Kinae, and H. Mori. 2000. The dietary effect of monoglucosylrutin on azomethane-induced colon carcinogenesis. *Asian Pacific Journal of Cancer Prevention* 1:213–218.

Miyagi, Y., A. S. Om, K. M. Chee, and M. R. Bennink. 2000. Inhibition of azoxymethane-induced colon cancer by orange juice. *Nutrition and Cancer* 36:224–229.

Miyamoto, S., Y. Yasui, H. Ohigashi, T. Tanaka, and A. Murakami. 2010. Dietary flavonoids suppress azoxymethane-induced colonic preneoplastic lesions in male C57BL/KsJ-db/db mice. *Chemico-Biological Interactions* 183:276–283.

Miyamoto, S., Y. Yasui, T. Tanaka, H. Ohigashi, and A. Murakami. 2008. Suppressive effects of nobiletin on hyperleptinemia and colitis-related colon carcinogenesis in male ICR mice. *Carcinogenesis* 29:1057–1063.

Moore, G. A. 2001. Oranges and lemons: Clues to the taxonomy of *Citrus* from molecular markers. *Trends in Genetics* 17:536–540.

Moyler, D. 2002. Terpeneless and sesquiterpeneless oils. In Dugo, G., and A. Di Giacomo (eds.), *Citrus: The genus citrus*, pp. 391–401. London: Taylor & Francis.

Murakami, A., Y. Nakamura, Y. Ohto, M. Yano, T. Koshiba, K. Koshimizu, H. Tokuda, H. Noshino, and H. Ohigashi. 2000b. Suppressive effects of citrus fruits on free radical generation and nobiletin, an anti-inflammatory polymethoxyflavonoid. *Biofactors* 12:187–192.

Murakami, A., Y. Nakamura, K. Torikai, T. Tanaka, T. Koshiba, K. Koshimizu, S. Kuwahara, Y. Takahashi, K. Ogawa, M. Yano, H. Tokuda, H. Nishino, Y. Mimaki, Y. Sashida, S. Kitanaka, and H. Ohigashi. 2000a. Inhibitory effect of citrus nobiletin on phorbol ester-induced skin inflammation, oxidative stress, and tumor promotion in mice. *Cancer Research* 60:5059–5066.

Murphy, E. A., J. M. Davis, J. L. McClellan, and M. D. Carmichael. 2011. Quercetin's effects on intestinal polyp multiplicity and macrophage number in the $Apc^{min/+}$ mouse. *Nutrition and Cancer* 63:421–426.

Murthy, K. N. C., G. K. Jayaprakasha, and B. S. Patil. 2011a. Obacunone and obacunone glucoside inhibit human colon cancer (SW480) cells by the induction of apoptosis. *Food and Chemical Toxicology* 49:1616–1625.

Murthy, K. N. C., G. K. Jayaprakasha, and B. S. Patil. 2012. D-limonene rich volatile oil from blood oranges inhibits angiogenesis, metastasis and cell death in human colon cancer cells. *Life Sciences* 9:429–439.

Murthy, K. N. C., G. K. Jayaprakasha, and B. S. Patil. 2013. Citrus limonoids and curcumin additively inhibit human colon cancer cells. *Food and Function* 4:803–810.

Murthy, K. N. C., G. K. Jayaprakasha, V. Kumar, K. S. Rathore, and B. S. Patil. 2011b. Citrus limonin and its glucoside inhibit colon adenocarcinoma cell proliferation through apoptosis. *Journal of Agricultural and Food Chemistry* 59:2314–2323.

Nakae, D., Y. Kobayashi, H. Akai, N. Andoh, H. Satoh, K. Ohashi, M. Tsutsumi, and Y. Konishi. 1997. Involvement of 8-hydroxyguanine formation in the initiation of rat liver carcinogenesis by low dose levels of N-nitrosodiethylamine. *Cancer Research* 57:1281–1287.

Nambiar, P. R., R. R. Gupta, and V. Misra. 2010. An "omics" based survey of human colon cancer. *Mechanisms of Mutagenesis* 693:3–18.

Narisawa, T., Y. Fukaura, M. Hasebe, M. Ito, R. Aizawa, M. Murakoshi, S. Uemura, F. Khachik, and H. Nishino. 1996. Inhibitory effects of natural carotenoids, α-carotene, β-carotene, lycopene and lutein, on colonic aberrant crypt foci formation in rats. *Cancer Letter* 107:137–142.

Narisawa, T., Y. Fukaura, S. Oshima, T. Inakuma, M. Yano, and H. Nishino. 1999. Chemoprevention by the oxygenated carotenoid β-cryptoxanthin of N-methylnitrosourea-induced colon carcinogenesis in F344 rats. *Cancer Science* 90:1061–1065.

Nguyen, H., E. M. Campi, W. R. Jackson, and A. F. Patti. 2009. Effect of oxidative deterioration on flavour and aroma components of lemon oil. *Food Chemistry* 112:388–393.

Niho, N., M. Mutoh, M. Takahashi, K. Tsutsumi, T. Sugimura, and K. Wakabayashi. 2005. Concurrent suppression of hyperlipidemia and intestinal polyp formation by NO-1886, increasing lipoprotein lipase activity in Min mice. *Proceedings of the National Academy of Sciences of the United States of America* 102:2970–2974.

Niho, N., M. Takahashi, T. Kitamura, Y. Shoji, and M. Itoh. 2003a. Concomitant suppression of hyperlipidemia and intestinal polyp formation in *Apc* deficient mice by peroxisome proliferator-activated receptor ligands. *Cancer Research* 63:6090–6095.

Niho, N., M. Takahashi, Y. Shoji, Y. Takeuchi, and S. Matsubara. 2003b. Dose-dependent suppression of hyperlipidemia and intestinal polyp formation in Min mice by pioglitazone, a PPAR gamma ligand. *Cancer Science* 94:960–964.

Nirmala, P., and M. Ramanathan. 2011. Effect of kaempferol on lipid peroxidation and antioxidant status in 1,2-dimethyl hydrazine induced colorectal carcinoma in rats. *European Journal of Pharmacology* 654:75–79.

Nishino, H., H. Tokuda, M. Murakoshi, Y. Satomi, M. Masuda, M. Onozuka, S. Yamaguchi, J. Takayasu, J. Tsuruta, M. Okuda, and F. Khachik. 2000. Cancer prevention by natural carotenoids. *Biofactors* 13:89–94.

Nogata, Y., K. Sakamoto, H. Shiratsuchi, T. Ishii, M. Yano, and H. Ohta. 2006. Flavonoid composition of fruit tissues of *Citrus* species. *Bioscience, Biotechnology, and Biochemistry* 70:178–192.

Pan, M. H., Y. S. Lai, C. S. Lai, Y. J. Wang, S. Li, C. Y. Lo, S. Dushenkov, and C. T. Ho. 2007. 5-Hydroxy-3,6,7,8,3′,4′-hexamethoxyflavone induces apoptosis through reactive oxygen species production, growth arrest and DNA damage-inducible gene 153 expression, and caspase activation in human leukemia cells. *Journal of Agriculture and Food Chemistry* 55:5081–5091.

Pandurangan, A. K., and S. Ganapasam. 2008. Protective role of luteolin on the status of lipid peroxidation and antioxidant defense against azoxymethane-induced experimental colon carcinogenesis. *Biomedicine & Pharmacotherapy* 62:590–597.

Pandurangan, A. K., and S. Ganapasam. 2011. Luteolin inhibits cell proliferation during azoxymethane-induced experimental colon carcinogenesis via Wnt/β-catenin pathway. *Investigational New Drugs* 29:273–284.
Pandurangan, A. K., and S. Ganapasam. 2013a. Luteolin induces apoptosis in azoxymethane-induced colon carcinogenesis through the involvement of Bcl-2, bax and caspase. *Journal of Chemical and Pharmaceutical Research* 5:143–148.
Pandurangan, A. K., and S. Ganapasam. 2013b. Luteolin modulates cellular thiols on azoxymethane-induced colon carcinogenesis. *Asian Journal of Experimental Biological Sciences* 4:245–250.
Pandurangan, A. K., and S. Ganapasam. 2013c. Cytotoxic effect of luteolin on human colorectal cancer cell line (HCT-15): Crucial involvement of reactive oxygen species. *Middle East Journal of Cancer* 4:175–180.
Pandurangan, A. K., S. K. Anandasadagopan, P. Dharmalingam, and S. Ganapasam. 2013b. Luteolin, a bioflavonoid inhibits azoxymethane-induced colorectal cancer through activation of Nrf2 signaling. *Toxicology Mechanisms and Methods* 24:13–20.
Pandurangan, A. K., P. Dharmalingam, S. K. Anandasadagopan, and S. Ganapasam. 2012. Effect of luteolin on the levels of glycoproteins during azoxymethane-induced colon carcinogenesis in mice. *Asian Pacific Journal of Cancer Prevention* 13:1569–1573.
Pandurangan, A. K., P. Dharmalingam, S. K. A. Sadagopan, and S. Ganapasam. 2014b. Luteolin inhibits matrix metalloproteinase 9 and 2 in azoxymethane-induced colon carcinogenesis. *Human & Experimental Toxicology* 33:1176–1185.
Pandurangan, A. K., P. Dharmalingam, S. K. A. Sadagopan, M. Ramar, A. Munusamy, and S. Ganapasam, S. 2013a. Luteolin induces growth arrest in colon cancer cells through involvement of Wnt/β-catenin/GSK-3β signaling. *Journal of Environmental Pathology, Toxicology and Oncology* 32:131–139.
Pandurangan, A. K., S. Kumar, P. Dharmalingam, and S. Ganapasam. 2014a. Luteolin, a bioflavonoid inhibits azoxymethane-induced colon carcinogenesis: Involvement of iNOS and COX-2. *Pharmacognosy Magazine* 10:306–310.
Pandurangan, A. K., S. K. A. Sadagopan, P. Dharmalingam, and S. Ganapasam. 2013c. Inhibitory effect of luteolin on the status of membrane bound ATPases against azoxymethane-induced colorectal cancer. *Journal of Chemical & Pharmaceutical Research* 5:123–127.
Pao, S., and P. J. Fellers. 2003. Citrus fruits. Oranges. *Encyclopedia of Food Sciences and Nutrition* 2:1341–1346.
Park, H. J., M. J. Kim, E. Ha, and J. H. Chung. 2008. Apoptotic effect of hesperidin through caspase3 activation in human colon cancer cells, SNU-C4. *Phytomedicine* 15:147–151.
Park, S. S., S. K. Park, J. H. Lim, Y. H. Choi, W. J. Kim, and S. K. Moon. 2011. Esculetin inhibits cell proliferation through the Ras/ERK1/2 pathway in human colon cancer cells. *Oncology Reports* 25:223–230.
Patel, S., and P. Gogna. 2015. Tapping botanicals for essential oils: Progress and hurdles in cancer mitigation. *Industrial Crops and Products* 76:1148–1163.
Patil, J. R., G. K. Jayaprakasha, K. N. C. Murthy, S. E. Tichy, M. B. Chetti, and B. S. Patil. 2009. Apoptosis-mediated proliferation inhibition of human colon cancer cells by volatile principles of *Citrus aurantifolia*. *Food Chemistry* 114:1351–1358.
Patil, J. R., G. K. Jayaprakasha, J. Kim, K. N. C. Murthy, M. B. Chetti, S. Y. Nam, and B. S. Patil. 2013. 5-Geranyloxy-7-methoxycoumarin inhibits colon cancer SW480 cells growth by inducing apoptosis. *Planta Medica* 79:219–226.
Patil, J. R., G. K. Jayaprakasha, K. N. Murthy, M. B. Chetti, and B. S. Patil. 2010. Characterization of *Citrus aurantifolia* bioactive compounds and their inhibition of human pancreatic cancer cells through apoptosis. *Microchemical Journal* 94:108–117.

Perez, J. L., G. K. Jayaprakasha, A. Cadena, E. Martinez, H. Ahmad, and B. S. Patil. 2010. *In vivo* induction of phase II detoxifying enzymes, glutathione transferase and quinone reductase by citrus triterpenoids. *BMC Complementary and Alternative Medicine* 10:1–8.

Perez, J. L., G. K. Jayaprakasha, V. Valdivia, D. Munoz, D. V. Dandekar, H. Ahmad, and B. S. Patil. 2009. Limonin methoxylation influences the induction of glutathione S-transferase and quinone reductase. *Journal of Agricultural and Food Chemistry* 57:5279–5286.

Peterson, J. J., G. R. Beecher, S. A. Bhagwat, J. T. Dwyer, S. E. Gebhardt, D. B. Haytowitz, and J. M. Holden. 2006a. Flavanones in grapefruit, lemons, and limes: A compilation and review of the data from the analytical literature. *Journal of Food Composition and Analysis* 19:74–80.

Peterson, J. J., J. T. Dwyer, G. R. Beecher, S. A. Bhagwat, S. E. Gebhardt, D. B. Haytowitz, and J. M. Holden. 2006b. Flavanones in oranges, tangerines (mandarins), tangors, and tangelos: A compilation and review of the data from the analytical literature. *Journal of Food Composition and Analysis* 19:66–73.

Poulose, S. M., E. D. Harris, and B. S. Patil. 2005. Citrus limonoids induce apoptosis in human neuroblastoma cells and have radical scavenging activity. *Journal of Nutrition* 135:870–377.

Poulose, S. M., E. D. Harris, and B. S. Patil. 2006. Antiproliferative effects of citrus limonoids against human neuroblastoma and colonic adenocarcinoma cells. *Nutrition and Cancer* 56:103–112.

Qiu, P., P. Dong, H. Guan, S. Li, C. T. Ho, M. H. Pan, D. J. McClements, and H. Xiao. 2010. Inhibitory effects of 5-hydroxy polymethoxyflavones on colon cancer cells. *Molecular Nutrition and Food Research* 54:S244–S252.

Qiu, P., H. Guan, P. Dong, S. Guo, J. Zheng, S. Li, Y. Chen, C. T. Ho, M. H. Pan, D. J. McClements, and H. Xiao. 2011b. The inhibitory effects of 5-hydroxy-3,6,7,8,3′,4′-hexamethoxyflavone on human colon cancer cells. *Molecular Nutrition and Food Research* 55:1523–1532.

Qiu, P., H. Guan, P. Dong, S. Li, C. T. Ho, M. H. Pan, D. J. McClements, and H. Xiao. 2011a. The p53-, Bax-and p21-dependent inhibition of colon cancer cell growth by 5-hydroxy polymethoxyflavones. *Molecular Nutrition & Food Research* 55:613–622.

Rajabi, F., A. Feiz, and R. Luque. 2015. An efficient synthesis of coumarin derivatives using a SBA-15 supported cobalt (II) nanocatalyst. *Catalysis Letters* 145:1621–1625.

Ramon, S. D. R. V., B. Somoza, T. Ortega, A. M. Villar, and T. Tejerina. 1999. Vasodilatory effect in rat aorta of eriodictyol obtained from *Satureja obovata*. *Planta Medica* 65(3):234–238.

Ramos, A. A., C. Pereira-Wilson, and A. R. Collins. 2010. Protective effects of ursolic acid and luteolin against oxidative DNA damage include enhancement of DNA repair in *Caco*-2 cells. *Mutation Research/Fundamental and Molecular Mechanisms of Mutagenesis* 692:6–11.

Raphael, T. J., and G. Kuttan, G. 2003. Effect of naturally occurring triterpenoids glycyrrhizic acid, ursolic acid, oleanolic acid and nomilin on the immune system. *Phytomedicine* 10:483–489.

Raut, J. S., and S. M. Karuppayil. 2014. A status review on the medicinal properties of essential oils. *Industrial Crops and Products* 62:250–264.

Rohr, A. C., C. K. Wilkins, P. A. Clausen, M. Hammer, G. D. Nielsen, P. Wolkoff, and J. D. Spengler. 2002. Upper airway and pulmonary effects of oxidation products of (+)-α-pinene, D-limonene, and isoprene in BALB/c mice. *Inhalation Toxicology* 14:663–684.

Rokkaku, T., R. Kimura, C. Ishikawa, T. Yasumoto, M. Senba, F. Kanaya, and N. Mori. 2013. Anticancer effects of marine carotenoids, fucoxanthin and its deacetylated product, fucoxanthinol, on osteosarcoma. *International Journal of Oncology* 43:1176–1186.

Rossato, M. F., G. Trevisan, C. I. B. Walker, J. Z. Klafke, A. P. de Oliveira, J. G. Villarinho, and J. Ferreira. 2011. Eriodictyol: A flavonoid antagonist of the TRPV1 receptor with antioxidant activity. *Biochemical Pharmacology* 81:544–551.

Roy, A., and S. Saraf. 2006. Limonoids: Overview of significant bioactive triterpenes distributed in plants kingdom. *Biological and Pharmaceutical Bulletin* 29:191–201.

Saidu, N. E. B., S. Valente, E. Bana, G. Kirsch, D. Bagrel, and M. Montenarh. 2012. Coumarin polysulfides inhibit cell growth and induce apoptosis in HCT116 colon cancer cells. *Bioorganic and Medicinal Chemistry* 20:1584–1593.

Saiprasad, G., P. Chitra, R. Manikandan, and G. Sudhandiran. 2013. Hesperidin alleviates oxidative stress and downregulates the expressions of proliferative and inflammatory markers in azoxymethane-induced experimental colon carcinogenesis in mice. *Inflammation Research* 62:425–440.

Saito, T., D. Abe, and K. Sekiya. 2007. Nobiletin enhances differentiation and lipolysis of 3T3-L1 adipocytes. *Biochemical and Biophysical Research Communications* 357:371–376.

Sakai, R., S. Yokobe, N. Abe, N. Miyoshi, Y. Murata, and Y. Nakamura. 2012. Luteolin overcomes resistance to benzyl isothiocyanate-induced apoptosis in human colorectal cancer HCT-116 cells. *Journal of Food and Drug Analysis* 20:389–393.

Saud, S. M., M. R. Young, Y. L. Jones-Hall, L. Ileva, M. O. Evbuomwan, J. Wise, N. H. Colburn, Y. S. Kim, and G. Bobe. 2013. Chemopreventive activity of plant flavonoid isorhamnetin in colorectal cancer is mediated by oncogenic Src and β-catenin. *Cancer Research* 73:5473–5484.

Sawabe, A., M. Morita, T. Kiso, H. Kishine, Y. Ohtsubo, T. Minematsu, Y. Matsubara, and T. Okamoto. 1999. Isolation and characterization of new limonoid glycosides from *Citrus unshiu* peels. *Carbohydrate Research* 315:142–147.

Sequetto, P. L., T. T. Oliveira, I. R. Maldonado, L. E. F. Augusto, V. J. Mello, V. R. Pizziolo, M. R. Almeida, M. E. Silva, and R. D. Novaes. 2014. Naringin accelerates the regression of pre-neoplastic lesions and the colorectal structural reorganization in a murine model of chemical carcinogenesis. *Food and Chemical Toxicology* 64:200–209.

Shan, B. E., M. X. Wang, and R. Q. Li. 2009. Quercetin inhibits human SW480 colon cancer growth in association with inhibition of cyclin D1 and survivin expression through Wnt/β-catenin signaling pathway. *Cancer Investigation* 27:604–612.

Shao, H., K. Jing, E. Mahmoud, H. Huang, X. Fang, and C. Yu. 2013. Apigenin sensitizes colon cancer cells to antitumor activity of ABT-263. *Molecular Cancer Therapeutics* 12:2640–2650.

Shigeshiro, M., S. Tanabe, and T. Suzuki. 2013. Dietary polyphenols modulate intestinal barrier defects and inflammation in a murine model of colitis. *Journal of Functional Foods* 5:949–955.

Shin, Y. W., S. H. Bok, T. S. Jeong, K. H. Bae, N. H. Jeoung, M. S. Choi, S. H. Lee, and Y. B. Park. 1999. Hypocholesterolemic effect of naringin associated with hepatic cholesterol regulating enzyme changes in rats. *International Journal for Vitamin and Nutrition Research* 69:341–347.

Singhal, J., L. D. Nagaprashantha, R. Vatsyayan, S. Awasthi, and S. S. Singhal. 2012. Didymin induces apoptosis by inhibiting N-Myc and upregulating RKIP in neuroblastoma. *Cancer Prevention Research* 5:473–483.

Sivagami, G., R. Vinothkumar, C. P. Preethy, A. Riyasdeen, M. A. Akbarsha, V. P. Menon, and N. Nalini. 2012. Role of hesperetin (a natural flavonoid) and its analogue on apoptosis in HT-29 human colon adenocarcinoma cell line—A comparative study. *Food and Chemical Toxicology* 50:660–671.

Smith, D. C., S. Forland, E. Bachanos, M. Matejka, and V. Barrett. 2001. Qualitative analysis of citrus fruit extracts by GC/MS: An undergraduate experiment. *Chemical Educator* 6:28–31.

Sporn, M. B., and N. Suh. 2002. Chemoprevention: An essential approach to controlling cancer. *Nature Reviews Cancer* 2:537–543.

Stanley, W. L., and L. Jurd. 1971. Citrus coumarins. *Journal of Agricultural and Food Chemistry* 19:1106–1110.

Suzuki, R., H. Kohno, Y. Yasui, K. Hata, S. Sugie, S. Miyamoto, K. Sugawara, T. Sumida, Y. Hirose, and T. Tanaka. 2007. Diet supplemented with citrus unshiu segment membrane suppresses chemically induced colonic preneoplastic lesions and fatty liver in male db/db mice. *International Journal of Cancer* 120:252–258.

Szarka, C. E., G. R. Pfeiffer, S. T. Hum, L. C. Everley, A. M. Balshem, D. F. Moore, S. Litwin, E. B. Goosenberg, H. Frucht, P. F. Engstrom, and M. L. Clapper. 1995. Glutathione S-transferase activity and glutathione S-transferase μ expression in subjects with risk for colorectal cancer. *Cancer Research* 55:2789–2793.

Takemura, H., H. Nagayoshi, T. Matsuda, H. Sakakibara, M. Morita, A. Matsui, T. Ohura, and K. Shimoi. 2010. Inhibitory effects of chrysoeriol on DNA adduct formation with benzo[a]pyrene in MCF-7 breast cancer cells. *Toxicology* 274:42–48.

Tan, K. W., Y. Li, J. W. Paxton, N. P. Birch, and A. Scheepens. 2013. Identification of novel dietary phytochemicals inhibiting the efflux transporter breast cancer resistance protein (BCRP/ABCG2). *Food Chemistry* 138:2267–2274.

Tanaka, T. 1997. Chemoprevention of human cancer: Biology and therapy. *Critical Reviews in Oncology/Hematology* 25:139–174.

Tanaka, T., M. de Azevedo, N. Duran, J. B. Alderete, F. Epifano, S. Genovese, M. Tanaka, T. Tanaka, and M. Curini. 2010. Colorectal cancer chemoprevention by 2β-cyclodextrin inclusion compounds of auraptene and 4′-geranyloxyferulic acid. *International Journal of Cancer* 126:830–840.

Tanaka, T., T. Kawamori, M. Ohnishi, H. Makita, H. Mori, K. Satoh, and A. Hara. 1995. Suppression of azoxymethane-induced rat colon carcinogenesis by dietary administration of naturally occurring xanthophylls astaxanthin and canthaxanthin during the postinitiation phase. *Carcinogenesis* 16:2957–2963.

Tanaka, T., H. Kohno, and H. Mori. 2001. Chemoprevention of colon carcinogenesis by dietary non-nutritive compounds. *Asian Pacific Journal of Cancer Prevention* 2:165–177.

Tanaka, T., H. Kohno, M. Murakami, R. Shimada, S. Kagami, T. Sumida, Y. Azuma, and H. Ogawa. 2000. Suppression of azoxymethane-induced colon carcinogenesis in male F344 rats by mandarin juices rich in β-cryptoxanthin and hesperidin. *International Journal of Cancer* 88:146–150.

Tanaka, T., H. Makita, K. Kawabata, H. Mori, M. Kakumoto, K. Satoh, A. Hara, T. Sumida, T. Tanaka, and H. Ogawa. 1997. Chemoprevention of azoxymethane-induced rat colon carcinogenesis by the naturally occurring flavonoids, diosmin and hesperidin. *Carcinogenesis* 18:957–965.

Tanaka, T., M. Shnimizu, and H. Moriwaki. 2012. Cancer chemoprevention by carotenoids. *Molecules* 17:3202–3242.

Tanaka, T., Y. Yasui, R. Ishigamori-Suzuki, and T. Oyama. 2008. Citrus compounds inhibit inflammation-and obesity-related colon carcinogenesis in mice. *Nutrition and Cancer* 60:70–80.

Tang, M., K. Ogawa, M. Asamoto, N. Hokaiwado, A. Seeni, S. Suzuki, S. Takahashi, T. Tanaka, K. Ichikawa, and T. Shirai. 2007. Protective effects of citrus nobiletin and auraptene in transgenic rats developing adenocarcinoma of the prostate (TRAP) and human prostate carcinoma cell. *Cancer Science* 98:471–477.

Tang, M. X., K. Ogawa, M. Asamoto, T. Chewonarin, S. Suzuki, T. Tanaka, and T. Shirai. 2011. Effects of nobiletin on PhIP-induced prostate and colon carcinogenesis in F344 rats. *Nutrition and Cancer* 63:227–233.

Thakur, A., R. Singla, and V. Jaitak. 2015. Coumarins as anticancer agents: A review on synthetic strategies, mechanism of action and SAR studies. *European Journal of Medicinal Chemistry* 101:476–495.

Thompson, C. L., E. K. Larkin, S. Patel, N. A. Berger, S. Redline, and L. Li. 2011. Short duration of sleep increases risk of colorectal adenoma. *Cancer* 117:841–847.

Tian, Q., E. G. Miller, H. Ahmad, L. Tang, and B. S. Patil. 2001. Differential inhibition of human cancer cell proliferation by citrus limonoids. *Nutrition and Cancer* 40:180–184.

Tsimplouli, C., L. Mahaira, G. Rassidakis, and K. Dimas. 2010. Anticancer activity of a novel kaempferol glucoside, Tac, is mediated by a mechanism involving RSK inhibition. *European Journal of Cancer Supplements* 8:67–68.

Turktekin, M., E. Konac, H. I. Onen, E. Alp, A. Yilmaz, and S. Menevse. 2011. Evaluation of the effects of the flavonoid apigenin on apoptotic pathway gene expression on the colon cancer cell line (HT29). *Journal of Medicinal Food* 14:1107–1117.

Turner, N. D., K. J. Paulhill, C. A. Warren, L. A. Davidson, R. S. Chapkin, J. R. Lupton, R. J. Carroll, and N. Wang. 2007. Quercetin suppresses early colon carcinogenesis partly through inhibition of inflammatory mediators. In *II International Symposium on Human Health Effects of Fruits and Vegetables: FAVHEALTH 2007 841*, pp. 237–242.

Ugocsai, K., A. Varga, P. Molnar, S. Antus, and R. J. Molna. 2005. Effects of selected flavonoids and carotenoids on drug accumulation and apoptosis induction in multidrug-resistant colon cancer cells expressing MDR1/LRP. *In Vivo* 19:433–438.

Urquiaga, I., and F. Leighton. 2000. Plant polyphenol antioxidants and oxidative stress. *Biological Research* 33:55–64.

Vaiyapuri, M., and N. Namasivayam. 2009. Effect of luteolin on glycoproteins metabolism in 1,2-dimethylhydrazine induced experimental colon carcinogenesis. *Journal of Biochemical Technology* 1:57–61.

Vanamala, J., T. Leonardi, B. S. Patil, S. S., Taddeo, M. E. Murphy, L. M. Pike, R. S. Chapkin, J. R. Lupton, and N. D. Turner. 2006. Suppression of colon carcinogenesis by bioactive compounds in grapefruit. *Carcinogenesis* 27:1257–1265.

Van de Waterbeemd, H., H. Lennernas, and P. Artursson. 2003. *Methods and principles in medicinal chemistry* (vol. 18). Weinheim, Germany: Wiley-VCH Verlag.

Villegas, I., C. Alarcón de la Lastra, A. Orjales, and C. La Casa. 2003. A new flavonoid derivative, dosmalfate, attenuates the development of dextran sulphate sodium-induced colitis in mice. *International Immunopharmacology* 3:1731–1741.

Walle, T. 2007. Methoxylated flavones, a superior cancer chemopreventive flavonoid subclass? *Seminars in Cancer Biology* 17:354–362.

Warren, C. A., K. J. Paulhill, L. A. Davidson, J. R. Lupton, S. S. Taddeo, M. Y. Hong, R. J. Carroll, R. S. Chapkin, and N. D. Turner. 2009. Quercetin may suppress rat aberrant crypt foci formation by suppressing inflammatory mediators that influence proliferation and apoptosis. *Journal of Nutrition* 139:101–105.

Wei, E. K., G. A. Colditz, E. L. Giovannucci, C. S. Fuchs, and B. A. Rosner. 2009. Cumulative risk of colon cancer upto age 70 years by risk factor status using data from the Nurses Health Study. *American Journal of Epidemiology* 170:863–872.

Wijnen, J. T., H. F. Vasen, and P. M. Khan. 1998. Clinical findings with implications for genetic testing in families with clustering of colorectal cancer. *New England Journal of Medicine* 339:511–518.

Wilkinson, J. I., and M. L. Clapper. 1997. Detoxication enzymes and chemoprevention. *Proceedings of Society of Experimental Biology and Medicine* 216:192–200.

Xavier, C. P., C. F. Lima, A. Preto, R. Seruca, M. Fernandes-Ferreira, and C. Pereira-Wilson. 2009. Luteolin, quercetin and ursolic acid are potent inhibitors of proliferation and inducers of apoptosis in both KRAS and BRAF mutated human colorectal cancer cells. *Cancer Letters* 281:162–170.

Xavier, C. P., C. F. Lima, M. Rohde, and C. Pereira-Wilson. 2011. Quercetin enhances 5-fluorouracil-induced apoptosis in MSI colorectal cancer cells through p53 modulation. *Cancer Chemotherapy and Pharmacology* 68:1449–1457.

Xie, Y. Y., D. Yuan, J. Y. Yang, L. H. Wang, and C. F. Wu. 2009. Cytotoxic activity of flavonoids from the flowers of *Chrysanthemum morifolium* on human colon cancer Colon205 cells. *Journal of Asian Natural Products Research* 11:771–778.

Yamada, K., S. Araki, M. Tamura, I. Sakai, and Y. Takahashi. 1998. Relation of serum total cholesterol, serum triglycerides and fasting plasma glucose to colorectal carcinoma *in situ*. *International Journal of Epidemiology* 27:794–798.

Yang, K., S. A. Lamprecht, Y. Liu, H. Shinozaki, K. Fan, D. Leung, H. Newmark, V. E. Steele, G. J. Kelloff, and M. Lipkin. 2000. Chemoprevention studies of the flavonoids quercetin and rutin in normal and azoxymethane-treated mouse colon. *Carcinogenesis* 21:1655–1660.

Yang, Y., X. Zhou, M. Xiao, Z. Hong, Q, Gong, L. Jiang, and J. Zhou. 2010. Discovery of chrysoeriol, a PI3K-AKT-mTOR pathway inhibitor with potent antitumor activity against human multiple myeloma cells *in vitro*. *Journal of Huazhong University of Science and Technology [Medical Sciences]* 30:734–740.

Yin, F., L. Cheng, and F. C. Lou. 2004. Studies on the constituents of *Citrus medica* L. var. sarcodactylis (Noot.) Swingle. *Chinese Journal of Natural Medicines* 2:149–151.

Yoo, H. H., M. Lee, H. J. Chung, S. K. Lee, and D. H. Kim. 2007. Effects of diosmin, a flavonoid glycoside in citrus fruits, on P-glycoprotein-mediated drug efflux in human intestinal *Caco*-2 cells. *Journal of Agricultural and Food Chemistry* 55:7620–7625.

Yoon, H., T. W. Kim, S. Y. Shin, M. J. Park, Y. Yong, D. W. Kim, T. Islam, Y. H. Lee, K. Y. Jung, and Y. Lim. 2013. Design, synthesis and inhibitory activities of naringenin derivatives on human colon cancer cells. *Bioorganic & Medicinal Chemistry Letters* 23:232–238.

Yoon, J. H., and S. J. Baek. 2005. Molecular targets of dietary polyphenols with anti-inflammatory properties. *Yonsei Medical Journal* 46:85–96.

Yoshida, T., M. Konishi, M. Horinaka, T. Yasuda, A. E. Goda, H. Taniguchi, K. Yano, M. Wakada, and T. Sakai. 2008. Kaempferol sensitizes colon cancer cells to TRAIL-induced apoptosis. *Biochemical and Biophysical Research Communications* 375: 129–133.

Yu, J., L. Wang, R. L. Walzem, E. G. Miller, L. M. Pike, and B. S. Patil. 2005. Antioxidant activity of citrus limonoids, flavonoids, and coumarins. *Journal of Agricultural and Food Chemistry* 53:2009–2014.

Zhang, H., M. Zhang, L. Yu, Y. Zhao, N. He, and X. Yang. 2012. Antitumor activities of quercetin and quercetin-5′,8-disulfonate in human colon and breast cancer cell lines. *Food and Chemical Toxicology* 50:1589–1599.

Zhong, Y., C. Krisanapun, S. H. Lee, T. Nualsanit, C. Sams, P. Peungvicha, and S. J. Baek. 2010. Molecular targets of apigenin in colorectal cancer cells: Involvement of p21, NAG-1 and p53. *European Journal of Cancer* 46:3365–3374.

Zhu, X., F. Luo, Y. Zheng, J. Zhang, J. Huang, C. Sun, and K. Chen. 2013. Characterization, purification of poncirin from edible citrus ougan (*Citrus reticulate* cv. suavissima) and its growth inhibitory effect on human gastric cancer cells SGC-7901. *International Journal of Molecular Sciences* 14:8684–8697.

4 Citrus Polysaccharides and Their Biofunctions

Shiguo Chen, Junhui Li, Shan Li, Lufeng Yan, and Donghong Liu

CONTENTS

4.1 INTRODUCTION

Polysaccharides are one of the main components in citrus fruits. A number of polysaccharides have been identified in citrus peel, endocarps, segments, and pulps (Figure 4.1). Most studies on structural polysaccharides in citrus fruits have dealt with the polysaccharides found in the peel, because these cell wall polysaccharides affect the growth, ripening, and storage properties of the fruit. It has been reported that the peel contained about 55 mol% pectic polysaccharides, including homogalacturonan (HG), rhamnogalacturonan I (RG-I), and rhamnogalacturonan II (RG-II)

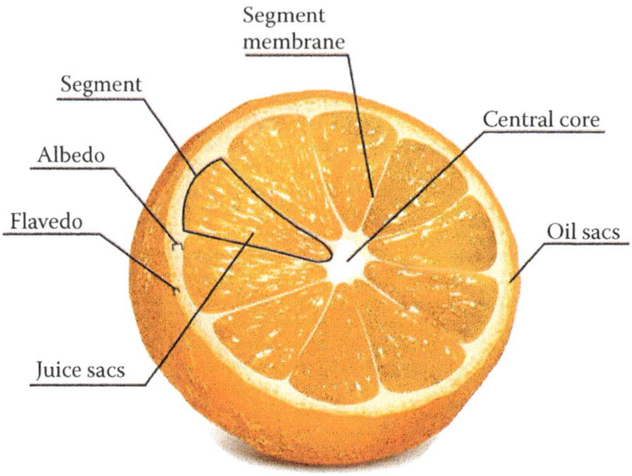

FIGURE 4.1 The structure of citrus fruit.

together with type I arabinogalactan (AG) and arabinan. It also contained cellulose (22 mol%) and other noncellulosic polysaccharides (14 mol%), including xyloglucans (XGs) (10 mol%), heteromannans (2 mol%), and heteroxylans (2 mol%), although the proportions were different among different species (Prabasari et al. 2011).

The citrus fruit is considered to be one of the most important sources of pectin in the whole world and the citrus peel represents the main raw material used in the industrial extraction of pectins, which are utilized as a valuable food ingredient. Although commercial citrus pectin (CP) is usually prepared from citrus peels, other parts of the citrus fruit have been employed and these parts include the endocarp, segment membrane, and pulp (Table 4.1). As these fruit parts are usually considered to be processing waste and are produced in large quantities during citrus fruit processing, it would be very advantageous to recover the valuable pectin from them.

The extracted pectin represents an important processed product that has huge potential applications in the food and health care industries (Lattimer and Haub 2010). Humans consume pectin as a dietary fiber, which is not digested by the enzymes in the small intestine but can be degraded by microbes in the colon. It maintains its gelling action in the digestive tract, so that it slows digestion. This is very beneficial in patients with Dumping syndrome whose stomachs undergo very rapid emptying (Lawaetz et al. 1983). Pectin is also capable of diminishing blood cholesterol levels and stimulating lipid excretion (Brown et al. 1999). Also, several studies have shown that orally administered pectin decreases the risk of intestinal infection and diarrhea in children by favoring the growth of "good" bacteria in the colon to the demise of pathogenic bacteria (Olano-Martin, Gibson, and Rastell 2002).

Due to its high molecular weight, pectin can't be directly absorbed by the intestine and performs important functions in the gastrointestinal tract. Modified citrus pectin (MCP) is low molecular pectin prepared by chemical, physical or enzymatic methods. MCP has been shown to play a significant inhibitory role in cancer cell metastasis, invasion, angiogenesis, and survival. The interaction and inactivation of oncogenes by MCP in prostate, breast, liver, lung, melanoma, and multiple myeloma

TABLE 4.1
Chemical Composition of the Pectic Polysaccharides from Different Part of Citrus

Sample	Extraction Methods	Monosaccharide Composition (mol%)									
		Rha	Fuc	Ara	Xyl	Man	Gal	Glc	GalA	DM	DA
Citrus peel[a]											
WEP	Hot water	1.4	0.2	8.2	1.4	nd	4.7	4.1	80.1	76.5	5.5
OEP	Oxalate	1.3	nd	6.2	0.7	nd	3.3	2.0	86.5	73.7	2.3
HEP	Hot dilute HCl	2.3	0.3	21.9	0.2	0.8	4.6	3.6	66.3	65.1	3.0
OHEP	Cold dilute NaOH	5.9	nd	12.1	2.5	0.3	7.0	0.8	71.3	10.0	nd
Endocarp	Hot dilute HCl	0.94	nd	1.53	nd	nd	10.3		100	66.2	
Pulp[b]	Hot dilute HCl										
P-0	0 M NaCl	8.2	11.83	17.81	6.09	0.68	15.8	0.81	35.5	28.06	5.12
P-1	0.1 M NaCl	8.53	19.78	11.72	4.45	nd	10.67	0.93	42.37	49.30	6.42
P-1	0.3 M NaCl	11.12	25.04	8.98	3.50	nd	6.43	0.59	42.88	24.71	5.02

[a] Pectins of different structures were extracted from citrus peels with water, oxalate, hot dilute hydrochloric acid, and cold dilute sodium hydroxide. Homogalacturonans (HGs) were isolated from the four pectins by mild acid hydrolysis after deesterification.

[b] Pectins of different structures were purified from hot diluted acid extracts of citrus pulp using an anion exchange column and eluted with 0 M, 0.1 M, and 0.3 M NaCl, and the collected fractions were precipitated using ethanol for monosaccharide analysis.

cancers suggest that MCP could play an important role in cancer chemotherapy and chemoprevention. After several days of consuming PectaSol®, a modified form of pectin, there was improved clearance of toxic elements like arsenic and cadmium via the urinary track where the heavy metals appeared to be chelated by the modified pectin and then eliminated in the urine (Eliaz et al. 2006).

Cellulose is a tough and water-insoluble substance, which is found in the protective cell walls of plants, particularly in the stalks, stems, trunks, and all woody portions of the plant. Cellulose nanofibrils in citrus are embedded with pectin (Marin et al. 2007) and its isolation requires the implementation of pectin removal processes (Habibi, Mahrouz, and Vignon 2009; Ifuku et al. 2011). The research on the utilization and bioactivity of cellulose as well as the semicellulose like polysaccharides in citrus are quite limited, which is due to the low solubility of these polysaccharides as well as the shortage of suitable enzymatic or chemical isolation methods. However, these polysaccharides, because of a very complex structure with interrelationships that makes the isolation and characterization of these polysaccharides quite difficult, needs to be further investigated.

The present chapter reviews the advances in the extraction, structural characterization, and applications, as well as the health benefits of citrus polysaccharides and their modified products.

4.2 PECTIN AND ITS MODIFIED PRODUCTS FROM CITRUS

Pectin takes its name from the Ancient Greek and means "congealed, curdled." These compounds are the most complex polysaccharides found in plant and make up the structural components of the middle lamellae. The middle lamella function as a hydrating agent and cements together the primary cell walls of two adjoining cells. They are commonly produced during the initial stages of primary cell wall growth and make up one-third of the cell wall of dry substances of dicotyledonous and some monocotyledonous plants. At present, citrus peels and apple pomace are the main raw materials used to produce commercial pectins. Other potential sources of pectins that have been considered are sugar beet and the residues from the seed heads of sunflowers and other plants.

Citrus fruits are particularly rich in pectin, especially in the albedo tissue (Liu, Shi, and Langrish 2006) and the large quantities of citrus (e.g., lemon, lime, orange, and grapefruit and so on) waste generated by the fruit juice industry have become the most important raw material for the production of commercial pectin (May, Phillips, and Williams 1997) The pectins obtained from citrus fruits were reported to have predominantly HG (with variable degrees of methyl/acetyl esterification) and smaller quantities of RG-I and RG-II regions. This may be because the currently used harsh acidic extraction procedures resulted in the stripping away of the side chains, particularly with RG-I and arabinose containing polymers (Ralet and Thibault 1994; Ros, Schols, and Voragen 1996; Yapo et al. 2007). These side chains, however, are important for their bioactivity as well as for their rheology properties.

4.2.1 Citrus Pectin and Its Structure

The pectic heteropolysaccharides extracted from citrus peel with hot acid usually contain several different structural regions (Table 4.2). The most abundant region is HG, a linear homopolymer with α-1,4-linked galacturonic acids that comprise about 65%

TABLE 4.2
Structural Elements Present in Pectic Substances and Possible Variations within an Individual Element

Structural Elements	Diversity Based On
Homogalacturonan	• Length of the homogalacturonan segment between individual rhamnose residues • Degree and distribution of methyl esterification and acetyl esterification
Rhamnogalacturonan I (RG-I)	• Nature of neutral and acidic sugars present in side chains • Length, sugar, and linkage composition and degree of branching of side chains • Distribution of side chains over the alternating rhamnose-(1→4)-galacturonic acid backbone • Rhamnose:galacturonic acid ratio • Degree and distribution of acetyl esterification, methyl esters present
Rhamnogalacturonan II (RG-II)	• Very conserved structure • Proportion of rate sugars such as O-me-xylose, KDO, DHA • Distribution of neutral sugars in the side chains • Number, type, and distribution of uronic acids in the (side) chains • Attachment and distribution of RG-II chains over the pectin molecule
Rylogalacturonan	• Degree of xylose substitution, length of short xylose side chains • Other sugars (e.g., frucose) present in side chains • Degree of methyl esterification and acetylation • Distribution of substituents over the backbone
Arabinan	• Size, degree, and type of branching, distribution of branches over backbone • Polymer attached to (pectin, arabinogalactan)
Arabinogalactan I and arabinogalactan II	• Size, ratio of arabinose to galactose, other sugars in chains • Linkage present • Distribution of substituent over the backbone • Attachment to other pectin (or protein) and distribution over molecule
Apiogalacturonan	• Degree of apiose substitution • Length of apiose chains • Methyl esterification • Distribution of substituent over the backbone

of the chain. The most structurally complex region is RG-II and it makes up 10% of the chain. RG-I represents the rest and makes up 20% to 35% of the chain; it contains a backbone of repeating disaccharide [-α-D-GalA-1,2-α-*L*-Rha-1-4-]n with a high degree of branched oligosaccharides attached to its backbone (Figure 4.2) (Mohnen 2008). In most reports on pectin structure, these three regions are considered to be critical structural elements, however, there is still no consensus on how these regions link to one another or how they interact with other polymers in the cell wall. A supported model with HG as backbone and RG-I and RGII linked via their backbones is the most supported structure (Coenen et al. 2007; Ishii, Matsunaga, and Hayashi 2001). Other components such as Rylogalacturonan, Arabinan, Arabinogalactan I, Arabinogalactan II, and Apiogalacturonan have been also reported to present in the *Citrus species.*

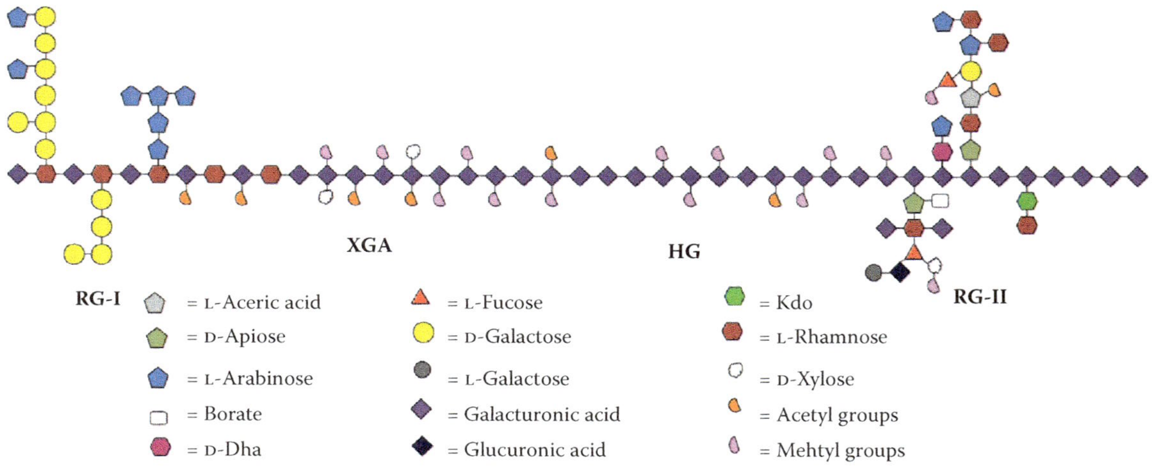

FIGURE 4.2 Schematic structure of pectin showing the four pectic polysaccharides homogalacturonan (HG), xylogalacturonan (XGA), rhamnogalacturonan I (RG-I), and rhamnogalacturonan II (RG-II) linked to one another. The representative pectin structure shown is not quantitatively accurate. HG should be increased 12.5-fold and RG-I increased 2.5-fold to approximate the amounts of these polysaccharides in walls.

The bioactivities and rheology properties of citrus pectin are related to the GalA and neutral sugar content, the amounts and distribution of substituents (methoxyl and acetyl groups), and to its molar mass and distribution (Axelos and Thibault 1991). The length of HG and the proportion of HG, RG-I, and RG-II in the molecule might also influence the properties of pectin (Bonnin et al. 2002) and affect its use as a food additive or functional ingredient. The pectin from different parts of the citrus fruit has totally different structures with different proportions of HG, RG-I, and RG-II, and a different monosaccharide composition (Table 4.1).

Pectins from citrus peel consist of HG stretches with similar lengths when extracted by hot water, oxalate, hot dilute acid or cold diluted alkaline (Yapo et al. 2007). It is concluded that citrus peel pectins consist predominantly of HG with a few RG-I regions and only minor amounts of RG-II. With the exception of alkaline extraction, the other three extraction methods yielded pectins with similar HG backbones, but the RG-I and RG-II branches varied among the different methods. Further isolation of the different structural regions indicated that RG-I was composed mainly of arabinose, galactose, galacturonic acid, and rhamnose, indicating the presence of arabinan and/or (arabino) galactan side chains. Debranched RG-I was obtained by treating the RG-I arising from acid-extracted pectin with endo-1,5-α-L-arabinanase, endo-1,4-β-D-galactanase, α-L-arabinofuranosidase, and β-D-galactosidase in admixture. The molar ratios of galacturonic acid to rhamnose in RG-I and debranched RG-I were very close to 1/1, indicating a strict repeating [GalA–Rha]n pattern in the backbone. RG-II, however, was only found in acid extracted pectin. The different extraction methods also caused changes in the degree of methylation (DM) and the degree of acylation (DA).

The endocarp is another source of pectin that comes from citrus processing waste. The pectin isolated from the endocarp of *Citrus depressa* has been studied by Yukihiro (Tamaki et al. 2008). The proposed structure of this pectin is shown in Figure 4.3.

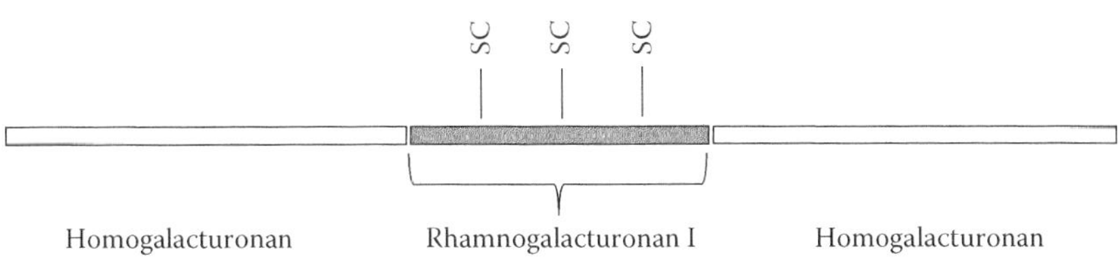

Side chain (SC)

R1 → 3: R2[→4Gal1→4Gal1→4Gal1→4Gal1→4Gal1→4Gal1]→ (R1 attached at position 3)

R3 → 6 (three times): R4[→4Gal1→4Gal1→4Gal1→4Gal1]→ (R3 attached at position 6)

R1 = Gal→, Ara→, Ara1→5Ara1→

R2 = Ga→, Ara→, Ara1→5Ara1→

R3 = R4→6Gal1→6Gal1→, R4→6Gal1→6Gal1→6Gal1→

R4 = Gal→, Ara→, Ara1→3Ara1→

FIGURE 4.3 Schematic structure model of the pectin of endocarp from *C. depressa*. Galacturonic acid residues were methoxylated (65%) and acetylated (2%).

This pectin was less complex than other pectins and was classified as a high-methoxyl type. The main sugar was galacturonic acid, which accounted for 88.7% of the total sugars. The main neutral sugar was galactose, which represented approximately 80% of the neutral sugars with small amounts of arabinose and rhamnose. Other sugars such as xylose and glucose, which are often present in citrus pectins, were not detected. The high galactose content in endocarp pectin may make it a good source of MCP, as the galactose is one of the main monosaccharides of the RG-I regions, which is the most important content for MCP and the result in bioactivities.

Citrus segment membrane, which is inside the citrus peel and endocarp, also contains high amounts of pectin. However, the structures of these polysaccharides have not been studied in detail. Unlike to the pectin extracted from citrus peels by hot acid, the pectin for citrus segments can be conveniently obtained as acidic and basic processes during citrus canning for removing of the segments, which also means no extra hot-acid extraction, which is applied in commercial pectin producing, is needed. However, those suspension liquids are usually discharged directly down the drain and can cause environmental pollution. Presently, there are few publications or industrial pilot studies on the recovery of pectic polysaccharides from processing water or on the mitigation of the chemical oxygen demand (COD) value. Edashige, Murakami, and Tsujita (2008) isolated several pectins containing both RG-I and RG-II from the segment membrane of four species of citrus using a lab scale, using hot acid extraction method. The results showed that galacturonic acid was the main component of the backbone of the pectin from citrus segments, but the content was much lower than pectin from citrus peel extracted by harsh acid treatments. The detection of rhamnose indicated the presence of a ramified region and the presence of arabinose suggests the presence of arabinan or side chains. Pectin from different species possessed different glycosyl compositions and their inhibitory effect on pancreatic lipase activity also varied, but all of them had relatively high inhibitory properties. Therefore, segment membranes are also a promising functional food resource and the recovery of segment membrane pectin from the canning process water could have both economical and environmental importance. Recently, we have established a pilot recovery system for recovery of the citrus segments polysaccharides and reduced the COD of effluent decreased dramatically through polysaccharide recovery, and find these polysaccharides might be potential healthy food thickening and gelling agents (Chen et al. 2017).

4.2.2 Extraction of Citrus Pectin

There are many procedures used for extraction of citrus pectin. However, most of these procedures typically involve two principal steps. The first step is the extraction of the starting plant material with hot diluted acid, cold diluted sodium hydroxide, or chelating agents (Whistler and Bemiller 1993). The second step involves the isolation of the extracted pectin from the liquid through alcohol precipitation. However, the preparation conditions differ based on the desired functional properties of the final product.

4.2.2.1 Commercial Preparation of High-Ester Pectin

In summary, commercial high-ester pectin is prepared using the following procedures (Figure 4.4): the raw material is suspended in hot water with an appropriate amount

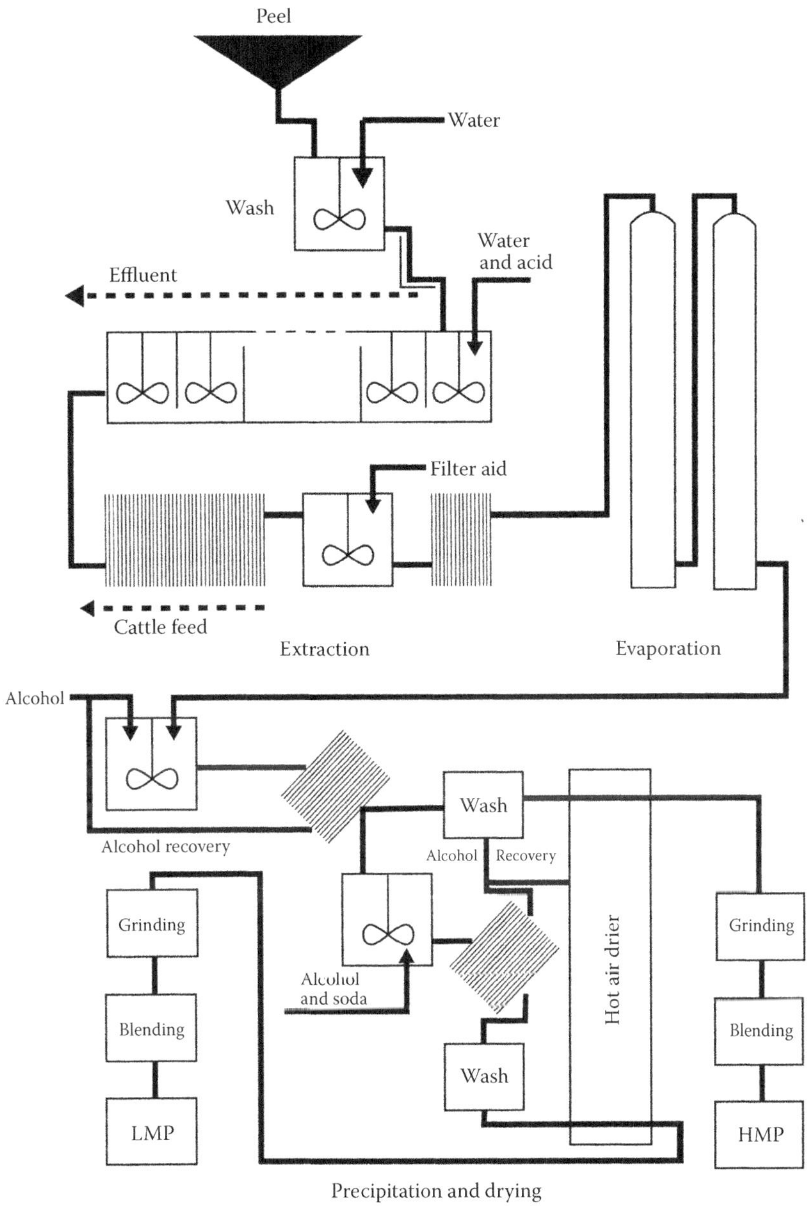

FIGURE 4.4 Typical process for pectin production from dried citrus peel.

of strong acid, typically within a pH 1–2 window, a temperature of 70°C to 90°C and a duration of 30 minutes to 6 hours. The undissolved solids are removed by filtration, and the resulting solution is ion-exchanged, concentrated, and then mixed with alcohol to precipitate the pectin. The precipitate is squeezed and pressed to remove the alcohol, purified by washing with more alcohol, and finally dried and milled.

Nitric acid is commonly used for acidifying the raw plant material because it is less corrosive than hydrochloric acid. The peel-to-water ratio is chosen so that the resulting solution contains about 6 to 10 g/L of pectin. The process turns protopectin into water-soluble pectin; it further dissolves the pectin and leaches it from the undissolved plant material. The modification of protopectin involves limited hydrolysis of branch chains in the molecular parts and the degree of esterification (DE) can be controlled within an interval of about 55% to 80% by choosing extraction conditions that hydrolyze the desired number of methyl ester groups. Furthermore, since various molecular parts in the protopectin molecule differ with respect to their rate of hydrolysis as a function of pH and temperature, subtle tailoring of pectin functionality is possible by selecting specific combinations of raw material and extraction conditions. Some plant materials are very different from those in citrus fruit and behave differently with respect to the extraction conditions.

The undissolved materials are removed in a series of filtrations, which are difficult because of the high solution viscosity, and that is the reason why the pectin concentration cannot exceed 10 g/L. Insoluble filter aids such as wood cellulose and perlite may be used to facilitate this process. The undissolved plant material is typically used as cattle feed. The extract is then optionally modified by ion exchange to remove Ca^{2+} ions, so that the resulting pectin becomes more soluble. Next, the solution can be concentrated by evaporation or membrane filtration before the pectin is precipitated by pouring the extract into an alcohol, such as 2-propanol. The stringy precipitate is squeezed by pressing to remove alcohol and then washed in a fresh batch of alcohol. After thorough squeezing, the precipitate is dried. The dried precipitate is finally milled and sieved to a powder of controlled particle size.

4.2.2.2 Production Process for Low-Ester Pectin

The processing steps for low-ester pectin is almost the same as for high-ester pectin, except for a separate deesterification treatment that usually takes place after filtration. Since the glycosidic bonds of the main chain of pectin are reasonably resistant to hydrolysis, it is possible to deesterify dissolved pectin with acid while keeping the loss of molecular weight at an acceptable level. It is also possible to deesterify pectin by using a strong alkali (e.g., NaOH), but the main chain of high-ester pectin is vulnerable to alkaline conditions, so it is necessary to perform this reaction at low temperature and/or while the pectin is not in aqueous solution. In industrial production, such treatments can be done on the pectin precipitate obtained after alcohol precipitation by reaction with alkali while the precipitate is suspended in the alcohol.

4.2.2.3 Other Extraction Methods

Microwave and ultrasound technologies appear to be the most likely candidates, when compared to other novel extraction methods that are able to perform with greater efficiency and produce more consistent products. The enzymatic augmentation of the extraction process and the use of subcritical water to replace acidified water as the solvent are also useful methods to improve the traditional pectin extraction technologies. Although these nonconventional technologies produce citrus pectin more efficiently and with higher purity, the high cost and the scale-up limitations hindered their application in industry. However, the knowledge gained from the optimization of processing parameters in experimental scale are crucial to better understanding of the probability of a positive outcome from a scale-up procedure (Adetunji et al. 2017).

4.2.3 Modification of Citrus Pectin

Many methods are available for the modification of pectin that will extend its functional effects to humans. The major starting material for the preparation of modified pectin is commercial citrus pectin. Modifications can be made by chemical, enzymatic, heat, and a few other treatments. A summary of the changes made to the pectin molecule by the different modification methods and the resulting product quality, including molecular weight, esterification degree, and other molecular changes, are shown in Table 4.3.

Acidic degradation of pectin is considered to be the dominant method for preparing low molecular weight pectin (Khotimchenko et al. 2012). This procedure is able to generate pectin fractions with the desired molecular-mass distribution by controlling acid concentration, temperature, and hydrolysis time. The reduction in the molecular weight of pectin improves its absorption in the gastrointestinal tract. Besides reducing pectin's molecular weight, acid degradation is also associated with deesterification of pectin when the acidic degradation conditions are extremely harsh.

Modification of pH is another chemical-based procedure used to prepare low molecular weight pectins that have special functions, by subjecting them to acid and alkaline conditions. Hao et al. (2013) found that the monosaccharide composition of pH-modified citrus pectin was similar to untreated citrus pectin, with molecular weight reduction to ~61 and 9 kDa by different hydrolysis time. Fourier transform infrared spectroscopy (FTIR) analysis gave supporting evidence for the successful removal of methyl esters by the pH modification procedure. Leclere, Van Cutsem, and Michiels (2013) showed that alkaline treatment initiated the β-elimination reaction, which caused depolymerization of the polysaccharide backbone and deesterification of the HG regions, whereas acid treatment cleaved neutral sugars and released the branched regions from the pectin backbone, which generated low molecular weight pectin with high amounts of arabinogalactans and galactans.

The chemical sulfation of pectin is another modification procedure that confers anticoagulant activities to pectin (Cipriani et al. 2009; Maas et al. 2012; Vityazev et al. 2010). However, the conventional sulfation reaction is performed under harsh conditions using chlorosulfonic acid as the sulfating agent. These harsh conditions produce severe changes in the pectin structure and alter the bioactivity of the sulfated derivatives. Y. Hu et al. (2015) described a milder sulfation method using the pyridine-sulfur-trioxide complex in dimethyl sulfoxide (DMSO). The results indicated only a small decrease in the GalA content and molecular weight after sulfation. Structural characterization by infrared (IR) and nuclear magnetic resonance (NMR) spectra indicated sulfation occurred mainly at positions C-2, C-3 of the GalA (located in backbone region of the CP). Anticoagulant assays demonstrated that sulfated CP (TBA-3) could prolong activated partial thromboplastin time and thrombin time, with an activity of 51.96 IU/mg and 15.2 IU/mg, respectively. Further

TABLE 4.3
Pectin Modification Methods and the Resulting Product Quality

	Result Products Quality			
Modification Methods	**Molecular Weight**	**Etherification Degree**	**Other Molecule Changes**	**Reference**
Acidic degradation	Decrease, wide distribution	Obvious de-esterification	Not mentioned	Khotimchenko et al. 2012
pH modification	Wild distribution, from 9 to 61 kDa	Methylesters-removed widely	Not mentioned	Hao et al. 2013
Sulfation modification	About 47 kDa	Decreased	Sulfated at positions C-2, C-3 of the GalA; Decrease in the GalA content and particle size	Y. Hu et al. 2015
Ultrasound treatment	From 464 kDa to 296 kDa	Slightly decreased	Not mentioned	Zhang et al. 2013
High temperature modification	About 24 kDa	No obvious change	Not mentioned	Hao et al. 2013
Microwave-assisted solvent-free modification	No obvious change	No obvious change	Etherification with several fatty acids	Calce et al. 2012
Cross-linked with adipic acid	No obvious change	No obvious change	Cross-linked with adipic acid	Li et al. 2007
Graft poly-acrylamide	No obvious change	No obvious change	Graft polyacrylamide at positions C-2 of the GalA	Sutar et al. 2008

investigation on coagulation factors indicated TBA-3 could achieve inactivation of thrombin with both heparin cofactor II and antithrombin.

Enzymic degradation of pectin is a very specific target modification method for pectin, although the sources of the enzymes are quite limited (Yadav et al. 2009). Cameron et al. (2011) reported on the mode of action of a unique citrus thermally tolerant pectin methylesterase (TT-PME) and the nanostructural modifications that it produced. As the methyl ester distribution in pectin homogalacturonan has a major influence on the rheology property and biofunction of pectin, the enzyme was used to produce a controlled demethylesterification series from a model homogalacturonan and the resulting fragments were further digested by endopolygalacturonase to produce oligogalacturonides.

High-temperature MCP was prepared by heating and cooling the pectin several times (Hao et al. 2013). The monosaccharide composition of high-temperature citrus pectin was similar to that of citrus pectin, whereas its molecular weight was greatly reduced to ~24 kDa, which was even lower than that of pH-modified heparin. FTIR analysis indicated this method had no obvious influence on the esterified group, although the molecular size of the pectin was greatly reduced.

Ultrasonic treatment is an effective, energy-saving, and environmentally friendly way to prepare and process polymer particles, and is particularly effective in breaking up aggregates and reducing particle size and molecular weight (Guo et al. 2014). Zhang et al. (2013) found that the average molecular weight of citrus pectin reduced from 464 to 296 kDa when exposed to sonication for 30 minutes. The degree of methylation of citrus pectin changed slightly and its monosaccharide components remained unchanged when high-intensity ultrasound was applied. The reduced (Gal + Ara)/Rha ratio after ultrasonication suggested degradation in the neutral sugar side chains of citrus pectin. Atomic force microscopy analysis confirmed that the citrus pectin chains were degraded by ultrasonic treatment at the nano level. However, the single use of ultrasound had limitations in that the final molecular weight was still very high. Thus, a combination of ultrasonic treatment with other methods, such as enzymic degradation or subcritical water, is sometimes used to generate pectic oligosaccharides with a molecular size lower than 10 kDa (Ma et al. 2016).

Because natural pectins have low water resistance and poor mechanical properties, their use in manufactured products is limited. The chemical modification with fatty acids improved their water resistance and barrier properties, and as a consequence the impetus for initiating new packaging applications was created. The reaction was successfully accomplished with a microwave-assisted solvent-free modification of the pectin system (Calce et al. 2012). This reaction involved the esterification of several fatty acids with pectin's alcoholic functions. The reaction was performed by simply mixing the reagents with a catalytic amount of the inorganic base (potassium carbonate) and irradiating the obtained mixture with microwaves for a short time.

Similarly, pectin cross-linked with adipic acid was synthesized and used for heavy metal removal from wastewater (Li et al. 2007). The modified pectin had a rough, porous phase covered with carboxyl group that conferred it with a high adsorption capacity. In another study, polyacrylamide was grafted onto pectin, producing polymers cross-linked with varying amounts of glutaraldehyde. The final products showed improved film forming properties and better gelling properties than unmodified pectin.

4.2.4 Health Benefits of Citrus Pectin and Modified Citrus Pectin

4.2.4.1 Health Benefits of Citrus Pectin

4.2.4.1.1 Bioabsorption of Heavy Metals

Pectin is the important component of citrus peels and has been found to have great affinity for metal ions such as Pb^{2+}, AS^{2+}, Zn^{2+}, and Cu^{2+}. Biosorption of metal ions depends on parameters such as pH, temperature, ionic strength, metal concentration, co-ion presence, and sorbent properties. Kartel, Kupchik, and Veisov (1999) determined that the sequence $Pb^{2+} > Cu^{2+} > Co^{2+} > Ni^{2+} > Zn^{2+} > Cd^{2+}$ was typical for cation binding to pectin. Balaria and Schiewer (2008) found that carboxylic acid groups are active participants in Pb^2 binding by citrus pectin. Because pectin and other polysaccharide solutions contain much more carboxyl anions and need more balanced cations. Some divalent cations help citrus pectin form a stable structure. Dronnet et al. (1996) found the presence of acetyl groups decreased the affinity of some divalent metal ions for sugar-beet pectins, while citrus pectins displayed cooperative interactions for all metal ions, such as Ca^{2+}, Cu^{2+}, Ni^{2+}, Cd^{2+}, Pb^{2+}, and Zn^{2+}.

The metal chelating property of pectin makes it an ideal candidate for use in chelation therapy. The standard Western medical approach for removing heavy metals from the body is chelation therapy. This procedure, when performed with harsh chelators and introduced intravenously, can cause multiple side effects while potentially robbing the body of some of its essential nutrients. Citrus pectin, however, works as a gentle chelator in the bloodstream and is very useful for ongoing detoxification with little side effects. Two recent clinical studies have found that citrus pectin is a very safe and effective supplement for reducing the heavy metal load in the body. A clinical study was recently conducted where citrus pectin was administered to a group of volunteers, and their baseline levels of total body mercury burden was compared before and after 4 months of citrus pectin treatment (Eliaz, Weil, and Wilk 2007). The results showed a significant average decrease (over 60%) in the total body mercury burden after treatment with citrus pectin. In other studies, patients who had been given citrus pectin were proven to increase urinary secretion of heavy metals such as lead, mercury, cadmium, and arsenic (Eliaz et al. 2006; Zhao et al. 2008). All these studies supported that citrus pectin is a viable alternative to the harsher intravenous chelating therapies, as it was proven to be effective and to have negligible side effects.

4.2.4.1.2 Antidiabetes and Control Blood Lipids

Diabetes and high blood lipids have dramatically risen at an alarming rate globally over the past few decades and are expected to hit about 439 million people by 2030 (Shaw, Sicree, and Zimmet 2010). Meanwhile, diabetes has become the sixth most common cause of morbidity and mortality affecting young and middle-aged people. Citrus pectin has been shown to decrease plasma and hepatic cholesterol concentrations in rats and guinea pigs. It has been shown to have a more significant hypocholestenolemic effect than guar gum or methylcellulose in rats. Previous studies in guinea pigs fed a hypercholesterolemic diet have shown that intake of citrus pectin isolated from prickly pear (*Opuntia* sp.) resulted in a significant decrease in plasma low-density lipoprotein (LDL). The presence of a pectic-like gel in the lumen of the small intestine could interfere with the equilibrium between the micellar phase and the molecular phase, which passes into the unstirred layer on the brush border, and in this way reduces the absorption of lipids (Brouns et al. 2011; Kay and Truswell 1977). Studies in rats have demonstrated that the molecular weight and the extent of methylation of pectin affect some of its physicochemical properties resulting in different metabolic responses.

4.2.4.1.3 Intestinal Health

Citrus pectin is a soluble fiber with powerful water-holding abilities. It can effectively bind to water in the intestine and form a gel to help prevent constipation and remove toxins. It also plays an important role in balancing the pH of the intestine. In the colon, the pectin ferments and forms butyric acid, and also binds to mutagens to assist in their removal. These effects help reduce the risk of colon cancer and other types of cancer.

Citrus pectins can also be fermented by the microbiotas. In the human intestine, pectins are fermented by the resident microbiota in the cecum and colon. The presence of pectins affects microbial composition and activity, and thereby the production of short-chain fatty acids (SCFAs) by the microbes in the colon (Dongowski, Lorenz, and Proll 2002; Onumpai et al. 2011). SCFA production as a result of pectin consumption provides the most direct link between pectin intake and associated health benefits (Wong et al. 2006). During *in vitro* fermentation using human fecal microbiota, pectin can significantly stimulate growth and activity of the genera *Bifidobacterium* and *Lactobacillus* (Onumpai et al. 2011), members of which have shown the potential to protect enterocytes from an acute inflammatory response (Candela et al. 2008). In rats, citrus pectins tended to increase not only the total SCFA concentration, but also the proportions of propionate and butyrate (Dongowski et al. 2002). Propionate has the potential to reduce blood cholesterol concentrations, while butyrate is an important energy source for intestinal epithelial cells and plays a role in the maintenance of colonic homeostasis (Wong et al. 2006).

Several recent clinical studies (Rabbani et al. 2001; Triplehorn and Millard 2002) have also demonstrated that oral pectin supplementation to children and infants reduced acute intestinal infections and significantly slowed diarrhea. It is thought that this is due to the reduction of pathogenic bacteria such as *Shigella*, *Salmonella*, *Klebsiella*, *Enterobacter*, *Proteus*, and *Citrobacter* from the intestine, and stimulated the growth of certain strains of *Bifidobacteria* and *Lactobacillus* (Olano-Martin et al. 2002). These bacteria are considered to be directly related to the health of the large intestine and their concentrations depict a healthy microflora population (Niture and Refai 2013).

A more recent study showed that dietary supplementation with pectin is a potential strategy to modulate the location of fermentation of dietary fibers, and in so doing control microbiota composition and SCFA production (Tian et al. 2017).

4.2.4.2 Health Benefits of Modified Citrus Pectin

MCPs can be made by chemical, enzymatic, or heat treatment as previously described. There are two main types of modified pectin that have been investigated in detail: GCS-100 and Pectasol-C (Table 4.4). GCS-100 is patented as a mammalian anticancer agent (Raz and Pienta 1998). The modification involves chemical modification of citrus pectin, chiefly intended to lower the molecular weight: the pectin is first alkali-treated and then acid-treated to lower the molecular weight to 10 kD (Raz and Pienta 1998). The resulting low molecular weight fragments are enriched with RG-I regions but very few RG-II regions (Figure 4.2), and was able to compete

TABLE 4.4
Modified Pectin and Its Biofunctions

Pectin	Modification	Effect	Reference
pH-MCP	pH	Decreasing prostate cancer MAT-LyLu cells	Leclere et al. 2013
PectaSol-C	Enzymatic treatment	Activation of the immune system	Ramachandran et al. 2011
Heat treated citrus pectin	Heat	Induced apoptosis of prostate cancer	Jackson et al. 2007
Heat-MCP	Heat	Inhibition of breast cancer, colon cancer, liver cancer, cervical cancer	Hao et al. 2013
HG-rich pectin	HG-rich	Inflammatory cell activation	Ye et al. 2015
Irradiation-MCP	Irradiation	Proliferation inhibition of cancer cells	Jeong et al. 2015
Galactose-rich MCP	Galactose-rich	Inhibition of metastatic B16-F1 and MLL cells	Kidd 1996
GCS-100	Chemical or enzymatic modification	Reduce lymphadenopathy	Morris et al. 2013
MCP	–	Inhibitory effect of MCP on liver metastasis	H. Y. Liu et al. 2008
GCS-100	Chemical or enzymatic modification	Overcome Bortezomib resistance and enhance dexamethasone-induced apoptosis	Chauhan et al. 2005
MCP	High pH and temperature treatment	Antimetastatic properties	Glinsky and Raz 2009
Heat-modified citrus pectin	Heat	Identification of a cytotoxic molecule	Leclere et al. 2015
Heat-modified citrus pectin	Heat	Induce apoptosis-like cell death and autophagy in HepG2 and A549 cancer cells	Leclere et al. 2015
Pectin	Acidic pH	Inhibitory activity on galectin-3	Gao et al. 2012

with galectin-3 (Gal-3) (Gao et al. 2012). Gal-3 is a lectin that possesses a conserved carbohydrate-binding domain that specifically recognizes galactose-contain oligosaccharides. Thus, the MCP can bind Gal-3 and possess a number of bioactivities related to Gal-3.

4.2.4.2.1 Activates Immune Responses

Citrus pectin exerts a favorable immunomodulatory response in human peripheral blood cells through its effect on cytokine production (Salman et al. 2008). The carbohydrate composition of pectin is very important in determining the different immune responses. The RG-I-arabinan fraction of pectin can enhance secretion of granulocyte colony-stimulating factor (G-CSF) by murine colonic MCE 301 cells (Matsumoto et al. 2008). RG-I-arabinogalactan can also activate macrophages and dendritic cells (Inngjerdingen et al. 2008). Methyl-esterified pectic oligosaccharides with 4,5-unsaturated nonreducing ends enhanced the T-helper 1 (Th1) dependent, delayed-type hypersensitivity in a murine influenza vaccine model, reduced the Th2 cytokine production in splenocytes *in vitro* (Vos et al. 2007b), and decreased allergic asthma in mice (Vos et al. 2007a).

A more recently study indicated MCP acts as a substance with immunostimulatory properties in human blood samples, including the activation of functional NK cells against K562 leukemic cells in culture. The selective immunostimulatory properties are proposed to be attributed to the presence of a low degree of methyl esterification and unsaturated oligogalacturonic acids. Further research is necessary to determine if changing the degree of esterification of oligogalacturonic acid in MCP can alter the immune response. *In vivo* studies are necessary to better understand the applications of MCP as an immune enhancer (Ramachandran 2011).

4.2.4.2.2 Anticancer of Modified Citrus Pectin from Citrus

Despite enormous efforts that have been made in the search for novel drugs and treatments, cancer continues to be a major public health problem. Moreover, the emergence of resistance to cancer chemotherapy often prevents complete remission. Researchers have thus turned to natural products mainly from plant origin to circumvent resistance. More than 60% of anticancer drugs in use nowadays are derived from natural products (Leclere et al. 2015). Modified pectins have demonstrated chemopreventive and antitumoral activities against some aggressive and recurrent cancers. It has been reported to exert antitumor activity on different cell lines and in different mice models. This range of activities is probably due to the different methods used to prepare MCP and their structural differences. A summary of the different kinds of MCP and their effects are shown in Table 4.4.

MCP is useful in the prevention and treatment of metastatic cancer, especially in solid tumors like melanoma and cancers of the prostate, colon, and breast (Nicholas 2009). The possible role of MCP in the suppression of metastasis is presented in Figure 4.5. Using various cancer types in animal models proved that MCP inhibited cancer progression by suppressing angiogenesis and metastasis (Glinskii et al. 2005; Glinsky and Raz 2009; Johnson et al. 2007). Pienta et al. (1995) were the first to show that MCP reduced prostate cancer metastasis in rats. Their results showed only

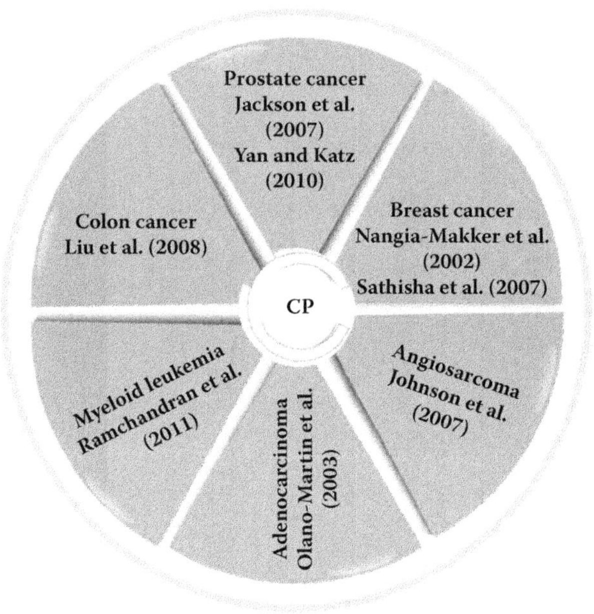

FIGURE 4.5 Possible inhibitory role of CP and MCP in suppression of different cancers.

50% of the rats that drank water with MCP (0.1% weight/volume) had any metastases, while 94% of the rats that drank plain water had cancer metastasize to their lungs. Similarly, several animal studies have found that MCP helps reduce the spread of not only prostate cancer but also breast, skin, and liver cancer (H. Y. Liu et al. 2008; Nangia-Makker et al. 2002; Platt and Raz 1992). MCP was fed to mice with these types of cancer and the mice were found to have a much lower chance of the tumor spreading to the lungs. Human studies have also shown that MCP increased the prostate-specific antigen (PSA) doubling time (Guess et al. 2003). In the phase II pilot study of 10 men whose prostate cancer had returned after an initial treatment with surgery or radiation, the prostate-specific antigen doubling time (PSADT) increased in 8 (80%) of the 10 men after taking MCP for 12 months. These studies appear to show that the presence of MCP makes it difficult for cancer cells to break off from the main tumor, aggregate, and spread to other organs.

4.2.4.2.3 Modified Pectin and Galectin-3 (Gal-3)

Gal-3 is a 30 kD chimeric protein, consisting of a serine-6-phosphorylation site at the N-terminus, a long thin "collagen-like" tail, and a globular structure at the C-terminus (Salman et al. 2008). It can be expressed in the cell nucleus within the cell cytoplasm, at the cell surface, and also released into the extracellular regime. The diverse location of Gal-3 allows it to participate in a wide range of biological processes that include heterotypic and homotypic cell adhesion, cell migration, invasion. The biofunctions of Gal-3 are linked to its ability to bind to galactooligosaccharides present on cell surfaces. Such bindings are implicit in the important roles played by Gal-3 in various stages of cancer progression and metastasis (Morris et al. 2013).

MCP is rich in galactose and antagonizes the binding protein Gal-3, resulting in suppression of cancer cell metastasis. It can interact with Gal-3 and inhibits its activity, which is required for cell-to-cell adhesion/aggregation of the cancer cells. The possible role of MCP and its interaction with galectin and cancer cell metastasis is suggested in Figure 4.6. Inhibition of Gal-3 by pectin polysaccharide or MCP not only possesses antimetastatic effects on cancer cells *in vitro* and *in vivo* (Glinsky et al. 2000; Inohara and Raz 1994; H. Y. Liu et al. 2008; Nangia-Makker et al. 2002; Pienta et al. 1995; Platt and Raz 1992; Yan and Katz 2010), but also inhibits cancer cell invasion in MDA-MB-231 human metastatic breast cancer cells and human buccal metastatic cells (Sathisha et al. 2007) and angiogenesis in endothelial cells (Nangia-Makker et al. 2002). Human clinical trials with MCP showed an increase in PSADT, a marker of slowing the progression of prostate cancer (Guess et al. 2003), and a significant improvement in the quality of life and stabilization of the disease for patients with advanced solid tumors (Azémar et al. 2007). In addition to the therapeutic role against cancer, MCP has been shown to remove toxic metals from the body (Eliaz et al. 2006; Zhao et al. 2008), and reduce experimentally induced kidney injury and fibrosis *in vivo* by reducing Gal-3 levels (Kolatsijoannou et al. 2011). The specific binding of a pectin galactan to the recombinant form of human Gal-3

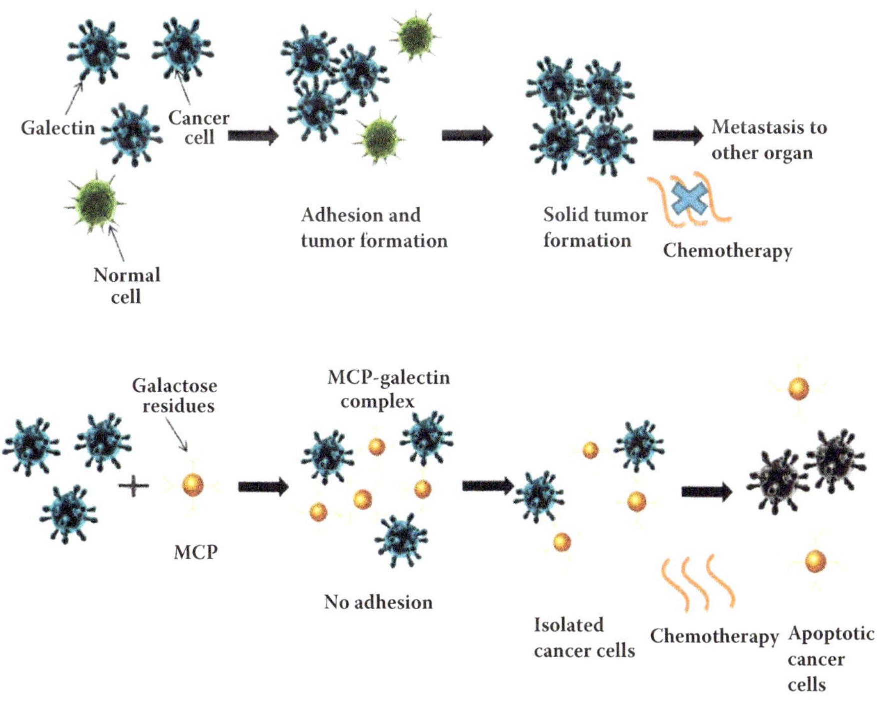

FIGURE 4.6 (Top) A possible mechanism of cancer metastasis. (Bottom) The inhibitory role of MCP-galectin complex interaction in cancer cell metastasis.

has been physically observed by a combination of fluorescence microscopy, flow cytometry, and atomic force microscopy (Glinsky and Raz 2009). All these studies suggested MCP inactivates key transcriptional factors, cell cycle regulators, and cancer metastatic factor Gal-3, which leads to the inhibition of cancer cell growth, angiogenesis, and metastasis. Compared to MCP, CP has limited solubility in water and is unable to interact with Gal-3, but in its modified form MCP can form small, linear, water-soluble fibers in water.

4.2.4.2.4 Digestion and Absorption of Modified Pectin

Regular ingestion of citrus pectin supports digestive health, but the molecules are too large to enter the circulation, which means that the benefits are restricted to the GI tract. PectaSol-C® solves this limitation by an advanced modification process, which reduces the size as well as the structure of the pectin, which makes it absorb into the body circulation and with benefits related to health.

Animal studies have suggested that MCP was able to act upon Gal-3 at its sites of action after absorbed into the bloodstream. Orally administered MCP has been found to assist with excretion of toxic elements by urine and reduce the toxicities in the bloodstream (Eliaz et al. 2006). Amidated pectin and the monomer D-galacturonic acid were stable in human saliva and simulated gastric juice, but can only be broken down by gut microbes (Knaup et al. 2008). However, the experimental evidence from human and animal studies on orally consumed MCP does suggest that MCP can be absorbed in the small intestine (Sakurai et al. 1996). It is still not apparent whether the pectin was modified in the body by endogenous enzymes and/or by other factors, it appears that a balance is needed: MCP has to be small enough to be absorbed but not too disruptive that it is unable to exert its effects. A possible answer to the mechanism of systemic uptake of MCP may be found in the analogous studies on the uptake of β-glucans, which exerts it function by binding to specific glucose/mannose receptors, suggesting a direct microbial-independent immunomodulating effect.

Although the mechanism remains to be established, it is well known that macrophages abundantly express Gal-3, both intra- and extracellularly, and that Gal-3 may modulate the function of the cell by binding to glycoconjugates (F. T. Liu et al. 1995). Considering the effect of pectin fractions on macrophages and the specific binding of Gal-3 to β-D-(1-4)-galactans, this could facilitate an uptake mechanism by macrophages, and also suggest an additional mechanism whereby MCP could exert its effects against disease.

4.2.4.2.5 Other Benefits of Modified Citrus Pectin

Besides the anticancer effects, MCP also functions effectively in decreasing cholesterol, bioabsorption of heavy metals (Eliaz et al. 2007), growing bone cells (Kokkonen et al. 2007), transporting drugs (L. Liu et al. 2008), and anticoagulation and antithrombosis (Cipriani et al. 2009).

4.2.5 Use of Citrus Pectin in the Food Industry

Pectin is a popular ingredient worldwide for the promotion of health (Wicker et al. 2014), and today it is in short supply despite having a global market value of more than $850 million in 2013 and a more than 5% annual growth rate (Ciriminna, Chavarría-Hernández, and Pagliaro 2015). It is an important polysaccharide with multiple applications in foods, pharmaceuticals, and other industries. Its importance in the food sector lies in its ability to form a gel in the presence of Ca^{2+} ions or a solute at low pH. It is widely used as a texturizer, stabilizer, and emulsifier in a variety of foods (Luo, Kang, and Zhong 2015; Thakur, Singh, and Handa 1997; Zhang, Lin, and Zhong 2015). A mixture of low methoxy (LM) pectin, aluminum hydroxide, and magnesium oxide has been reported to be useful in the treatment of gastric and duodenal ulcers (Harsha et al. 2015). Pectin alone or in combination with gelatin is used as an encapsulating agent for the controlled release of drugs. Furthermore, modified pectin significantly inhibits the growth of colon cancer cells and possesses obvious antitumor properties (Almeida et al. 2015). Thus, pectin is referred to as the balance for human health. However, the pectin from different sources exhibited a different gelling ability due to variations in parameters of structures, and as a result the commercial applications of pectin are quite limited (Thakur et al. 1997).

Recently, pectin has been used as a fat or sugar replacer in low-calorie foods and to make edible films. The joint FAO/WHO Committee on Food Additives recommended pectin as a safe additive with no limits on acceptable daily intake except as dictated by good manufacturing practice. According to the degree of esterification (DE) with methanol, which is the ratio of esterified galacturonic acid groups to total galacturonic acid groups, pectin can be classified as high methoxyl pectin (HMP) and low methoxyl pectin (LMP). HMP has over 50% of their carboxyl groups esterified (DE >50), while LMP have a DE <50. The DE affects the gelling properties of pectins. In this way, LMP forms gel in the presence of multivalent ions, which acts as a bridge between pairs of carboxyl groups of different pectin chains. On the other hand, HMP forms gel in acidic media with the addition of different sugars such as sucrose or glucose (Mishra, Banthia, and Majeed 2012; Videcoq et al. 2011). Among the different sources, the pectin from citrus is the most used pectin in the food industry.

4.2.5.1 Fruit Jelly and Jam

In traditional jelly production, sugar contributes to the product's structure, given that gels with HMP are formed only if a cosolute is present (typically sucrose at a concentration greater than 55%). Since sugar consumption is directly related to diabetes and other illnesses, an alternative for gel production without sucrose is to use LMP in the presence of calcium. Khouryieh, Aramouni, and Herald (2005) prepared three jelly formulations using sucralose, LMP, and maltodextrin with either xanthan gum or locust bean gum to make sugar-free jelly. The overall acceptability based on aroma, taste, texture, and spreadability for no-sugar-added grape jelly averaged 5.8 to 6.4 on a 9-point hedonic scale (Acosta, Víquez, and Cubero 2008). The jelly provided less than 12 calories per serving, which allowed the product to be labeled "low calorie."

Jam is another food product using large amounts of pectin. The addition of pectin adds a lot of benefits to jam products for calorie-conscious consumers and to partly fill the need for sugar-free products for diabetics. The demand for jams with less or even without sugar is also increasing. LMP is also used to form pectin-calcium gels in different products (Abdullah and Cheng 2001; Poiana, Alexa, and Mateescu 2012), and the addition of a small amount of HMP will improve the texture, gel formation, and spreadability of the jam (Afoakwa et al. 2006; Gajar and Badrie 2002).

4.2.5.2 Beverages

Pectin is usually used as a beverage stabilizer. Good suspension stability was achieved for juices by mixing with pectin and other stabilizers, such as xanthan gum, carboxymethyl cellulose, and agar, in different proportions. A good stabilizer also makes it possible to suspend the juice sacs (Ren et al. 2015). Pectin can be used to prepare a stable emulsion by mixing flaxseed oil in carrot juice with pectin as the emulsifier. The pectin stabilized the emulsion by creating an electrostatic repulsion between the emulsion droplets and these emulsions were stable up to 8 days at 4°C ± 2°C (Shanmugam and Ashokkumar 2015).

Pectins also play an important role in dietetic soft drinks and ready-made nutritive drinks, and enjoy a significant share of the beverage market. The addition of 5 g (1.1% w/v) of HM pectin that enhanced the satiety of orange juices (Tiwary, Ward, and Jackson 1997) and other low-calorie beverages were effective in reducing subsequent energy intake (Perrigue et al. 2010). Recently, a novel, two-part beverage, consisting of alginate-pectin and calcium that formed a stable, fibrous gel in the stomach was developed. It was used to determine the effects on subjective satiety and food intake in overweight and obese women (Pelkman et al. 2007). Sanaka et al. (2007) prepared a ready-made nutritive drink with maltodextrin, fat, protein, agar, pectin, and water. The authors reported that 5.2 g of pectin (1% w/v) in the drink delayed gastric emptying but had no effect on postprandial glucose response when compared with the control.

4.2.5.3 Use of Citrus Pectin as a Fat Replacer

A legitimate fat replacer cannot only reduce the fat content in the food, but also maintain or even improve the sensory quality of the original food. As a water-soluble dietary fiber, pectin has valuable biological activity, which makes it a potential and high-quality fat replacer for use in different food systems. It has been used in spreads, mayonnaise, salad dressing, manufactured meat, ice cream, cheese, soups, sauces, desserts, and baked foods. Pectin can be used not only to replace fat but also to decrease the content of sugar in food. Pectin (20%) was used to prepare low-fat Frankfurter sausages and the resulting products showed a decrease in hardness with good emulsification stability (Candogan and Kolsarici 2003). A mixture of HMP and LMP from orange peels was also used to replace 80% fat in mayonnaise and 40% fat in pound cake. The addition of pectin in margarine made from palm oil was also reported to give negative correlation to the hardness, cohesiveness, elasticity modulus, and toughness modulus characteristics, but gave positive correlation to adhesiveness and relaxation time characteristics (Ayuningtias et al. 2013).

4.2.5.4 Use of Citrus Pectin as Edible Films

Due to pectin's biodegradability, biocompatibility, edibility, and versatile chemical and physical properties (such as gelation and selective gas permeability), it gained widespread support for the development and application of edible films intended for active food packaging (Espitia et al. 2014). The structure of films with a continuous network using high methoxyl pectin and soy flour was unaffected by surface density. Water vapor permeability increased with the rise in surface density, while kP_{O2} and kP_{CO2} decreased (Giancone et al. 2009). Farris et al. (2011) prepared composite films formed with gelatin and LMP from simultaneous reversible and permanent polyion-complex hydrogels, which created a unique new structure with improved properties and offer potential for tailoring them to a wide range of targeted applications. Galus and Lenart (2013) developed and characterized composite films based on sodium alginate and pectin. Pectin edible films incorporated with natural bioactive compounds have also exhibited antimicrobial properties against foodborne pathogens (Tripathi, Mehrotra, and Dutta 2010). Nisin incorporated into a pectin/polylactic acid film was effective in reducing *Listeria monocytogenes* in a typical growth medium (e.g., orange juice and liquid egg) (Jin et al. 2009).

Vartiainen et al. (2010) made films from fluidized pectin and nanoclay. These hybrid films showed significantly improved barrier properties against oxygen. In addition, the ability to prevent the water vapor transmission was slightly increased and the films were totally impermeable to grease. In a study by Çokaygil, Banar, and Seyhan (2014), orange-peel-derived pectin jelly/cornstarch-based biocomposite films with and without layered silicates were produced. The analysis of x-ray diffraction (XRD), FTIR, differential scanning calorimetry (DSC), thermo-gravimetric analysis (TGA), and scanning electron microscopy (SEM) showed that of all the films tested, pectin jelly/15% modified starch-based biodegradable composite films (54/46 w/w) containing 0.25 wt% of Na-LS emerged as the most promising in terms of texture structure and mechanical integrity. The analysis indicated that biocomposite film had remarkable specifications in terms of elasticity and barrier properties when compared with LDPE film, and can be safely used in some food packaging applications.

4.2.5.5 Other Applications and Future Use

In the food industry, pectin is also use in frozen food, baked food, confectionary products, and dairy products. Pectin (0.2%) can reduce the drip losses up to 36% in the frozen-thawed vegetable curry. It can also increase the storage modulus G′ (elasticity) and loss modulus G″ (viscosity) of treated bread dough through the interactions with bread components such as wheat proteins during dough development and baking. HMP can also be used to stabilize acidified milk drinks against flocculation of milk protein.

In the medical industry, pectin is part of an extensively used natural biodegradable polymer formulations for drug delivery, wound dressing, and tissue engineering. Pectin has numerous benefits in formulations because it can be easily tailored into hydrogels, films, scaffolds, microparticles, and nanoparticles (Mishra et al. 2012). Complex films constructed with pectin as an anionic polyelectrolyte and chitosan as a cationic species showed encouraging potential for a sigmoidal drug delivery system with an initial, controllable slow release followed by a burst release immediately after a change in pH (Ghaffari et al. 2007). Similar work using two different low methoxy citrus pectins and one high-calcium-sensitive pectin as a shell and a hydrophobic oil as core to make microcapsules that would function to control the release of drugs (Muhiddinov et al. 2004). Biopolymer core–shell nanoparticles were fabricated using zein as the core and pectin as the shell (Yusuf 2015). These core–shell biopolymer nanoparticles would be useful for incorporating curcumin into functional foods and beverages, as well as dietary supplements and pharmaceutical products (K. Hu et al. 2015).

Citrus pectin is also a good material for food printing. Traditional foods like rice, meat, fruit, and vegetables, largely consumed by people every day, are not printable by nature. For them to be extrudable, hydrocolloids have to be added to these solid materials. To enable their extrusion capabilities, adding hydrocolloids in these solid materials has been utilized in many culinary fields (Lipton et al. 2010). HMP and LMP, with their different gelling mechanisms, could be potential hydrocolloid material for 3D printing. However, the relevant research using pectin as material in 3D food printing is still quite limited. I. H. Liu, Chang, and Lin (2015) used a 3D plotting system to make chitosan-based tissue scaffolds with interconnected pores using pure chitosan and chitosan cross-linked with pectin and genipin. The scaffolds were stronger, less likely to degrade and better at promoting osteoblast cell proliferation *in vitro* compared to the freeze-dried scaffolds. The results of this study suggested pectin could be a good 3D printing material when combined with other gels.

4.3 CELLULOSE

Cellulose is a tough and water-insoluble substance, which is found in the protective cell walls of plants, particularly in stalks, stems, trunks, and all woody portions of plant. Cellulose, which consists of β-(1→4)-linked glucose repeating units, is the largest renewable biological resources. It is used in a wide range of industrial domains such as fiber, paper, polymer, textile, and food. Cellulose is a highly functionalized, linear stiff-chain homopolymer that is characterized by its chirality, biodegradability, and broad chemical modifying capacity.

The production of citrus juice on an industrial level leads to a considerable quantity of solid and liquid residues (globally around 8–20 million tons per year), which is considered waste or is used as a supplement in agricultural practices. During citrus juice production, only about half of the fresh citrus weight is transformed into juice (Pascual and Carmona 1980), generating a great amount of residue (peel, pulp, seeds, leaves, and whole citrus fruits that do not meet quality requirements), which has a moisture content of approximately 78 wt% (Dolores et al. 1993; Garcia-Castello et al. 2006). In general, citrus juice residues have no economic value, even though they are rich in soluble sugars, cellulose, hemicellulose, pectin, and essential oils that could form the basis of several industrial processes (Rezzadori and Benedett 2012). This huge amount of waste is, in most cases, burned or disposed in landfills (Magín et al. 2008).

It has been reported that the cellulose content in citrus peels (orange peels and lemon peels) ranges from 12.7% to 13.6% and the hemicellulose from 5.3% to 6.1% (Ververis et al. 2007). The insoluble solids from the albedo tissue of orange have 22% cellulose (Prabasari et al. 2011). Extraction of dietary fiber from citrus junos peels contains 29.04% cellulose, 32.21% pectin, and 11.70% hemicellulose. Based on these numbers, citrus and the by-products of citrus processing represent a very good resource for cellulose and cellulose derivatives.

It has been reported that the yield percentage of cellulose production from dried grapefruit peel is 20.04%. Cellulose from grapefruit peel is converted to CMC by alkalization and etherification, and exhibits a thixotropic flow behavior (Karatas and Arslan 2016). Pomelo (*Citrus grandis*) peel is one of the underutilized waste materials that has potential in the production of functional ingredients, due to its high fiber content. The cellulose content in pomelo albedo is about 21.29% and possesses a high level of purity, low crystallinites, and good water-holding capabilities (Zain, Yusop, and Ahmad 2015).

4.3.1 Extraction of Citrus Cellulose

Since the cellulose nanofibrils in fruit tissues are embedded with pectin (Marin et al. 2007), which is mainly composed of galacturonic acid, its isolation requires the implementation of a pectin removal procedure (Habibi et al. 2009; Ifuku et al. 2011). To remove pectin from the citrus juice residue, a multistep chemical treatment consists of three processes: inorganic substance removal, pectin depolymerization, and dissolution. It was reported that this procedure was effective in removing pectin (May 1990). This multistep treatment however was not as useful in the purification of cellulose, since a significant amount of chemical reagents and long reaction times were needed. On the other hand, hydrothermal treatment (Martínez et al. 2010) has emerged as an alternative method to purify cellulose and is a simpler procedure. Most noncellulosic materials such as pectin and hemicellulose were first collected from fruits by these treatments, but residues from pectin and hemicellulose were detected. A more recent study combined multistep and hydrothermal treatments to extract cellulose from mandarin (*Citrus unshiu*) peel waste. The cellulose nanofibrils from mandarin peel waste were uniform at 2 to 3 nm and smaller than that of wood nanofibrils (Hiasa et al. 2014).

A more recent investigation was initiated to study the use of citrus peel waste for the production of bacterial cellulose (BC) by *Komagataeibacter xylinus* CICC No. 10529 and to study the structural properties of the BC films in both citrus peel and pomace enzymolysis (CPPE) and Hestrin-Schramm (HS) media (Fan et al. 2016). The average diameters of BC, obtained from CPPE and HS mediums, were 50 nm and 60 nm, respectively. The crystallinity index of BC from the CPPE medium was approximately 63%, which was lower than BC produced from the HS medium (65%). The two varieties of BC showed no significant differences in relation to their color parameters. Thus, BC production from CPPE medium had similar properties to BC from HS medium, but it is more environmentally friendly and cheaper to produce.

4.3.2 Health Benefits and Applications in Functional Foods

Natural cellulose has a high crystalloid, strong inter- and intramolecular hydrogen bonded structure built by the hydroxy groups and is insoluble in water and common organic solvents. Knowledge of these properties highlighted the need for research and development and many methods were developed to modify cellulose and to expand its usage. Cellulose derivatives (monocarboxy cellulose [MCC], carboxymethyl cellulose [CMC], and so on) have greatly expanded the application of cellulose in the food industry.

Fiber is often classified as soluble dietary fiber (SDF) and insoluble dietary fiber (IDF). Typically, solubility refers simply to fibers that are dispersible in water. The SDF/IDF ratio is important for both dietary and functional properties. It is generally accepted that those fiber sources suitable for use as a food ingredient should have an SDF/IDF ratio close to 1:2. As a kind of dietary IDF, cellulose has been proven to show a lipid-regulating ability (Chou, Chien, and Chau 2008; Krzysik et al. 2011), which is the typical function of dietary fiber. Cellulose binds with bile salts in the duodenum and prevents lipid absorption and reduce lipid uptake (Torcello-Gomez et al. 2015).

The physicochemical properties of dietary fiber can improve the viscosity, texture, sensory characteristics, and shelf life of food products. Cellulose can effectively stabilize oil-in-water emulsions and has considerable potential for use as a new hydrocolloid in the food industry (Jia et al. 2015). In addition, cellulose can also be used to reduce the fat content of food products (Martinez-Cervera et al. 2015). Baked biscuits still have an acceptable texture even after replacing the fat with a cellulose emulsion (Tarancon et al. 2015).

The functional and nutritional properties of cellulose and cellulose derivatives is a growing area of interest for researchers, especially as it applies to the development of healthier foods. Cellulose is not only desirable for its nutritional properties, but also for its functional and technological properties, and because of those properties it can be used to upgrade agricultural products and by-products for use as food ingredients. Furthermore, cellulose can also be used in developing green materials for food packing.

CMC is a cellulose derivative, widely used as a thickener, water binder, extrusion aid, and film former in the pharmaceutical, cosmetics, and food industries. CMC is capable of improving the consistency and flow properties of many beverages (Yasar, Togrul, and Arslan 2007). CMC exerts its effect by increasing viscosity and creating a well-ordered structure in milk-juice drinks, and acts as a moisture binder, emulsion stabilizer, and texture modifier (Yasar et al. 2007). Other applications of CMC are the emulsification of curcumin in CMC for oral administration for its potent anti-inflammatory effect, the incorporation of lysozyme and lactoferrin in cellulose-based food packaging for its antimicrobial activity, and the inclusion of CMC in paper packing for protein payload improvement (Barbiroli et al. 2012).

4.4 CONCLUSIONS

Polysaccharides are the main components belonging to the group of citrus polysaccharides that includes pectin, cellulose, semicellulose, arabinan, and arabinogalactan.

These compounds have been extracted from citrus peel, segments, and pulps. The utilization of natural pectic substances in the medicinal field is always encouraged because of their lack of side effects. In this review, we focused on the polysaccharides from citrus. These polysaccharides are not only important for their wide usage in the food industry as texturizers, stabilizers, and emulsifiers in food products but also for a number of biological reasons, such as the removal of toxic compounds, anticancer activity, antidiabetes activity, and activities that strengthen the immune system.

Pectin is one the most interesting and widely used polysaccharides from citrus. A number of studies have shown that CP and MCP may have genuine anticarcinogenic abilities and suppress the activity of human cancer cells. By analyzing the molecular mechanisms, it is now clear that CP and MCP target and inactivate galectin-3, a cancer cell survival factor; Nf-kβ, a transcription factor involved in tumorigenesis; and many kinases and apoptotic markers, which lead to suppression of numerous human cancers. Animal studies and human clinical studies show that CP and MCP play a major role in detoxification of metals/carcinogens from the body, which could prevent early stages of human malignancies. Future studies should focus on the isolation and evaluation of more of these bioactive pectin components.

Although citrus pectin and its derivatives are usually considered to be safe, and the U.S. Food and Drug Administration generally regards modified citrus pectin as safe, as with any kind of dietary supplement a few adverse effects will be reported. For example, a few reports have stated that modified citrus pectin may cause hypersensitivity and some reports have indicated that the ingredients contained in citrus pectin products have caused an allergic response.

In summary, citrus polysaccharides, because of their wide distribution and high content, as well as their special use in food industry and multiple health benefits, should be given greater consideration and support not only for the development of new products and new applications, but also for expanding the understanding of the biological and physiological mechanisms controlling the activity and efficacies of these bioactive compounds.

ACKNOWLEDGMENT

Financial support from the China Nature Sciences Foundation (Project C200501) is acknowledged.

REFERENCES

Abdullah, A., and Cheng, T. C. 2001. Optimization of reduced calorie tropical mixed fruits jam. *Food Qual Prefer* 12: 63–68.

Acosta, O., Víquez, F., and Cubero, E. 2008. Optimisation of low calorie mixed fruit jelly by response surface methodology. *Food Qual Prefer* 19: 79–85.

Adetunji, L. R., Adekunle, A., Orsat, V., and Raghavan, V. 2017. Advances in the pectin production process using novel extraction techniques: A review. *Food Hydrocoll* 62: 239–250.

Afoakwa, E. O., Nartey, D., Ashong, J., and Annor, G. 2006. Effect of sugar, pectin and acid balance on the quality characteristics of pineapple (*Ananas comosus*) jam. Paper presented at 13th World Food Congress.

Almeida, E. A. M. S., Facchi, S. P., Martins, A. F., Nocchi, S., Schuquel, I. T. A., Nakamura, C. V., Rubira, A. F., and Muniz, E. C. 2015. Synthesis and characterization of pectin derivative with antitumor property against *Caco*-2 colon cancer cells. *Carbohydr Polym* 115: 139–145.

Axelos, M. A. V., and Thibault, J. F. 1991. Influence of the substituents of the carboxyl groups and of the rhamnose content on the solution properties and flexibility of pectins. *Int J Biol Macromol* 13: 77–82.

Ayuningtias, A., Nurhasanah, S., Nurhadi, B., and Subroto, E. 2013. Textural characteristic of margarine enriched with pectin fiber by using blending method. *Int J Advanced Sci Eng Inform Tech* 3: 268–272.

Azémar, M., Hildenbrand, B., Haering, B., and Heim, M. E. 2007. Clinical benefit in patients with advanced solid tumors treated with modified citrus pectin: A prospective pilot study. *Clin Med Oncol* 1: 73–80.

Balaria, A., and Schiewer, S. 2008. Assessment of biosorption mechanism for Pb binding by citrus pectin. *Sep Purif Tech* 63: 577–581.

Barbiroli, A., Bonomi, F., Capretti, G., Iametti, S., Manzoni, M., Piergiovanni, L., and Rollini, M. 2012. Antimicrobial activity of lysozyme and lactoferrin incorporated in cellulose-based food packaging. *Food Control* 26: 387–392.

Bonnin, E., Dolo, E., Goff, A. L., and Thibault, J. F. 2002. Characterisation of pectin subunits released by an optimised combination of enzymes. *Carbohydr Res* 337: 1687–1696.

Brouns, F., Theuwissen, E., Adam, A., Bell, M., Berger, A., and Mensink, R. P. 2011. Cholesterol-lowering properties of different pectin types in mildly hyper-cholesterolemic men and women. *Eur J Clin Nutr* 66: 591–599.

Brown, L., Rosner, B., Willett, W. W., and Sacks, F. M. 1999. Cholesterol-lowering effects of dietary fiber: A meta-analysis. *Am J Clin Nutr* 69: 30–42.

Calce, E., Bugatti, V., Vittoria, V., and De Luca, S. 2012. Solvent-free synthesis of modified pectin compounds promoted by microwave irradiation. *Molecules* 17: 12234–12242.

Cameron, R. G., Luzio, G. A., Vasu, P., Savary, B. J., and Williams, M. A. 2011. Enzymatic modification of a model homogalacturonan with the thermally tolerant pectin methylesterase from citrus: 1. Nanostructural characterization, enzyme mode of action, and effect of pH. *J Agr Food Chem* 59: 2717–2724.

Candela, M., Perna, F., Carnevali, P., Vitali, B., Ciati, R., Gionchetti, P., Rizzello, F., Campieri, M., and Brigidi, P. 2008. Interaction of probiotic *Lactobacillus* and *Bifidobacterium* strains with human intestinal epithelial cells: Adhesion properties, competition against enteropathogens and modulation of IL-8 production. *Int J Food Microbiol* 125: 286–292.

Candogan, K., and Kolsarici, N. 2003. The effects of carrageenan and pectin on some quality characteristics of low-fat beef frankfurters. *Meat Sci* 64: 199–206.

Chauhan, D., Li, G. L., Podar, K., Hideshima, T., Neri, P., He, D. L., Mitsiades, N., Richardson, P., Chang, Y., Schindler, J., Carver, B., and Anderson, K. C. 2005. A novel carbohydrate-based therapeutic GCS-100 overcomes bortezomib resistance and enhances dexamethasone-induced apoptosis in multiple myeloma cells. *Cancer Res* 65: 8350–8358.

Chen, J., Cheng, H., Wu, D., Linhardt, R., Zhi, Z., Yan, L., Chen, S., and Ye, X. 2017. Green recovery of pectic polysaccharides from citrus canning processing water. *J Clean Prod* 144: 459–469.

Chou, S. Y., Chien, P. J., and Chau, C. F. 2008. Particle size reduction effectively enhances the cholesterol-lowering activities of carrot insoluble fiber and cellulose. *J Agr Food Chem* 56: 10994–10998.

Cipriani, T. R., Gracher, A. H., Souza, L. M., Fonseca, R. J., Belmiro, C. L., Gorin, P. A., Sassaki, G. L., and Iacomini, M. 2009. Influence of molecular weight of chemically sulfated citrus pectin fractions on their antithrombotic and bleeding effects. *Thromb Haemost* 101: 860–866.

Ciriminna, R., Chavarría-Hernández, N., and Pagliaro, M. 2015. Pectin: A new perspective from the biorefinery standpoint. *Biofuel Bioprod Bior* 9: 368–377.

Coenen, G. J., Bakx, E. J., Verhoef, R. P., Schols, H. A., and Voragen, A. G. J. 2007. Identification of the connecting linkage between homo- or xylogalacturonan and rhamnogalacturonan type I. *Carbohydr Polym* 70: 224–235.

Çokaygil, Z., Banar, M., and Seyhan, A. T. 2014. Orange peel-derived pectin jelly and corn starch-based biocomposite film with layered silicates. *J Appl Polym Sci* 131: 1–12.

Dolores, M., Martínez-Teruel, A., Gallego, J., and Núñez, J. 1993. Chemical changes during the ensiling of orange peel. *Anim Feed Scie Tech* 43: 269–274.

Dongowski, G., Lorenz, A., and Proll, J. 2002. The degree of methylation influences the degradation of pectin in the intestinal tract of rats and in vitro. *J Nutr* 132: 1935–1944.

Dronnet, V. M., Renard, C. M. G. C., Axelos, M. A. V., and Thibault, J. F. 1996. Characterisation and selectivity of divalent metal ions binding by citrus and sugar-beet pectins. *Carbohydr Polym* 30: 253–263.

Edashige, Y., Murakami, N., and Tsujita, T. 2008. Inhibitory effect of pectin from the segment membrane of citrus fruits on lipase activity. *J Nutr Scie Vitaminol* 54: 409–415.

Eliaz, I., Hotchkiss, A. T., Fishman, M. L., and Rode, D. 2006. The effect of modified citrus pectin on urinary excretion of toxic elements. *Phytother Res* 20: 859–864.

Eliaz, I., Weil, E., and Wilk, B. 2007. Integrative medicine and the role of modified citrus pectin/alginates in heavy metal chelation and detoxification: Five case reports. *Forsche Komp* 14: 358–364.

Espitia, P. J. P., Du, W.-X., de Jesús Avena-Bustillos, R., Soares, N. d. F. F., and McHugh, T. H. 2014. Edible films from pectin: Physical-mechanical and antimicrobial properties—A review. *Food Hydrocoll* 35: 287–296.

Fan, X., Gao, Y., He, W., Hu, H., Tian, M., Wang, K., and Pan, S. 2016. Production of nano bacterial cellulose from beverage industrial waste of citrus peel and pomace using *Komagataeibacter xylinus*. *Carbohydr Polym* 151: 1068–1072.

Farris, S., Schaich, K. M., Liu, L., Cooke, P. H., Piergiovanni, L., and Yam, K. L. 2011. Gelatin–pectin composite films from polyion-complex hydrogels. *Food Hydrocoll* 25: 61–70.

Gajar, A. M., and Badrie, N. 2002. Processing and quality evaluation of a low-calorie Christophene Jam (*Sechium edule* (Jacq.) Swartz. *J Food Sci* 67: 341–346.

Galus, S., and Lenart, A. 2013. Development and characterization of composite edible films based on sodium alginate and pectin. *J Food Eng* 115: 459–465.

Gao, X., Yuan, Z., Tao, Z., Xue, H., Xiao, W., Foday, A. D., Tai, G., and Zhou, Y. 2012. Analysis of the neutral polysaccharide fraction of MCP and its inhibitory activity on galectin-3. *Glycoconjugate J* 29: 159–165.

Garcia-Castello, E., Lora-Garcia, J., Garcia-Garrido, J., and Rodriguez-Lopez, A. D. 2006. International congress on membranes and membrane processes energetic comparison for leaching waste liquid from citric juice production using both reverse osmosis and multiple-effect evaporation. *Desalination* 191: 178–185.

Ghaffari, A., Navaee, K., Oskoui, M., Bayati, K., and Rafiee-Tehrani, M. 2007. Preparation and characterization of free mixed-film of pectin/chitosan/Eudragit® RS intended for sigmoidal drug delivery. *Eur J Pharm Biophar* 67: 175–186.

Giancone, T., Torrieri, E., Masi, P., and Michon, C. 2009. Protein–polysaccharide interactions: Phase behaviour of pectin–soy flour mixture. *Food Hydrocoll* 23: 1263–1269.

Glinskii, O. V., Huxley, V. H., Glinsky, G. V., Pienta, K. J., Raz, A., and Glinsky, V. V. 2005. Mechanical entrapment is insufficient and intercellular adhesion is essential for metastatic cell arrest in distant organs. *Neoplasia* 7: 522–527.

Glinsky, V. V., Huflejt, M. E., Glinsky, G. V., Deutscher, S. L., and Quinn, T. P. 2000. Effects of Thomsen-Friedenreich antigen-specific peptide P-30 on beta-galactoside-mediated homotypic aggregation and adhesion to the endothelium of MDA-MB-435 human breast carcinoma cells. *Cancer Res* 60: 2584–2588.

Glinsky, V. V., and Raz, A. 2009. Modified citrus pectin anti-metastatic properties: One bullet, multiple targets. *Carbohydr Res* 344: 1788–1791.

Guess, B. W., Scholz, M. C., Strum, S. B., Lam, R. Y., Johnson, H. J., and Jennrich, R. I. 2003. Modified citrus pectin (MCP) increases the prostate-specific antigen doubling time in men with prostate cancer: A phase II pilot study. *Prostate Cancer Prostatic Dis* 6: 301–304.

Guo, X., Ye, X., Sun, Y., Wu, D., Wu, N., Hu, Y., and Chen, S. 2014. Ultrasound effects on the degradation kinetics, structure, and antioxidant activity of sea cucumber fucoidan. *J Agr Food Chem* 62: 1088–1095.

Habibi, Y., Mahrouz, M., and Vignon, M. R. 2009. Microfibrillated cellulose from the peel of prickly pear fruits. *Food Chem* 115: 423–429.

Hao, M., Yuan, X., Cheng, H., Xue, H., Zhang, T., Zhou, Y., and Tai, G. 2013. Comparative studies on the anti-tumor activities of high temperature-and pH-modified citrus pectins. *Food Funct* 4: 960–971.

Harsha, M. R., Chandra, P. S. V., and Dharmesh, S. M. 2015. Modified pectic polysaccharide from turmeric (*Curcuma longa*): A potent dietary component against gastric ulcer. *Carbohydr Polym* 138: 143–155.

Hiasa, S., Iwamoto, S., Endo, T., and Edashige, Y. 2014. Isolation of cellulose nanofibrils from mandarin (*Citrus unshiu*) peel waste. *Ind Crop Prod* 62: 280–285.

Hu, K., Huang, X., Gao, Y., Huang, X., Xiao, H., and McClements, D. J. 2015. Core-shell biopolymer nanoparticle delivery systems: Synthesis and characterization of curcumin fortified zein-pectin nanoparticles. *Food Chem* 182: 275–281.

Hu, Y., Ye, X., Yin, X., and Chen, S. 2015. Sulfation of citrus pectin by pyridine-sulfurtrioxide complex and its anticoagulant activity. *LWT—Food Sci Tech* 60: 1162–1167.

Ifuku, S., Adachi, M., Morimoto, M., and Saimoto, H. 2011. Fabrication of cellulose nanofibers from parenchyma cells of pears and apples. *Fiber* 67: 86–90.

Inngjerdingen, M., Inngjerdingen, K. T., Patel, T. R., Allen, S., Chen, X., Rolstad, B., Morris, G. A., Harding, S. E., Michaelsen, T. E., and Diallo, D. 2008. Pectic polysaccharides from Biophytum petersianum Klotzsch, and their activation of macrophages and dendritic cells. *Glycobiology* 18: 1074–1084.

Inohara, H., and Raz, A. 1994. Effects of natural complex carbohydrate (citrus pectin) on murine melanoma cell properties related to galectin-3 functions. *Glycoconjugate J* 11: 527–532.

Ishii, T., Matsunaga, T., and Hayashi, N. 2001. Formation of rhamnogalacturonan II-borate dimer in pectin determines cell wall thickness of pumpkin tissue. *Plant Physiol* 126: 1698–1705.

Jackson, C. L., Dreaden, T. M., Theobald, L. K., Tran, N. M., Beal, T. L., Eid, M., Gao, M. Y., Shirley, R. B., Stoffel, M. T., Kumar, M. V., and Mohnen, D. 2007. Pectin induces apoptosis in human prostate cancer cells: Correlation of apoptotic function with pectin-structure. *Glycobiology* 17(8): 805–819.

Jeong, K. Y., Lee, E. J., Kim, S. J., Yang, S. H., Sung, Y. C., and Seong, J. 2015. Irradiation-induced localization of IL-12-expressing mesenchymal stem cells to enhance the curative effect in murine metastatic hepatoma. *Int J Cancer* 137: 721–730.

Jia, X. J., Xu, R. R., Shen, W., Xie, M. X., Abid, M., Jabbar, S., Wang, P., Zeng, X. X., and Wu, T. 2015. Stabilizing oil-in-water emulsion with amorphous cellulose. *Food Hydrocoll* 43: 275–282.

Jin, T., Liu, L., Zhang, H., and Hicks, K. 2009. Antimicrobial activity of nisin incorporated in pectin and polylactic acid composite films against *Listeria monocytogenes*. *Int J Food Sci Tech* 44: 322–329.

Johnson, K. D., Glinskii, O. V., Mossine, V. V., Turk, J. R., Mawhinney, T. P., Anthony, D. C., Henry, C. J., Huxley, V. H., Glinsky, G. V., and Pienta, K. J. 2007. Galectin-3 as a potential therapeutic target in tumors arising from malignant endothelia 1. *Neoplasia* 9: 662–670.

Karatas, M., and Arslan, N. 2016. Flow behaviours of cellulose and carboxymethyl cellulose from grapefruit peel. *Food Hydrocoll* 58: 235–245.

Kartel, M. T., Kupchik, L. A., and Veisov, B. K. 1999. Evaluation of pectin binding of heavy metal ions in aqueous solutions. *Chemosphere* 38: 2591–2596.

Kay, R. M., and Truswell, A. S. 1977. Effect of citrus pectin on blood lipids and fecal steroid excretion in man. *Am J Clin Nutr* 30: 171–175.

Khotimchenko, M., Kovalev, V., Kolenchenko, E., and Khotimchenko, Y. 2012. Acidic method for the low molecular pectin preparation. *Int J Pharm Pharm Sci* 4: 279–283.

Khouryieh, H. A., Aramouni, F. M., and Herald, T. J. 2005. Physical, chemical and sensory properties of sugar-free jelly. *J Food Quality* 28: 179–190.

Kidd, P. M. 1996. A new approach to metastatic cancer prevention: Modified citrus pectin (MCP), a unique pectin that blocks cell surface lectins. *Altern Med Rev* 1: 4–10.

Knaup, B., Kempf, M., Fuchs, J., Valotis, A., Kahle, K., Oehme, A., Richling, E., and Schreier, P. 2008. Model experiments mimicking the human intestinal transit and metabolism of D-galacturonic acid and amidated pectin. *Mol Nutr Food Res* 52: 840–848.

Kokkonen, H., Ilversao, J. M., Morra, M., Schols, H. A., and Tuukkanen, J. 2007. Effect of modified pectin molecules on the growth of bone cells. *Biomacromolecules* 8: 509–515.

Kolatsijoannou, M., Price, K. L., Winyard, P. J., and Long, D. A. 2011. Modified citrus pectin reduces galectin-3 expression and disease severity in experimental acute kidney injury. *Plos One* 6: e18683.

Krzysik, M., Grajeta, H., Prescha, A., and Weber, R. 2011. Effect of cellulose, pectin and chromium(III) on lipid and carbohydrate metabolism in rats. *J Trace Elem Med Bio* 25: 97–102.

Lattimer, J. M., and Haub, M. D. 2010. Effects of dietary fiber and its components on metabolic health. *Nutrients* 2: 1266–1289.

Lawaetz, O., Blackburn, A. M., Bloom, S. R., Aritas, Y., and Ralphs, D. N. 1983. Effect of pectin on gastric emptying and gut hormone release in the dumping syndrome. *Scand J Gastroenterol* 18: 327–336.

Leclere, L., Fransolet, M., Cambier, P., Bkassiny, S. E., Tikad, A., Dieu, M., Vincent, S. P., Cutsem, P. V., and Michiels, C. 2015. Identification of a cytotoxic molecule in heat-modified citrus pectin. *Carbohydr Polym* 137: 39–51.

Leclere, L., Van Cutsem, P., and Michiels, C. 2013. Anti-cancer activities of pH-or heat-modified pectin. *Front Pharmacol* 4: 128: 1–8.

Li, F. T., Yang, H., Zhao, Y., and Xu, R. 2007. Novel modified pectin for heavy metal adsorption. *Chinese Chem Lett* 18: 325–328.

Lipton, J., Arnold, D., Nigl, F., Lopez, N., Cohen, D., Norén, N., and Lipson, H. 2010. Multi-material food printing with complex internal structure suitable for conventional post-processing. *Solid Freeform Fabrication Symposium* (pp. 809–815).

Liu, F. T., Hsu, D. K., Zuberi, R. I., Kuwabara, I., Chi, E. Y., and Henderson, W. R. 1995. Expression and function of galectin-3, a beta-galactoside-binding lectin, in human monocytes and macrophages. *Am J Pathol* 147: 1016–1028.

Liu, H. Y., Huang, Z. L., Yang, G. H., and Lu, W. Q. 2008. Inhibitory effect of modified citrus pectin on liver metastases in a mouse colon cancer model. *World J Gastroenterol* 14: 7386–7391.

Liu, I. H., Chang, S. H., and Lin, H. Y. 2015. Chitosan-based hydrogel tissue scaffolds made by 3D plotting promotes osteoblast proliferation and mineralization. *Biomed Mater* 10: 035004.

Liu, L., Fishman, M. L., Hicks, K. B., Kende, M., and Ruthel, G. 2008. Pectin/zein beads for potential colon-specific drug delivery: Synthesis and in vitro evaluation. *Drug Deliver* 13: 417–423.

Liu, Y., Shi, J., and Langrish, T. A. G. 2006. Water-based extraction of pectin from flavedo and albedo of orange peels. *Chem Eng J* 120: 203–209.

Luo, Y., Kang, P., and Zhong, Q. 2015. Casein/pectin nanocomplexes as potential oral delivery vehicles. *Int J Pharm* 486: 59–68.

Ma, X., Zhang, L., Wang, W., Zou, M., Ding, T., Ye, X., and Liu, D. 2016. Synergistic effect and mechanisms of combining ultrasound and pectinase on pectin hydrolysis. *Food Bioprocess Tech* 9: 1249–1257.

Maas, N. C., Gracher, A. H. P., Sassaki, G. L., Gorin, P. A., Iacomini, M., and Cipriani, T. R. 2012. Sulfation pattern of citrus pectin and its carboxy-reduced derivatives: Influence on anticoagulant and antithrombotic effects. *Carbohydr Polym* 89: 1081–1087.

Magín, L. J., Rodriguez-Fernandez, J., Agudelo, J. R. 2008. Diesel particulate emissions from used cooking oil biodiesel. *Bioresource Technol* 99: 731–740.

Marin, F. R., Soler-Rivas, C., Benavente-Garcia, O., Castillo, J., and Perez-Alvarez, J. A. 2007. By-products from different citrus processes as a source of customized functional fibres. *Food Chem* 100: 736–741.

Martínez, M., Gullón, B., Yáñez, R., Alonso, J. L., and Parajó, J. C. 2010. Kinetic assessment on the autohydrolysis of pectin-rich by-products. *Chem Eng J* 162: 480–486.

Martinez-Cervera, S., Salvador, A., and Sanz, T. 2015. Cellulose ether emulsions as fat replacers in muffins: Rheological, thermal and textural properties. *LWT—Food Sci Technol* 63: 1083–1090.

Matsumoto, T., Moriya, M., Sakurai, M. H., Kiyohara, H., Tabuchi, Y., and Yamada, H. 2008. Stimulatory effect of a pectic polysaccharide from a medicinal herb, the roots of *Bupleurum falcatum* L., on G-CSF secretion from intestinal epithelial cells. *Int Immunopharmacol* 8: 581–588.

May, C. D. 1990. Industrial pectins: Sources, production and applications. *Carbohydr Polym* 12: 79–99.

May, C. D., Phillips, G. O., and Williams, P. A. 1997. Pectins. In *Thickening and Gelling Agents for Food*, A. Imeson, ed. London: Blackie Academic & Professional, 124–152.

Mishra, R., Banthia, A., and Majeed, A. 2012. Pectin based formulations for biomedical applications: A review. *Asian J Pharm Clin Res* 5: 1–7.

Mohnen, D. 2008. Pectin structure and biosynthesis. *Curr Opin Plant Biol* 11: 266–277.

Morris, V. J., Belshaw, N. J., Waldron, K. W., and Maxwell, E. G. 2013. The bioactivity of modified pectin fragments. *Bioact Carbohydr Diet Fibre* 1: 21–37.

Muhiddinov, Z., Khalikov, D., Speaker, T., and Fassihi, R. 2004. Development and characterization of different low methoxy pectin microcapsules by an emulsion–interface reaction technique. *J Microencapsul* 21: 729–741.

Nangia-Makker, P., Hogan, V., Honjo, Y., Baccarini, S., Tait, L., Bresalier, R., and Raz, A. 2002. Inhibition of human cancer cell growth and metastasis in nude mice by oral intake of modified citrus pectin. *J Nat Cancer Ins* 94: 1854–1862.

Nicholas, J. 2009. Fighting cancer metastasis and heavy metal toxicities with modified citrus pectin. *Life Extension* 5.

Niture, S. K., and Refai, L. 2013. Plant pectin: A potential source for cancer suppression. *Am J Pharmacol Toxicol* 8: 9–19.

Olano-Martin, E., Gibson, G. R., and Rastell, R. A. 2002. Comparison of the in vitro bifidogenic properties of pectins and pectic-oligosaccharides. *J Appl Microbiol* 93: 505–511.

Onumpai, C., Kolida, S., Bonnin, E., and Rastall, R. A. 2011. Microbial utilization and selectivity of pectin fractions with various structures. *Appl Environ Microbiol* 77: 5747–5754.

Pascual, J. M., and Carmona, J. F. 1980. Composition of citrus pulp. *Anim Feed Sci Technol* 5: 1–10.

Pelkman, C. L., Navia, J. L., Miller, A. E., and Pohle, R. J. 2007. Novel calcium-gelled, alginate-pectin beverage reduced energy intake in nondieting overweight and obese women: Interactions with dietary restraint status. *Am J Clin Nutr* 86: 1595–1602.

Perrigue, M., Carter, B., Roberts, S. A., and Drewnowski, A. 2010. A low-calorie beverage supplemented with low-viscosity pectin reduces energy intake at a subsequent meal. *J Food Sci* 75: 300–305.

Pienta, K. J., Naik, H., Akhtar, A., Yamazaki, K., Replogle, T. S., Lehr, J., Donat, T. L., Tait, L., Hogan, V., and Raz, A. 1995. Inhibition of spontaneous metastasis in a rat prostate cancer model by oral administration of modified citrus pectin. *J Nat Cancer Inst* 87: 348–353.

Platt, D., and Raz, A. 1992. Modulation of the lung colonization of B16-F1 melanoma cells by citrus pectin. *J Nat Cancer Inst* 84: 438–442.

Poiana, M. A., Alexa, E., and Mateescu, C. 2012. Tracking antioxidant properties and color changes in low-sugar bilberry jam as effect of processing, storage and pectin concentration. *Chem Cent J* 6: 1–11.

Prabasari, I., Pettolino, F., Liao, M. L., and Bacic, A. 2011. Pectic polysaccharides from mature orange (*Citrus sinensis*) fruit albedo cell walls: Sequential extraction and chemical characterization. *Carbohydr Polym* 84: 484–494.

Rabbani, G. H., Teka, T., Zaman, B., Majid, N., Khatun, M., and Fuchs, G. J. 2001. Clinical studies in persistent diarrhea: Dietary management with green banana or pectin in Bangladeshi children. *Gastroenterology* 121: 554–560.

Ralet, M. C., and Thibault, J. F. 1994. Extraction and characterisation of very highly methylated pectins from lemon cell walls. *Carbohydr Res* 260: 283–296.

Ramachandran, C., Wilk, B. J., Hotchkiss, A., Chau, H., Eliaz, I., and Melnick, S. J. 2011. Activation of human T-helper/inducer cell, T-cytotoxic cell, B-cell, and natural killer (NK)-cells and induction of natural killer cell activity against K562 chronic myeloid leukemia cells with modified citrus pectin. *BMC Complem Altern Med* 11: 1–9.

Raz, A., and Pienta, K. J. 1998. Method for inhibiting cancer metastasis by oral administration of soluble modified citrus pectin. US Patent US5834442 A.

Ren, J. N., Song, S. Y., Tai, Y. N., Xu, M., Dong, M., Yang, Z. Y., Wu, H. J., Pan, S. Y., and Fan, G. 2015. Suspension stability of an orange juice beverage containing sacs. *Modern Food Sci Technol* 31: 277–283.

Rezzadori K., and Benedett, S. 2012. Proposals for the residues recovery: Orange waste as raw material for new products. *Food Bioprod Process* 90: 606–614.

Ros, J. M., Schols, H. A., and Voragen, A. G. J. 1996. Extraction, characterisation, and enzymatic degradation of lemon peel pectins. *Carbohydr Res* 282: 271–284.

Sakurai, M. H., Matsumoto, T., Kiyohara, H., and Yamada, H. 1996. Detection and tissue distribution of anti-ulcer pectic polysaccharides from *Bupleurum falcatum* by a polyclonal antibody. *Planta Med* 62: 341–346.

Salman, H., Bergman, M., Djaldetti, M., Orlin, J., and Bessler, H. 2008. Citrus pectin affects cytokine production by human peripheral blood mononuclear cells. *Biomed Pharmacother* 62: 579–582.

Sanaka, M., Yamamoto, T., Anjiki, H., Nagasawa, K., and Kuyama, Y. 2007. Effects of agar and pectin on gastric emptying and post-prandial glycaemic profiles in healthy human volunteers. *Clin Exp Pharmacol Physiol* 34: 1151–1155.

Sathisha, U. V., Jayaram, S., Nayaka, M. A. H., and Dharmesh, S. M. 2007. Inhibition of galectin-3 mediated cellular interactions by pectic polysaccharides from dietary sources. *Glycoconjugate J* 24: 497–507.

Shanmugam, A., and Ashokkumar, M. 2015. Characterization of ultrasonically prepared flaxseed oil enriched beverage/carrot juice emulsions and process-induced changes to the functional properties of carrot juice. *Food Bioprocess Technol* 8: 1258–1266.

Shaw, J. E., Sicree, R. A., and Zimmet, P. Z. 2010. Global estimates of the prevalence of diabetes for 2010 and 2030. *Diabetes Res Clin Pract* 87: 4–14.

Sutar, P. B. M., R. K., Pal, K., Banthia, A. K. 2008. Development of pH sensitive polyacrylamide grafted pectin hydrogel for controlled drug delivery system. *J Mater Sci* 19: 2247–2253.

Tamaki, Y., Konishi, T., Fukuta, M., and Tako, M. 2008. Isolation and structural characterisation of pectin from endocarp of *Citrus depressa*. *Food Chem* 107: 352–361.

Tarancon, P., Hernandez, M. J., Salvador, A., and Sanz, T. 2015. Relevance of creep and oscillatory tests for understanding how cellulose emulsions function as fat replacers in biscuits. *LWT—Food Sci Technol* 62: 640–646.

Thakur, B. R., Singh, R. K., and Handa, A. K. 1997. Chemistry and uses of pectin: A review. *Critical Rev Food Sci Nutr* 37: 47–73.

Tian, L., Bruggeman, G., Berg, M. V. D., Borewicz, K., Scheurink, A. J. W., Bruininx, E., Vos, P. D., Smidt, H., Schols, H. A., and Gruppen, H. 2017. Effects of pectin on fermentation characteristics, carbohydrate utilization and microbial community composition in the gastrointestinal tract of weaning pigs. *Mol Nutri Food Res* 61: 1–10.

Tiwary, C. M., Ward, J. A., and Jackson, B. A. 1997. Effect of pectin on satiety in healthy US army adults. *J Am Coll Nutr* 16: 423–428.

Torcello-Gomez, A., Fraguas, C. F., Ridout, M. J., Woodward, N. C., Wilde, P. J., and Foster, T. J. 2015. Effect of substituent pattern and molecular weight of cellulose ethers on interactions with different bile salts. *Food Funct* 6: 730–739.

Tripathi, S., Mehrotra, G., and Dutta, P. 2010. Preparation and physicochemical evaluation of chitosan/poly(vinyl alcohol)/pectin ternary film for food-packaging applications. *Carbohydr Polym* 79: 711–716.

Triplehorn, C., and Millard, P. S. 2002. A rice-based diet with green banana or pectin reduced diarrhea in infants better than a rice-alone diet. *ACP J Club* 136: 67.

Vartiainen, J., Tammelin, T., Pere, J., Tapper, U., and Harlin, A. 2010. Biohybrid barrier films from fluidized pectin and nanoclay. *Carbohydr Polym* 82: 989–996.

Ververis, C., Georghiou, K., Danielidis, D., Hatzinikolaou, D. G., Santas, P., Santas, R., and Corleti, V. 2007. Cellulose, hemicelluloses, lignin and ash content of some organic materials and their suitability for use as paper pulp supplements. *Bioresource Technol* 98: 296–301.

Videcoq, P., Garnier, C., Robert, P., and Bonnin, E. 2011. Influence of calcium on pectin methylesterase behaviour in the presence of medium methylated pectins. *Carbohydr Polym* 86: 1657–1664.

Vityazev, F., Golovchenko, V., Patova, O., Drozd, N., Makarov, V., Shashkov, A., and Ovodov, Y. S. 2010. Synthesis of sulfated pectins and their anticoagulant activity. *Biochemistry (Moscow)* 75: 759–768.

Vos, A. P., Esch, B. C. V., Stahl, B., M'Rabet, L., Folkerts, G., Nijkamp, F. P., and Garssen, J. 2007a. Dietary supplementation with specific oligosaccharide mixtures decreases parameters of allergic asthma in mice. *Int Immunopharmacol* 7: 1582–1587.

Vos, A. P., Haarman, M., Van Ginkel, J. W., Knol, J., Garssen, J., Stahl, B., Boehm, G., and M'Rabet, L. 2007b. Dietary supplementation of neutral and acidic oligosaccharides enhances Th1-dependent vaccination responses in mice. *Pediatr Allergy Immunol* 18: 304–312.

Whistler, R. L., and Bemiller, J. N., eds. 1993. *Industrial gums: Polysaccharides and their derivatives*. San Diego, CA: Academic Press.

Wicker, L., Kim, Y., Kim, M.-J., Thirkield, B., Lin, Z., and Jung, J. 2014. Pectin as a bioactive polysaccharide—Extracting tailored function from less. *Food Hydrocoll* 42: 251–259.

Wong, J. M., De Souza, R., Kendall, C. W., Emam, A., and Jenkins, D. J. 2006. Colonic health: Fermentation and short chain fatty acids. *J Clin Gastroenterol* 40: 235–243.

Yadav, S., Yadav, P. K., Yadav, D., and Yadav, K. D. S. 2009. Pectin lyase: A review. *Process Biochem* 44: 1–10.

Yan, J., and Katz, A. 2010. PectaSol-C modified citrus pectin induces apoptosis and inhibition of proliferation in human and mouse androgen-dependent and -independent prostate cancer cells. *Integ Cancer Ther* 9: 197–203.

Yapo, B. M., Lerouge, P., Thibault, J. F., and Ralet, M. C. 2007. Pectins from citrus peel cell walls contain homogalacturonans homogenous with respect to molar mass, rhamnogalacturonan I and rhamnogalacturonan II. *Carbohydr Polym* 69: 426–435.

Yasar, F., Togrul, H., and Arslan, N. 2007. Flow properties of cellulose and carboxymethyl cellulose from orange peel. *J Food Eng* 81: 187–199.

Ye, X. Q., Chen, J. L., Jin, M. R., Cheng, H., Chen, S. G., and Liu, D. H. 2015. Advances in Pectin Modification Methods and Biological Mechanisms. *Chin J Food Sci* 15: 1–9.

Yusuf, K. 2015. Novel use of pectin as a microneedle base. *Chem Pharm Bull* 63: 300–304.

Zain, N., Yusop, S. M., and Ahmad, I. 2015. Preparation and characterization of cellulose and nanocellulose from pomelo (*Citrus grandis*) albedo. *J Nutr Food Sci* 5: 334.

Zhang, L., Ye, X., Xue, S. J., Zhang, X., Liu, D., Meng, R., and Chen, S. 2013. Effect of high-intensity ultrasound on the physicochemical properties and nanostructure of citrus pectin. *J Sci Food Agr* 93: 2028–2036.

Zhang, Y., Lin, J., and Zhong, Q. 2015. The increased viability of probiotic *Lactobacillus salivarius* NRRL B-30514 encapsulated in emulsions with multiple lipid-protein-pectin layers. *Food Res Int* 71: 9–15.

Zhao, Z., Liang, L., Fan, X., Yu, Z., Hotchkiss, A. T., Wilk, B. J., and Eliaz, I. 2008. The role of modified citrus pectin as an effective chelator of lead in children hospitalized with toxic lead levels. *Altern Ther Health Med* 14: 34–38.

5 Phenolic Compounds and Bioactive Agents in Citrus Fruits

Guihua Xu, Donghong Liu, Yaqin Ma, and Tian Ding

CONTENTS

5.1 INTRODUCTION: POLYPHENOLS IN CITRUS FRUITS

It is widely accepted that fruits and vegetables can have a strong beneficial impact on human health. This is based in part on overwhelming findings from epidemiological and clinical studies that have shown a strong relationship between high consumption of fruits and vegetables and a lower incidence of degenerative diseases, including cancer, heart disease, inflammation, arthritis, immune system decline, brain dysfunction, and cataracts (Kaur and Kapoor 2001; Scalbert and Williamson 2000). The positive influence of a diet rich in fruits and vegetables is attributed to the naturally occurring antioxidants in them, including vitamin C, vitamin E, phenolic compounds, and carotenoids. These phytochemicals have antioxidant activities which help protect cells against oxidative damage caused by free radicals (Kris-Etherton et al. 2002).

More than 8000 naturally occurring phenolic compounds have been identified (Balasundram et al. 2006), and they can be divided into several classes according to their structure (Table 5.1). It should be pointed out that phenolic compounds are usually present in the form of esters or glycosides rather than as free phenols. Phenolic compounds in the diet generally include flavonoids, phenolic acids, and tannins (King and Young 1999). As for citrus fruits, the major phenolic compounds are normally flavonoids and phenolic acids.

TABLE 5.1
Classes and Structures of Phenolic Compounds

Class	Structure
Simple phenolics, benzoquinones	C_6
Hydroxybenzoic acids	C_6-C_1
Phenylacetic acids	C_6-C_2
Hydroxycinnamic acids	C_6-C_3
Napthoquinones	C_6-C_4
Xanthones	C_6-C_1-C_6
Anthraquinones	C_6-C_2-C_6
Flavonoids	C_6-C_3-C_6
Lignans, neolignans	$(C_6$-$C_3)_2$
Biflavonoids	$(C_6$-C_3-$C_6)_2$
Lignins	$(C_6$-$C_3)_n$
Condensed tannins	$(C_6$-C_3-$C_6)_n$

5.2 PHENOLIC ACIDS

Phenolic acids (Figure 5.1) are widely distributed in fruits and vegetables and carry out many physiological functions, such as antioxidant, anti-inflammatory, and anti-mutation activities; they also impede or prevent cardiovascular disease (Morton et al. 2000). Fruits and vegetables are a rich source of phenolic compounds, with flavonoids accounting for about two-thirds and phenolic acids one-third of total phenolic compound content. It is estimated that the daily intake of phenolic acids ranges from

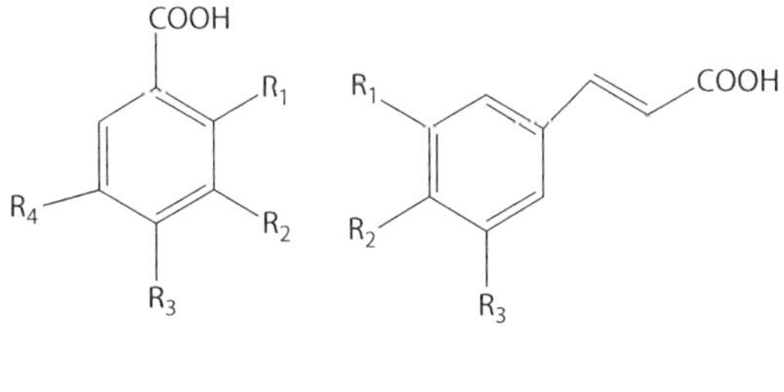

Benzoic acid derivatives

Benzoic acid: R_1=R_2=R_3=R_4=H
Salicylic acid: R_2=R_3=R_4=H R_1=OH
Gentisic acid: R_2=R_3=H R_1=R_4=OH
Vanillic acid: R_1=R_4=H R_2=OH R_3=OMe
Gallic acid: R_1=H R_2=R_3=R_4=OH
Syringic acid: R_1=H R_3=OH R_2=R_4=OMe

Cinnamic acid derivatives

Cinamic acid: R_1=R_2=R_3=H
Coumaric acid: R_2=OH R_1=R_3=H
Caffeic acid: R_1=R_2=OH R_3=H
Ferulic acid: R_1=OMe R_2=OH R_3=H
Sinapic acid: R_1=R_3=OMe R_2=O H

FIGURE 5.1 Chemical structures of phenolic acids in citrus fruits.

25 mg to 1 g; the intake is mainly dependent on the type and quantity of the fruits and vegetables consumed (Clifford 1999).

5.2.1 Metabolism and Physiological Functions of Phenolic Acids

A great deal of recent attention has been given to the metabolism of phenolic acid and its physiological functions. Rechner et al. investigated the absorption and metabolism of hydroxycinnamates from artichokes (Rechner et al. 2001). Their results indicated that the caffeic acid esters present in encapsulated artichoke extract were absorbed, metabolized, and excreted as methylated phenolic acids, that is, ferulic, isoferulic, dihydroferulic, and vanillic acids. Andreasen et al. (2001) reported that diferulic acids were potent antioxidants and were abundant structural components of the plant cell wall, but these phenolics were ester linked to cell wall polysaccharides and could not be absorbed in that form. Their study showed that the diferulic acids could be released from cereal brans via digestion by intestinal enzymes and that free diferulic acid was absorbed and entered the circulatory system. Plumb et al. (1999) suggested that chlorogenic acid ingested by humans was most likely cleaved into caffeic acid and quinic acid by an esterase enzyme(s) provided by the colonic microflora. Olthof et al. (2001) suggested that one-third of chlorogenic acid and almost all of the caffeic acid were absorbed in the small intestine of humans. Koshihara et al. (1984) found that caffeic acid was a selective inhibitor for 5-lipoxygenase and as a consequence inhibited leukotriene biosynthesis. Leukotrienes are significantly involved in immunoregulation and in a variety of disease conditions, including asthma, inflammation, and various allergies. Rao et al. (1993) indicated that caffeic acid esters were potent inhibitors of human colon tumor cell growth, which suggested that these compounds might possess antitumor activity against colon cancer. Jayaprakasam et al. (2006) synthesized a series of ferulic and caffeic acid esters and tested the compounds for tumor cell proliferation, cyclooxygenase enzymes (COX-1 and COX-2), and lipid peroxidation inhibitory activities *in vitro*. Dietary supplementation of caffeic acid (0.2 and 0.8%, wt/wt) in rats resulted in a statistically significant increase of α-tocopherol in both plasma and lipoprotein. Caffeic acid was not detectable in plasma under fasting conditions. The results demonstrated the physiological relevance of caffeic acid and its antioxidant action *in vivo*, through a direct contribution to the antioxidant defense system and a sparing effect on α-tocopherol (Nardini et al. 1997).

The antioxidant capacity of phenolic acids has been extensively investigated. Gulcin (2006) reported that caffeic acid was an effective antioxidant when they evaluated it for 2,2′-azino-bis(3-ethylbenzothiazoline-6-sulfonic acid) [ABTS(+)] scavenging, α,α-diphenyl-β-picrylhydrazyl (DPPH) scavenging, superoxide anion radical scavenging, total reducing power, metal chelating activities, and when tested via the ferric thiocyanate method. Chen and Ho (1997) compared the antioxidant and free radical scavenging activities of caffeic acid, caffeic acid phenethyl ester, ferulic acid, ferulic acid phenethyl ester, rosmarinic acid, and chlorogenic acid with those of α-tocopherol and butylated hydroxytoluene (BHT). They found that the antioxidant capacity of chlorogenic acid was lower than that of caffeic acid, indicating that esterification reduced the antioxidant activity of caffeic acid.

Saija et al. suggested that caffeic and ferulic acids would be good candidates for employment as topical protective agents against UV irradiation-induced skin damage (Saija et al. 1999, 2000). Castelluccio et al. (1996) suggested that hydroxycinnamates were effective in enhancing the resistance of low-density lipoprotein (LDL) to oxidation, in the order caffeic acid > ferulic acid > *p*-coumaric acid, and ferulic acid was a more effective antioxidant against LDL oxidation than the hydrophilic antioxidant ascorbic acid. Lodovici et al. (2001) examined the antioxidant activity of 3-OH-benzoic acid, 4-OH-benzoic acid, 2,3-dihydroxybenzoic acid, 3,4-dihydroxybenzoic acid, ferulic acid, caffeic acid, and 2-coumaric, 3-coumaric, and 4-coumaric acids. Their results suggested that some natural phenolic acids had interesting protective activities against DNA oxidation *in vitro*.

5.3 FLAVANONES

5.3.1 Types and Chemical Structures of Citrus Flavonoids

Flavonoids are important biological substances in oranges, limes, lemons, citrons, mandarins, grapefruits, and other citrus fruits. According to their chemical structures, citrus flavonoids can be divided into four categories: flavanones, flavones, flavonols (Figure 5.2), and anthocyanin (found only in blood oranges). More than 60 flavonoids have been identified in citrus fruits.

Flavanones are flavonoids that are glycosylated with a disaccharide at position 7, thus creating flavanone glycosides. Flavanones are present in the largest amounts, with naringenin and hesperetin representing the major contributers. They exist in the form of glycosides (and not the aglycone form) in all citrus fruits. Among the flavanone-*O*-glycoside forms, two predominant types have been classified: neohesperidosides and rutinosides. Naringin, neohesperidin, and neoeriocitrin belong to the neohesperidosides and have a bitter taste. Hesperidin, narirutin, and didymin belong to rutinosides and do not possess a bitter taste. It is worth mentioning that naringin and neohesperidin can be hydrogenated into dihydrochalcone, which has a sweetness level 300 to 1000 times higher than sucrose (Del Río et al. 1997). Peterson et al. (2006) summarized the content and composition of flavanone glycosides in citrus fruits.

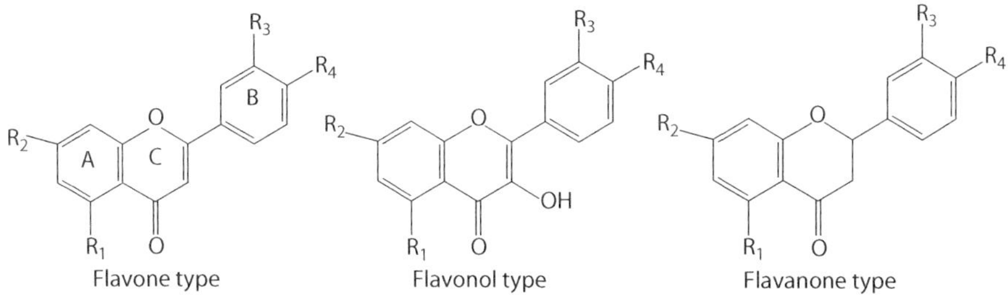

FIGURE 5.2 Chemical structures of flavone, flavonol, and flavanone.

5.3.2 Antioxidant Capacity of Flavanone Glycosides

In recent years, many researchers have been actively studying the antioxidant capacities of flavanone glycosides. Di Majo et al. (2005) elucidated the antioxidant and pro-oxidant behaviors of some common flavanones and determined their structure–activity relationships by using the crocin bleaching inhibition assay. The data revealed that the replacement of the 7-OH group of the flavanones with an *O*-neohesperidose influenced the relationship between structure and antioxidant activity. When in the glycosylated form, the 3′,4′-catechol structure noticeably increased the antioxidant power, and *O*-methylation decreased the antioxidant activity. Wilmsen et al. (2005) found that hesperidin significantly reduced the level of the free radical DPPH with an efficacy of similar to that of trolox (the positive control). Yu et al. (2005) adopted a variety of *in vitro* models, measuring compounds such as β-carotene–linoleic acid, DPPH, superoxide, and hamster LDL to measure the antioxidant activity of 11 citrus bioactive compounds: 2 limonoids, 8 flavonoids, and 1 coumarin. Their results indicated that flavonoids containing a chromanol ring system had stronger antioxidant activities compared to limonoids and coumarin, which lack the hydroxy groups. Kim and Lee (2004) investigated a wide range of natural and synthetic polyphenolics in a vitamin C equivalent antioxidant capacity (VCEAC) assay by using free blue/green ABTS radicals. The tested polyphenolics were grouped into the following categories: vitamins (β-carotene, α-tocopherol, vitamin A, and vitamin C), phenolic acids (benzoic acid, phenylacetic acid, cinnamic acid, and their derivatives), flavonoids (anthocyanidin, flavanol, chalcone, flavanone, flavone, flavonol, isoflavone, and their derivatives), synthetic food additives (butylated hydroxyanisole, BHT, tert-butylhydroquinone, and propylene glycol), and other miscellaneous polyphenolics (ellagic acid, sesamol, eugenol, thymol, and so on). Their results showed a positive linear relationship between VCEAC and the number of free OH groups around the flavonoid framework. The polyphenolics typically had higher VCEAC than monophenolics, and the glycosylated flavonoids showed less-potent antioxidant capacities than their aglycone forms.

5.3.3 Metabolism and Physiological Functions of Flavanones

Hsiu et al. (2002) compared the metabolic pharmacokinetics of naringin and naringenin in rabbits. Naringenin was administered intravenously and orally to rabbits, and naringin was administered orally. Their results showed that the absolute bioavailability of oral naringenin was only 4%, whereas after taking the conjugated naringenin into account, it increased to 8%. When naringin was administered orally, very little naringenin was detected, but its glucuronides/sulfates were found circulating in the plasma.

Miyake et al. (2000) reported on flavonoid metabolites in plasma and urine of eriocitrin-treated rats. Their results suggested that eriocitrin was metabolized by intestinal bacteria and transformed into eriodictyol and 3,4-dihydroxyhydrocinnamic metabolites before being absorbed. Following administration of eriocitrin, plasma in the rats exhibited greater resistance to lipid peroxidation. Abe et al. (1993) investigated the structures of biliary metabolites of orally administered hesperetin in rats. Four isolated metabolites were identified (hesperetin 3′-*O*-β-D-glucuronide, hesperetin 7-*O*-β-D-glucuronide, hesperetin 7-*O*-sulfate-3′-*O*-β-D-glucuronide, and hesperetin

7,3′-di-*O*-sulfate) on the basis of chemical and spectroscopic evidence. Manach et al. (2003) studied the bioavailability of the flavanones hesperidin and narirutin in humans after the ingestion of two doses of orange juice. Flavanone metabolites appeared in plasma 3 h after ingesting the juice, reached a maximum level between 5 and 7 h and then returned to baseline at 24 h. Nielsen et al. (2006) demonstrated that the bioavailability of hesperidin was modulated by its enzymatic conversion to hesperetin-7-glucoside, which changed the absorption site from the colon to the small intestine. These findings may affect future interventions concerning the health benefits of citrus flavonoids.

Flavanones have many other important physiological functions, such as anti-inflammatory, antiatherosclerosis, anticancer, and antibacterial functions.

5.3.3.1 Anti-Inflammation

Da Silva Emim et al. (1994) reported that the pretreatment of rats with hesperidin (50 and 100 mg/kg of body weight) reduced carrageenan-induced paw edema by 47% and 63%, respectively, within 5 h. At 100 mg/kg, hesperidin decreased the dextran-induced rat paw edema by 33% without influencing the histamine-induced paw edema. Hesperidin also helped to control carrageenan-induced pleurisy by reducing the volume of exudate and the number of migrating leucocytes by 48% and 34%, respectively, compared to control values. Equal doses of duartin and claussequinone were ineffective in all of the above tests. Pretreatment of mice with hesperidin (100 mg/kg, subcutaneously) reduced acetic acid-induced abdominal constriction by 50% but did not affect the tail flick response. These results indicated that hesperidin obtained from citrus cultures may represent potential therapeutic use as a mild anti-inflammatory agent. Guardia et al. (2001) reported that intraperitoneal administration of rutin, quercetin (flavonols), and hesperidin (flavanone), given at daily doses equivalent to 80 mg/kg, inhibited both acute and chronic phases of inflammation in this experimental model. Rutin was the most active in the chronic phase. The intestinal anti-inflammatory activities of two flavonoids, hesperidin and diosmin, were evaluated in the the trinitrobenzenesulfonic acid acute-stage model of rat colitis (Crespo et al. 1999). The results showed that pretreatment with diosmin (10 mg/kg) or hesperidin (10 and 25 mg/kg) reduced colonic damage, compared to the response in trinitrobenzenesulfonic acid-treated control rats.

5.3.3.2 Antiatherosclerosis

Cha et al. (2001) demonstrated that the addition of 1% dietary hesperetin can reduce hepatic triacylglycerol accumulation induced by orotic acid, and the hesperetin was also associated with reduced activity of the triacylglycerol synthetic enzyme phosphatidate phosphohydrolase. Kim et al. (2003) demonstrated that hesperetin metabolites played as potent a role as hesperetin in plasma lipid-lowering activities *in vivo*, and their results further suggested that cholesterol biosynthesis and esterification were concomitantly reduced by hesperetin and its metabolites, as indicated by decreased hydroxymethylglutaryl coenzyme A (CoA) reductase and acyl-CoA:cholesterol acyltransferase activities.

5.3.3.3 Anticancer

In an investigation of the anticancer activities of citrus flavonoids, So et al. (1996) reported that citrus flavonoids were effective inhibitors of human breast cancer cell proliferation *in vitro*, especially when paired with quercetin. Koyuncu et al. (1999) suggested that hesperidin has potential as a chemopreventive agent against tumor promoter-induced inflammation and hyperplasia.

5.3.3.4 Antibacterial

Kawaguchi et al. (2004) reported that the administration of 1 mg of naringin at 3 h before infection resulted in protection from lethal shock. Treatment with naringin resulted not only in a significant decrease in bacterial numbers in the spleen and liver, but also in a decrease in plasma lipopolysaccharide levels. Z. Yi et al. (2007) reported that hesperidin displayed a broad spectrum of antimicrobial activity and exerted antimicrobial effects in antimicrobial tests with *Escherichia coli*, *Staphylococcus aureus*, *Staphylococcus epidermidis*, *Salmonella enterica* serovar Typhi, and *Enterobacter cloacae*.

5.4 POLYMETHOXY FLAVONES

5.4.1 Chemical Structures of Several PMFs

Polymethoxy flavones (PMFs) exist almost exclusively in citrus fruits. PMFs are abundant in cold-pressed orange peel oil after a long-term freezing step to remove waxes. Fruit juice and pulp contain only a trace amount of PMFs. More than 20 PMFs have been isolated and identified, and among these, nobiletin, tangeretin, and sinensetin are the most common. The composition of PMFs can be used as a citrus marker. Like other flavonoids, flavones contain the chromone backbone with a C_6-C_3-C_6 carbon skeleton, but their phenyl rings are substituted with many methoxy groups, particularly in the A-ring. Figure 5.3 shows the chemical structures of several PMFs.

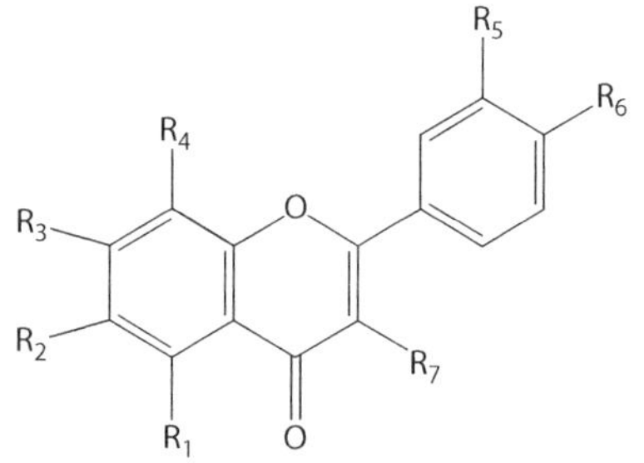

	R_1	R_2	R_3	R_4	R_5	R_6	R_7
Sinensetin	OCH_3	OCH_3	OCH_3	H	OCH_3	OCH_3	H
Tetramethylscutellarein	OCH_3	OCH_3	OCH_3	H	H	OCH_3	H
Isosinensetin	OCH_3	H	OCH_3	OCH_3	OCH_3	OCH_3	H
Nobiletin	OCH_3	OCH_3	OCH_3	OCH_3	OCH_3	OCH_3	H
Tangeretin	OCH_3	OCH_3	OCH_3	OCH_3	H	OCH_3	H
Heptamethoxyflavone	OCH_3	OCH_3	OCH_3	OCH_3	OCH_3	OCH_3	OCH_3

FIGURE 5.3 Chemical structures of several PMFs.

5.4.2 Metabolism and Physiological Functions of PMFs

Yasuda et al. (2003) reported that the urinary metabolite of nobiletin is 4′-hydroxy-3′,5,6,7,8-pentamethoxyflavone. Nielsen et al. (2000) investigated the *in vivo* biotransformation and excretion of tangeretin. Rats were treated with tangeretin at a daily rate of 100 mg/kg of body weight. Ten major metabolites with the flavonoid structure still intact were identified. The metabolites identified were either demethylated or hydroxylated derivatives of the parent compound, and the changes were primarily in the 4′-position of the B-ring. The total urinary excretion of tangeretin metabolites with an intact flavan nucleus was about 11% of the administered daily dose.

The chemical structures of PMFs are different from those of other citrus flavonoid in that they are not glycosylated, and therefore the polarities are rather weak, which is important for many physiological activities, such as biofilm permeability, metabolism products, and so on (Manthey and Guthrie 2002). Although their levels are very low, they still show anticancer, antiviral, and anti-inflammatory activities, and also inhibition of platelet aggregation and other important physiological functions.

5.4.2.1 Anticancer

The anticancer effects of PMFs have attracted increasing attention in recent years. Many *in vitro* experiments have shown that PMFs have considerable anti-cancer cell proliferation, antimutagenic, and anti-tumor-promoting effects. Kandaswami et al. (1991, 1992) examined the effects of four plant flavonoids (quercetin, taxifolin, nobiletin, and tangeretin) on the *in vitro* growth of a human squamous cell carcinoma cell line (HTB43). Nobiletin and tangeretin clearly inhibited cell growth at all concentrations tested on days 5 and 7. However, quercetin and taxifolin exhibited no significant inhibition at any of the concentrations tested.

Hirano et al. (1995) found that tangeretin inhibited the growth of HL-60 cells *in vitro*, partially through induction of apoptosis and by not causing serious side effects on immune cells. Kawaii et al. (1999a) investigated 27 citrus flavonoids for their antiproliferative activities against several tumor and normal human cell lines. Based on their results, seven flavonoids were shown to be active against the tumor cell lines while exhibiting weak antiproliferative activity against the normal human cell lines. The rank order of potency was luteolin > natsudaidain > quercetin > tangeretin > eriodictyol > nobiletin > 3,3′,4′,5,6,7,8-heptamethoxyflavone. Iwase et al. (2001) investigated nobiletin and 3,5,6,7,8,3′,4′-heptamethoxyflavone for anti tumor-initiating activity on two-stage carcinogenesis of mouse skin tumors. 3,5,6,7,8,3′,4′-Heptamethoxyflavone exhibited a remarkable anti tumor-initiating effect on mouse skin.

5.4.2.2 Antibacterial

Del Rio et al. (1998) found that PMFs inhibited the growth of *Penicillium digitatum*, *Phytophthora citrophthora*, and *Geotrichum* spp. Almada-Ruiz et al. (2003) reported that four PMFs had antifungal activity against *Colletotrichum gloeosporioides*.

5.4.2.3 Hypolipidemic Effect

Kurowska and Manthey (2004) investigated the hypolipidemic effects of flavanones and PMFs. Diets containing 1% PMF significantly reduced serum total and very-low-density lipoprotein (VLDL) and also LDL cholesterol (by 19%–27% and 32%–40%, respectively) and either reduced or tended to reduce serum triacylglycerols. Comparable reductions were achieved by feeding a 3% mixture of hesperidin and naringin (1:1, wt/wt), implying a lower hypolipidemic potency of hesperidin/naringin compared to PMFs. High-performance liquid chromatography–mass spectrometry (HPLC-MS) analysis identified high serum, liver, and urine concentrations of tangeretin metabolites, including dihydroxytrimethoxy flavone and monohydroxytetramethoxy flavone glucuronides and aglycones. Malterud and Rydland (2000) reported inhibitory activities of PMFs toward soybean 15-lipoxygenase. Eguchi et al. (2006) reported that nobiletin appeared to be a promising phytochemical for regulating atherosclerosis, with reasonable mechanisms of action.

5.4.2.4 Other Effects

Li et al. (2006) suggested that PMFs can ameliorate hypertriglyceridemia, and their antidiabetic effects may occur as a consequence of adipocytokine regulation and peroxisome proliferator-activated receptor alpha (PPARα) and (PPARγ) activation. Itoigawa et al. (1994) found that 3,5,6,7,8,3′,4′-heptamethoxyflavone (HEPTA) and natsudaidain produced a positive inotropic effect on guinea pig papillary muscle. Lin et al. (2003) suggested that nobiletin is a candidate requiring characterization as a novel immunomodulatory and anti-inflammatory drug.

5.4.2.5 Safety

Delaney et al. (2001, 2002) isolated PMFs from orange oil and reported the following composition of PMFs: nobiletin, 30.7%; 3,3′,4′,5,6,7,8-heptamethoxyflavone, 27.9%; trimethylscutellarein, 14.5%; tangeretin, 10.4%; sinensetin, 5.8%; 5-demethyl-nobiletin, 2.0%; hexa-*O*-methylquercetagetin, 1.3%; 5-demethyltetramethylscutellarein, 0.6%; other flavonoids, 2.7%. To assess the effect of the PMF mixture on humoral immune responses, female $B6C3F_1$ mice ($n = 8$) were exposed to the PMF mixture by gavage at 5, 50, 150, and 500 mg/kg/day for 28 days. The results showed that long-term, high-dose exposure to a standardized mixture of citrus PMFs caused mild suppression of NK cell activity; however, humoral immunity was not sensitive to suppression at the same exposure levels. Similarly, the results demonstrated that the PMF mixture was not genotoxic in *in vitro* assay systems.

5.5 APPLICATION OF CITRUS PHENOLIC COMPOUNDS

5.5.1 Application of Citrus Phenolic Compounds in Taxonomy

The botanical classification of citrus fruits is difficult. The two extensively used systems are named Swingle (16 species) and Tanaka (162 species). The two systems are based on the composition of phenolic compounds (mainly flavonoids) in citrus fruits, which can provide useful information for the taxonomy of citrus fruits.

Nogata et al. (2006) reported that the flavonoid composition of citrus fruits was approximately the same within each section of Tanaka's classification system, except for the species in the Aurantium section and those with a peculiar flavonoid composition, such as bergamot (*Citrus bergamia*), marsh grapefruit (*Citrus paradist*), sour orange (*Citrus aurantium*), and shunkokan (*Citrus shunkokan*). Mizuno et al. (1991) investigated the composition of seven PMFs in citrus fruit peels, and their results supported the morphological classification systems presented by Swingle and Tanaka, except for several taxa. In work reported by Di Mauro et al. (2002), hydroxycinnamic acids (ferulic, *p*-coumaric, sinapic, and caffeic acids) were detected in 113 orange juices, and their concentrations were utilized to develop a data bank of Italian juices that can be employed to check the variety and geographical origin of commercial products.

5.5.2 Application of Citrus Phenolic Compounds in Juice Adulteration

The adulteration of fruit juices is a serious problem that has beleaguered the fruit and vegetable industries. However, by using the inherent differences in flavonoid composition among citrus varieties, it should be possible to develop and apply appropriate procedures to detect unwanted adulteration practices.

Ooghe and Detavernier (1999) reported that FG and PMF patterns are characteristic of the different citrus juices and can be used to detect the addition of non-*C. sinensis* juices, such as grapefruit, mandarin, sour/bitter oranges, or bergamot, to sweet orange juice. Reverse-phase HPLC has been applied to quantify levels of PMFs and carotenoids in orange and tangelo juices. Using canonical discriminant analysis, the addition of 10% tangelo to orange juice can be detected (Pan et al. 2002). The determination of 4-hydroxycinnamic acids (caffeic, *p*-coumaric, ferulic, and sinapic acids) was performed on 82 orange juices derived from the most important blood and blond cultivars grown in Italy. Discriminant analysis of the experimental results showed that these acids could be used as markers of blood and blond cultivars. The statistical model was used to recognize some mixtures of blood and blond juices (Rapisarda et al. 1998).

5.5.3 Application of Citrus Phenolic Compounds in Fingerprints

Many citrus fruits are used as raw materials in Chinese medicine, including the compounds chenpi, qingpi, huajuhong, Zhishi, zhiqiao, and so on. Fingerprints of citrus fruits based on their phenolic compound content can be used to control the quality of traditional Chinese medicines. HPLC fingerprints of pericarpium citri reticulatae (PCR; chenpi) and pericarpium citri reticulatae viride (PCRV; qingpi) were studied. The results suggested that fingerprints as a quality control measure for PCR and PCRV not only distinguish mixed peels but also discriminate authentic PCR and PCRV (L.-Z. Yi et al. 2007).

5.6 HEALTH BENEFITS OF CITRUS PHENOLIC COMPOUNDS

Epidemiological and animal studies point to a possible protective effect of flavonoids against cardiovascular diseases and some types of cancer. The possible beneficial effects of citrus fruits are due not only to the high amounts of vitamins and minerals but also to their phenolic compounds (flavonoids). Citrus flavonoids have many health benefits, as described previously; however, the most valuable health-promoting effects are the prevention of atheroscerosis and cancer.

5.6.1 Citrus Flavonoids in the Prevention of Atherosclerosis

Flavonoids were once named vitamin P, for their effects on capillary fragility and bleeding. The regular consumption of flavonoids can reduce the risk of coronary diseases in older men. Researchers have suggested that hesperetin limits the rise in hepatic triacylglycerol and cholesterol content induced by orotic acid, and PMFs are novel flavonoids with cholesterol- and triacylglycerol-lowering potentials (Cha et al. 2001; Kurowska and Manthey 2004). Epidemiological studies have reported that the intake of citrus flavonoid-containing foods attenuates cardiovascular diseases, and a limited number of clinical studies have revealed lipid-lowering, insulin-sensitizing, antihypertensive and anti-inflammatory properties for these compounds. The mechanisms underlying flavonoid-induced metabolic regulations have not been fully established (Assini et al. 2013; Mulvihill and Huff 2012).

5.6.2 Citrus Flavonoids in the Prevention of Cancer

As mentioned previously, citrus phenolic compounds, especially flavonoids and PMFs, have a role in the prevention and treatment of some human cancers. However, the exact cellular and molecular mechanisms by which citrus flavonoids play their roles in human cancer prevention and treatment need further clarification. Meanwhile, long-term clinical trials on the intake, metabolism, and cytotoxicity of citrus flavonoids are required before a dietary recommendation for an increase in the intake of citrus flavonoids can be endorsed for the prevention or treatment of cancers (Benavente-Garcia and Castillo 2008).

Citrus fruits contain sugar, organic acids, and a number of physiologically functional components, such as citric acid, ascorbic acid, minerals, coumarins, and the flavonoids naringin, hesperidin, neohesperidin, rutin, naringenin, hesperetin, narirutin, and tangeretin (Kawaii et al. 1999b). China is one of the major producers of citrus fruits (of various species), especially South China, such as Zhejing, Fujian and Jiangxi Provinces. The following is a summary of the authors' latest research involving phenolic compounds in citrus fruits.

5.7 JUICE COMPONENTS AND ANTIOXIDANT CAPACITIES OF CITRUS VARIETIES CULTIVATED IN CHINA

5.7.1 Contents of Flavanone Glycosides and Phenolic Acids

The levels of four major FGs (narirutin, naringin, hesperidin, and neohesperidin) in citrus fruits have been determined (Table 5.2). Hesperidin and narirutin are considered the major FGs in mandarin and orange juices, whereas naringin and neohesperidin were not detected in these fruits. For mandarin and orange juices, the hesperidin content ranged from 304.46 mg/L (Zhuhong) to 533.64 mg/L (Yinzhaocheng), and narirutin content ranged from 10.28 mg/L (Skaggs Bonanza) to 288.12 mg/L

TABLE 5.2
FG Contents of Citrus Juices

	FG Content (mg/L)			
Fruit	**Narirutin**	**Hesperidin**	**Naringin**	**Neohesperidin**
Wase satsuma	169.4	337.44	ND	ND
Satsuma	288.12	450.60	ND	ND
Ponkan	42.63	379.92	ND	ND
Bendizao	42.44	417.94	ND	ND
Manju	43.70	315.88	ND	ND
Hybrid 439	119.80	501.44	ND	ND
Zhuhong	24.42	304.46	ND	ND
Skaggs bonanza	10.28	489.64	ND	ND
Hamlin	136.74	427.76	ND	ND
Liubencheng	89.49	506.40	ND	ND
Yinzhaocheng	84.12	533.64	ND	ND
Lemon	ND	237.96	ND	ND
Huyou	94.04	38.26	348.53	265.25
Miyou	ND	42.17	108.52	6.71
Sijiyou	ND	21.81	125.79	ND

Source: Modified from Xu, G. et al., *Food Chemistry* 106(2):545–551, 2008.
Abbreviation: ND, not detected.

(satsuma). Only hesperidin was detected in lemon juice, while Huyou had all four of the FGs and the highest levels of naringin (348.53 mg/L), neohesperidin (265.25 mg/L), and total FGs (746.08 mg/L). Narirutin was not detectable in Miyou juice, and narirutin and neohesperidin were not detectable in Sijiyou juice. Generally, mandarin, orange, and grapefruit had higher levels of FGs, while the levels in lemon and pomelo were lower. Furthermore, among grapefruits, Huyou had the highest content of total FGs (746.08 mg/L). The amount of flavonone aglycones was 359.48 mg/L, which was higher than the average aglycone value (270 mg/kg) in grapefruits reported by Peterson et al. (2006). Therefore, future study and more attention should be paid to this variety.

Seven phenolic acids, which included four cinnamic acids (caffeic, *p*-coumaric, ferulic, and sinapic acids) and three benzoic acids (protocatechuic, *p*-hydroxybenzoic, and vanillic acids) were quantified by HPLC with photodiode array (PDA) (Xu et al. 2008a). The results are shown in Table 5.3. Generally, ferulic was dominant in the citrus juices, with the exception of Miyou and Sijiyou, in which sinapic acid (4.55 mg/L Miyou) and *p*-coumaric (8.79 mg/L Sijiyou) levels were higher than ferulic acid levels. Total phenolic acids ranged from 14.00 mg/L (Miyou) to 72.61 mg/L (Liubencheng). Generally, mandarins (except Manju) and oranges had higher phenolic acid contents than did grapefruits or pomelos. The phenolic acid results of Xu et al.'s analysis were lower to some extent than previous reports (Rapisarda et al. 1998, 2003), which might be the result of the variety diversity.

TABLE 5.3
Phenolic Acid Contents of Citrus Juices

	Phenolic Acid Content (mg/L)							
	Cinnamic Acids				Benzoic Acids			
Fruit	Caffeic	*p*-Coumaric	Ferulic	Sinapic	Protocatechuic	*p*-Hydroxybenzoic	Vanillic	Total
Wase satsuma	2.71	6.19	36.49	2.90	0.86	1.69	3.40	54.24
Satsuma	2.74	3.66	40.07	2.78	0.71	1.16	2.71	53.83
Ponkan	5.24	2.79	26.07	3.36	0.57	0.90	0.94	39.85
Bendizao	5.39	7.24	45.00	6.05	0.55	0.74	0.69	65.66
Manju	2.54	1.32	18.11	2.78	0.55	0.86	0.94	27.10
Hybrid 439	5.50	3.47	16.53	9.12	0.82	1.77	3.65	40.87
Zhuhong	6.55	3.06	45.91	4.39	0.58	1.01	1.39	62.90
Skaggs bonanza	5.02	8.15	32.14	5.09	0.61	1.04	1.14	53.19
Hamlin	3.26	6.17	39.94	7.88	0.70	0.99	1.17	60.10
Liubencheng	5.68	13.49	43.20	6.83	1.02	1.17	1.21	72.61
Yinzhaocheng	4.79	9.17	40.13	6.24	0.75	0.87	0.87	62.82
Lemon	2.07	11.57	35.77	6.75	0.72	0.79	0.85	58.50
Huyou	2.54	2.60	11.13	3.88	0.71	1.07	2.86	24.79
Miyou	2.02	3.75	1.63	4.55	0.76	0.67	0.63	14.00
Sijiyou	7.23	8.79	6.77	3.77	0.81	0.81	1.17	29.35

Source: Modified from Xu, G. et al., *Food Chemistry* 106(2):545–551, 2008.

5.7.2 Antioxidant Capacities of Citrus Juices

Total phenolics were measured by the Folin–Ciocalteu method, and the antioxidant capacity of the citrus juice was evaluated in the ferric reducing antioxidant power (FRAP) and DPPH assays (Table 5.4). Total phenolics and scavenging of DPPH by Hybrid 439 achieved the highest values, 1555.49 mg/L and 61.62%, respectively, which suggested that Hybrid 439 is a valuable variety with high antioxidant capacity that may be beneficial to health. For the FRAP assay, Hamlin had the highest value, 899.31 ascorbic acid equivalent antioxidant capacity (AEAC) mg/L, while lemons had the lowest level, 307.43 AEAC mg/L. It seems that many factors, such as different citrus varieties, maturity, material preparation, and analysis methods, could have caused the divergent results. Generally, oranges had higher antioxidant capacities than the other citrus varieties.

Correlation coefficients of total phenolics (gallic acid equivalents, or GAE), FRAP (AEAC), DPPH (percent inhibition), total FGs, and total phenolic acids are shown in Table 5.5. Total FG values correlate highly ($p < 0.01$) with total phenolics and FRAP (AEAC), but correlation with DPPH was not significant. Since FGs are the major phenolic compounds, they obviously exhibited a high correlation with total phenolics. Correlation coefficients of total phenolic acids with FRAP (AEAC) and DPPH were not significant, which indicated that phenolic acids play a minimal role in the antioxidant capacity of citrus juices.

TABLE 5.4
Antioxidant Capacities of Citrus Juices

	Antioxidant Capacity Based on Method		
Fruit	**FRAP (AEAC, mg/L)**	**Total Phenolics (GAE, mg/L)**	**% Inhibition of DPPH**
Wase satsuma	454.72	863.38	26.31
Satsuma	598.48	1109.23	33.65
Ponkan	476.19	830.32	29.67
Bendizao	482.98	972.88	25.39
Manju	361.24	774.54	23.69
Hybrid 439	875.93	1555.49	61.62
Zhuhong	541.14	1043.12	36.75
Skaggs bonanza	765.33	1173.28	50.92
Hamlin	899.31	1499.71	60.24
Liubencheng	886.26	1462.52	60.13
Yinzhaocheng	712.61	1245.59	47.82
Lemon	307.43	751.82	24.50
Huyou	617.50	1241.46	39.83
Miyou	510.16	863.38	37.71
Sijiyou	442.22	801.40	35.79

Source: Modified from Xu, G. et al., *Food Chemistry* 106(2):545–551, 2008.

TABLE 5.5
Agreement between Detection Methods for Activity Levels of Total Phenolic Acids, FRAP, DPPH, Total FGs, and Total Phenolic Acids

	Correlation Coefficient *r* (Method of Detection)			
Parameter Measured	Total Phenolic Acids (GAE)	FRAP (AEAC)	DPPH (% Inhibition)	Total FGs
FRAP	0.904			
DPPH	0.845	0.962		
Total FGs	0.659	0.643	0.459	
Total phenolic acids	0.472	0.336	0.227	0.341

Source: Modified from Xu, G. et al., *Food Chemistry* 106(2):545–551, 2008.
Note: Data are based on 15 samples for each method.

5.8 COMPOSITION OF MAJOR FLAVANONE GLYCOSIDES AND ANTIOXIDANT CAPACITIES OF THREE CITRUS VARIETIES

5.8.1 Contents of Individual Flavonone Glycosides in the Peel and Flesh of Three Citrus Types: Ponkan, Satsuma Mandarin, and Huyou

The main flavonoid compounds, narirutin, naringin, hesperidin, and neohesperidin, that are generally found in citrus fruits were separated and quantified by reverse-phase HPLC (Table 5.6). Ponkan and Satsuma are from the species *C. reticlata* and are rich in narirutin and have an exceptionally high hesperidin content, while Huyou contains all four FGs, most likely because it is a hybrid of *C. sinensis* and *C. grandis.* The data in Table 5.6 clearly show that the content of all FGs decreases a great deal with maturity, which is in agreement with the report by Ortuno et al. (1997). For instance, the content of hesperidin from the flesh of Ponkan decreased from 3204 mg/100 g fresh weight (FW) in early season to 164.30 mg/100 g FW in late season. Based on these findings, it would appear to be more profitable to utilize the early dropped fruits to extract specific flavonone glycosides. Generally, there is a higher content of FGs in the peel than in the flesh, which suggests that citrus peel is a better source of phenolic compounds. The peel of Ponkan can be used to obtain hesperidin because it occurs in these fruits at remarkably high levels, the peel of Satsuma can be utilized to obtain narirutin and hesperidin, based on the mandarins higher narirutin/hesperidin ratio, and the peel of Huyou can be used to extract neohesperidin and naringin for its noticeably higher flavonone glycoside content.

5.8.2 Total Phenolic and Total Flavonoid Contents and AEAC Levels, Determined in DPPH, ABTS, Reducing Power, and FRAP Assays

As shown in Table 5.7, the content of total phenolics (GAE, in mg/100 g FW) varies greatly between varieties and during maturity. The values ranged from 70.24

TABLE 5.6
Contents of Individual Flavonones in Peel and Flesh of Ponkan, Satsuma Mandarin, and Huyou Fruits

	Flavonone Content (mg/100 g FW)				
Sample	**Narirutin**	**Naringin**	**Hesperidin**	**Neohesperidin**	**Total FGs**
In flesh					
Ponkan					
Unripe	341.49	ND	3204.49	ND	3545.98
Half-ripe	44.63	ND	374.08	ND	418.71
Ripe	20.04	ND	164.30	ND	184.33
Satsuma					
Unripe	340.06	ND	415.29	ND	755.35
Half-ripe	79.66	ND	135.88	ND	215.53
Ripe	31.29	ND	50.40	ND	81.69
Hunyou					
Unripe	285.12	849.01	160.55	472.24	1766.92
Half-ripe	63.30	175.32	13.69	99.73	352.04
Ripe	24.38	62.43	Trace	32.02	118.83
In peel					
Ponkan					
Unripe	124.05	ND	6172.71	ND	6296.77
Half-ripe	50.78	ND	1928.89	ND	1979.67
Ripe	26.95	ND	1233.73	ND	1260.68
Satsuma					
Unripe	859.31	ND	2951.50	ND	3810.80
Half-ripe	366.83	ND	1718.29	ND	2085.12
Ripe	258.33	ND	1469.20	ND	1727.53
Hunyou					
Unripe	77.80	1110.33	145.16	1037.11	2370.41
Half-ripe	38.67	721.44	32.39	592.43	1384.92
Ripe	19.07	279.13	14.28	313.05	625.54

Source: Modified from Xu, G. et al., *Journal of Food Biochemistry* 33(5):453–469, 2009.
Abbreviation: ND, not detected.

(ripe Satsuma) to 447.70 (unripe Huyou) in the flesh and 409.88 (ripe Satsuma) to 1080.11 (unripe Ponkan) in the peel respectively. As for Ponkan, there was a clear decline with maturity. The total phenolics in the flesh of unripe Ponkan was 399.38 mg/100 g FW, which decreased to 98.73 mg/100 g FW in the flesh of ripe Plonkan. In the Ponkan peels, the content of total phenolics decreased from 1080.11 to 417.95 mg/100 g FW. A similar tendency was seen in the other two varieties. Citrus peels have been widely studied because they contain numerous biologically active compounds, including natural antioxidants, such as phenolic acids and flavonoids (Bocco et al. 1998; Manthey and Grohmann 2001). Phenolic acids and flavonoids act as free radical scavengers and have shown beneficial health-promoting

TABLE 5.7
Total Phenolic Acids, Total Flavonoids, and Antioxidant Capacities of Ponkan, Satsuma Mandarin, and Huyou Fruits, Determined by Four Different Antioxidant Methods

	Content or Capacity (mg/100 g FW)					
Sample	Total Phenolics (GAE)	Total Flavonoids (RE)	DPPH (AEAC)	ABTS (AEAC)	Reducing Power (AEAC)	FRAP (AEAC)
In flesh						
Ponkan						
Unripe	399.38	481.994	131.84	542.58	170.56	196.49
Half-ripe	123.41	73.03	38.71	52.78	53.97	44.86
Ripe	98.73	37.39	40.11	73.20	47.38	44.84
Satsuma						
Unripe	225.66	150.52	61.93	138.94	101.63c	79.38
Half-ripe	100.76	39.43	26.91	73.81	38.06	41.25
Ripe	70.24	28.83	28.68	79.05	39.09	27.13
Huyou						
Unripe	447.70	321.96	59.45	949.84	157.96	215.49
Half-ripe	193.79	104.74	53.59	266.63	83.11	70.80
Ripe	99.71	46.87	41.83	118.01	62.84	30.89
In peel						
Ponkan						
Unripe	1080.11	1356.52	525.03	1262.56	611.42	440.84
Half-ripe	561.52	599.29	214.97	725.18	277.54	268.14
Ripe	417.95	346.55	161.98	572.02	226.51	210.58
Satsuma						
Unripe	510.36	306.52	140.50	713.40	143.16	177.28
Half-ripe	422.58	208.05	113.98	597.05	129.82	169.46
Ripe	409.88	156.96	103.20	589.78	169.68	192.00
Huyou						
Unripe	955.45	739.43	139.51	1655.67	280.39	326.03
Half-ripe	671.87	440.64	110.01	1171.66	151.52	219.19
Ripe	453.64	266.67	98.14	793.94	168.62	147.50

Source: Modified from Xu, G. et al., *Journal of Food Biochemistry* 33(5):453–469, 2009.

effects for chronic and degenerative diseases (Kim and Lee 2004). As can be seen in the data, the content of total phenolics in peels is significantly higher than in flesh. Taking ripe Ponkan as an example, the content of total phenolics in peels was 417.95 (GAE mg/100 g), while in flesh it was 98.73 (GAE mg/100 g), showing that citrus peels are a good source of dietary phenolics. Similar trends were observed when examining the total flavonoid results.

The scavenging models utilizing the stable DPPH radical and the ABTS radical are widely used to evaluate radical scavenging ability in a relatively short time compared with other methods, and reducing power may serve as a significant indicator of potential antioxidant activity. In the FRAP assay yellow Fe^{3+}-TPTZ [2,4,6-Tris(2-pyridyl)-*s*-triazine] is reduced to blue Fe^{2+}-TPTZ via electron-donating compounds under acidic conditions. Since the DPPH, ABTS, reducing power, and FRAP assays are all commonly used methods to determine the *in vitro* antioxidant capacities of fruits, it was decided to evaluate the antioxidant capacity with all four assay methods. The results indicated that, in general, the AEAC measured in each assay decreased with maturity on all fruits studied, and the citrus peel showed greater antioxidant potential than flesh. As shown in Table 5.7, similar results or trends can be obtained using the four methods. However, absolute values of antioxidant capacities varied greatly according to the methodology used. Therefore, we suggest that the total antioxidant capacities of samples be measured with at least two different methods.

5.8.3 Correlations between Total Phenolics, Total Flavonoids, DPPH, ABTS, Reducing Power, FRAP Assays, and Total FGs

Significant correlations were found when we compared the assay methods (Table 5.8). Briefly, the total phenolics determined by the Folin–Ciocalteu method correlated strongly with antioxidant capacity measured by the FRAP method, which had the highest r value ($r = 0.969$), followed by ABTS ($r = 0.949$), reducing power ($r = 0.880$), and DPPH ($r = 0.798$). There was a strong relationship that suggested that the higher the total phenolics, the greater the antioxidant capacity. The same relationship appeared to be present between antioxidant capacity and both total FGs and total flavonoids. In addition, the correlation coefficients in all cases were very high, which

TABLE 5.8
Correlation Coefficients for Total Phenolics, Total Flavonoids, DPPH, ABTS, Reducing Power, FRAP, and Total FGs

	Correlation Coefficient (r) for					
Parameter Measured	Total Phenolics	Total Flavonoids	DPPH	ABTS	Reducing Power	FRAP
Total flavonoids	0.921					
DPPH	0.798	0.930				
ABTS	0.949	0.786	0.588			
Reducing power	0.880	0.971	0.968	0.719		
FRAP	0.969	0.935	0.850	0.896	0.929	
Total FGs	0.795	0.859	0.864	0.640	0.849	0.824

Source: Modified from Xu, G. et al., *Journal of Food Biochemistry* 33(5):453–469, 2009.
Abbreviation: The following detection methods were used for each parameter (n = 18 in each group): total phenolics, GAE; total flavonoids, RE; DPPH, AEAC; ABTS, AEAC; reducing power, AEAC; FRAP, AEAC; total FGs, relative fresh weight (in mg/100 g FW).

signifies that all the methods used are capable of determining the antioxidant capacity of citrus extracts and that the total phenolics assay (an efficient and simply assay) can serve as a key indicator of antioxidant ability.

The results clearly showed that total FGs, determined by HPLC, had a strong correlation with all the methods that measure antioxidant capacity, which indicated that FGs could be the the main antioxidant component of the methanol extracts of citrus samples and that they may play a key role in the total antioxidant capacity of citrus fruits. Although total flavonoids content was highly correlated with total FGs ($r = 0.859$), the absolute values of total FGs determined by HPLC were much higher than the values for total flavonoids. The total flavonoids assay is a nonspecific test and estimates the total flavonoid content in citrus fruits. The key reason for our result may due to that rutin was selected as the standard sample instead of an FG (such as hesperidin).

5.9 EFFECTS OF HEAT TREATMENT ON PHENOLIC COMPOUNDS AND ANTIOXIDANT CAPACITIES OF CITRUS PEEL EXTRACTS

Phenolic acids account for about one-third of phenolic compounds in plant foods, and they are present in free and bound forms in the plant material. Bound phenolics may be linked to various plant components through ester, ether, or acetal bonds (Robbins 2003). It is generally believed that many antioxidative phenolic compounds in plants are in a covalently bound form. Based on this knowledge, some processing methods were employed to liberate them and enhance the antioxidant capacity of citrus peel extracts (Seok-Moon et al. 2004; Seung-Cheol et al. 2005). For instance, it has been reported that heat treatment may liberate some low-molecular-weight phenolic compounds and thereby increase the antioxidant capacity of citrus peel (Seok-Moon et al. 2004). However, the effect of heat treatment on citrus flavonoids was not examined in that study, and no clear quantitative relationships were elucidated.

5.9.1 Phenolic Acid Composition and Distribution in the Huyou Peel

Phenolic acids in Huyou peel (HP) were divided into four fractions: free, ester, glycoside, and ester-bound forms (Ayaz et al. 2005). Seven phenolic acids in HP were detected and quantified, namely, benzoic acids (*p*-hydroxybenzoic acid and vanillic acid), cinnamic acids (caffeic acid, *p*-coumaric acid, ferulic acid, and sinapic acid), and chlorogenic acid (as a representative of phenolic esters). The comparative analyses of phenolic acids in the four fractions indicated that there were lower amounts of phenolic acids in the free fraction than in the other three parts (Table 5.9). Similarly, it has been reported that the majority of phenolic acids in citrus fruits are present in the bound form (Gorinstein et al. 2004; Peleg et al. 1991). As well, since the ester-bound fraction accounts for a considerable amount of the phenolic acids, in this study the phenolic acids were divided into four parts and analyzed separately.

The content of phenolic acids varied in the four different fractions, and in general the hydroxycinnamic acid content decreased in the following order: ferulic

TABLE 5.9
Phenolic Acid Composition and Distribution of Untreated and Heat-Treated Huyou Peel

		Phenolic Acid Composition (µg/g DW)							
		Cinnamics				Benzoics			
Phenolic Acid Form	Heat Treatment	Caffeic	*p*-Coumaric	Ferulic	Sinapic	*p*-Hydroxybenzoic	Vanillic	TCB	Chlorogenic
Free	Untreated		5.58	4.56	3.69	Trace	54.18	68.64	126.58
	20°C, 30 min		21.77	49.40	5.76	4.04	72.54	153.51	65.96
	120°C, 60 min		35.30	68.88	6.37	7.76	109.55	227.86	65.10
	120°C, 90 min		45.05	69.98	10.66	10.16	160.71	296.56	35.27
	90°C, 30 min		6.33	6.85	4.30	Trace	63.32	83.71	104.02
	150°C, 30 min		54.10	44.88	2.26	14.00	82.00	197.24	17.47
Ester	Untreated	4.11	24.10	142.68	53.39	10.25	76.05	310.59	
	120°C, 30 min	4.57	16.77	106.85	44.53	8.77	68.83	250.32	
	120°C, 60 min	3.77	10.27	66.35	32.01	5.58	47.96	144.99	
	120°C, 90 min	3.05	7.26	44.99	25.05	4.53	42.80	127.68	
	90°C, 30 min	5.60	26.86	138.35	48.41	10.19	75.10	304.51	
	150°C, 30 min	5.20	11.72	57.36	25.61	6.58	58.88	165.35	
Glycoside	Untreated	2.13	52.26	37.35	2.34	6.38	53.70	154.16	
	120°C, 30 min	2.38	50.55	40.66	2.87	5.97	52.74	155.17	
	120°C, 60 min	1.94	50.69	30.38	1.69	5.38	44.18	134.27	
	120°C, 90 min	Trace	43.89	16.25	Trace	4.60	26.87	91.91	
	90°C, 30 min	1.58	65.96	34.61	1.80	7.08	44.22	155.25	
	150°C, 30 min	1.11	55.96	25.90	1.15	6.51	40.51	132.13	

(Continued)

TABLE 5.9 (CONTINUED)
Phenolic Acid Composition and Distribution of Untreated and Heat-Treated Huyou Peel

Phenolic Acid	Heat	Phenolic Acid Composition (μg/g DW)							
		Cinnamics				Benzoics			
Form	Treatment	Caffeic	*p*-Coumaric	Ferulic	Sinapic	*p*-Hydroxybenzoic	Vanillic	TCB	Chlorogenic
Ester-bound	Untreated	Trace	26.04	77.65	5.72	4.19	5.54	119.15	
	120°C, 30 min	Trace	14.84	55.92	3.55	2.56	3.22	80.2	
	120°C, 60 min	Trace	11.98	46.19	2.32	2.03	1.28	66.12	
	120°C, 90 min	Trace	9.92	39.34	1.81	1.94	1.48	54.96	
	90°C, 30 min	Trace	20.65	72.01	7.21	3.50	5.85	110.08	
	150°C, 30 min	Trace	10.31	34.02	4.24	1.65	5.72	57.85	

Source: Modified from Xu, G. et al., *Journal of Agricultural and Food Chemistry* 55(2):330–335, 2007.
Note: "Trace" indicates content was lower than 1 μg/g DW.

acid > sinapic acid > coumaric acid > caffeic acid, which was in accordance with findings in other reports (Bocco et al. 1998; Peleg et al. 1991; Rapisarda et al. 1998). Within the four hydroxycinnamic acids analyzed, caffeic acid had the lowest concentration and was only present in the ester and glycoside forms. It should be noted that chlorogenic acid, which can be hydrolyzed to caffeic acid under alkaline conditions (Luthria and Mukhopadhyay 2006), was present primarily in the free fraction. Furthermore, the recovery of caffeic acid was very low after alkaline hydrolysis, so the calculated caffeic acid content here was obviously underestimated.

As shown in Table 5.9, cinnamic acids occur most frequently as esters, while the benzoic acids are present mainly in the form of esters and glycosides. These results are in agreement with the findings reported by Herrmann (1989).

5.9.2 Effect of Heat Treatment on the Phenolic Acid Distribution of HP

The effect of heat treatment with different heating times and temperatures on the phenolic acids distribution of HP is shown in Table 5.9. In the free phenolic acids fraction, the content of benzoic acids and cinnamic acids significantly increased after heat treatment ($p < 0.05$). For example, after heat treatment at 120°C for 90 min, the content of *p*-coumaric acid increased from 5.58 to 45.05 μg/g dry weight (DW), ferulic acid increased from 4.56 to 69.98 μg/g DW, vanillic acid increased from 54.18 to 160.71 μg/g DW, and total cinnamics and benzoics (TCB) increased from 68.64 to 296.56 μg/g DW. In addition, chlorogenic acid (an ester of caffeic and quinic acid) was determined in its free form. Nevertheless, its content decreased with heating time and temperature, which indicated that the ester bond is cleaved by heat treatment. In the ester, glycoside, and ester-bound fractions, the content of phenolic acids also decreased after heat treatment. For example, after being heated at 120°C for 90 min, the content of ferulic acid in the ester fraction decreased from 142.68 to 44.99 μg/g DW, from 37.35 to 16.25 μg/g DW in the glycosides fraction, and from 77.65 to 39.34 μg/g DW in the ester-bound fraction.

The effect of heat treatment on the content of total phenolic acids, the sum of all seven detected phenolic acid in the four fractions, is shown in Figure 5.4. In general, there was a drop in total phenolic acids with increasing time and temperature, which indicated that some phenolic acids are probably destroyed by heat treatment even though there is an increase in the free phenolic acid fraction.

These results suggest that after heat treatment, the distribution of phenolic acids is changed due to the hydrolysis of esterified and glycosylated bonds. To sum up, the free fraction increased while the ester, glycoside, and ester-bound fractions decreased. Likewise, there was a decrease in total phenolic acid content.

5.9.3 Effect of Heat Treatment on the FG Composition of HP

Four major FGs in Huyou peel (HP), namely, narirutin, naringin, hesperidin, and neohesperidin, were determined by HPLC, and the effect of heat treatment on the FG composition of HP is shown in Table 5.10. Clearly, after heat treatment the content of all four FGs declines with increasing time and temperature. The effect of heat

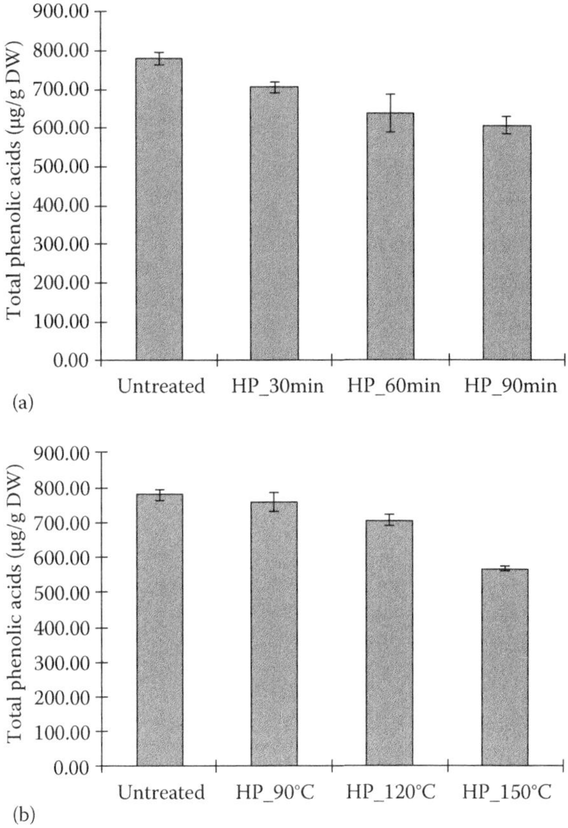

FIGURE 5.4 Total phenolic acid contents of untreated and heat-treated Huyou peel (HP). (a) HP was heated at 120°C for 30, 60, or 90 min. (b) HP was heated at 90°C, 120°C, or 150°C for 30 min. (Modified from Xu, G. et al., *Journal of Agricultural and Food Chemistry* 55(2):330–335, 2007.)

TABLE 5.10
FG Content of Untreated and Heat-Treated Huyou Peel

	FG Content (mg/g DW)			
Heat Treatment	**Narirutin**	**Naringin**	**Hesperidin**	**Neohesperdin**
Untreated	2.81	31.57	2.04	24.09
120°C, 30 min	2.74	31.14	1.83	24.10
120°C, 60 min	2.31	31.00	1.87	23.58
120°C, 90 min	2.24	27.82	1.49	21.34
90°C, 30 min	2.61	30.94	1.83	23.64
150°C, 30 min	2.40	28.72	1.70	21.33

Source: Modified from Xu, G. et al., *Journal of Agricultural and Food Chemistry* 55(2):330–335, 2007.

treatment on the content of total FGs is shown in Figure 5.5. The total FG content decreased from 60.50 to 52.89 mg/g DW after 90 min at 120°C and to 54.15 mg/g DW after 30 min at 150°C.

It appears that FGs are destroyed when subjected to high temperatures and for extended time periods. Since FGs are the predominate citrus flavonoids, it may not be beneficial to use fairly high temperatures to enhance the antioxidant capacity of citrus peels. These opinions are not in agreement with the views of Seok-Moon et al. (2004). Chenpi, which are dried tangerine peels used as a seasoning and in traditional Chinese medicine, is usually prepared by sun drying and is said to have health-promoting effects because of its high citrus flavonoid content

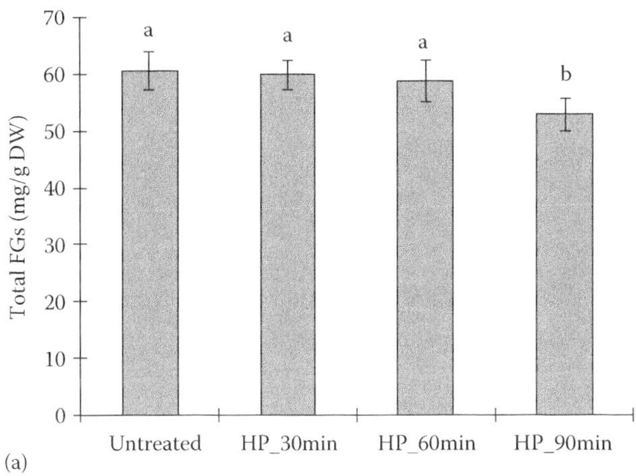

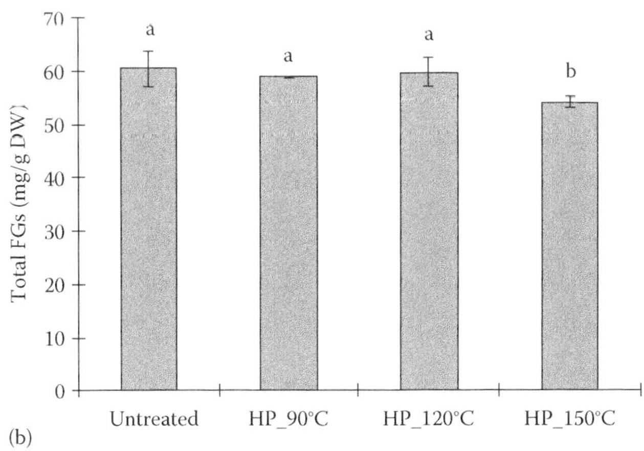

FIGURE 5.5 Total FG content of untreated and heat-treated Huyou peel (HP) samples. (a) HP was heated at 120°C for 30, 60, or 90 min. (b) HP was heated at 90°C, 120°C, or 150°C for 30 min. (Modified from Xu, G. et al., *Journal of Agricultural and Food Chemistry* 55(2):330–335, 2007.)

(Chinese Pharmacopoeia Commission 2010). Therefore, we would like to suggest that heat treatment at a proper and resonable temperature (for example, 100°C or so) should be used to prepare chenpi.

5.9.4 Effect of Heat Treatment on the Antioxidant Capacity of Huyou Peel Extract

The antioxidant capacity of the methanol extract of HP was evaluated by the ABTS decoloration method and the FRAP assay. The total phenolic content was also determined on the methanol extract. The results are shown in Figure 5.6. Obviously, the antioxidant capacity and phenolic content of the HP extract increased with heating time and temperature. For example, after being heated at 120°C for 90 min, total phenolic content (TPC) increased from 37.33 to 47.20 GAE mg/g DW when measured by the Folin–Ciocalteu method, the total antioxidant capacity (TAC) increased from 43.66 to 58.21 trolox equivalent antioxidant capacity (TEAC) mg/g DW when measured by the ABTS method, and the TAC increased from 19.66 to 33.14 TEAC mg/g DW when measured by the FRAP assay. According to the results of the ABTS radical scavenging assay, the maximum TAC of HP extract was produced after heat treatment of 150°C for 30 min (59.95 TEAC mg/g DW). This was in accordance with the results of analysis via FRAP (38.2 TEAC mg/g DW) and TPC (50.07 GAE mg/g DW). However, the effect of heat treatment at a lower temperature (90°C for 30 min) on TAC was minimal.

The correlation coefficients among TPC, ABTS assay, FRAP assay, and TCB in free fraction results are shown in Figure 5.7. As shown, the correlation coefficients in each case were significant ($p < 0.05$), which indicated that the increase of TAC of HP extract could be due at least in part to the increase of TCB in the free fraction. It is possible that the TCB in the free fraction serves as an indicator of the increase of lower-molecular-weight phenolic compounds liberated from HP after heat treatment.

There are many plants and herbs that have potential medical uses for humans, so it is necessary to pay more attention to the methods which can enhance the antioxidant capacity of their extracts. As reported before, the antioxidant capacity of citrus peel can be enhanced after heat treatment (Seok-Moon et al. 2004). Here, we employed HPLC-PDA instead of gas chromatography (GC)-MS to investigate the changes of phenolic acid content and FGs after heat treatment in order to achieve more precise and comprehensive results. Furthermore, a more complicated extraction method was adopted; as a result, more specific information could be obtained about heat treatment on the antioxidant capacity of citrus peel. Finally, we must be mindful of the possibility that perhaps there are other reasons responsible for the enhancement of antioxidant capacity of HP after heat treatment, which were not considered in this research.

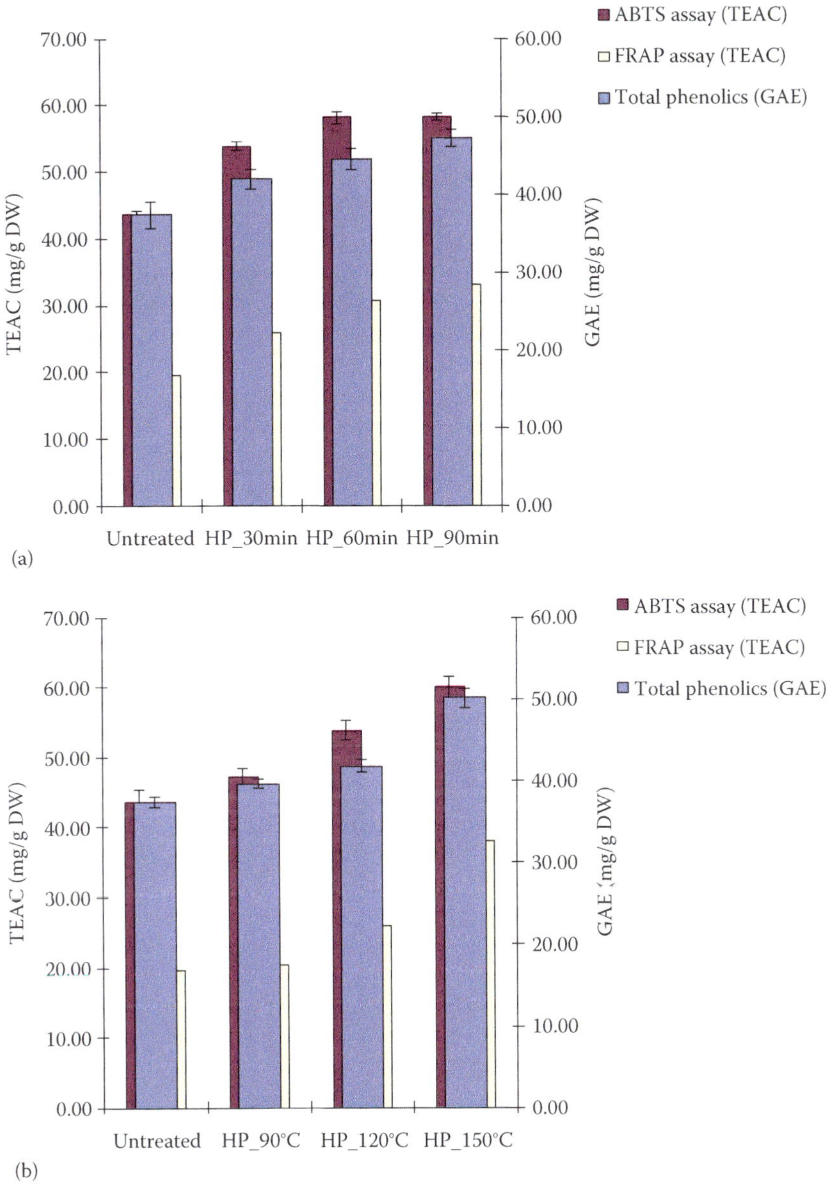

FIGURE 5.6 TPC, ABTS, and FRAP assay results with methanol extracts of untreated and heat-treated Huyou peel (HP) samples. (a) HP was heated at 120°C for 30, 60, or 90 min. (b) HP was heated at 90°C, 120°C, or 150°C for 30 min. (Modified from Xu, G. et al., *Journal of Agricultural and Food Chemistry* 55(2):330–335, 2007.)

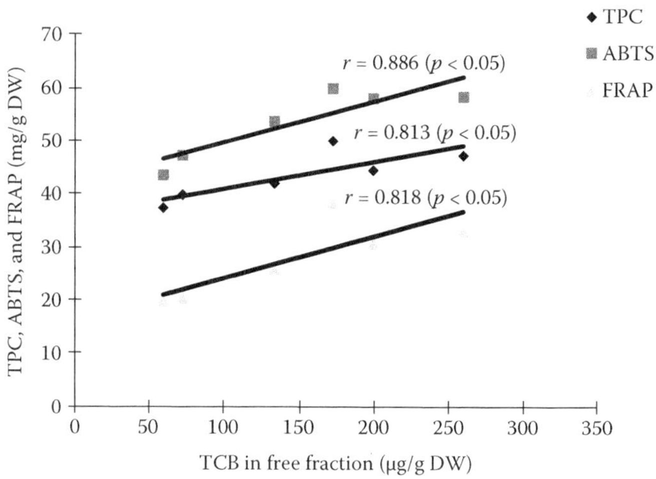

FIGURE 5.7 Correlations among TCB in the free fraction, TPC fraction, ABTS assay product, and FRAP assay product ($n = 6$). (Modified from Xu, G. et al., *Journal of Agricultural and Food Chemistry* 55(2):330–335, 2007.)

5.10 PHENOLIC COMPOUNDS AND ANTIOXIDANT CAPACITIES OF HOT WATER EXTRACTS OF CITRUS PEELS

China is one of the major producers of citrus fruits, and some *Citrus* species are recorded explicitly in the *Chinese Pharmacopoeia*, such as qingpi (dried immature fruit of tangerine, *C. reticulata*), chenpi (dried peels of mature *C. reticulata*), zhiqiao (dried mature fruit of sour orange, *C. aurantium* L.), and Zhishi (dried immature fruit of sour orange, *C. aurantium* L.), for having genuine medical functions. These traditional Chinese medicines are used to activate vital energy and circulation, eliminate phlegm, overcome physical stagnation, and so on. The primary active biological constituents present in citrus fruits are flavonoids, which include the following three types: flavanones, flavones, and flavonols. Phenolic acids are also present in considerable amounts in citrus fruits, and this has triggered some scientific interest (Rapisarda et al. 2003; Wang et al. 2007; Xu et al. 2007).

To prepare some of these traditional medicines in China, infusions are made by seeping selected herbal medicines in boiling water to extract their bioactive components. Hesperidin was found in high amounts in the infused extract and was considered the main functional component. However, citrus peels also contain many other naturally occurring flavonoids with biological properties, such as nobiletin and tangeretin. These flavonoids are attracting a great deal of interest for their potential anticancer, antiviral, anti-inflammatory, and antiatherogenic activities (Manthey et al. 2001; Middleton et al. 2000; Whitman et al. 2005). In south China, dried citrus peels were infused in boiling water and consumed just like tea, and this was deemed to be beneficial to one's health.

Satsuma mandarin (*Citrus unshiu* Marc.) and Ponkan (*C. poonensis* Hort. ex Tanaka) are two very popular citrus varieties in south China and have been examined to determine the efficiency of infusion cooking in extracting phenolic compounds

(FGs, PMFs, and phenolic acids) and also to determine the antioxidant activity of the peels. Most published papers have focused on the infusion of teas (Giulian et al. 2007; Kyle et al. 2007; Langley-Evans 2000; Mehra and Baker 2007; Mossion et al. 2008), but so far there have only been a few papers on the hot water extraction of chenpi.

Traditionally in China, traditional medicines were extracted by boiling water for 30 min or more, and compounds were extracted twice in most cases. In our study (Xu et al. 2008b), we investigated three infusion temperatures (40°C, 70°C, and 100°C), three extraction times (30, 60, and 90 min) and three reextractions (first, second, and third) on the yield of phenolic compounds and the antioxidant capacity of the hot water extracts of citrus peel.

5.10.1 Hot Water Extraction of Flavonoids from Satsuma Mandarin and Ponkan

Four citrus flavonoids consisting of two FGs (narirutin and hesperidin) and two PMFs (nobiletin and tangeretin) were detected simultaneously by HPLC-PDA. As seen in Table 5.11, Satsuma mandarin peels have a much higher FG content than

TABLE 5.11
Flavonoids in Hot Water Extracts and Peels of Satsuma Mandarin and Ponkan

		Flavonoids in Extract (mg/g DW)			
		FGs		PMFs	
Citrus Variety	Extraction Treatment	Narirutin	Hesperidin	Nobiletin	Tangeretin
Satsuma	40°C, 30 min	2.25	1.77	0.25	0.09
	70°C, 30 min	2.45	1.83	0.19	0.06
	100°C, 30 min	2.69	2.47	0.10	0.04
	100°C, 60 min	2.56	2.10	0.09	0.03
	100°C, 90 min	2.62	2.44	0.09	0.03
	100°C, 30 min (second)	0.93	1.55	0.04	0.02
	100°C, 30 min (third)	0.42	1.05	0.03	0.02
	Peel	7.66	62.01	0.31	0.16
Ponkan	40°C, 30 min	0.32	2.88	4.49	1.31
	70°C, 30 min	0.36	2.48	4.45	1.45
	100°C, 30 min	0.30	3.55	4.16	1.39
	100°C, 60 min	0.23	3.21	4.40	1.39
	100°C, 90 min	0.29	3.25	4.27	1.39
	100°C, 30 min (second)	0.12	2.45	0.79	0.15
	100°C, 30 min (third)	0.08	1.96	0.52	0.14
	Peel	1.10	38.94	8.70	4.30

Source: Modified from Xu, G. et al., *Journal of Food Science* 73(1):C11–C18, 2008.
Note: A 2-mL aliquot of water extract was mixed with 6 mL of methanol-dimethyl sulfoxide (1:1) for 12 h and shaken vigorously every 1 h in order to extract the FGs and PMFs. Similarly, 200 mg of citrus peel powder was extracted using 8 mL of methanol-dimethyl sulfoxide under the same conditions.

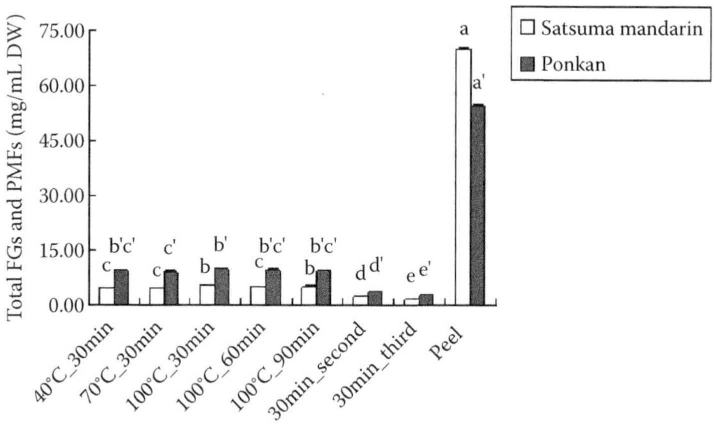

FIGURE 5.8 Total FGs and PMFs of hot water extracts of citrus peel. (Modified from Xu, G. et al., *Journal of Food Science* 73(1):C11–C18, 2008.)

Ponkan; however, Ponkan is more abundant in PMFs. These results are comparable with those reported by other researchers (Green et al. 2007; Lu et al. 2006). Although hesperidin was selected as the standard criteria of chenpi (Chinese Pharmacopoeia Commission 2010), it seems that FGs, especially hesperidin, are difficult to extract by hot water, while PMFs, especially nobiletin, are extracted in greater amounts in hot water. These results are in accordance with the report by Friedman et al. (2006), which indicated that significantly greater amounts of flavonoids were extracted from tea leaves with aqueous ethanol than with boiled water. Increasing the extraction temperature appeared to increase the yield of hesperidin to some extent but had little impact on the yield of narirutin, nobiletin, and tangeretin. It was noticed that the yield of PMFs at lower temperatures was even higher, and considerable portions of PMFs were extracted with hot water, compared to low extraction rates of hesperidin. A multiple extraction seemed more effective in extracting FGs than PMFs, and there were considerable amounts of FGs and PMFs in the third extraction.

As shown in Figure 5.8, very small portions of FGs and PMFs are extracted by using hot water, and the yields of total FGs and PMFs in Ponkan are much higher than from Satsuma mandarins. Since flavonoids were considered to be the main biological compounds in chenpi, Ponkan maybe a more suitable source of chenpi, even though Satsuma mandarin has a much higher content of hesperidin than Ponkan.

5.10.2 Phenolic Acid Contents of Hot Water Extracts of Satsuma Mandarin and Ponkan

Seven phenolic acids consisting of four hydrocinnamics (caffeic, *p*-coumaric, sinapic, and ferulic) and three hydrobenzoics (protocatechuic, *p*-hydroxybenzoic, and vanillic) were determined by HPLC-PDA, and the results are shown in Table 5.12. For both

TABLE 5.12
Phenolic Acid Contents of Hot Water Extracts and Peels of Satsuma Mandarin and Ponkan

		Phenolic Acid Content (mg/g DW)						
		Cinnamics					Benzoics	
Citrus Variety	Extraction Treatment	Caffeic	*p*-Coumaric	Ferulic	Sinapic	Protocatechuic	*p*-Hydroxybenzoic	Vanillic
Satsuma	40°C, 30min	134.1	244.5	2265.2	125.7	209.4	46.5	28.2
	70°C, 30 min	122.1	243.2	2285.5	123.7	90.4	48.6	37.7
	100°C, 30 min	120.4	248.0	2287.3	128.2	1140.3	46.4	31.9
	100°C, 60 min	113.6	235.4	2190.3	120.2	1118.8	44.3	27.3
	100°C, 90 min	110.1	225.9	2091.6	126.9	767.0	43.6	26.7
	100°C, 30 min (second)	33.6	60.9	530.4	39.1	113.8	16.1	9.9
	100°C, 30 min (third)	12.5	18.1	117.3	17.8	92.6	8.0	4.6
	Peel	143.7	299.7	2,755.6	194.9	16.6	62.4	50.3
Ponkan	40°C, 30 min	811.2	626.3	2652.8	148.9	82.8	37.4	21.9
	70°C, 30 min	790.8	604.8	2582.3	136.8	29.6	37.1	22.6
	100°C, 30 min	766.4	581.2	2464.9	139.7	343.3	35.7	15.6
	100°C, 60 min	761.4	585.4	2477.0	148.1	466.2	35.7	16.5
	100°C, 90 min	703.9	542.2	2285.4	141.4	348.8	32.6	13.0
	100°C, 30 min (second)	312.7	180.2	808.1	49.3	249.5	16.4	5.3
	100°C, 30 min (third)	121.7	52.4	255.1	17.4	15.4	10.0	2.9
	Peel	1,089.0	682.9	3,099.2	179.6	33.1	48.3	26.3

Source: Modified from Xu, G. et al., *Journal of Food Science* 73(1):C11–C18, 2008.

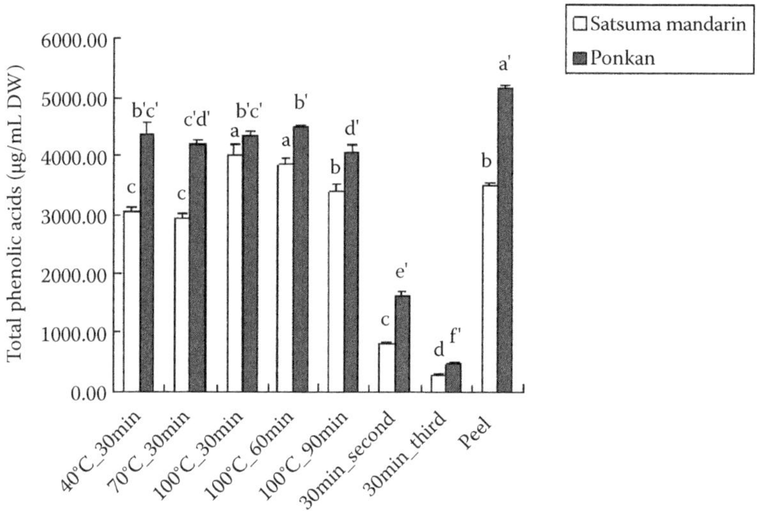

FIGURE 5.9 Total phenolic acids of hot water extracts of citrus peel. (Modified from Xu, G. et al., *Journal of Food Science* 73(1):C11–C18, 2008.)

Satsuma mandarin and Ponkan, ferulic was the dominant acid, while Ponkan had a much higher content of caffeic and *p*-coumaric than Satsuma mandarin. It appears that increasing the extraction temperature affects the change of each phenolic acid differently. However, when increasing the extraction time at 100°C, the level of individual phenolic acids exhibited a general declining trend, and we believe that the phenolic acids are broken down when subjected to 100°C hot water conditions. Surprisingly, the content of protocatechuic acid was much higher using hot water extraction than from conventional citrus peel extraction. We suggest that protocatechuic could be transferred from other phenolic acids under high temperatures, since its content was rather low in the citrus peels; more work should be done in this field. After two consecutive extractions, the yields of individial phenolic acid and total phenolic acids are higher than their contents in citrus peels; therefore, we supposed that hot water could make the bound form of phenolic acids easier to be hydrolyzed, while alkaline conditions could not hydrolyze citrus peel as completely as the hot water extracts. As shown in Figure 5.9, Ponkan has a higher total phenolic acid content than Satsuma mandarin based on the hot water extraction results. From this perspective, Ponkan appears to be a more suitable source of chenpi than Satsuma mandarin, since phenolic acids are also compounds with multiple functional properties.

5.10.3 Antioxidant Capacities of Hot Water Extracts and Methanol Extracts of Citrus Peels via TPC, FRAP, and DPPH Assays

The TPC of hot water and methanol extracts of citrus peel were determined by the Folin–Ciocalteu assay, and their antioxidant capacities were evaluated by the FRAP and DPPH assays, as shown in Figure 5.10. According to the results for total phenolic

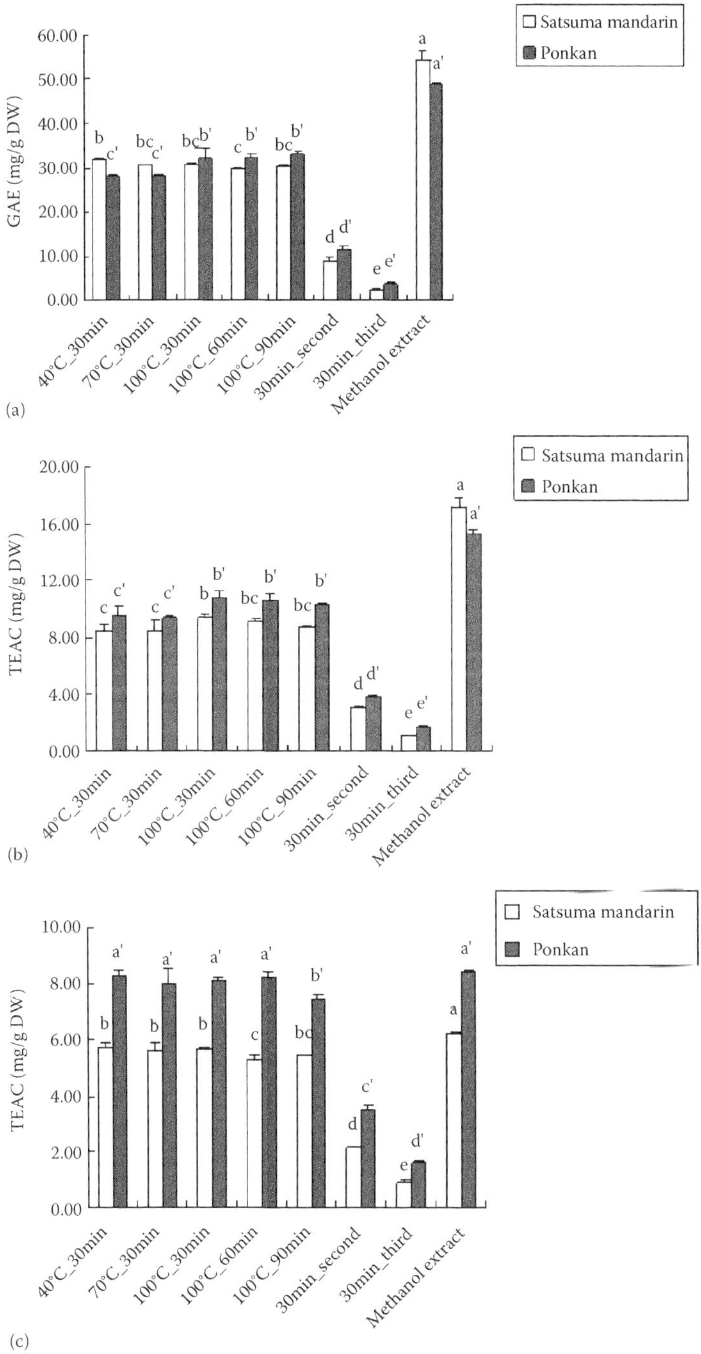

FIGURE 5.10 TPC (a), FRAP (b) assay results with hot water extracts and methanol extracts of citrus peel. DPPH (c) assay results with hot water extracts and methanol extracts of citrus peel. (Modified from Xu, G. et al., *Journal of Food Science* 73(1):C11–C18, 2008.)

content and FRAP, higher prolonged extraction temperatures yield higher total phenolic content and TEAC values for Ponkan, but these conditions had little influence on Satsuma mandarin extraction. Increasing the extraction time did not increase the total phenolic content and TEAC values. As for the DPPH assay, both extraction temperature and time had very little impact on the antioxidant capacity of the hot water extracts. Based on the above observations, we believe that increasing the extraction time or temperature does not increase the antioxidant capacity of the water extract. When the antioxidant capacities of water extracts and methanol extracts were compared, we found that the second extraction in hot water yielded a similar total phenolic content (Folin–Ciocalteu assay) and TEAC (FRAP assay) as the methanol extraction. Furthermore, the first infusion of hot water extract had an equivalent TEAC to that via DPPH of the methanol extract. It has been reported that the water extract of dried Satsuma mandarin peels has a higher antioxidant capacity than the methanol extract (Higashi-Okai et al. 2002). As for Ponkan and Satsuma mandarin, the hot water extract of Ponkan has a higher TEAC than Satsuma mandarin, while the TPC are similar. In conclusion, hot water extraction was effective in extracting antioxidant compounds in citrus peels and could acquire an antioxidant capacity comparable to those from the methanol extract. We feel that the high antioxidant capacity observed in chenpi will make them effective agents in fighting some diseases.

5.11 SUMMARY

Citrus fruits are one of the major agricultural products in China. The phenolic compounds represent a class of secondary metabolites within citrus that exihibit important biological properties. Juices from 15 citrus varieties (seven mandarins, four sweet oranges, one lemon, one grapefruit, and two pomelos) grown in China were investigated. Huyou and Hybrid 439 were considered valuable varieties based on their antioxidant capacities and high nutritional values. The composition of the major phenolic compounds in the flesh and peels of Ponkan, Satsuma mandarin, and Huyou during maturity were determined by HPLC. It was found that their contents decreased with maturity, and a similar tendency was also observed for the antioxidant capacities of the methanol extracts. Unripe citrus fruits were especially rich in flavonoids; Ponkan was a suitable source of hesperidin, nobiletin, and tangeretin, and Satsuma mandarin could be used to extract hesperidin, while Huyou was a good source of naringin and neohesperidin. The effects of heat treatment on Huyou peel in terms of phenolic compounds and antioxidant capacity were investigated. HPLC-PDA was used to determine the levels of phenolic acids. The analysis divided the phenolic compounds into four fractions: free, ester, glycoside, and ester-bound plus FGs in Huyou peel before and after heat treatment. The results showed that after heat treatment, the free phenolic acid fraction increased, while the ester, glycoside, and ester-bound fractions decreased, and the content of total FGs declined ($p < 0.05$). Furthermore, antioxidant activity of the methanol extracts of Huyou peel increased ($p < 0.05$). The correlation coefficients among total phenolics content, ABTS, FRAP assay, and TCB in the free fraction were significantly higher ($p < 0.05$), which means that the increase of total antioxidant capacity of HP extract was due at least in part to the increase of TCB in the free fraction. In addition, the FGs may have been degraded when heated at higher temperatures and longer times. Therefore, we suggested that an appropriate heat treatment be used to enhance the antioxidant capacity of citrus peel extracts. The efficiency of infusion cooking on extracting phenolic compounds and also the antioxidant activity of hot water extracts of citrus peels were investigated. Peels of two citrus varieties, namely, Satsuma mandarin and Ponkan, which belong to *C. reticulata*, were selected. The results showed that hot water extraction was efficient in extracting phenolic acids. As for the citrus flavonoids (narirutin, nobiletin, and tangeretin), they were easier to extract than hesperidin. The results of the antioxidant capacity assays indicated that for citrus peels, hot water extraction produced almost the same antioxidant capacity as methanol extraction. We suggest that Ponkan is a more suitable source of chenpi, since its hot water extract has a much higher content of phenolic acids, FGs, and PMFs and a higher antioxidant capacity than Satsuma mandarin. Finally, we also suggest that two successive extractions at 100°C for 30 min is a more efficient method of extracting the phenolic compounds in chenpi.

ACKNOWLEDGMENT

Financial support from the China Nature Sciences Foundation (Project C200501) is acknowledged.

REFERENCES

Abe, K., Y. Nakada, A. Suzuki, and E. Yumioka. 1993. Biliary metabolites of hesperetin in rats. *Japanese Journal of Pharmacognosy* 47(1):43–46.

Almada-Ruiz, E., M. A. Martinez-Tellez, M. M. Hernandez-Alamos, S. Vallejo, E. Primo-Yufera, and I. Vargas-Arispuro. 2003. Fungicidal potential of methoxylated flavones from citrus for in vitro control of *Colletotrichum gloeosporioides*, causal agent of anthracnose disease in tropical fruits. *Pest Management Science* 59(11):1245–1249.

Andreasen, M. F., P. A. Kroon, G. Williamson, and M. T. Garcia-Conesa. 2001. Intestinal release and uptake of phenolic antioxidant diferulic acids. *Free Radical Biology and Medicine* 31(3):304–314.

Assini, J. M., E. E. Mulvihill, and M. W. Huff. 2013. Citrus flavonoids and lipid metabolism. *Current Opinion in Lipidology* 24(1):34–40.

Ayaz, F. A., S. Hayırlıoglu-Ayaz, J. Gruz, O. Novak, and M. Strnad. 2005. Separation, characterization, and quantitation of phenolic acids in a little-known blueberry (*Vaccinium arctostaphylos* L.) fruit by HPLC-MS. *Journal of Agricultural and Food Chemistry* 53(21):8116–8122.

Balasundram, N., K. Sundram, and S. Samman. 2006. Phenolic compounds in plants and agri-industrial by-products: Antioxidant activity, occurrence, and potential uses. *Food Chemistry* 99(1):191–203.

Benavente-Garcia, O., and J. Castillo. 2008. Update on uses and properties of citrus flavonoids: New findings in anticancer, cardiovascular, and anti-inflammatory activity. *Journal of Agricultural and Food Chemistry* 56(15):6185–6205.

Bocco, A., M. E. Cuvelier, H. Richard, and C. Berset. 1998. Antioxidant activity and phenolic composition of citrus peel and seed extracts. *Journal of Agricultural and Food Chemistry* 46(6):2123–2129.

Castelluccio, C., G. P. Bolwell, C. Gerrish, and C. Rice-Evans. 1996. Differential distribution of ferulic acid to the major plasma constituents in relation to its potential as an antioxidant. *Biochemical Journal* 316(2):691–694.

Cha, J. Y., Y. S. Cho, I. Kim, T. Anno, S. M. Rahman, and T. Yanagita. 2001. Effect of hesperetin, a citrus flavonoid, on the liver triacylglycerol content and phosphatidate phosphohydrolase activity in orotic acid-fed rats. *Plant Foods for Human Nutrition* 56(4):349–358.

Chen, J. H., and C. T. Ho. 1997. Antioxidant activities of caffeic acid and its related hydroxycinnamic acid compounds. *Journal of Agricultural and Food Chemistry* 45(7):2374–2378.

Chinese Pharmacopoeia Commission. 2010. *Pharmacopoeia of the People's Republic of China*. Beijing: Chemical Industry Press.

Clifford, M. N. 1999. Chlorogenic acids and other cinnamates: Nature, occurrence and dietary burden. *Journal of the Science of Food and Agriculture* 79(3):362–372.

Crespo, M. E., J. Galvez, T. Cruz, M. A. Ocete, and A. Zarzuelo. 1999. Anti-inflammatory activity of diosmin and hesperidin in rat colitis induced by TNBS. *Planta Medica* 65(7):651–653.

Da Silva Emim, J. A., A. S. Oliveira, and A. J. Lapa. 1994. Pharmacological evaluation of the anti-inflammatory activity of a citrus bioflavonoid, hesperidin, and the isoflavonoids, duartin and claussequinone, in rats and mice. *Journal of Pharmacy and Pharmacology* 46(2):118–122.

Del Rio, J. A., M. C. Arcas, O. Benavente-Garcia, and A. Ortuno. 1998. Citrus polymethoxylated flavones can confer resistance against *Phytophthora citrophthora*, *Penicillium digitatum*, and *Geotrichum* species. *Journal of Agricultural and Food Chemistry* 46(10):4423–4428.

Del Río, J. A., M. D. Fuster, F. Sabater, I. Porras, A. García-Lidón, and A. Ortuño. 1997. Selection of citrus varieties highly productive for the neohesperidin dihydrochalcone precursor. *Food Chemistry* 59(3):433–437.

Delaney, B., K. Phillips, D. Buswell, B. Mowry, D. Nickels, D. Cox, H. B. Wang, and J. Manthey. 2001. Immunotoxicity of a standardized citrus polymethoxylated flavone extract. *Food and Chemical Toxicology* 39(11):1087–1094.

Delaney, B., K. Phillips, C. Vasquez, A. Wilson, D. Cox, H. B. Wang, and J. Manthey. 2002. Genetic toxicity of a standardized mixture of citrus polymethoxylated flavones. *Food and Chemical Toxicology* 40(5):617–624.

Di Majo, D., M. Giammanco, M. La Guardia, E. Tripoli, S. Giammanco, and E. Finotti. 2005. Flavanones in citrus fruit: Structure–antioxidant activity relationships. *Food Research International* 38(10):1161–1166.

Di Mauro, A., A. Passerini, P. Rapisarda, and E. Maccarone. 2002. Distribution of hydroxycinnamic acids as a criterion to evaluate variety and geographical origin of Italian orange juices. *Italian Journal of Food Science* 14(3):301–315.

Eguchi, A., A. Murakami, and H. Ohigashi. 2006. Nobiletin, a citrus flavonoid, suppresses phorbol ester-induced expression of multiple scavenger receptor genes in THP-1 human monocytic cells. *FEBS Letters* 580(13):3321–3328.

Friedman, M., C. E. Levin, S. H. Choi, E. Kozukue, and N. Kozukue. 2006. HPLC analysis of catechins, theaflavins, and alkaloids in commercial teas and green tea dietary supplements: Comparison of water and 80% ethanol/water extracts. *Journal of Food Science* 71(6):C328–C337.

Giulian, R., C. E. I. dos Santos, S. D. Shubeita, L. M. da Silva, J. F. Dias, and M. L. Yoneama. 2007. Elemental characterization of commercial mate tea leaves (*Ilex paraguariensis* A. St.-Hil.) before and after hot water infusion using ion beam techniques. *Journal of Agricultural and Food Chemistry* 55(3):741–746.

Gorinstein, S., M. Cvikrova, I. Machackova, R. Haruenkit, Y. S. Park, S. T. Jung, K. Yamamoto, A. L. Martinez Ayala, E. Katrich, and S. Trakhtenberg. 2004. Characterization of antioxidant compounds in Jaffa sweeties and white grapefruits. *Food Chemistry* 84(4):503–510.

Green, C. O., A. O. Wheatley, A. U. Osagie, E. Morrison, and H. N. Asemota. 2007. Determination of polymethoxylated flavones in peels of selected Jamaican and Mexican citrus (*Citrus* spp.) cultivars by high-performance liquid chromatography. *Biomedical Chromatography* 21(1):48–54.

Guardia, T., A. E. Rotelli, A. O. Juarez, and L. E. Pelzer. 2001. Anti-inflammatory properties of plant flavonoids. Effects of rutin, quercetin and hesperidin on adjuvant arthritis in rat. *Farmaco* 56(9):683–687.

Gulcin, I. 2006. Antioxidant activity of caffeic acid (3,4-dihydroxycinnamic acid). *Toxicology* 217(2–3):213–220.

Herrmann, K. 1989. Occurrence and content of hydroxycinnamic and hydroxybenzoic acid compounds in foods. *Critical Reviews in Food Science and Nutrition* 28(4):315–347.

Higashi-Okai, K., K. Kamimoto, A. Yoshioka, Y. Okai. 2002. Potent suppressive activity of fresh and dried peels from Satsuma mandarin (Citrus unshiu Marcorv.) on hydroperoxide generation from oxidized linoleic acid. *Phytotherapy Research* 16(8):781–784.

Hirano, T., K. Abe, M. Gotoh, and K. Oka. 1995. Citrus flavone tangeretin inhibits leukaemic HL-60 cell growth partially through induction of apoptosis with less cytotoxicity on normal lymphocytes. *British Journal of Cancer* 72(6):1380–1388.

Hsiu, S. L., T. Y. Huang, Y. C. Hou, D. H. Chin, and P. D. L. Chao. 2002. Comparison of metabolic pharmacokinetics of naringin and naringenin in rabbits. *Life Sciences* 70(13):1481–1489.

Itoigawa, M., K. Takeya, and H. Furukawa. 1994. Cardiotonic flavonoids from citrus plants (*Rutaceae*). *Biological and Pharmaceutical Bulletin* 17(11):1519–1521.

Iwase, Y., Y. Takemura, M. Ju-ichi, M. Yano, C. Ito, H. Furukawa, T. Mukainaka, M. Kuchide, H. Tokuda, and H. Nishino. 2001. Cancer chemopreventive activity of 3,5,6,7,8,3,4′-heptamethoxyflavone from the peel of citrus plants. *Cancer Letters* 163(1):7–9.

Jayaprakasam, B., M. Vanisree, Y. Zhang, D. L. Dewitt, and M. G. Nair. 2006. Impact of alkyl esters of caffeic and ferulic acids on tumor cell proliferation, cyclooxygenase enzyme, and lipid peroxidation. *Journal of Agricultural and Food Chemistry* 54(15):5375–5381.

Kandaswami, C., E. Perkins, G. Drzewiecki, D. S. Soloniuk, and E. Middleton, Jr. 1992. Differential inhibition of proliferation of human squamous cell carcinoma, gliosarcoma and embryonic fibroblast-like lung cells in culture by plant flavonoids. *Anticancer Drugs* 3(5):525–530.

Kandaswami, C., E. Perkins, D. S. Soloniuk, G. Drzewiecki, and E. Middleton, Jr. 1991. Antiproliferative effects of citrus flavonoids on a human squamous cell carcinoma in vitro. *Cancer Letters* 56(2):147–152.

Kaur, C., and H. C. Kapoor. 2001. Antioxidants in fruits and vegetables: The millennium's health. *International Journal of Food Science and Technology* 36(7):703–725.

Kawaguchi, K., S. I. Kikuchi, R. Hasunuma, H. Maruyama, R. Ryll, and Y. Kumazawa. 2004. Suppression of infection-induced endotoxin shock in mice by a citrus flavanone naringin. *Planta Medica* 70(1):17–22.

Kawaii, S., Y. Tomono, E. Katase, K. Ogawa, and M. Yano. 1999a. Antiproliferative activity of flavonoids on several cancer cell lines. *Bioscience Biotechnology and Biochemistry* 63(5):896–899.

Kawaii, S., Y. Tomono, E. Katase, K. Ogawa, and M. Yano. 1999b. Quantitation of flavonoid constituents in citrus fruits. *Journal of Agricultural and Food Chemistry* 47(9):3565–3571.

Kim, D. O., and C. Y. Lee. 2004. Comprehensive study on vitamin C equivalent antioxidant capacity (VCEAC) of various polyphenolics in scavenging a free radical and its structural relationship. *Critical Reviews in Food Science and Nutrition* 44(4):253–273.

Kim, H. K., T. S. Jeong, M. K. Lee, Y. B. Park, and M. S. Choi. 2003. Lipid-lowering efficacy of hesperetin metabolites in high-cholesterol fed rats. *Clinica Chimica Acta* 327(1–2):129–137.

King, A., and G. Young. 1999. Characteristics and occurrence of phenolic phytochemicals. *Journal of the American Dietetic Association* 99(2):213–218.

Koshihara, Y., T. Neichi, and S. Murota. 1984. Caffeic acid is a selective inhibitor for leukotriene biosynthesis. *Biochimica et Biophysica Acta* 792(1):92–97.

Koyuncu, H., B. Berkarda, F. Baykut, G. Soybir, C, Alatli, H. Gül, and M. Altun. 1999. Preventive effect of hesperidin against inflammation in CD-1 mouse skin caused by tumor promoter. *Anticancer Research* 19(4 Part B):3237–3241.

Kris-Etherton, P. M., K. D. Hecker, A. Bonanome, S. M. Coval, A. E. Binkoski, K. F. Hilpert, A. E. Griel, and T. D. Etherton. 2002. Bioactive compounds in foods: Their role in the prevention of cardiovascular disease and cancer. *American Journal of Medicine* 113(Supplement 2):71–88.

Kurowska, E. M., and J. A. Manthey. 2004. Hypolipidemic effects and absorption of citrus polymethoxylated flavones in hamsters with diet-induced hypercholesterolemia. *Journal of Agricultural and Food Chemistry* 52(10):2879–2886.

Kyle, J. A. M., P. C. Morrice, G. McNeill, and G. G. Duthie. 2007. Effects of infusion time and addition of milk on content and absorption of polyphenols from black tea. *Journal of Agricultural and Food Chemistry* 55(12):4889–4894.

Langley-Evans, S. C. 2000. Antioxidant potential of green and black tea determined using the ferric reducing power (FRAP) assay. *International Journal of Food Sciences and Nutrition* 51(3):181–188.

Li, R. W., A. G. Theriault, K. Au, T. D. Douglas, A. Casaschi, E. M. Kurowska, and R. Mukherjee. 2006. Citrus polymethoxylated flavones improve lipid and glucose homeostasis and modulate adipocytokines in fructose-induced insulin resistant hamsters. *Life Sciences* 79(4):365–373.

Lin, N., T. Sato, Y. Takayama, Y. Mimaki, Y. Sashida, M. Yano, and A. Ito. 2003. Novel anti-inflammatory actions of nobiletin, a citrus polymethoxy flavonoid, on human synovial fibroblasts and mouse macrophages. *Biochemical Pharmacology* 65(12):2065–2071.

Lodovici, M., F. Guglielmi, M. Meoni, and P. Dolara. 2001. Effect of natural phenolic acids on DNA oxidation in vitro. *Food and Chemical Toxicology* 39(12):1205–1210.

Lu, Y., C. Zhang, P. Bucheli, and D. Wei. 2006. Citrus flavonoids in fruit and traditional Chinese medicinal food ingredients in China. *Plant Foods for Human Nutrition* 61(2):57–65.

Luthria, D. L., and S. Mukhopadhyay. 2006. Influence of sample preparation on assay of phenolic acids from eggplant. *Journal of Agricultural and Food Chemistry* 54(1):41–47.

Malterud, K. E., and K. M. Rydland. 2000. Inhibitors of 15-lipoxygenase from orange peel. *Journal of Agricultural and Food Chemistry* 48(11):5576–5580.

Manach, C., C. Morand, A. Gil-Izquierdo, C. Bouteloup-Demange, and C. Remesy. 2003. Bioavailability in humans of the flavanones hesperidin and narirutin after the ingestion of two doses of orange juice. *European Journal of Clinical Nutrition* 57(2):235–242.

Manthey, J. A., and K. Grohmann. 2001. Phenols in citrus peel byproducts. Concentrations of hydroxycinnamates and polymethoxylated flavones in citrus peel molasses. *Journal of Agricultural and Food Chemistry* 49(7):3268–3273.

Manthey, J. A., and N. Guthrie. 2002. Antiproliferative activities of citrus flavonoids against six human cancer cell lines. *Journal of Agricultural and Food Chemistry* 50(21):5837–5843.

Manthey, J. A., N. Guthrie, and K. Grohmann. 2001. Biological properties of citrus flavonoids pertaining to cancer and inflammation. *Current Medicinal Chemistry* 8(2):135–153.

Mehra, A., and C. L. Baker. 2007. Leaching and bioavailability of aluminium, copper and manganese from tea (Camellia sinensis). *Food Chemistry* 100(4):1456–1463.

Middleton E, Jr., C. Kandaswami, and T. C. Theoharides. 2000. The effects of plant flavonoids on mammalian cells: Implications for inflammation, heart disease, and cancer. *Pharmacological Reviews* 52(4):673–751.

Miyake, Y., K. Shimoi, S. Kumazawa, K. Yamamoto, N. Kinae, and T. Osawa. 2000. Identification and antioxidant activity of flavonoid metabolites in plasma and urine of eriocitrin-treated rats. *Journal of Agricultural and Food Chemistry* 48(8):3217–3224.

Mizuno, M., M. Iinuma, M. Ohara, T. Tanaka, and M. Iwamasa. 1991. Chemotaxonomy of the genus *Citrus* based on polymethoxyflavones. *Chemical and Pharmaceutical Bulletin* 39(4):945–949.

Morton, L. W., R. A. A. Caccetta, I. B. Puddey, and K. D. Croft. 2000. Chemistry and biological effects of dietary phenolic compounds: Relevance to cardiovascular disease. *Clinical and Experimental Pharmacology and Physiology* 27(3):152–159.

Mossion, A., M. Potin-Gautier, S. Delerue, I. L. Hecho, and P. Behra. 2008. Effect of water composition on aluminium, calcium and organic carbon extraction in tea infusions. *Food Chemistry* 106(4):1467–1475.

Mulvihill, E. E., and M. W. Huff. 2012. Citrus flavonoids and the prevention of atherosclerosis. *Cardiovascular and Hematological Disorders Drug Targets* 12(2):84–91.

Nardini, M., F. Natella, V. Gentili, M. D. Felice, and C. Scaccini. 1997. Effect of caffeic acid dietary supplementation on the antioxidant defense system in rat: An in vivo study. *Archives of Biochemistry and Biophysics* 342(1):157–160.

Nielsen, I. L. F., W. S. S. Chee, L. Poulsen, E. Offord-Cavin, S. E. Rasmussen, H. Frederiksen, M. Enslen, D. Barron, M. N. Horcajada, and G. Williamson. 2006. Bioavailability is improved by enzymatic modification of the citrus flavonoid hesperidin in humans: A randomized, double-blind, crossover trial. *Journal of Nutrition* 136(2):404–408.

Nielsen, S. E., V. Breinholt, C. Cornett, and L. O. Dragsted. 2000. Biotransformation of the citrus flavone tangeretin in rats. Identification of metabolites with intact flavane nucleus. *Food and Chemical Toxicology* 38(9):739–746.

Nogata, Y., K. Sakamoto, H. Shiratsuchi, T. Ishii, M. Yano, and H. Ohta. 2006. Flavonoid composition of fruit tissues of *Citrus* species. *Bioscience, Biotechnology, and Biochemistry* 70(1):178–192.

Olthof, M. R., P. C. H. Hollman, and M. B. Katan. 2001. Chlorogenic acid and caffeic acid are absorbed in humans. *Journal of Nutrition* 131(1):66–71.

Ooghe, W., and C. Detavernier. 1999. Flavonoids as authenticity markers for *Citrus sinensis* juice. *Fruit Processing* 9(8):308–313.

Ortuno, A., I. Reynaldo, M. D. Fuster, J. Botia, D. Garcia Puig, F. Sabater, A. Garcia Lidon, I. Porras, and J. A. Del Rio. 1997. *Citrus* cultivars with high flavonoid contents in the fruits. *Scientia Horticulturae* 68(14):231–236.

Pan, G. G., P. A. Kilmartin, B. G. Smith, and L. D. Melton. 2002. Detection of orange juice adulteration by tangelo juice using multivariate analysis of polymethoxylated flavones and carotenoids. *Journal of the Science of Food and Agriculture* 82(4):421–427.

Peleg, H., M. Naim, R. L. Rouseff, and U. Zehavi. 1991. Distribution of bound and free phenolic acid in oranges (*Citrus sinensis*) and grapefruits (*Citrus paradisi*). *Journal of the Science of Food and Agriculture* 57(3):417–426.

Peterson, J. J., G. R. Beecher, S. A. Bhagwat, J. T. Dwyer, S. E. Gebhardt, D. B. Haytowitz, and J. M. Holden. 2006. Flavanones in grapefruit, lemons, and limes: A compilation and review of the data from the analytical literature. *Journal of Food Composition and Analysis* 19:S74–S80.

Plumb, G. W., M. T. Garcia-Conesa, P. A. Kroon, M. Rhodes, S. Ridley, and G. Williamson. 1999. Metabolism of chlorogenic acid by human plasma, liver, intestine and gut microflora. *Journal of the Science of Food and Agriculture* 79(3):390–392.

Rao, C. V., D. Desai, B. Simi, N. Kulkarni, S. Amin, and B. S. Reddy. 1993. Inhibitory effect of caffeic acid esters on azoxymethane-induced biochemical changes and aberrant crypt foci formation in rat colon. *Cancer Research* 53(18):4182–4188.

Rapisarda, P., G. Carollo, B. Fallico, F. Tomaselli, and E. Maccarone. 1998. Hydroxycinnamic acids as markers of Italian blood orange juices. *Journal of Agricultural and Food Chemistry* 46(27):464–470.

Rapisarda, P., P. Pannuzzo, G. Romano, and G. Russo. 2003. Juice components of a new pigmented citrus hybrid *Citrus sinensis* (L.) Osbeck × *Citrus clementina* Hort. ex Tan. *Journal of Agricultural and Food Chemistry* 51(6):1611–1616.

Rechner, A. R., A. S. Pannala, and C. A. Rice-Evans. 2001. Caffeic acid derivatives in artichoke extract are metabolised to phenolic acids in vivo. *Free Radical Research* 35(2):195–202.

Robbins, R. J. 2003. Phenolic acids in foods: An overview of analytical methodology. *Journal of Agricultural and Food Chemistry* 51(10):2866–2887.

Saija, A., A. Tomaino, R. Lo Cascio, D. Trombetta, A. Proteggente, A. De Pasquale, N. Uccella, and F. Bonina. 1999. Ferulic and caffeic acids as potential protective agents against photooxidative skin damage. *Journal of the Science of Food and Agriculture* 79(3):476–480.

Saija, A., A. Tomaino, D. Trombetta, A. De Pasquale, N. Uccella, T. Barbuzzi, D. Paolino, and F. Bonina. 2000. In vitro and in vivo evaluation of caffeic and ferulic acids as topical photoprotective agents. *International Journal of Pharmaceutics* 199(1):39–47.

Scalbert, A., and G. Williamson. 2000. Dietary intake and bioavailability of polyphenols. *Journal of Nutrition* 130(8):2073S–2085S.

Seok-Moon, J., K. So-Young, K. Dong-Ryul, J. Seong-Chun, K. C. Nam, D. U. Ahn, and L. Seung-Cheol. 2004. Effect of heat treatment on the antioxidant activity of extracts from citrus peels. *Journal of Agricultural and Food Chemistry* 52(11):3389–3393.

Seung-Cheol, L., J. Seok-Moon, K. So-Young, K. C. Nam, and D. U. Ahn. 2005. Effect of far-infrared irradiation on the antioxidant activity of defatted sesame meal extracts. *Journal of Agricultural and Food Chemistry* 53(5):1495–1498.

So, F. V., N. Guthrie, A. F. Chambers, M. Moussa, and K. K. Carroll. 1996. Inhibition of human breast cancer cell proliferation and delay of mammary tumorigenesis by flavonoids and citrus juices. *Nutrition and Cancer* 26(2):167–181.

Wang, Y.-C., Y.-C. Chuang, and Y.-Y. Ku. 2007. Quantitation of bioactive compounds in citrus fruits cultivated in Taiwan. *Food Chemistry* 102(4):1163–1171.

Whitman, S. C., E. M. Kurowska, J. A. Manthey, and A. Daugherty. 2005. Nobiletin, a citrus flavonoid isolated from tangerines, selectively inhibits class A scavenger receptor-mediated metabolism of acetylated LDL by mouse macrophages. *Atherosclerosis* 178(1):25–32.

Wilmsen, P. K., D. S. Spada, and M. Salvador. 2005. Antioxidant activity of the flavonoid hesperidin in chemical and biological systems. *Journal of Agricultural and Food Chemistry* 53(12):4757–4761.

Xu, G., J. Chen, D. Liu, Y. H. Zhang, P. Jiang, and X. Ye. 2008a. Minerals, phenolic compounds and antioxidant capacity of citrus peel extract by hot water. *Journal of Food Science* 73(1):C11–C18.

Xu, G., D. Liu, J. Chen, X. Ye, Y. Ma, and J. Shi. 2008b. Juice components and antioxidant capacity of citrus varieties cultivated in China. *Food Chemistry* 106(2):545–551.

Xu, G., D. Liu, J. Chen, X. Ye, and J. Shi. 2009. Composition of major flavonone glycosides and antioxidant capacity of three citrus varieties. *Journal of Food Biochemistry* 33(5):453–469.

Xu, G., X. Ye, J. Chen, and D. Liu. 2007. Effect of heat treatment on the phenolic compounds and antioxidant capacity of citrus peel extract. *Journal of Agricultural and Food Chemistry* 55(2):330–335.

Yasuda, T., Y. Yoshimura, H. Yabuki, T. Nakazawa, K. Ohsawa, Y. Mimaki, and Y. Sashida. 2003. Urinary metabolites of nobiletin orally administered to rats. *Chemical and Pharmaceutical Bulletin* 51(12):1426–1428.

Yi, L.-Z., D.-L. Yuan, Y.-Z. Liang, P.-S. Xie, and Y. Zhao. 2007. Quality control and discrimination of pericarpium citri reticulatae and pericarpium citri reticulatae viride based on high-performance liquid chromatographic fingerprints and multivariate statistical analysis. *Analytica Chimica Acta* 588(2):207–215.

Yi, Z., Y. Yu, Y. Liang, and B. Zeng. 2007. In vitro antioxidant and antimicrobial activities of the extract of pericarpium citri reticulatae of a new *Citrus* cultivar and its main flavonoids. *LWT Food Science and Technology* 41(4):597–603.

Yu, J., L. M. Wang, R. L. Walzem, E. G. Miller, L. M. Pike, and B. S. Patil. 2005. Antioxidant activity of citrus limonoids, flavonoids, and coumarins. *Journal of Agricultural and Food Chemistry* 53(6):2009–2014.

6 Carotenoids in Citrus

Yujing Sun, Huan Cheng, Yaqin Ma, and Qianying (Sophia) Ye

CONTENTS

6.1 INTRODUCTION

In nature, there are more than 700 identified carotenoids; they are divided into two groups, carotenes, which have a hydrocarbon structure, and xanthophylls, which contain oxygen atoms in their structure. Carotenoids are a group of natural pigments present in fruits and vegetables that give them their yellow to red colors. All carotenoids possess some common chemical features: a polyisoprenoid structure, a long conjugated chain of double bonds, and a near-bilateral symmetry around the central double bond. Different carotenoids are derived essentially by modification of the base structure via cyclization of the end groups and by the incorporation of oxygen, functions that give them their characteristic colors and antioxidant properties.

For human health benefits, their importance comes from their activities related to provitamin A, antioxidants, cell differentiation and proliferation regulators, cell-cell communication stimulators, immune function modulators, carcinogen metabolism modulators, and blue light filters. Many studies have shown strong correlations between carotenoid intake and improvement of the immune system and the reduction of risk of some degenerative diseases, such as cancer, cardiovascular disease, cataracts, and macular degeneration (Fernández-García et al. 2012; Chatterjee et al. 2012; Kaulmann and Bohn 2014).

Citrus fruits are a good source of carotenoids, with high β-cryptoxanthin, lutein, and zeaxanthin contents, and also some special carotenoids, such as β-citraurin and violaxanthin. Additionally, the concentration of carotenoids in citrus are higher in peels than in pulp, so citrus by-products (peels, and so on) could potentially be a good source of carotenoids for the production of nutraceuticals.

6.2 CAROTENOIDS IN CITRUS

Citrus fruits contain a complex mixture of carotenoids and have the largest number of carotenoids found in any type of fruit. Approximately 115 different carotenoids have been reported in citrus fruits, including a large number of isomers (Figure 6.1) (Xu et al. 2006). The composition of the carotenoids vary greatly with growing conditions, fruit maturity, and citrus variety (Tables 6.1 and 6.2).

The caratoneoid contents in the peel are about 2.5–15 times higher than their respective concentrations in pulp (Tao 2002). Hence, the peel is the principal location of carotenoids in citrus fruits. With development and maturation of citrus fruit, α-carotene and β-carotene, which are known to be in the upstream portion of the carotenoid biosynthesis pathway, decrease. On the other hand, β-cryptoxanthin, β-citraurin, and zeaxanthin, which are found in the downstream portion of the carotenoid biosynthesis pathway, increase steadily and become the principal carotenoid components (Tao 2002).

Citrus cultivars can be roughly divided into three groups, according to the type of carotenoids they contain (Ikoma et al. 2016): those that are β-cryptoxanthin abundant, are violaxanthin abundant, or are carotenoid poor. Most mandarin cultivars, including Satsuma mandarin and Ponkan (*C. reticulata* Blanco), are classified into the β-cryptoxanthin-abundant category in terms of both their flavedo and juice sacs. In contrast, orange cultivars, including the common sweet orange (*C. sinensis* L. Osbeck Trovita) and navel orange (*C. sinensis* L. Osbeck Washington), are classified into the violaxanthin-abundant category in terms of both their flavedo and juice sacs. Other cultivars, including lime (*C. aurantifolia* Cristm. Swingle), lemon (*C. limon* L. Burm. f.), grapefruit (*C. paradisi* Macfad.), and pomelo (*C. grandis* L. Osbeck), can be separated from oranges and mandarins because of the low violaxanthin and β-cryptoxanthin contents in both the flavedo and juice sacs. The main carotenoids in citrus are as follows.

6.2.1 CRYPTOXANTHIN

Cryptoxanthin (Figure 6.2), a type of xanthophyll, is abundant in Satsuma mandarin orange (*Citrus unshiu* Marc.). In terms of structure, cryptoxanthin is closely related to β-carotene, with only the addition of a hydroxyl group.

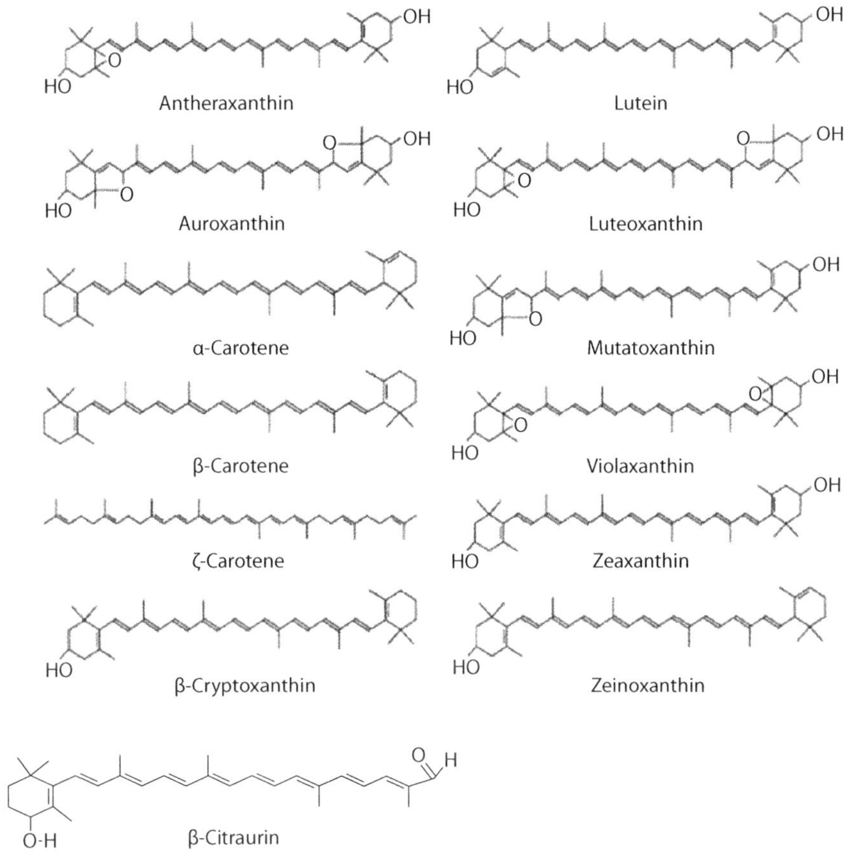

FIGURE 6.1 Chemical structures of carotenoids that occur in citrus fruits.

TABLE 6.1
Carotenoid Content in Pulp of Various Citrus Types

	Carotenoid Content (µg/g FW)				
Citrus Type	**Lutein**	**Zeaxanthin**	**Cryptoxanthin**	**Carotene**	**Carotene**
Citrus reticulate Blanco	2.71	1.28	7.38	0.05	0.77
Citrus sinensis Osbeck	2.35	0.69	0.58	0.01	0.17
Citrus aurantium Linn	0.83	0.37	0.67	0.02	0.08
Citrus sp. × *Citrus* sp.	2.17	0.69	4.00	0.01	0.42
Citrus grandis Osbeck	0.40	0.40	0.30	0	0
Citrus limon Burm.	0.09	0.12	0.43	0	0
Fortunella wingle	0.65	0.29	1.90	0	0
Poncirus trifoliata Raf.	0.08	0.30	0	0	0

Source: Tao, J., PhD dissertation, Zhejian University, 2002.
Abbreviation: FW, fresh weight.

TABLE 6.2
Carotenoid Content in Peels of Various Citrus Types

Citrus Type	Carotenoid Content (μg/g FW)				
	Lutein	Zeaxanthin	Cryptoxanthin	Carotene	Carotene
Citrus reticulate Blanco	27.37	5.76	22.89	0.03	0.45
Citrus sinensis Osbeck	12.31	2.51	2.05	0	0
Citrus aurantium L.	15.37	5.66	4.29	0	0
Citrus sp × *Citrus* sp.	15.53	3.31	9.79	0	0.21
Citrus grandis Osbeck	0.77	0.59	0.12	0	0
Citrus limon Burm.	1.17	0.08	1.55	0	0
Fortunella wingle	5.51	1.28	4.50	0	0
Poncirus trifoliata Raf.	7.31	2.06	0.40	0	0.64

Source: Tao, J., PhD dissertation, Zhejian University, 2002.
Abbreviation: FW, fresh weight.

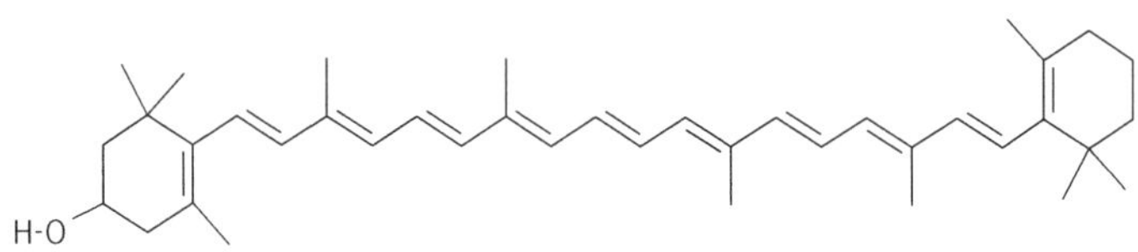

FIGURE 6.2 Structure of all-*trans*-β-cryptoxanthin.

6.2.2 β-Citraurin

Citraurin (Figure 6.3), a C_{30} apocarotenoid, is an uncommon carotenoid that accumulates specifically in the peel of some *Citrus* cultivars (Oberholster et al. 2001). It is a color-imparting pigment responsible for the reddish color of citrus fruits (Farin et al. 1983). It was first discovered in 1936 in Sicilian oranges. In citrus fruits, the accumulation of β-citraurin is low and is only observed in the flavedos of some varieties during fruit ripening. The citrus varieties that accumulate β-citraurin are considered more attractive because of their red-orange color (Ríos et al. 2010). β-Citraurin is a degradation product of β-cryptoxanthin and zeaxanthin (Ikoma et al. 2016).

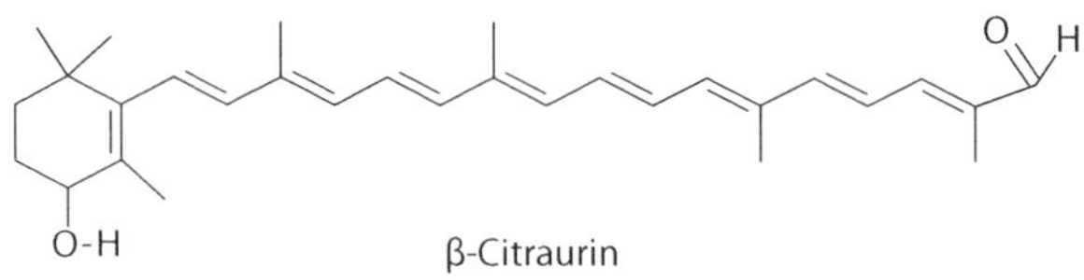

FIGURE 6.3 Structure of all-*trans*-β-citraurin.

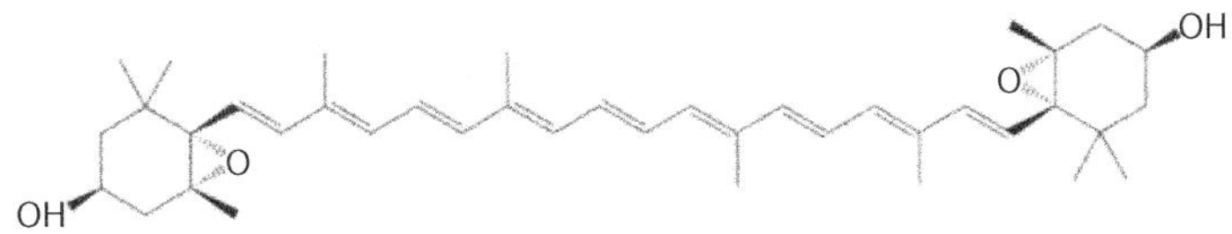

FIGURE 6.4 Structure of all-*trans*-β-violaxanthin.

6.2.3 Violaxanthin

Violaxanthin (Figure 6.4) is a natural xanthophyll pigment with an orange color and is found in a variety of plants, including pansies. It is biosynthesized from zeaxanthin via epoxidation. Mature sweet orange (*Citrus sinensis* Osbeck) accumulates violaxanthin isomers predominantly in fruits that contain 9-*cis*-violaxanthin as the principal carotenoid (Lee and Castle 2001).

6.2.4 Lutein and Zeaxanthin

The xanthophyll lutein (β,ε-carotene-3,30-diol) generally coexists in nature with its stereoisomer, zeaxanthin (β,β-carotene-3,30-diol) (Shegokar and Mitri 2012). They are the major xanthophylls, besides β-cryptoxanthin, in citrus pulp and peel. Chemically, lutein (Figure 6.5) contains the basic C_{40} isoprenoid structure characteristic of carotenoids, as well as 10 conjugated double bonds (9 conjugated double bonds in the polyene chain and 1 double bond in the β-ionone ring) (Sparrow and Kim 2010). Comparatively, in addition to its C_{40} isoprenoid structure, zeaxanthin (Figure 6.6) contains 11 conjugated double bonds consisting of 9 conjugated double bonds in the polyene chain and 2 double bonds in the β-ionone rings (Sparrow and Kim 2010).

6.2.5 Lycopene

Lycopene is a highly unsaturated straight-chain hydrocarbon consisting of 11 conjugated and 2 unconjugated double bonds. It is responsible for the red color of

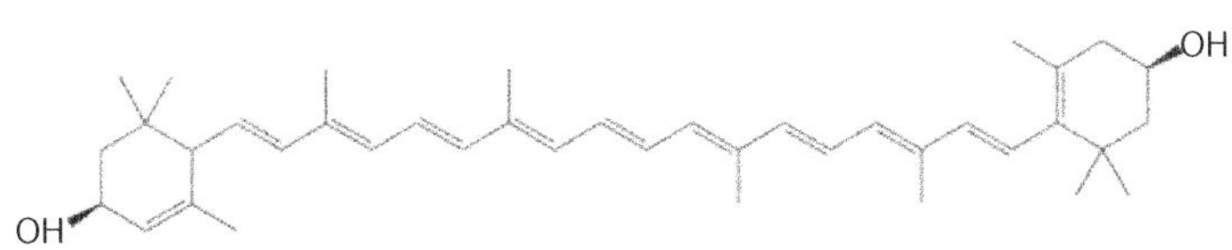

FIGURE 6.5 Structure of all-*trans*-β-lutein.

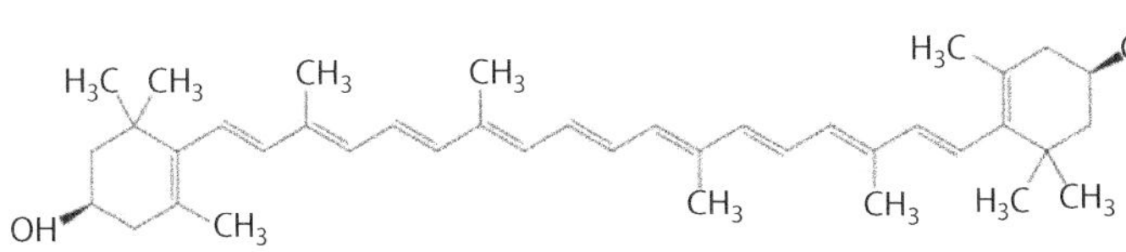

FIGURE 6.6 Structure of all-*trans*-β-zeaxanthin.

many fruits and vegetables, such as tomatoes. Unlike some other carotenoids, lycopene lacks the terminal ionic ring in its structure and has no provitamin A activity. Tomatoes and tomato-based foods account for more than 85% of all dietary sources of lycopene. Lycopene is an uncommon carotenoid in citrus fruits. Most lycopene-accumulating *Citrus* cultivars are mutant grapefruits, pomelos, and oranges (Xu et al. 2006). Grapefruits have the highest number of lycopene-accumulating mutants, including the famous cultivars Marsh Pink, Ruby Red, and Star Ruby (Gmitter 1993; Xu et al. 2006). Lycopene is also found in the lycopene-accumulating navel orange mutant Cara Cara and sweet orange mutant Hong Anliu (Alquézar et al. 2008; Fanciullino et al. 2008; Liu et al. 2007; Xu et al. 2009).

6.3 EXTRACTION OF CITRUS CAROTENOIDS

The peels of citrus fruits represent 30%–40% of the total fruit weight. The peels are an abundant source of natural carotenoids. However, most citrus peels are by-products in conventional food processing and are considered a waste product, which results in undesirable environmental pollution. Based on these facts, it became apparent that the development of an efficient procedure to extract carotenoids from citrus peels would be a worthwhile undertaking.

Today, various extraction methods have been employed for the extraction of carotenoids from citrus: solvent extraction, soxhlet extraction, and nonconventional methods of extraction, such as ultrasound-assisted, microwave-assisted, or enzymatic methods and an innovative technique that uses supercritical carbon dioxide (SC-CO_2) for extraction (Singh et al. 2015). Organic solvents such as hexane, dichloromethane, tetrahydrofuran, and ethyl acetate are frequently used; however, the newer technologies that use novel environmentally friendly solvents (nontoxic) hold promise for safer products and a cleaner environment (Wang et al. 2008).

The green and environmentally friendly techniques for extraction of carotenoids are SC-CO_2 extraction and ultrasound-assisted extraction (UAE) with ethanol. The SC-CO_2 extraction procedure (Lim et al. 2003) extracts pure compound with a high yield without the use of harmful organic solvents, and because it operates at lower temperatures it can be used to extract thermolabile compounds. This procedure uses SC-CO_2 as a green solvent and other solvents as modifiers that are generally recognized as safe (GRAS) compounds.

An investigation of UAE of citrus carotenoids (Sun et al. 2011) showed that use of ethanol produced higher carotenoid extraction yields than classical solvent extraction (CE) and was similar to the yields achieved when hexane and tetrahydrofuran were used in UAE (Figure 6.7). The higher extraction yield of ethanol under UAE in comparison to solvent extraction may be due to its physical properties. The physical properties (surface tension, viscosity, vapor pressure) of the solvent are the main factors that affect extraction yield under UAE. The vapor pressure of the solvent is the most important factor among these properties. Moreover, the vapor pressure is directly linked to its boiling point. The boiling point of ethanol is the highest of all the solvents that were used in this study. Thus, the relatively high boiling point of ethanol, combined with its molecular affinity with carotenoids, make

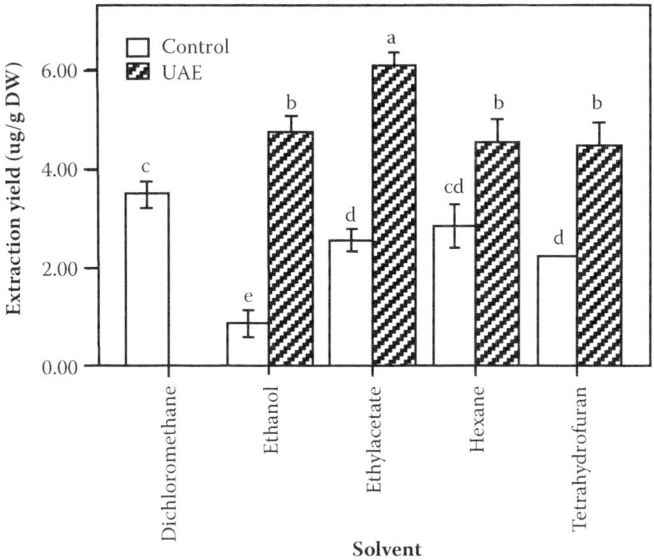

FIGURE 6.7 Effect of solvents on extraction yield of all-*trans*-β-carotene from UAE and CE. Different letters on bars show significant differences ($p < 0.05$). (From Sun, Y. J. et al., *Ultrasonics Biochemistry* 18:243–249, 2011.)

ethanol extractability similar to hexane and tetrahydrofuran in UAE yields, while the molecular affinity of the solvent is the main factor affecting extraction yield under CE. Ethanol had the weakest extraction yield with CE, because of the weak affinity between ethanol and carotenoids. The four nonpolar solvents hexane, tetrahydrofuran, ethyl acetate, and dichloromethane are routinely used for the extraction of carotenoids, but they all have toxicity concerns, while ethanol, which is a GRAS compound, may be a safe and effective solvent for UAE of carotenoids.

6.4 STABILITY OF CITRUS CAROTENOIDS

In practice, it is very important to take into account the sensitivities of carotenoids to oxidation, degradation, adduct formation, and isomerization, which are due to their polyisoprenoid structures that consist of a long chain of conjugated double bonds. The selection of suitable extraction conditions must consider the ramifications of all these damaging reactions. However, most studies have been carried out with β-carotene. These studies have reported the impact of processing, storage, heating, and extraction on the stability of β-carotene, but only a few studies have considered the stability of the many other carotenoids in citrus. Therefore, we focused our review of carotenoid stability here on β-carotene, and we have used β-carotene stability as a model for all citrus carotenoids. It is well known that carotenoids can undergo autoxidation, thermos degradation, photodegradation, adduct formation, isomerization, and other modifications when exposed to atmospheric oxygen, heat, light, acid, or ultrasound.

6.4.1 Autoxidation

Reactions of carotenoids with atmospheric oxygen occur with relative ease, especially in systems consisting of purified carotenoids in organic solvents. Autoxidation of β-carotene in benzene or tetrachloromethane in the dark at 30°C under 1 atm of oxygen or by bubbling oxygen through the solvent was found to occur with an induction period of less than 1 h, followed by rapid production of oxidation products. Under these conditions, β-carotene was completely consumed within 30 h. A combination of high-performance liquid chromatography (HPLC), Fourier transform infrared spectroscopy (FTIR), and gas chromatography-mass spectrometry (GC-MS) was used to monitor the reactions and identify over 20 oxidation products, such as β-carotene-5,6-epoxide, β-carotene-5,6,5′,6′-diepoxide, β-apo-13-carotene, β-apo-14′-carotene, retinal, β-ionone acetaldehyde, β-ionone, 4-oxo-β-ionone, 2,2,6-trimethylcyclohexanone, 9-*cis*-β-carotene, 13-*cis*-β-carotene, and so on (Henry et al. 2000; Clark et al. 1997).

6.4.2 Thermal Degradation

Thermal treatment of carotenoids in the presence of oxygen results in the formation of volatile compounds and larger nonvolatile components (Bonnie and Choo 1999). Kanasawud and Crouzet (1990) proposed a mechanism for β-carotene degradation based on the products found during heating β-carotene at 97°C for up to 3 h in the presence of air. GC-MS and absorption spectrophotometry were used to identify the degradation products. The results suggested that β-carotene reacts with oxygen to form 5,6-epoxy-β-carotene, which can then be converted to mutatochrome, 5,6,5,6-diepoxy-β-carotene, or luteochrome. Luteochrome may be converted to aurochrome, which may then be cleaved to form dihydroactinidiolide. 2,5,6-Epoxy-β-carotene may be cleaved to form 5,6-epoxy-β-ionone, which may be converted to β-ionone, 2-hydroxy-2,6,6-trimethylcyclohexanone, 2,6,6-trimethylcyclohexanone, and 2-hydroxy-2,6,6-trimethylcyclohexane-1-carboxaldehyde. 2-Hydroxy-2,6,6-trimethylcyclohexane-1-carboxaldehyde can form β-cyclocitral, while 2-hydroxy-2,6,6-trimethylcyclohexanone can form 2,6,6-trimethyl-2-cyclohexen-1-one (Kanasawud and Crouzet 1990). Marty and Berset (1990) determined via HPLC that heating pure β-carotene at 180°C for 2 h resulted in the formation of *cis* isomers as well as oxidation products.

6.4.3 Photodegradation

Light exposure degrades carotenoids, and several mechanisms of action have been proposed. Photooxidation produces chemical species thought to be carotenoid radical cations (Gao et al. 1997; Konovalova et al. 2001; Mortensen and Skibsted 1996). Laser flash photolysis studies have produced evidence to suggest that rapid bleaching of β-carotene in some solvents, like chloroform, can occur due to the light exciting the β-carotene molecules, which then instantly react with the solvent (chloroform in this case) to form either a carotenoid-solvent free radical adduct or a β-carotene radical. The same work also suggested that the β-carotene molecules in the excited state could return to ground state and be attacked by radical by-products created during the initial reaction; they could then undergo a slower degradation process that could contribute to β-carotene radical cation formation (Mortensen and Skibsted 1996).

6.4.4 Acids

Exposure to acids is thought to produce ion pairs, which can then dissociate into a carotenoid carbocation. This process can be seen in the following equation (Konovalov and Kispert 1999): $Car + AH \leftrightarrow (CarH^{+}...A^{-}) \leftrightarrow CarH^{+} + A^{-}$.

Optical spectra of carotenoid carbocations have been reported for β-carotene, 8′-apo-caroten-8′-al, and canthaxanthin upon exposure to trifluoroacetic acid in benzene, CH_2Cl_2, or acetonitrile solvents (Konovalov and Kispert 1999). These same researchers showed that canthaxanthin and 8′-apo-β-caroten-8′-al, when incorporated into sol-gels, could be degraded by sulfuric acid (pH 3 to 3.5) (He and Kispert 2001).

6.4.5 Ultrasonic Degradation

The degradation of β-carotene may also occur during its extraction from fruit and vegetables. UAE has been widely used for the extraction of carotenoids due to its high extraction efficiency and extraction rate, but the effects of cavitation of ultrasound energy may accelerate or trigger chemical reactions in the extraction medium. However, the stability of carotenoids under high-intensity ultrasound treatment is seldom reported. Zhao et al. (2006) found that (all-*E*)-astaxanthin under ultrasound treatment in a model system degraded to unidentified colorless compounds. Sun et al. (2006) extracted lutein from chicken liver and found that the extraction rate of lutein was improved under ultrasound treatment without saponification, but lutein degradation occurred under ultrasound treatment with saponification. Sun et al. (2010) examined the factors affecting the degradation kinetics and products of β-carotene decomposition caused by ultrasound treatment in a model system. The results indicated that the type of solvents and temperature were important factors in determining the outcome of the degradation reaction (Figures 6.8 and 6.9). The highest rate of β-carotene degradation was with dichloromethane. The degradation rate of β-carotene decreased as the temperature increased, which did not conform with the normal Arrhenius activation energy relationship between reaction rates and temperature. The degradation mechanism of β-carotene in dichloromethane at 5°C to 15°C was different from that at 25°C (Figure 6.10). Isomerization of β-carotene under ultrasound treatment occurred at 25°C, but isomerization, degradation, and oxidation of β-carotene under ultrasound treatment occurred at 5°C to 15°C. As for the degradation products, two isomers (15-*cis*-β-carotene, di-*cis*-β-carotene) were identified (Figures 6.11 and 6.12 and Table 6.3), while the other products were characterized by identifying their functional groups. The occurrence of a new polar function group (C–O) (Figure 6.13), a β-apo-carotene (Figure 6.14) may be one of the degradation products.

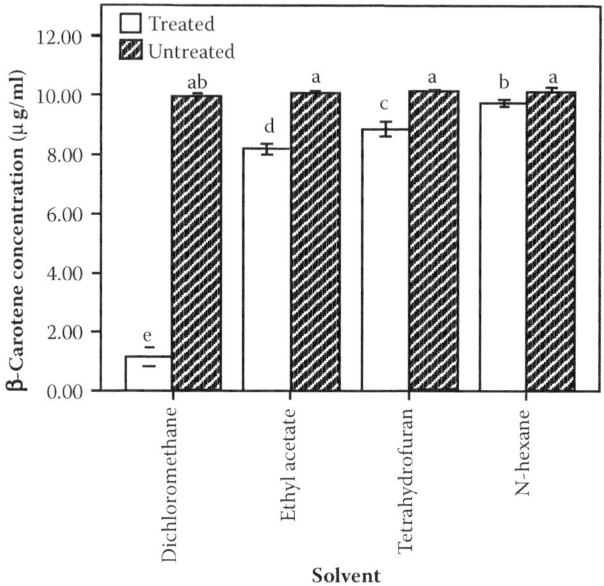

FIGURE 6.8 Effects of solvent on stability of all-*trans*-β-carotene from ultrasound treatment. Different letters on bars show significant differences ($p < 0.05$). (From Sun, Y. J. et al., *Ultrasonics Sonochemistry* 17:654–661, 2010.)

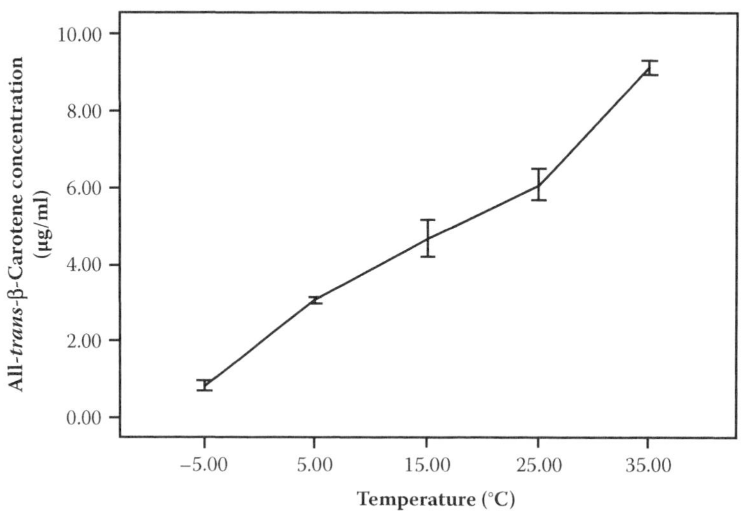

FIGURE 6.9 Effects of temperature on stability of all-*trans*-β-carotene under ultrasound treatment. (From Sun, Y. J. et al., *Ultrasonics Sonochemistry* 17:654–661, 2010.)

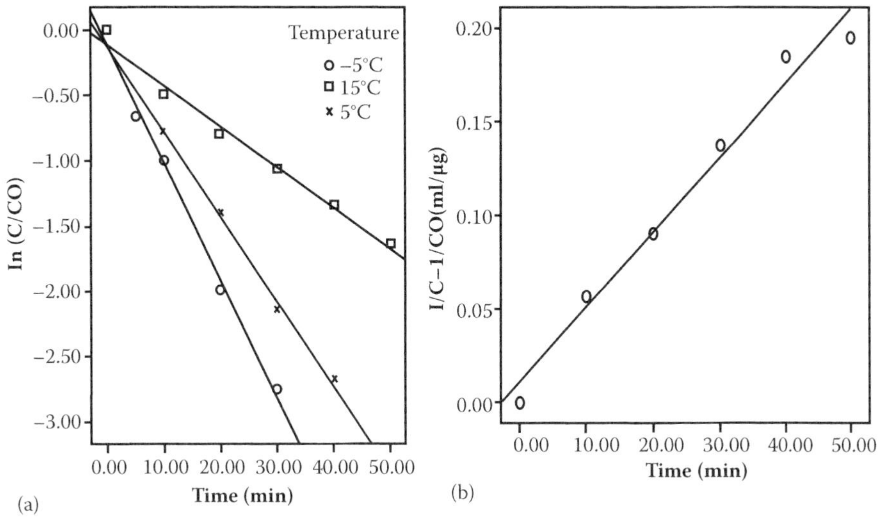

FIGURE 6.10 Degradation kinetics curve for all-*trans*-β-carotene under ultrasound treatment at −5°C, 5°C, or 15°C (a) or at 25°C (b). (From Sun, Y. J. et al., *Ultrasonics Sonochemistry* 17:654–661, 2010.)

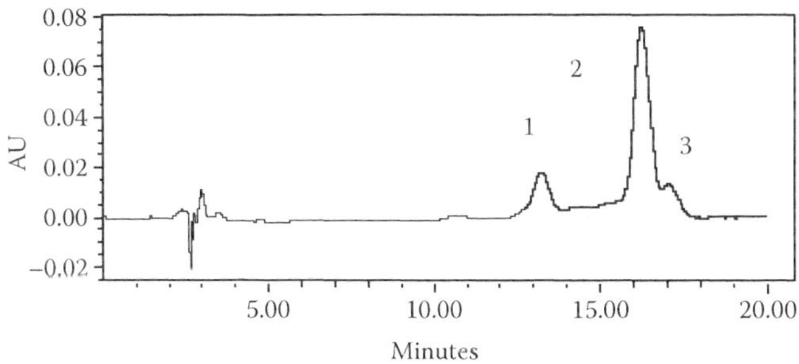

FIGURE 6.11 HPLC results showing peaks 1, 2, and 3 for all-*trans*-β-carotene treated by ultrasound at 455 nm (5°C). (From Sun, Y. J. et al., *Ultrasonics Sonochemistry* 17:654–661, 2010.)

6.4.6 Isomerization

Electron transfer reactions are also thought to play a role in isomerization of carotenoids. Some studies have suggested that isomerization is mediated by carotenoid radical cations and possibly dications formed via electron transfer from neutral carotenoids to radicals or compounds, such as iron, that can easily transfer charges (Wei et al. 1997; Gao et al. 2003).

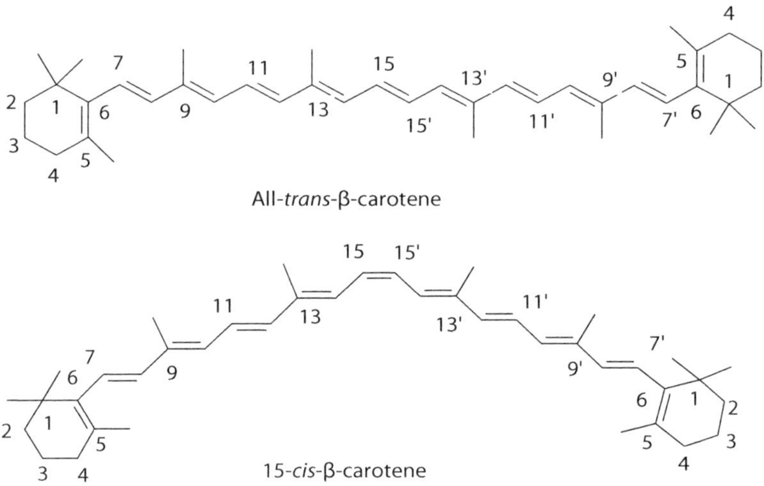

FIGURE 6.12 Structures of all-*trans*-β-carotene and 15-*cis*-β-carotene. (From Sun, Y. J. et al., *Ultrasonics Sonochemistry* 17:654–661, 2010.)

TABLE 6.3
Data for Isomers of β-Carotene in Dichloromethane after Ultrasound Treatment

Peak No.	β-Carotene Isomer(s)	Retention Time (min)	λ Found (nm)	λ Reported (nm)	Q Value	
					Found	Reported
1	15-*cis*-β-carotene	13.45	341, 450, 475	447, 471	0.49	0.44
2	All-*trans*-β-carotene	16.45	455, 483	453, 477		
3	*cis*-β-carotene	17.36	449	475		

Source: Sun, Y. J. et al., *Ultrasonics Sonochemistry* 17:654–661, 2010.

This is thought to occur because of the lower energy barriers for configural transformation of the cation species compared to those of neutral carotenoids (Wei et al. 1997; Gao et al. 1996). Wei et al. (1997) found that oxidation of canthaxanthin and 8-apo-β-caroten-8-al resulted in the formation of radical cations, and this was followed by the formation of *cis* isomers. Canthaxanthin and β-carotene in dichloromethane were also found to undergo this process, as determined by bulk electrolysis with simultaneous absorption spectroscopy (Gao et al. 1996).

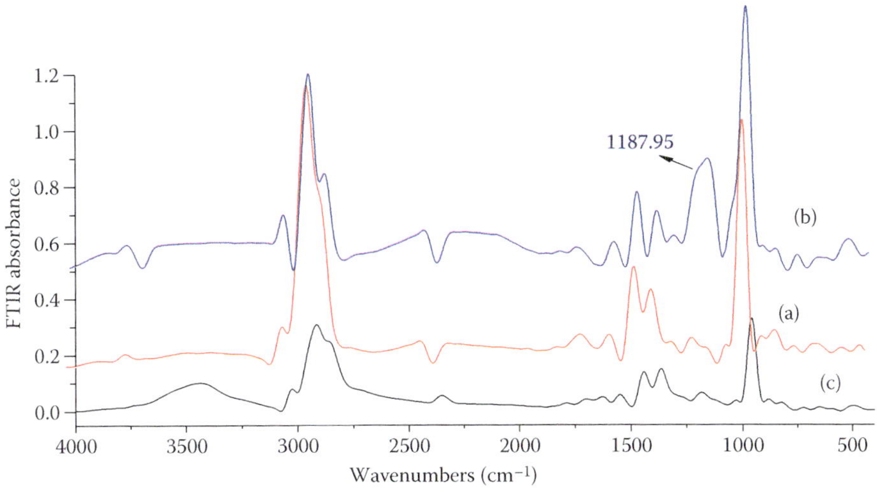

FIGURE 6.13 FTIR spectra at different wavelengths (cm^{-1}) corresponding to all-*trans*-β-carotene treated by ultrasound for 10 min (a) or 30 min (b) or not treated (c) (5°C). (From Sun, Y. J. et al., *Ultrasonics Sonochemistry* 17:654–661, 2010.)

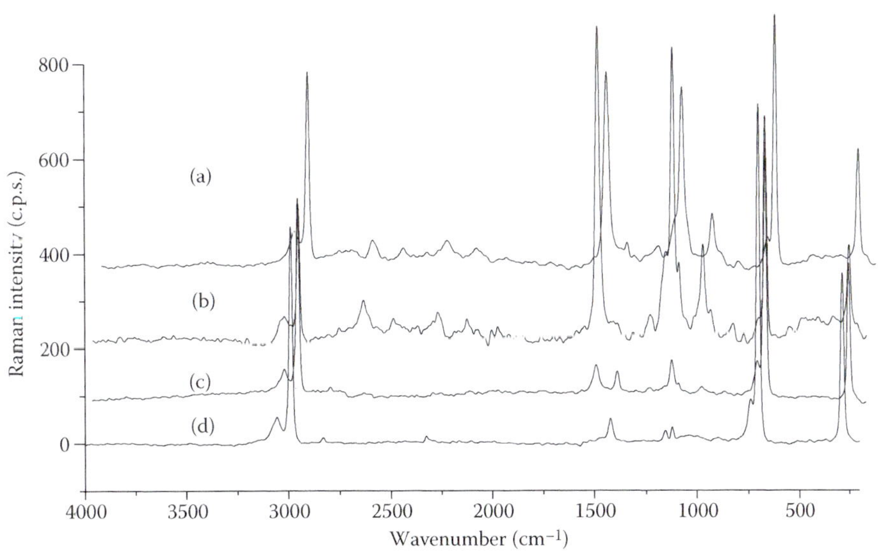

FIGURE 6.14 Raman spectra at different wavelengths (cm^{-1}) corresponding to the solvent, dichloromethane (d), and all-*trans*-β-carotene treated by ultrasound for 10 min (a), untreated (b), or treated for 30 min (c) (5°C). (From Sun, Y. J. et al., *Ultrasonics Sonochemistry* 17:654–661, 2010.)

The *cis* radical cations and dications are thought to undergo reactions with neutral *trans*-carotenoids still present in the system to form either neutral *cis*-carotenoid or *cis* radical cations, as well as new *trans* radical cations that can be recycled through the isomerization reaction sequence (Wei et al. 1997; Gao et al. 1999). Isomerization has also been found to occur during the oxidation of ethyl-8-apo-caroten-8-oate and β-carotene by ferric chloride. The study by Gao et al. proposed that if iron is present in the system, ferrous iron can react with the *cis* radical cations or dications and can be oxidized to ferric iron to form either the neutral *cis*-carotenoid or the *cis* radical cation (Gao et al. 2003). Isomerization is also thought to occur when carotenoids are exposed to acids. Similar to the carotenoid radical cation formed in electron transfer-based isomerization, a carotenoid carbocation ($CarH^+$) is believed to be an intermediate in *trans-cis* isomerization. This carbocation has been predicted by Austin model 1 calculations of rotation barriers to have a lower barrier to rotation than neutral carotenoids, facilitating the isomerization process (Konovalov and Kispert 1999).

6.5 HEALTH EFFECTS OF CAROTENOIDS IN CITRUS

Although over 700 carotenoids have been identified in nature, only about 20 carotenoids have been found in human blood and tissues. Close to 90% of the carotenoids found in humans and in their diet are β-carotene, α-carotene, lycopene, lutein, and cryptoxanthin. The biological properties of β-carotene and lycopene have been the center of research for many years, and recently, lutein, zeaxanthin, and cryptoxanthin have gained the attention of scientists; most of the other carotenoids are seldom studied. Here, only the unique health effects of lutein, zeaxanthin, and cryptoxanthin, which are abundant in citrus fruits and have been the focus of recent scientific investigation, are discussed.

6.5.1 Health Effects of Cryptoxanthin

6.5.1.1 Provitamin A Activity

Vitamin A is essential for the promotion of growth, embryonal development, and visual function. Carotenoids can also be divided into provitamin A and non-provitamin A compounds. The major provitamin A carotenoids in the diet are β-carotene and α-carotene, which are well-known, but recent studies have found that β-cryptoxanthin can also be converted to vitamin A and help prevent vitamin A deficiency. One cryptoxanthin molecule is converted into one vitamin A (retinol) molecule. The contribution of provitamin A carotenoids to the daily vitamin A intake depends on dietary habits and available food sources. It has been estimated that carotenoids from fruits and vegetables provide more than 70% of the vitamin A intake in developing countries; in Western societies, the contribution is a much lower percentage.

It was established many years ago that β-cryptoxanthin forms retinol (vitamin A). Despite this, research on the mechanism of retinol formation from carotenoids has focused almost exclusively on β-carotene. In fact, no recent research has focused on the mechanisms by which β-cryptoxanthin forms retinol. However, some studies on β-carotene metabolism have included data on β-cryptoxanthin. These studies suggest that the same enzymes involved in β-carotene cleavage can also cleave β-cryptoxanthin. Generally, the kinetics of these reactions differ, suggesting that β-carotene is the preferred substrate. However, because the same enzymes that cleave β-carotene to retinal also cleave β-cryptoxanthin with relatively high efficiencies, there is a good probability that β-cryptoxanthin forms retinol by the same mechanism.

6.5.1.2 Promotion of Bone Health

Among various carotenoids, β-cryptoxanthin has been found to have a unique anabolic effect on bone mass by stimulating osteoblastic bone formation and inhibiting osteoclastic bone resorption.

Bone is a dynamic tissue, constantly forming and reforming. Good nutrition is essential for bone homeostasis, with vitamin D and calcium playing essential roles (Rondanelli et al. 2013). A series of cell culture and rodent studies have suggested that β-cryptoxanthin may also be involved in bone health and homeostasis by promoting osteoclast formation and inhibiting osteoblast actions (Yamaguchi and Uchiyama 2003, 2004; Uchiyama and Yamaguchi 2004; Uchiyama et al. 2005; Ikeda et al. 2012; Iino et al. 2014). Most of these studies were summarized by Yamaguchi in 2012. Collectively, the results suggest that β-cryptoxanthin has an effect on bone health that is not duplicated by other carotenoids.

Moderate (probably physiological) concentrations of β-cryptoxanthin appear to increase calcium content, protein content, and alkaline phosphatase activity in bone *in vitro*. These changes can be inhibited by inhibitors of RNA polymerase II or by the protein synthesis inhibitor cycloheximide. Both calcium and alkaline phosphatase participate in the mineralization of bone. β-Cryptoxanthin also appears to stimulate gene expression for proteins involved in bone formation and mineralization in osteoblasts, such as insulin-like growth factor 1 and transforming growth factor β1, an effect possibly mediated by protein kinase C or mitogen-activated protein kinase (Yamaguchi and Weitzmann 2009). Transforming growth factor β1 is involved in the differentiation of preosteoblasts to osteoblasts.

Furthermore, β-cryptoxanthin stimulates Runt-related transcription factor 2 (Runx 2, also known as core binding factor a1, or CBF a1), a key transcription factor associated with osteoblast differentiation (Uchiyama and Yamaguchi 2005a). Although retinoic acid, the active form of vitamin A, can also mediate bone formation, it did not always duplicate the actions of β-cryptoxanthin on gene expression in these studies (Uchiyama and Yamaguchi 2005b).

In vitro, β-cryptoxanthin inhibits bone resorption induced by parathyroid hormone or prostaglandin E2 by preventing osteoclast cell formation via the Receptor activator of nuclear factor κB ligand (RANKL) (Uchiyama and Yamaguchi 2004). β-Cryptoxanthin decreased the number of mature osteoclasts in culture, an action that was inhibited by inhibitors of capase-3, suggesting that β-cryptoxanthin induces apoptotic cell death (Uchiyama and Yamaguchi 2006a). Zinc appears to have synergistic effects with β-cryptoxanthin, stimulating bone formation and inhibiting bone resorption (Uchiyama et al. 2005c; Yamaguchi et al. 2006; Yamaguchi and Uchiyama 2008).

Animal studies have provided some corroboration for these cell culture results. Rats fed moderately high doses of β-cryptoxanthin (50–100 mg/kg of body weight) in combination with zinc sulfate (zinc; 1–5 mg/kg) for 1 week showed increases of alkaline phosphatase activity and calcium concentrations in diaphyseal tissues (Yamaguchi et al. 2006). Additionally, young male or older female rats fed high doses of β-cryptoxanthin (100–500 mg/kg) for 1 week also showed increased alkaline phosphatase and calcium in diaphyseal and metaphyseal tissues (Uchiyama and Yamaguchi 2005c; Uchiyama et al. 2004). Furthermore, several laboratories found that feeding moderate doses of β-cryptoxanthin to ovariectomized rats inhibited bone loss (Uchiyama and Yamaguchi 2006b; Lino et al. 2014) and periodontal bone resorption (Matsumoto et al. 2013).

There have been a few human studies on the effects of foods rich in β-cryptoxanthin or β-cryptoxanthin on osteoporosis; most of these studies have been small, and the results have been inconclusive. No significant associations were found in a prospective cohort study of carotenoid intake and risk of hip fracture or other indices of bone health in the Framingham Osteoporosis Study (Sahni et al. 2009a,b). In another study conducted in the United States, β-cryptoxanthin intakes were higher in postmenopausal women with osteoporosis than in postmenopausal women without osteoporosis (Yang et al. 2008). On the other hand, epidemiological studies in Japan showed that high β-cryptoxanthin intake or the intake of Satsuma mandarins was associated with high bone mineral density in postmenopausal women (Sugiura et al. 2008, 2011; Sugiura 2014). Furthermore, comparison of the highest and lowest tertiles of dietary intakes of vitamin C and β-cryptoxanthin in menopausal female subjects from Mikkabi, Japan, showed that these antioxidants were associated with higher bone mineral density (odds ratio of 0.40, 95% confidence interval of 0.17–0.92) (Sugiura et al. 2011). In a 4-year follow-up of the Mikkabi study, 15 postmenopausal women developed osteoporosis. After adjustments for confounders, the odds ratio for osteoporosis in the highest tertiles of serum β-cryptoxanthin was 0.07 (95% confidence interval, 0.01–0.88). Serum β-cryptoxanthin was also inversely associated with osteopenia/osteoporosis ($p = 0.037$) (Sugiura et al. 2012). There are many differences between these studies that could explain the varied results, including study design, study population, and diets. One factor that may be important is that the Japanese have one of the highest intakes of β-cryptoxanthin, coupled with relatively low intakes of preformed vitamin A.

6.5.2 Eye Health Effects of Lutein and Zeaxanthin

Lutein has been shown to play a central role in reducing the incidence of eye diseases, such as age-related macular degeneration (AMD), cataracts, and retinitis pigmentosa (Aleman et al. 2001; Olmedilla et al. 2003). A frequently used parameter in determining the condition of the retina, especially while diagnosing AMD, is the macular pigment density (MPD), which is an indirect measurement of the concentration of lutein and zeaxanthin in the macula (Alves-Rodrigues and Shao 2004). The macular pigment, which principally absorbs harmful blue light (at 440 nm) and shields sensitive photoreceptors from damaging UV rays, is entirely composed of lutein, its metabolite, *meso*-zeaxanthin, and zeaxanthin and is thought to be critical in preserving visual health (Alves-Rodrigues and Shao 2004; Johnson 2014). High levels of MPD have been positively correlated with lower rates of AMD and retinitis pigmentosa (Dagnelie et al. 2000; Phelan and Bok 2000; Richer et al. 2002). For cataracts, a similar parameter is the lens optical density (Hammond et al. 1997) but, unlike the directly proportional relationship between MPD and the incidence of AMD, increased lens optical density is correlated with decreased lens function or health (Trumbo and Ellwood 2006).

A number of studies between 1987 and 1999 suggested that dietary carotenoids (lutein, zeaxanthin, lycopene, α-carotene, and β-carotene) from fruits and vegetables may play a role in reducing the risk of eye diseases (Goldberg et al. 1988; Sommerburg et al. 1998; Yannuzzi et al. 1993). Although it was relatively straightforward to hypothesize that lutein and zeaxanthin have a role in visual health, given the particularly high concentrations and exclusive presence of both xanthophylls in certain ocular tissues (Alves-Rodrigues and Shao 2004), demonstration of their contributions to vision improvement or in delaying the onset of ocular diseases was a more labyrinthine exercise. First, through various observational studies, the following were established: (a) depending on the outcome assessed, dietary intake of lutein significantly reduced the risk of AMD (Seddon et al. 1994) or reduced the risk of cataract extraction (Brown et al. 1999; Chasan-Taber et al. 1999; Hankinson et al. 1992); (b) pure crystalline lutein supplementation (Bernstein et al. 2002) as well as dietary intake (Curran-Celentano et al. 2001) resulted in significantly higher MPD; and (c) there was a negative correlation between increasing lutein levels in the retina and the risk of developing AMD (Bone et al. 2001).

Various observational studies have also examined the relationship between lutein consumption and cataract risk. Some key findings of these studies were (a) subjects consuming lutein-rich spinach at least 5 times a week had a 39% lower risk of cataract extraction than those whose spinach consumption was once a week (Hankinson et al. 1992); (b) subjects on lutein supplementation at 1.3 mg/day had a 50% lower risk of developing nuclear cataracts than those receiving 0.3 mg of lutein per day (Lyle et al. 1999); and (c) the prevalence of posterior subcapsular cataracts was 50% lower in 66- to 75-year-old subjects with a plasma lutein concentration greater than 0.20 mmol/L, compared to those whose lutein plasma levels were less than 0.14 mmol/L (Gale et al. 2001). Given the delicateness of the eye and the invasive nature of strategies for determining and quantifying metabolic products in the retina and the lens, it is impractical to directly measure the effects of eye lutein concentration on the incidence of ocular diseases in living subjects. A key study reported direct measurements of the actual concentrations of lutein and zeaxanthin in the macular pigment of donor eyes with and without AMD (Bone et al. 2001). The study concluded, after examining 56 retinas, from both AMD and control subjects, that those control subjects with the highest amount of lutein were 82% less inclined to develop AMD than those with the lowest levels of the xanthophylls. Observational studies are sufficient for the purpose of demonstrating links between nutrient supplementation and tissue concentrations of a particular compound with disease risk, but they fall short of establishing a direct cause–effect relationship between the consumption of a particular nutrient and a particular health advantage (Alves-Rodrigues and Shao 2004). Therefore, intervention studies are used to establish causality. For example, the Age-Related Eye Disease Study Research Group investigated the effect of lutein on AMD progression in nearly 4000 patients over a 6-year period (Kassoff et al. 2001). In this work, which investigated the impact of diet supplements on ocular disease progression, an oral antioxidant supplement was used (because lutein supplements were not yet commercially available at the time the study commenced). The study concluded that the antioxidant supplement, which contained lutein, significantly delayed the progression of AMD in patients. Another study, the Lutein Antioxidant Supplementation Trial, modeled after the Age-Related Eye Disease Study trial, administered to study subjects a daily regimen of 10 mg of lutein, 10 mg of lutein plus a mixed antioxidant formula, or placebo to 90 AMD patients for 1 year (Richer et al. 2002). The study results indicated that patients who received the pure lutein supplement showed marked improvements in several objective parameters of visual function, including contrast sensitivity, visual acuity, and glare recovery compared to those on the placebo supplement. In a similar study, it was reported that among three distinct groups of cataract subjects receiving a daily dose of 100 mg of α-tocopherol, 15 mg of lutein, or placebo for 24 months, significant improvements in glare sensitivity and visual acuity were recorded for the lutein supplementation group relative to the group that received either the placebo or α-tocopherol (Olmedilla et al. 2003). More recently, other intervention studies have established that (a) 10-mg daily lutein or zeaxanthin supplements over a 48-week period significantly improved retinal functions, as determined by means of multifocal electroretinograms, in early AMD patients (Ma et al. 2012); (b) lutein supplementation significantly increased visual acuity and macular function as measured by microperimetry after a 6-month administration period (Weigert et al. 2011); (c) a 42% and 41% lower risk of developing age-related nuclear cataracts was found for people with the highest tertiles of plasma lutein and zeaxanthin concentrations, respectively, compared to subjects in the lowest tertiles of both xanthophylls groups (Karppi et al. 2012); (d) 0.5 mg/kg daily lutein supplementation reduced the incidence and severity of cataracts in diabetic male Wistar rats but not in the controls that received a placebo (Arnal et al. 2009); and (e) diets rich in lutein and zeaxanthin moderately reduced the prevalence of cataracts in women aged 50–79 years, in contrast to women of the same age range who were placed on diets not containing the oxycarotenoids (Moeller et al. 2008).

Another human intervention study demonstrated that lutein supplementation resulted in an increase in MPD in patients with retinitis pigmentosa, although the contribution of the oxycarotenoid to vision improvement in this particular study did not seem to be as significant as the results recorded for AMD and cataracts (Aleman et al. 2001). It is important to mention that most of the observational and intervention investigations focused on lutein alone (due to the significantly higher amounts found in the serum, ocular tissues, and most food sources) or a combination of lutein and zeaxantin (Bone et al. 1997; Khachik et al. 1997).

6.6 APPLICATIONS OF CAROTENOIDS FROM CITRUS IN FUNCTIONAL FOODS

Citrus fruits have many different types of carotenoids, which makes them a good source of carotenoids from functional foods. However, only cryptoxanthin functional citrus juices or beverages have been described in the literature.

In Spain, Hernández-Alvarez et al. (2016) developed a β-cryptoxanthin-rich functional beverage composed of mandarin juice, banana puree, grape juice, and skimmed milk that was designed to improve cardiovascular and bone remodeling markers in postmenopausal women. In Japan, Iwamoto et al. (2012) found that supplementation of highly concentrated β-cryptoxanthin in a Satsuma mandarin beverage improved adipocytokine profiles in obese Japanese women. Tanaka et al. (2012) prepared a citrus pulp and citrus juice from Satsuma mandarin (*Citrus unshiu* Mar.) juice for cancer chemoprevention of colon, tongue, and lung neoplasms. Makoto et al. (2010) found that intake of β-cryptoxanthin-fortified Satsuma mandarin juice may be effective in improving liver function. Yamaguchi et al. (2006) reported that prolonged intake of juice fortified with β-cryptoxanthin had stimulatory effects on bone formation and inhibitory effects on bone resorption in humans.

6.7 SUMMARY

Carotenoids not only serve a vital function in plants but also have an important role in improving human health. Citrus fruits are an excellent source of carotenoids and can be readily consumed as part of the human diet, while citrus by-products from commercial processing plants and the preharvest fruit drops can be extracted for their carotenoids and other bioactive compounds. It is important that these nutraceuticals be given greater scientific scrutiny by more researchers, so that the true potential of these compounds can be realized. Knowledge of their structure, stability, and bioactivities and the relationship between bioactivity and human health of familiar carotenoids, such as β-cryptoxanthin, and also an expanded research scope to include the lesser-known carotenoids in citrus, such as β-citraurin and violaxanthin, will provide health benefits and at the same time promote the advancement of citrus fruit research.

ACKNOWLEDGMENT

Financial support from the China Nature Sciences Foundation (Project C200501) is acknowledged.

REFERENCES

Aleman, T. S., J. L. Duncan, M. L. Bieber, E. De Castro, D. A. Marks, L. M. Gardner et al. 2001. Macular pigment and lutein supplementation in retinitis pigmentosa and Usher syndrome. *Investigative Ophthalmology and Visual Science* 42(8):1873–1881.

Alquezar, B., M. J. Rodrigo, and L. Zacarias. 2008. Regulation of carotenoid biosynthesis during fruit maturation in the red-fleshed orange mutant Cara Cara. *Phytochemistry* 69(10):1997–2007.

Alves-Rodrigues, A., and A. Shao. 2004. The science behind lutein. *Toxicology Letters* 150(1):57–83.

Arnal, E., M. Miranda, I. Almansa, M. Muriach, J. M. Barcia, F. J. Romero et al. 2009. Lutein prevents cataract development and progression in diabetic rats. *Graefe's Archive for Clinical and Experimental Ophthalmology* 247(1):115–120.

Bernstein, P. S., D. Y. Zhao, S. W. Wintch, I. V. Ermakov, R. W. McClane, and W. Gellermann. 2002. Resonance Raman measurement of macular carotenoids in normal subjects and in age-related macular degeneration patients. *Ophthalmology* 109(10):1780–1787.

Bone, R. A., J. T. Landrum, S. T. Mayne, C. M. Gomez, S. E. Tibor, and E. E. Twaroska. 2001. Macular pigment in donor eyes with and without AMD: A case-control study. *Investigative Ophthalmology and Visual Science* 42(1):235–240.

Bonnie, T. P., and Y. M. Choo. 1999. Oxidation and thermal degradation of carotenoids. *Journal of Oil Palm Research* 2(1):62–78.

Brown, L., E. B. Rimm, J. M. Seddon, E. L. Giovannucci, L. Chasan-Taber, D. Spiegelman et al. 1999. A prospective study of carotenoid intake and risk of cataract extraction in US men. *American Journal of Clinical Nutrition* 70(4):517–524.

Chasan-Taber, L., W. C. Willett, J. M. Seddon, M. J. Stampfer, B. Rosner, G. A. Colditz et al. 1999. A prospective study of carotenoid and vitamin A intakes and risk of cataract extraction in US women. *American Journal of Clinical Nutrition* 70(4):509–516.

Chatterjee, M., K. Roy, M. Janarthan, S. Das, and M. Chatterjee. 2012. Biological activity of carotenoids: Its implications in cancer risk and prevention. *Current Pharmaceutical Biotechnology* 13(1):180–190.

Clark, K. B., J. A. Howard, and A. R. Oyler. 1997. Retinoic acid oxidation at high oxygen pressures: Evidence for spin-forbidden direct addition of triplet molecular oxygen. *Journal of the American Chemical Society* 119:9560–9561.

Curran-Celentano, J., B. R. J. Hammond, T. A. Ciulla, D. A. Cooper, L. M. Pratt, and R. B. Danis. 2001. Relation between dietary intake, serum concentrations, and retinal concentrations of lutein and zeaxanthin in adults in a Midwest population. *American Journal of Clinical Nutrition* 74(6):796–802.

Dagnelie, G., I. S. Zorge, and T. M. McDonald. 2000. Lutein improves visual function in some patients with retinal degeneration: A pilot study via the internet. *Optometry* 71(3):147–164.

Fanciullino, A. L., M. Cercos, C. Dhuique-Mayer, Y. Froelicher, M. Talon, P. Ollitrault, and R. Morillon. 2008. Changes in carotenoid content and biosynthetic gene expression in juice sacs of four orange varieties (*Citrus sinensis*) differing in flesh fruit color. *Journal of Agricultural and Food Chemistry* 56(10):3628–3638.

Farin, D., R. Ikan, and J. Gross. 1983. The carotenoid pigments in the juice and flavedo of a mandarin hybrid (*Citrus reticulata*) cv Michal during ripening. *Phytochemistry* 22:403–408.

Fernández-García, E., I. Carvajal-Lérida, M. Jarén-Galán, J. Garrido-Fernández, A. Pérez-Gálvez, and D. Hornero-Méndez. 2012. Carotenoids bioavailability from foods: From plant pigments to efficient biological activities. *Food Research International* 46(2):438–450.

Gale, C. R., N. F. Hall, D. I. W. Phillips, and C. N. Martyn. 2001. Plasma antioxidant vitamins and carotenoids and age-related cataract. *Ophthalmology* 108(11):1992–1998.

Gao, G. Q., Y. Deng, and L. D. Kispert. 1997. Photoactivated ferric chloride oxidation of carotenoids by near-UV to visible light. *Journal of Physical Chemistry B* 101(39):7844–7849.

Gao, G. Q., Y. Deng, and L. D. Kispert. 1999. Photovoltaic response of carotenoid-sensitized electrode in aqueous solution: ITO coated with a mixture of TiO_2 nanoparticles, carotenoid, and polyvinylcarbazole. *Journal of the Chemical Society-Perkin Transactions* 2(6):1225–1229.

Gao, G., C. C. Wei, A. S. Jeevarajan, and L. D. Kispert. 1996. Geometrical isomerization of carotenoids mediated by cation radical/dication formation. *Journal of Physical Chemistry B* 100:5362–5366.

Gao, Y. L., S. Webb, and L. D. Kispert. 2003. Deprotonation of carotenoid radical cation and formation of a didehydrodimer. *Journal of Physical Chemistry B* 107(47):13237–13240.

Gmitter, F. G. 1993. Marsh grapefruit. *Fruit Varieties Journal* 47(3):130–133.

Goldberg, J., G. Flowerdew, E. Smith, J. A. Brody, and M. O. M. Tso. 1988. Factors associated with age-related macular degeneration. An analysis of data from the first National Health and Nutrition Examination Survey. *American Journal of Epidemiology* 128(4):700–710.

Hammond, B. R., B. R. Wooten, and D. M. Snodderly. 1997. Density of the human crystalline lens is related to the macular pigment carotenoids, lutein and zeaxanthin. *Optometry and Vision Science* 74(7):499–504.

Hankinson, S. E., M. J. Stampfer, J. M. Seddon, G. A. Colditz, B. Rosner, F. E. Speizer et al. 1992. Nutrient intake and cataract extraction in women: A prospective study. *British Medical Journal* 305(6849):335–339.

He, Z. F., and L. D. Kispert. 2001. Carotenoids in sol-gels: Incorporation, stability, and sensitivity to oxidant and acid. *Chemistry of Materials* 13(1):227–231.

Henry, L. K., N. L. Puspitasari-Nienaber, M. Jaren-Galan, R. B. van Breemen, G. L. Catignani, and S. J. Schwartz. 2000. Effects of ozone and oxygen on the degradation of carotenoids in an aqueous model system. *Journal of Agricultural and Food Chemistry* 48:5008–5013.

Hernández-Alvarez, E., I. Blanco-Navarro, B. Pérez-Sacristán, L. S. Sánchez-Siles, and F. Granado-Lorencio. 2016. In vitro digestion-assisted development of a β-cryptoxanthin-rich functional beverage. In vivo validation using systemic response and faecal content. *Food Chemistry* 208:18–25.

Iino, M., Y. Kozai, R. Kawamata et al. 2014. Effects of β-cryptoxanthin on bone-formation parameters in the distal femoral epiphysis of ovariectomized mice. *Oral Radiology* 30:1–8.

Ikeda, N., T. Sugiyama, T. Suzuki et al. 2012. Effects of β-cryptoxanthin on bone metabolism in a rat model of osteoporosis. *Journal of Animal and Veterinary Advances* 11:30–35.

Ikoma, Y., H. Matsumoto, and M. Kato. 2016. Diversity in the carotenoid profiles and the expression of genes related to carotenoid accumulation among citrus genotypes. *Breeding Science* 66(1):139–147.

Iwamoto, M., K. Imai, H. Ohta, B. Shirouchi, and M. Sato. 2012. Supplementation of highly concentrated β-cryptoxanthin in a satsuma mandarin beverage improves adipocytokine profiles in obese Japanese women. *Lipids in Health and Disease* 11:52.

Johnson, E. J. 2014. Role of lutein and zeaxanthin in visual and cognitive function throughout the lifespan. *Nutrition Reviews* 72(9):605–612.

Kanasawud, P., and J. C. Crouzet. 1990. Mechanism of formation of volatile compounds by thermal degradation of carotenoids in aqueous medium. 1. Beta-carotene degradation. *Journal of Agricultural and Food Chemistry* 38:237–243.

Karppi, J., J. A. Laukkanen, and S. Kurl. 2012. Plasma lutein and zeaxanthin and the risk of age-related nuclear cataract among the elderly Finnish population. *British Journal of Nutrition* 108(1):148–154.

Kassoff, A., J. Kassoff, J. Buehler, M. Eglow, F. Kaufman, M. Mehu et al. 2001. A randomized, placebo-controlled, clinical trial of high-dose supplementation with vitamins C and E, beta carotene, and zinc for age-related macular degeneration and vision loss: AREDS report no. 8. *Archives of Ophthalmology* 119(10):1417–1436.

Kaulmann, A., and T. Bohn. 2014. Carotenoids, inflammation, and oxidative stress: Implications of cellular signaling pathways and relation to chronic disease prevention. *Nutrition Research* 34(11):907–929.

Khachik, F., C. J. Spangler, J. C. Smith, Jr., L. M. Canfield, A. Steck, and H. Pfander. 1997. Identification, quantification, and relative concentrations of carotenoids and their metabolites in human milk and serum. *Analytical Chemistry* 69(10):1873–1881.

Konovalov, V. V., and L. D. Kispert. 1999. AMI, INDO/S and optical studies of carbocations of carotenoidmolecules. Acid induced isomerization. *Journal of the Chemical Society* 2:901–909.

Konovalova, T. A., Y. L. Gao, R. Schad, L. D. Kispert, C. A. Saylor, and L. C. Brunei. 2001. Photooxidation of carotenoids in mesoporous MCM-41, Ni-MCM-41 and Al-MCM-41 molecular sieves. *Journal of Physical Chemistry B* 105(31):7459–7464.

Lee, H. S., and W. S. Castle. 2001. Seasonal changes of carotenoid pigments and color in Hamlin, Earlygold, and Budd blood orange juices. *Journal of Agricultural and Food Chemistry* 49(2):877–882.

Lim, S., S. K. Jung, and M. K. Jwa. 2003. Extraction of carotenoids from *Citrus unshiu* press cake by supercritical carbon dioxide. *Food Science and Biotechnology* 12(5):513–520.

Liu, Q., J. Xu, Y. Z. Liu, X. L. Zhao, X. X. Deng, L. L. Guo, and J. Q. Gu. 2007. A novel bud mutation that confers abnormal patterns of lycopene accumulation in sweet orange fruit (*Citrus sinensis* L. Osbeck). *Journal of Experimental Botany* 58(15–16):4161–4171.

Lyle, B. J., J. A. Mares-Perlman, B. E. K. Klein, R. Klein, and J. L. Greger. 1999. Antioxidant intake and risk of incident age-related nuclear cataracts in the Beaver Dam Eye Study. *American Journal of Epidemiology* 149(9):801–809.

Ma, L., H. Dou, Y. Huang, X. Lu, X. Xu, F. Qian et al. 2012. Improvement of retinal function in early age-related macular degeneration after lutein and zeaxanthin supplementation: A randomized, double-masked, placebo-controlled trial. *American Journal of Ophthalmology* 154(4):625–634.

Makoto, O., S. Minoru, and U. Kayo. 2010. Effects of β-cryptoxanthin-fortified Satsuma mandarin (*Citrus unshiu* Marc.) juice on liver function and the serum lipid profile. *Journal of the Japanese Society for Food Science and Technology* 57(3):114–120.

Marty, C., and C. Berset. 1990. Factors affecting the thermal degradation of all-trans-beta-carotene. *Journal of Agricultural and Food Chemistry* 38:1063–1067.

Matsumoto, C., N. Ashida, S. Yokoyama et al. 2013. The protective effects of β-cryptoxanthin on inflammatory bone resorption in a mouse experimental model of periodontitis. *Bioscience Biotechnology and Biochemistry* 77:860–862.

Moeller, S. M., R. Voland, L. Tinker, B. A. Blodi, M. L. Klein, K. M. Gehrs et al. 2008. Associations between age-related nuclear cataract and lutein and zeaxanthin in the diet and serum in the carotenoids in the Age-Related Eye Disease Study (CAREDS), an ancillary study of the Women's Health Initiative. *Archives of Ophthalmology* 126(3):354–364.

Mortensen, A., L. H. Skibsted. 1996. Kinetics of photobleaching of beta-carotene in chloroform and formation of transient carotenoid species absorbing in the near infrared. *Free Radical Research* 25(4):355–368.

Oberholster, R., A. K. Cowan, P. Molnár, and G. Tóth. 2001. Biochemical basis of color as an aesthetic quality in *Citrus sinensis*. *Journal of Agricultural and Food Chemistry* 49:303–307.

Olmedilla, B., F. Granado, I. Blanco, and M. Vaquero. 2003. Lutein, but not atocopherol, supplementation improves visual function in patients with age-related cataracts: A 2-y double-blind, placebo-controlled pilot study. *Nutrition* 19(1):21–24.

Phelan, J. K., and D. Bok. 2000. A brief review of retinitis pigmentosa and the identified retinitis pigmentosa genes. *Molecular Vision* 6(16):116–124.

Richer, S. P., W. Stiles, L. Statkute, K. Y. Pei, J. Frankowski, J. Nyland et al. 2002. The Lutein Antioxidant Supplementation Trial (LAST). *Investigative Ophthalmology and Visual Science* 43(12):2542.

Ríos, G., M. A. Naranjo, M. J. Rodrigo, E. Alós, L. Zacarías, M. Cercós, and M. Talón. 2010. Identification of a GCC transcription factor responding to fruit colour change events in citrus through the transcriptomic analyses of two mutants. *BMC Plant Biology* 10:276.

Rondanelli, M., A. Opizzi, S. Perna et al. 2013. Update on nutrients involved in maintaining healthy bone. *Endocrinologia y Nutricion* 60:197–210.

Sahni, S., M. T. Hannan, J. Blumberg et al. 2009a. Protective effect of total carotenoid and lycopene intake on the risk of hip fracture: A 17-year follow-up from the Framingham Osteoporosis Study. *Journal of Bone and Mineral Research* 24:1086–1094.

Sahni, S., M. T. Hannan, J. Blumberg et al. 2009b. Inverse association of carotenoid intakes with 4-y change in bone mineral density in elderly men and women: The Framingham Osteoporosis Study. *American Journal of Clinical Nutrition* 89:416–424.

Seddon, J. M., U. A. Ajani, R. D. Sperduto, R. Hiller, N. Blair, T. C. Burton et al. 1994. Dietary carotenoids, vitamins A, C, and E, and advanced age-related macular degeneration. *Journal of the American Medical Association* 272(18):1413–1420.

Shegokar, R., and K. Mitri. 2012. Carotenoid lutein: A promising candidate for pharmaceutical and nutraceutical applications. *Journal of Dietary Supplements* 9(3):183–210.

Singh, A., S. Ahmad, and A. Ahmad. 2015. Green extraction methods and environmental applications of carotenoids: A review. *RSC Advances* 5:62358.

Sommerburg, O., J. E. E. Keunen, A. C. Bird, and F. J. G. M. Van Kuijk. 1998. Fruits and vegetables that are sources for lutein and zeaxanthin: The macular pigment in human eyes. *British Journal of Ophthalmology* 82(8):907–910.

Sparrow, J. R., and S. R. Kim. 2010. The carotenoids of macular pigment and bis-retinoid lipofuscin precursors in photoreceptor outer segments, in *Carotenoids: Physical, Chemical, and Biological Functions and Properties*, ed. J. T. Landrum, pp. 355–363. Boca Raton, FL: CRC Press/Taylor & Francis.

Sugiura, M. 2014. Nutritional epidemiologic survey of the relationship between Satsuma mandarin intake and the risk for lifestyle-related diseases: Mikkabi Prospective Cohort Study. *Journal of the Japanese Society for Food Science and Technology* 61:373–381.

Sugiura, M., M. Nakamura, K. Ogawa et al. 2008. Bone mineral density in post-menopausal female subjects is associated with serum antioxidant carotenoids. *Osteoporosis International* 19:211–219.

Sugiura, M., M. Nakamura, K. Ogawa et al. 2011. Dietary patterns of antioxidant vitamin and carotenoid intake associated with bone mineral density: Findings from postmenopausal Japanese female subjects. *Osteoporosis International* 22(1):143–152.

Sugiura, M., M. Nakamura, K. Ogawa et al. 2012. High serum carotenoids associated with lower risk for bone loss and osteoporosis in post-menopausal Japanese female subjects: Prospective cohort study. *PLoS One* 7:e52643.

Sun, T., Z. M. Xu, and J. S. Godber. 2006. Ultrasound assisted extraction in quantifying lutein from chicken liver using high-performance liquid chromatography. *Journal of Chromatography B* 830:158–160.

Sun, Y. J., G. P. Ma, X. Q. Ye, Y. Kakuda, and R. F. Meng. 2010. Stability of all-trans-beta-carotene under ultrasound treatment in a model system: Effects of different factors, kinetics and newly formed compounds. *Ultrasonics Sonochemistry* 17:654–661.

Sun, Y. J., D. H. Liu, J. C. Chen, X. Q. Ye, and D. Yu. 2011. Effects of different factors of ultrasound treatment on the extraction yield of the all-trans-β-carotene from citrus peels. *Ultrasonics Sonochemistry* 18:243–249.

Tanaka, T., T. Tanaka, M. Tanaka, and T. Kuno. 2012. Cancer chemoprevention by citrus pulp and juices containing high amounts of β-cryptoxanthin and hesperidin. *Journal of Biomedicine and Biotechnology* http://dx.doi.org/10.1155/2012/516981.

Tao, J. 2002. Physiological studies on carotenoid formation and regulation in citrus fruit. Ph.D. dissertation, Zhejiang University, Hangzhou, China.

Trumbo, P. R., and C. K. Ellwood. 2006. Lutein and zeaxanthin intakes and risk of age-related macular degeneration and cataracts: An evaluation using the Food and Drug Administration's evidence-based review system for health claims. *American Journal of Clinical Nutrition* 84(5):971–974.

Uchiyama, S., K. Ishiyama, K. Hashimoto et al. 2005. Synergistic effect of β-cryptoxanthin and zinc sulfate on the bone component in rat femoral tissues in vitro: The unique anabolic effect with zinc. *Biological and Pharmaceutical Bulletin* 28:2142–2145.

Uchiyama, S., T. Summida, and M. Yamaguchi. 2004. Anabolic effect of β-cryptoxanthin on bone components in femoral tissues of aged rats in vivo and in vitro. *Journal of Health Science* 50:491–496.

Uchiyama, S., and M. Yamaguchi. 2004. Inhibitory effect of β-cryptoxanthin on osteoclast-like cell formation in mouse marrow cultures. *Biochemical Pharmacology* 67: 1297–1305.

Uchiyama, S., and M. Yamaguchi. 2005a. Oral administration of β-cryptoxanthin prevents bone loss in streptozotocin-diabetic rats in vivo. *Biological and Pharmaceutical Bulletin* 28:1766–1769.

Uchiyama, S., and M. Yamaguchi. 2005b. β-Cryptoxanthin stimulates cell proliferation and transcriptional activity in osteoblastic MC3T3-E1 cells. *International Journal of Molecular Medicine* 15:671–681.

Uchiyama, S., and M. Yamaguchi. 2005c. β-Cryptoxanthin stimulates cell differentiation and mineralization in osteoblastic MC3T3-E1 cells. *Journal of Cellular Biochemistry* 95:1224–1234.

Uchiyama, S., and M. Yamaguchi. 2006a. β-Cryptoxanthin stimulates apoptotic cell death and suppresses cell function in osteoclastic cells: Change in their related gene expression. *Journal of Cellular Biochemistry* 98:1185–1195.

Uchiyama, S., and M. Yamaguchi. 2006b. Oral administration of β-cryptoxanthin prevents bone loss in ovariectomized rats. *International Journal of Molecular Medicine* 17:15–20.

Wang, Y. C., Y. C. Chuang, and H. W. Hsu. 2008. The flavonoid, carotenoid and pectin content in peels of citrus cultivated in Taiwan. *Food Chemistry* 106(1):277–284.

Wei, C., G. Gao, and L. D. Kispert. 1997. Selected cis/trans isomers of carotenoids formed by bulk electrolysis and iron (III) chloride oxidation. *Journal of the Chemical Society* 2:783–786.

Weigert, G., S. Kaya, B. Pemp, S. Sacu, M. Lasta, R. M. Werkmeister et al. 2011. Effects of lutein supplementation on macular pigment optical density and visual acuity in patients with age-related macular degeneration. *Investigative Ophthalmology and Visual Science* 52(11):8174–8178.

Xu, C. J., P. D. Fraser, W. J. Wang, P. M. Bramley. 2006. Differences in the carotenoid content of ordinary citrus and lycopene-accumulating mutants. *Journal of Agricultural and Food Chemistry* 54(15):5474–5481.

Xu, Q., K. Q. Yu, A. D. Zhu, J. L. Ye, Q. Liu, J. C. Zhang, and X. X. Deng. 2009. Comparative transcripts profiling reveals new insight into molecular processes regulating lycopene accumulation in a sweet orange (*Citrus sinensis*) red-flesh mutant. *BMC Genomics* 10:540.

Yamaguchi, M. 2012. Role of carotenoid β-cryptoxanthin in bone homeostasis. *Journal of Biomedical Science* 19:36.

Yamaguchi, M., A. Igarashi, S. Uchiyama, K. Sugawara, T. Sumida, S. Morita, H. Ogawa, M. Nishitani, and Y. Kajimoto. 2006. Effect of β-cryptoxanthin on circulating bone metabolic markers: Intake of juice (*Citrus unishu*) supplemented with beta-cryptoxanthin has an effect in menopausal women. *Journal of Health Science* 52(6):758–768.

Yamaguchi, M., and S. Uchiyama. 2003. Effect of carotenoid on calcium content and alkaline phosphatase activity in rat femoral tissues in vitro: The unique anabolic effect of β-cryptoxanthin. *Biological and Pharmaceutical Bulletin* 26:1188–1191.

Yamaguchi, M., and S. Uchiyama. 2004. β-Cryptoxanthin stimulates bone formation and inhibits bone resorption in tissue culture in vitro. *Molecular and Cellular Biochemistry* 258:137–144.

Yamaguchi, M., and S. Uchiyama. 2008. Combination of β-cryptoxanthin and zinc has potent effects on apoptotic cell death and suppression of bone resorption-related gene expression in osteoclastic cells. *International Journal of Molecular Medicine* 22:221–228.

Yamaguchi, M., S. Uchiyama, K. Ishiyama et al. 2006. Oral administration in combination with zinc enhances β-cryptoxanthin-induced anabolic effects on bone components in the femoral tissues of rats in vivo. *Biological and Pharmaceutical Bulletin* 29:371–374.

Yamaguchi, M., and M. N. Weitzmann. 2009. The bone anabolic carotenoid β-cryptoxanthin enhances transforming growth factor-b1-induced SMAD activation in MC3T3 preosteoblasts. *International Journal of Molecular Medicine* 24:671–675.

Yang, Z., Z. Zhang, K. L. Penniston et al. 2008. Serum carotenoid concentrations in postmenopausal women from the United States with and without osteoporosis. *International Journal for Vitamin and Nutrition Research* 78:105–111.

Yannuzzi, L. A., J. A. Sorenson, R. S. Sobel, J. R. Daly, J. T. DeRosa, J. M. Seddon et al. 1993. Antioxidant status and neovascular age-related macular degeneration. *Archives of Ophthalmology* 111(1):104–109.

Zhao, L. Y., G. H. Zhao, F. Chen, Z. F. Wang, J. H. Wu, and X. S. Hu. 2006. Different effects of microwave and ultrasound on the stability of (all-E)-astaxanthin. *Journal of Agricultural and Food Chemistry* 54:8346–8351.

7 Properties and Use of Citrus Flavonoids

José Antonio del Río, Licinio Díaz, and Ana Ortuño

CONTENTS

7.1 INTRODUCTION

Plants synthesize a vast range of organic compounds that are traditionally classified as primary and secondary metabolites. Phenolics are important plant secondary metabolites, and structurally they are organic aromatic compounds that contain one or more hydroxyl groups attached to a benzene ring. Flavonoids are a widely distributed group of phenolic compounds that are structurally related and characterized by a common benzo-γ-pyrone structure; they act as antioxidants in various biological systems (Morel et al. 1993). Fruits and vegetables, due to the presence of these compounds, have many healthful properties. A considerable amount of epidemiological evidence has revealed an association between people who consume a diet rich in fresh fruit and vegetables and a decreased risk of cardiovascular diseases and certain forms of cancer (Salah et al. 1995). It is generally assumed that the active dietary constituents that contribute to these protective effects are the antioxidant nutrients, although more recent work has highlighted the additional roles of polyphenolic components of higher plants (Hertog et al. 1993), which may act as antioxidants or agents of other mechanisms that contribute to anticarcinogenic or cardioprotective actions. These compounds have applications in food stabilization due to their ability to protect against peroxidation of oxygen-sensitive foods. They have also been demonstrated to inhibit a variety of enzymes, like hydrolases, hyaluronidase, alkaline phosphatase, arylsulfatase, cyclic AMP (cAMP) phosphodiesterase, lipase, α-glucosidase, and kinase.

Flavonoids are present in a wide variety of edible plants and especially in *Citrus* species. In this chapter, the many health-related properties and uses of citrus flavonoids are described. These properties have been found to include antioxidant, anticancer, antimicrobial, and anti-inflammatory activities, and also effects on capillary fragility and an ability to inhibit human platelet aggregation. In addition, the uses of citrus flavonoids and their derivatives in the fields of pharmacology, neutraceuticals, cosmetics, and food technology are described.

7.2 CITRUS FLAVONOIDS

Citrus is the largest genus in the family Rutaceae and is the most highly traded horticultural product in the world. The genus is formed by different species: *C. sinensis* (sweet and blood oranges), *C. unshi* (mandarin), *C. tangerina* and *C. reticulata* (tangerine), *C. clementine* (clementine), *C. aurantium* (sour orange), *C. limon* (lemon), *C. aurantifolia* (lime), *C. paradisi* (grapefruit), *C. grandis* (pomelo), *C. medica* (citron), and hybrids, such as tangelos, tangors, and limequats.

The edible parts of citrus fruits have high contents of many bioactive compounds that make high-pulp juices recommended for consumption, such as simple sugars, vitamin C, carotenoids, flavonoids, limonoids, fiber, folic acid, and potassium, all of which have important effects on health. Almost half of the citrus fruits produced are processed into juices, concentrates, jams, and other food products.

Four types of flavonoids (flavanones, flavones, flavonols, and anthocyanins, the last of which is found only in blood oranges) occur in citrus (Figure 7.1 and Table 7.1). Flavanones are the most abundant, but the highly methoxylated flavones exhibit

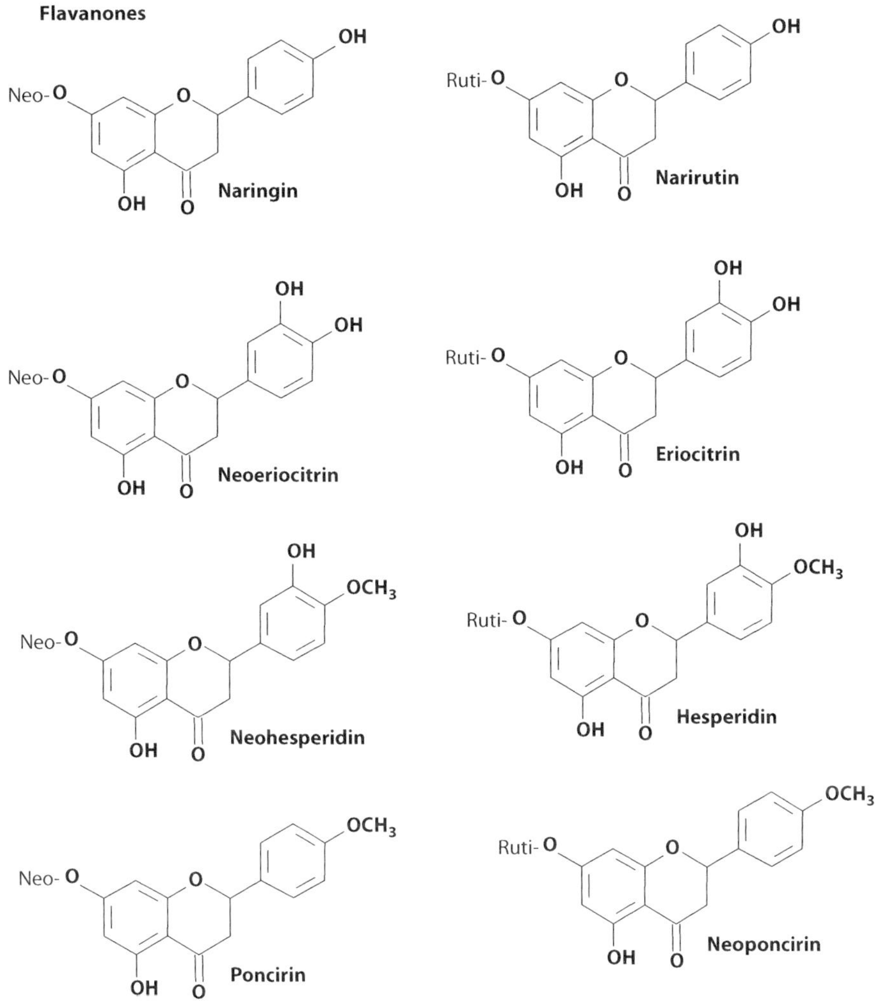

FIGURE 7.1 Structures of flavanone glycosides present in citrus fruits. Neo-, neohesperidoside; Ruti-, rutinoside.

higher biological activity, even though they occur in much lower concentrations. Most *Citrus* species accumulate substantial quantities of flavonoids during the development of their different organs (Castillo et al. 1992; Benavente-García et al. 1993).

Naringin (4′,5,7-trihydroxy-flavanone-7-rhamnoglucoside-glucopyranoside) is one of the phenolic flavonoids present in different citrus fruits and is the flavonoid most abundant in grapefruit and pomelos. Hesperetin (4′-methoxy-5,7,3′-trihydroxyflavanone) and its glycosides are also mainly present in citrus fruits. The aglycone is less dominant in nature than the glycosides. The most widely distributed glycosides of hesperetin are hesperidin and neohesperidin, which are conjugates with rhamnosyl-α-1,6-glucose (rutinoside) and rhamnosyl-α-1,2-glucose (neohesperidoside), respectively (Figure 7.2).

TABLE 7.1
Flavonoids Present in Citrus Fruits

Family	Relevant Species	Flavonoid Subclass	Flavonoid	References
Sweet orange (white and blood oranges)	*Citrus sinensis*	Flavanones	Naringenin 7-ß-rutinoside	Horowitz and Gentili (1977); Peterson et al. (2006); Gattuso et al. (2007); Khan et al. (2014)
			Naringenin 7-ß-rutinoside 4´-ß-D-glucoside	
			Didymin	
			Hesperidin	
			Naringin 4´-ß-D-glucoside	
			Neoeriocitrin	
			Neohesperidin	
			Poncirin	
		Flavones	Limocitrin 3-ß-D-glucoside	
			2″-*O*-ß-D-Xylosylvitexin	
			8-C-ß-D-Glucosyldiosmetin	
			Rhoifolin	
			Luteonin 7-ß-neohesperioside	
			Neodiosmin	
			Tetra-*O*-methylscutellarein	
			Tetra-*O*-methylisoscutellarein	
			Heptamethoxyflavone	
			Sinensetin	
			Isosinensetin	
			Tangeritin	
			Nobiletin	
			5-*O*-Desmethylnobiletin	
			Hexa-*O*-methylquercetagenin	
			Hexa-*O*-methylgossypetin	
			Taxifolin	
			Acacetin	
			6,8-di-C-Apigenin-glucoside	
			8-C-ß-D-Glucosyldiosmetin	
		Anthocyanins	Cyanidin 3-ß-D-glucoside	
			Cyanidin 3,5-di-ß-D-glucoside	
			Peonidin 5-ß-D-glucoside	
			Delphinidin 3-ß-D-glucoside	
			Petunidin 3-ß-D-glucoside	

(*Continued*)

TABLE 7.1 (CONTINUED)
Flavonoids Present in Citrus Fruits

Family	Relevant Species	Flavonoid Subclass	Flavonoid	References
Mandarins	*Citrus deliciosa*	Flavanones	Hesperidin	Horowitz and Gentili (1977); Peterson et al. (2006); Gattuso et al. (2007); Khan et al. (2014)
	Citrus unshiu		Narirutin	
	Citrus clementina		Naringin	
		Flavones	6,8-di-C-Apigenin-glucoside	
			8-C-ß-D-Glucosyl diosmetin	
			Sinensetin	
			Tangeritin	
			Nobiletin	
			Diosmin	
Sour orange	*Citrus aurantium*	Flavanones	Naringin	Horowitz and Gentili (1977); Peterson et al. (2006); Gattuso et al. (2007); Khan et al. (2014)
			Hesperidin	
			Neohesperidin	
			Eriocitrin	
			Neoeriocitrin	
			Poncirin	
			Didymin	
			Narirutin	
		Flavones	Diosmin	
			Tangeritin	
			Nobiletin	
			Kaempferol	
Tangerines	*Citrus tangerina*	Flavanones	Naringin	Horowitz and Gentili (1977); Peterson et al. (2006); Gattuso et al. (2007); Sun et al. (2010); Khan et al. (2014)
	Citrus reticulata		Hesperidin	
			Neohesperidin	
			Eriocitrin	
			Neoeriocitrin	
			Poncirin	
			Didymin	
			Narirutin	
		Flavones	Tangeritin	
			Nobiletin	
			Heptamethoxyflavone	
			Sinensetin	
			Quercetogenin	
			Acacetin	

(Continued)

TABLE 7.1 (CONTINUED)
Flavonoids Present in Citrus Fruits

Family	Relevant Species	Flavonoid Subclass	Flavonoid	References
Grapefruit	*Citrus paradisi*	Flavanone	Naringin Hesperidin Narirutin Neohesperidin Didymin Poncirin Neoeriocitrin Hesperetin Naringenin	Horowitz and Gentili (1977); Peterson et al. (2006); Ortuño et al. (2006); Gattuso et al. (2007); Zhang et al. (2011); Uckoo et al. (2012); Khan et al. (2014)
		Flavones	Vicenin-2 Lucenin-2,4´-methyl ester Diosmin Quercetin 3-*O*-sophroside Sinensetin Nobiletin Heptamethoxyflavone Tangeretin Rhoifolin	
		Flavonols	Rutin Quercetin	
Pummelo	*Citrus grandis*	Flavanones	*O*-triglicosyl naringenin Eriocitrin Neoeriocitrin Naringin Hesperidin Neohesperidin Acetylnaringin Melitidin	Horowitz and Gentili (1977); Peterson et al. (2006); Gattuso et al. (2007); Zhang et al. (2011); Khan et al. (2014)
		Flavones	Lucenin-2 Vicenin-2 Apigenin 6-*C*-glucosyl-7-*O*-glucoside Apigenin 6,8-di-*C*-(synapoil) glycoside Apigenin 6,8-di-C-(feruloyl) glycoside Vitexin Quecetin 3,7-triglucoside Kaempferol Rhoifolin Diosmin	

(*Continued*)

TABLE 7.1 (CONTINUED)
Flavonoids Present in Citrus Fruits

Family	Relevant Species	Flavonoid Subclass	Flavonoid	References
Lemon	*Citrus limon*	Flavanones	Hesperidin	Horowitz and Gentili (1977); Kawaii et al. (1999); Peterson et al. (2006); Gattuso et al. (2007); Khan et al. (2014)
			Eriocitrin	
		Flavones	Diosmin	
			Nobiletina	
			Heptamethoxiflavone	
			Natsudaidain	
			Luteonin	
		Flavonols	Rutin	
Limes	*Citrus latifolia* *Citrus aurantifolia* *Citrus limonia*	Flavanones	Neoponcirina	Horowitz and Gentili (1977); Peterson et al. (2006); Gattuso et al. (2007); Khan et al. (2014)
			Eriocitrin	
			Hesperidin	
			Isorhoifolin	
			Narirutin	
			Quecetin	
			Luteolin	
		Flavones	Diosmin	
			Heptamethoxiflavone	
			Natsudaidain	
			Tangeretin	
			Nobiletin	
			Taxifolin	
			Luteonin	
		Flavonols	Rutin	
Citrons	*Citrus medica*	Flavanones	Naringenin	Horowitz and Gentili (1977); Peterson et al. (2006); Gattuso et al. (2007); Khan et al. (2014); Venturini et al. (2014)
			Hesperidin	
			Hesperetin	
			Quercetin	
			Apigenin	
			Flavanomarein	
			Narirutin	
			Neohesperidin	
		Flavones	Diosmin	
			Luteolin	
			Sinensetin	
			Nobiletin	
			Tangeretin	
		Flavonols	Hyperoxyde	
			Rutin	
			Retusin	
			Robinetintrimethylether	
			Casticin	

(*Continued*)

TABLE 7.1 (CONTINUED)
Flavonoids Present in Citrus Fruits

Family	Relevant Species	Flavonoid Subclass	Flavonoid	References
Tangelos	*C. x tangelo*	Flavanones	Hesperidin	Horowitz and Gentili (1977); Peterson et al. (2006); Gattuso et al. (2007); Barreca et al. (2013); Khan et al. (2014)
			Narirutin	
			Eriocitrin	
			Dydimin	
			Naringin	
			Neoeriocitrin	
		Flavones	Lucenin-2	
			Vicenin-2	
Tangors	*C. x nobilis*	Flavanones	Narirutin	Horowitz and Gentili (1977); Peterson et al. (2006); Gattuso et al. (2007); Khan et al. (2014)
			Eriocitrin	
			Dydimin	
			Hesperidin	
			Naringin	

Hesperidin (hesperetin-7-rutinoside) is present in higher levels in lemons, limes, sweet oranges, tangerines, and tangor species of citrus fruits. On the other hand, neohesperidin (hesperetin-7-neohesperidoside) is the principal flavonoid of *C. aurantium*, and significant amounts also occur in grapefruits (Castillo et al. 1992; Gattuso et al. 2007).

The flavones in citrus are found in glycosylated (Figure 7.3) and aglycone states, the latter of which represent a greater variety of compounds, with their structures

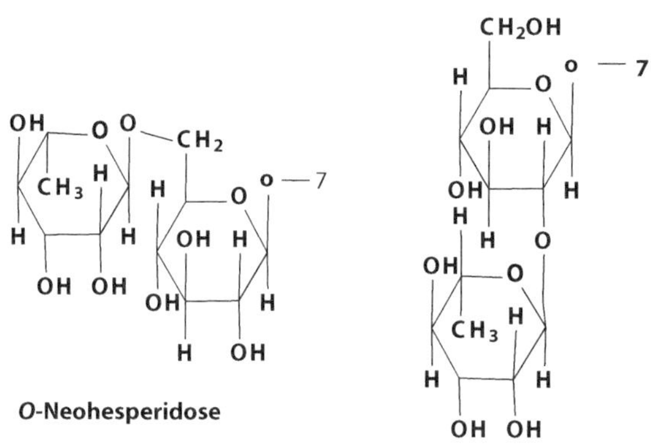

O-Neohesperidose

O-Rutinose

FIGURE 7.2 Structures of sugars present in flavanone glycosides. Rutinoside (rhamnopyranose-*R*-1,6-glucopyranose) and neohesperidoside (rhamnopyranose-*R*-1,2-glucopyranose) side groups are shown (O).

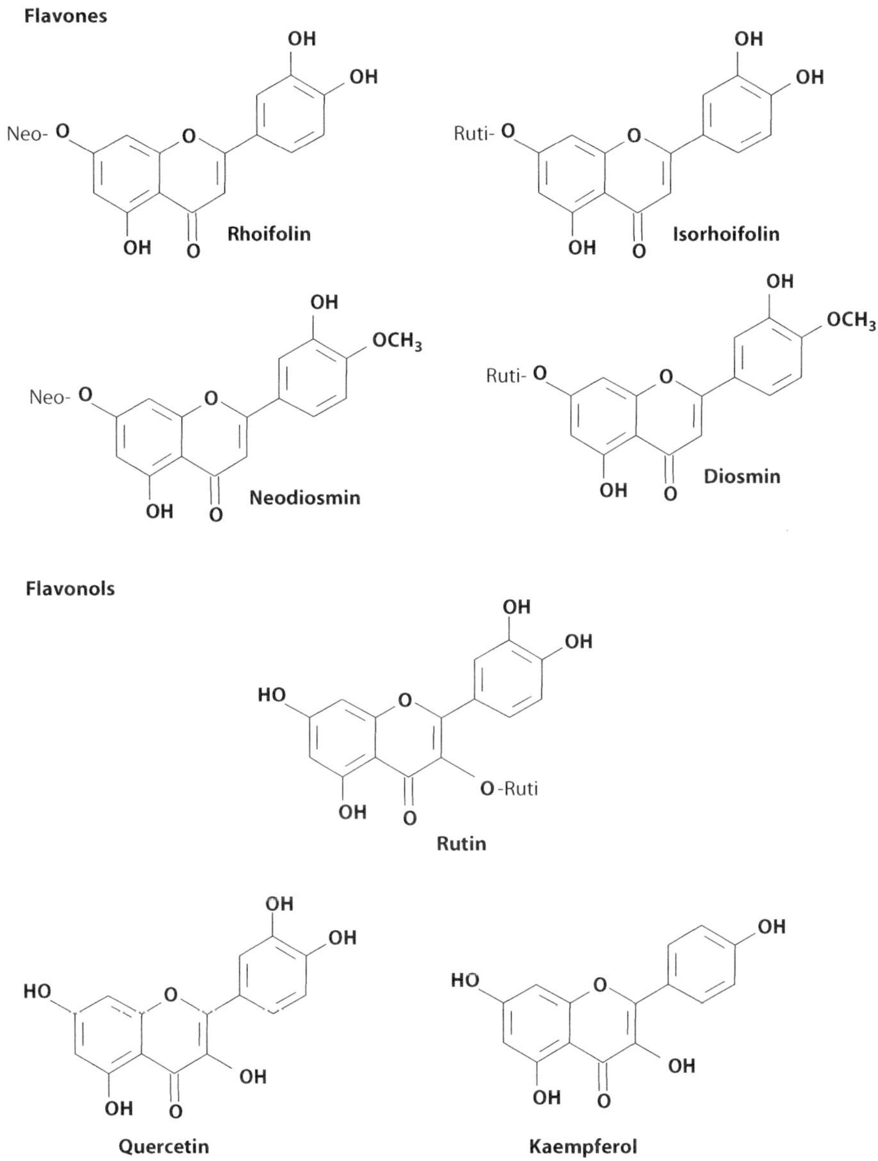

FIGURE 7.3 Structures of flavone glycosides and flavonols present in citrus fruits.

frequently multisubstituted with hydroxyl and/or methoxyl groups. Among these polymethoxylated flavones (PMFs) are scutellarein, sinensetin, tangeretin, quercetogetin, nobiletin, and heptamethoxyflavone (Figure 7.4) (Ooghe et al. 1994; Del Río et al. 2004). The glycosylated flavone diosmin is an important flavonoid in citrus. This flavone has pharmacological applications, being the active ingredient of certain drugs that are used in the treatment of several illnesses of the circulatory system.

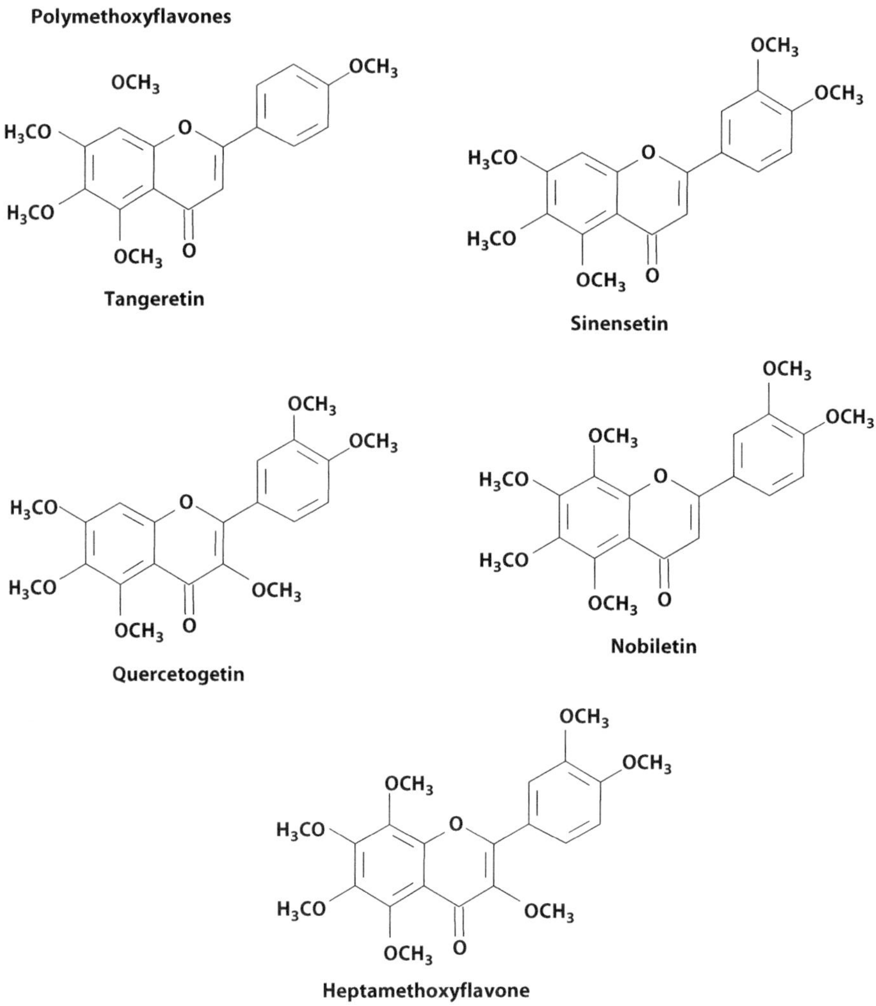

FIGURE 7.4 Structures of polymethoxyflavones present in citrus fruits.

Anthocyanins are flavonoids responsible for the colors of fruits and vegetables. All anthocyanins are based on the six most common anthocyanidins: pelargonidin, cyanidin, delphinidin, peonidin, petunidin, and malvidin, which only differ by the hydroxylation and methoxylation patterns on their B-rings. Structurally, anthocyanins are glycosides and acylglycosides of anthocyanidins, and the aglycone flavyliums (2-phenylbenzopyrilium) differ by their hydroxyl or methoxyl substitutions in their basic structures.

Several cultivars of red oranges (*Citrus sinensis* L. Osbeck), such as the cultivars Tarocco, Moro, and Sanguinello, are characterized by the presence of anthocyanins in both the rind and fruit juice vesicles. Cyanidin 3-glucoside and cyanidin 3-(6′-malonyl)-β-glucoside (Figure 7.5) (Hillebrant et al. 2004; Maccarone et al. 1998)

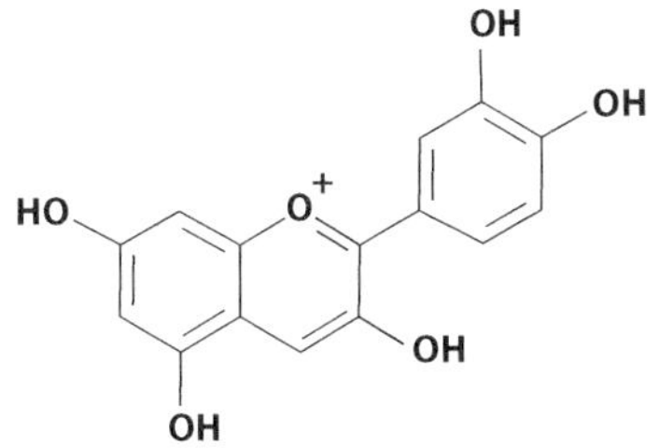

Cyanidin 3-glucoside

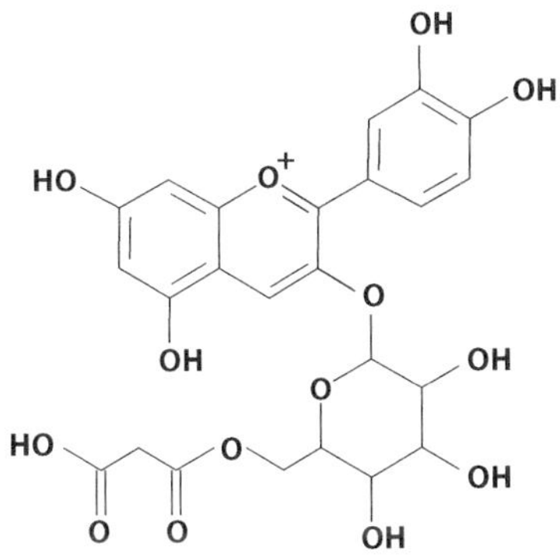

Cyanidin 3-(6"-malonyl)-ß-glucoside

FIGURE 7.5 Structures of principal anthocyanins present in red oranges.

are the main components of blood oranges, together with other minor constituents (Dugo et al. 2003). The amount and composition of anthocyanins in the pigmented orange cultivar vary greatly depending on variety, maturity, region of cultivation, and many other environmental conditions. Anthocyanins are currently under investigation for their ability to inhibit low-density lipoprotein (LDL) cholesterol, prevent blood clotting, and defend cells against dangerous carcinogens; they may prove to be significant compounds for human health.

Studies on the quantitative distribution of these flavonoids in citrus have shown that the 7-*O*-glycosylflavanones are the most abundant flavonoids in all species of the genus; aglycones are intermediates in its biosynthetic pathway (Benavente-García et al. 1995). Although flavones and flavonols have been found in low concentrations in citrus tissues, these types of flavonoids have been show to be powerful antioxidants and free radical scavengers.

The flavonoid contents of citrus fruits increase to a maximum during the early stage of fruit development and then remain constant (Figure 7.6). With an increase in fruit size, the flavonoid concentration decreases (Castillo et al. 1992). The hesperidin content of citrus fruits varies with species, with clementine mandarins containing higher concentration of this flavonoid (Gattuso et al. 2007). Grapefruits are extremely bitter when immature due to the high concentration of naringin in the juice vesicles; when this fruit ripens, the bitterness decreases, with a concomitant decrease of naringin.

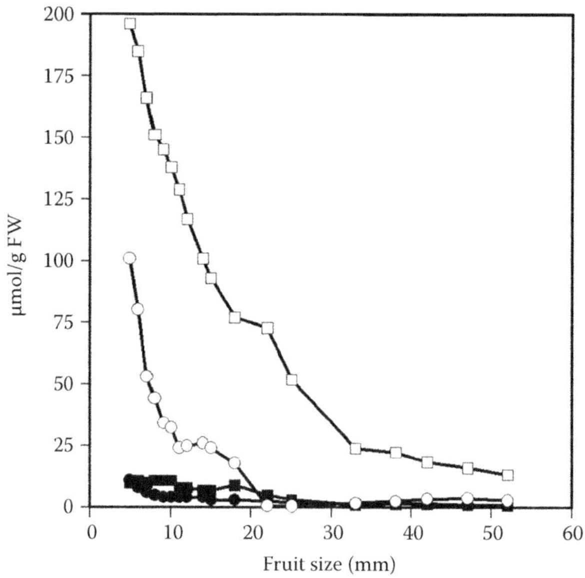

FIGURE 7.6 Flavonoid contents during development of *Citrus limon* cv. Fino 49 fruits. Hesperidin (□); eriocitrin (○); diosmin (■); and luteolin-7-rutinoside (●).

7.3 PROPERTIES OF CITRUS FLAVONOIDS

7.3.1 Antioxidant Capacity

The best-described property of almost every group of flavonoids is their capacity to act as antioxidants. Body cells and tissues are continuously threatened by the damage caused by free radicals and reactive oxygen species (ROS), which are produced during normal oxygen metabolism or are induced by exogenous damage (Finkel et al. 2000). When cells utilize oxygen to generate energy, free radicals are created as a consequence of respiratory metabolism. The primary reactive species in the human body are ROS and reactive nitrogen species (RNS). The different radicals formed (superoxide, hydroxyl, peroxides, alkoxyl, nitric oxide, nitrogen dioxide) may interact with nucleic acids and proteins and produce lipid peroxidation, DNA damage, and protein oxidation. Such oxidative stresses have been implicated in the etiology of many chronic and degenerative diseases, and they are also believed to play a major role in the aging process (Pham-Huy and Pham-Huy 2008; Valko et al. 2007). The mechanisms and the sequence of events by which free radicals interfere with cellular functions are not fully understood, but one of the most important events seems to be lipid peroxidation, which results in cellular membrane damage. This cellular damage causes a shift in the net charge of the cell, changing the osmotic pressure and leading to swelling and eventually cell death.

The implication of the role of oxidative stress in the etiology of chronic and degenerative diseases suggests that antioxidants, both endogenously and exogenously produced, are necessary to protect cellular components from oxidative damage. The endogenous antioxidant defense system is believed to play a primary role in modulating the redox balance and in reducing oxidative stress in the human body. This defense system can be largely divided into three groups: chain-breaking antioxidants, antioxidant enzymes, and metal binding proteins (Halliwell 1994). The exogenous antioxidants, which are obtained mostly through dietary sources, may act as important supplements to endogenous antioxidants.

The antioxidant defense mechanisms of the body include enzymes such as superoxide dismutase, catalase, and glutathione peroxidases, but also nonenzymatic counterparts, such as glutathione, ascorbic acid, and tocopherol. The increased production of ROS during injury results in consumption and depletion of the endogenous scavenging compounds. Phenolic compounds and particularly flavonoids have been shown to possess important antioxidant activities, as they can block the activities of these free radicals, and they may provide an additive effect to that of the endogenous scavenging compounds. Together with an ability to capture electrons, these characteristics import great stability to the flavonoid radicals formed, by means of a tautomeric dislocation which prevents the propagating chain reactions of these oxygen free radicals.

These antioxidant activities attenuate oxidative stress through their direct reactions with free radicals (direct scavenging free radicals) (Honzel et al. 2008) and chelation of metal ions (Kuhnau 1976). They stimulate the activities of important antioxidant enzymes and regulate key redox cell signaling, such as that through the nuclear factors erythroid-2-related factor (Nrf2) and nuclear factor kappa B (NF-κB) pathways (Speciale et al. 2011); thus, they also exert antioxidant activities through indirect mechanisms.

7.3.1.1 Free Radical Scavenging Activity

The antioxidant capacity of a flavonoid is closely linked to the particular structure of the flavonoid compound (Table 7.2). Three structural groups are important for determining the radical scavenging and/or antioxidative capacity (Bors et al. 1990): the *O*-dihydroxy (catechol) structure in the B-ring, the 2,3-double bond in conjugation with a 4-oxo function, and the presence of hydroxyl groups in both the 3 and 5

TABLE 7.2
TEACs for Some Citrus Flavonoids

Flavonoid	TEAC (mM)
Apigenin	1.45
Luteolin	201
Luteolin-7-glucoside	0.79
Naringin	0.24
Naringenin	1.53
Eriodictyol	1.8
Hesperidin	1.08
Cyanidin	4.42

Source: Rice-Evans, C. A. et al., *Free Radical Research* 22:375–383, 1995.

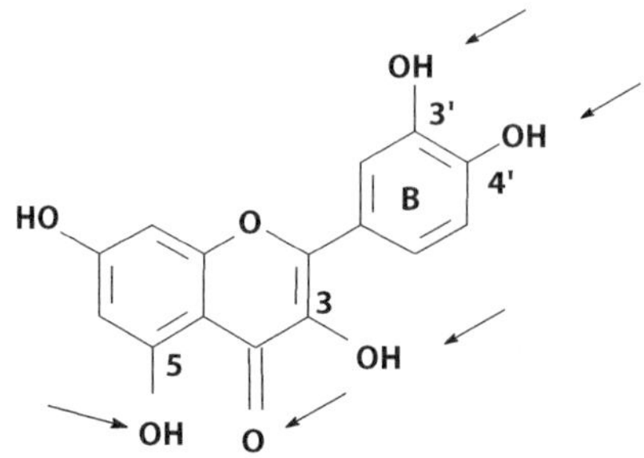

FIGURE 7.7 Functional groups of flavonoid structures implicated in antioxidant activity.

positions (Figure 7.7) (Benavente-García et al. 1997) for maximal radical scavenging capacity and the strongest absorption of radicals. The antioxidant capacity of any flavonoid is determined by a combination of these structural elements and others properties, such as the presence or absence of glycosidic moieties in the flavonoid skeleton (glycosides or aglycons), the glycosylation site, and the number and positions of the free hydroxyls and of the sterified hydroxyls.

Anthocyanins are powerful antioxidants by virtue of their electron-donating properties (Jung and Kwak 2010). In addition, they enhance the capacity for absorbing oxygen radicals in cells, stimulating the expression of detoxifying enzymes, minimizing the formation of oxidative adducts in DNA, and decreasing lipid peroxidation (Hashimoto et al. 2010). In general, the anthocyanidins (aglycones) have superior radical scavenging activity compared to the respective anthocyanins (glycosides); this activity decreases as the number of sugar moieties increases.

7.3.1.2 Quenchers of Singlet Oxygen

Flavonoids have been reported to act as quenchers of singlet oxygen, and based on structure–activity relationships, those citrus flavonoids with higher activity are in the order of quercetin > kaempferol > luteolin > tangeretin > eriodictyol > kaempferol > tangeretin > naringenin. For this activity, the structural factors are very important, where the basic element is the combination between the conjugation of the B-ring to the 4-oxo structure via a 2,3-double bond, with the presence of the 3-hydroxyl group (Sorata et al. 1984; Takahama et al. 1974).

7.3.1.3 Superoxide Anion Scavenging Activity

The superoxide anion scavenging activities of several citrus flavonoids have also been studied, using several methods and with different structural correlations (Darmon et al. 1990). The dominant structural element is the C-ring configuration, particularly the presence of a 3-hydroxyl group that activates the 2,3-double bond. The absence of a 3-hydroxyl group in flavanones and flavones weakens their scavenging capacity. However, the presence of the 2,3-double bond renders the flavonoid structure more reactive; for this reason, apigenin is a moderate scavenger while naringenin shows practically no activity. Xanthine oxidase is a source of oxygen free radicals that react with molecular oxygen, thereby releasing superoxide free radicals. Quercetin and luteolin inhibit xanthine oxidase activity, thereby resulting in decreased oxidative injury. In a study on structure–function relations, luteolin was reported to be the most potent inhibitor of xanthine oxidase (Chang et al. 1993).

7.3.1.4 Nitric Oxide Inhibition

Nitric oxide is produced by several different types of cells, including endothelial cells and macrophages. Although the nitric oxide is important in maintaining the dilation of blood vessels (Huk et al. 1998), higher concentrations of nitric oxide, produced by inducible nitric oxide synthase, may cause oxidative damage. In this case, activated macrophages greatly increase their simultaneous production of both nitric oxide and superoxide anions. Nitric oxide injury takes place for the most part through the peroxynitrite route, because peroxynitrite can directly oxidize LDLs, resulting in irreversible damage to the cell membrane. Several flavonoids reduce nitric oxide injury by interfering with inducible nitric oxide synthase activity (Shoskes 1998) and by interaction with nitric oxide molecules. In one study, the correlation between the flavonoid content and NO production inhibitory activity of fruit peel extracts of 20 *Citrus* species was analyzed (Choi et al. 2007). All citrus peel extracts inhibited lipopolysaccharide-induced NO production in a dose-dependent manner, and this inhibitory effect was significantly and positively correlated with the content of nobiletin and tangeretin. This result supports the premise that nobiletin-rich citrus may provide protection against human diseases resulting from excessive NO production (Choi et al. 2007).

7.3.1.5 Hydroxyl Radical Scavenging Activity

Flavonoids are excellent hydroxyl scavengers, and the degree of effectiveness depends on the binary substitution model in the B-ring, even when the hydroxyls are esterified with methyl groups, and the presence of sugar, with the 3-*O*-glycosides more active than their corresponding aglycones. On the other, a negative influence of the hydroxyl group in position 3 in monosubstituted B-ring compounds has been described (Pincemail et al. 1986).

7.3.1.6 Antilipoperoxidation Activity

The flavonoids are strong scavengers of lipoid radicals (Morel et al. 1993; Torel et al. 1986). The antilipoperoxidant activity of flavonoids depends on the nature of the organic substrate's susceptibility to oxidation, operational conditions, and even the method used to evaluate this potential. In the case of linoleic acid oxidation, the antioxidant activity of flavonoids is related to the inhibition of the formation of *trans,trans*-hydroperoxide isomers of this acid. The antioxidant efficiency depends on the B-ring substitution pattern. Similar results were obtained in the thermal autoxidation of palmitic acid.

7.3.1.7 Chelation of Transition Metals

Some authors have suggested that the activity for chelation of transition metals of the dihydroxylated compounds in the B-ring is partially based on the good correlation between iron-chelating ability and the antioxidant effect. Thus, the antiperoxidative capabilities of several flavonoids were ascribed to concomitant activities of scavenging of free radicals and of chelating iron (Morel et al. 1993). The transition metals iron and copper are essential cofactors of several proteins that are involved in oxygen metabolism, including hemoglobin, catalase, cytochrome P450, transferrin, lactoferrin, ferritin, and hemosiderin in the ferric case (Gutteridge and Halliwell 1994). Copper is present in several enzymes, such as Cu,Zn-superoxide dismutase and dopamine-β-hydroxylase. When the transition metals are present in the free state in biological systems, they can catalyze free radical reactions. Flavonoids, in the form of flavanones and flavones, possess the ability to form a complex with Cu^{2+} ions, and quercetin was found to form the most stable complex (Hudson and Lewis 1983). The ability of rutin to form a stable complex with Fe^{2+} has been demonstrated (Afanas'ev et al. 1989).

7.3.2 Anticarcinogenic Properties

Many of the pharmacological properties of citrus flavonoids can be linked to the abilities of these compounds to inhibit enzymes involved in cell activation. Different flavonoids have demonstrated a capacity to modify the activity of enzymatic systems in mammals, such as kinases, phospholipases, ATPase, lipooxygenases, cyclooxygenases, and phosphodiesterases (Yanez et al. 2004), and to interact with the nucleotide binding sites of regulatory enzymes (Manthey et al. 2001). Research has also shown that citrus flavonoids are potent radical scavengers and, thus, are able to prevent many age-related and degenerative events involving ROS (Benavente-García et al. 1997). Flavonoids showed inhibition of carcinogenesis *in vitro* as well as *in vivo*, and studies with animals and investigations using different cellular models suggest that certain flavonoids could inhibit tumor initiation as well as tumor progression (Manthey et al. 2001).

Cancer may be controlled by a variety of means, including suppression, blockage, and transformation. Agents of suppression prevent the formation of new cancers from procarcinogens, blocking agents prevent carcinogenic compounds from reaching critical initiation sites, and transformation agents act to facilitate the metabolism of carcinogenic components into less toxic materials or to prevent their biological actions. Flavonoids can act in all three ways (Manthey et al. 2001), and they may act in the different development stages of malignant tumors by protecting DNA against oxidative damage, inactivating carcinogens, inhibiting the expression of mutated genes and enzymes responsible for activating procarcinogenic substances, and by activating the systems responsible for xenobiotic detoxification (Bravo 1998).

7.3.2.1 Antimutagenesis

Due to their absorption of ultraviolet light, flavonoids can protect DNA from mutagenic damage. This effect is one of the physiological functions attributed to flavonoids in the plant kingdom (Stapleton and Walbot 1994), although it can be generalized to animal cells, particularly those of mammals. Recent experiments in template plasmid DNA irradiated with UV-B light showed the protective effect of naringenin and rutin against UV-induced DNA damage (Kooststra 1994). In parallel, flavonoids are able to quench free radicals, which may promote mutations when they are generated in the vicinity of DNA. This radical scavenging ability, in a direct or endogenous enzyme-mediated manner, is responsible for the protective effect of flavonoids in whole-body γ-ray-irradiated mice (Shimoi et al. 1994).

Flavonoids may also protect DNA by interacting directly with carcinogens that have escaped detoxification processes, as occurs with the chromosomal aberrations induced by bleomycin (Heo et al. 1994). Heo et al. (1994) showed that *in vitro* or *in vivo* treatment of lymphocytes with galangin, a flavonoid metabolic derivative, suppressed the induction of chromosome aberrations by bleomycin in a galangin dose-dependent manner.

7.3.2.2 Antiproliferation

Although most flavonoids appear to be nontoxic to humans and animals, they have been demonstrated to inhibit proliferation in many kinds of cancerous cell lines. It has been reported that citrus flavonoids (tangeretin, nobiletin, quercetin, and taxifolin) (Kandaswami et al. 1991) have antiproliferative effects on the squamous cell carcinoma cell line HTB43. Quercetin has demonstrated its antiproliferative activity against meningioma cells (Piantelli et al. 1993) and against colon cancer cells, for example, *Caco*-2 and HT-29 cells, with a dose-dependent effect (Kuo 1996). Diosmetin, another important citrus flavonoid and which is available in pharmacies as a venotonic, has shown antiproliferative activity in *Caco*-2 and HT-29 colon cancer cell lines, although with less efficacy than quercetin (Kuntz et al. 1999).

Flavonoids also have shown inhibitory effects on the growth of leukemia HL-60 cell via a nontoxic mechanism (Hirano 1994); these effects are almost equivalent to the effects of currently used anticancer agents. citrus flavones have shown their ability to inhibit the proliferation of different human cancer cells at low 50% inhibitory concentrations (IC_{50}s) (Table 7.3) (Manthey and Guthrie 2002). Most studies have found that IC_{50}

TABLE 7.3
Antiproliferative Activities of Citrus Flavonoids against Colon and Prostate Cancer Cell Lines

	IC_{50} (μM) for Cell Type	
Flavonoid	**Colon**	**Prostate**
Tangeretin	1.6	0.5
Nobiletin	4.7	1.0
Sinensetin	9.5	16.5
Apigenin	29.0	37.0
Luteolin	10.5	32.0
Eriodictyol	6.2	42.0
Hesperetin	149.0	181.0
Naringenin	154.0	150.0
Diosmin	>200	>200
Naringin	>200	>200
Hesperidin	77.0	101.0

Source: Manthey, E., and N. Guthrie, *Journal of Agricultural and Food Chemistry* 50:5837–5843, 2002.

for the inhibition of cell proliferation by active flavonoids are in the low-micromolar range, which are physiologically available concentrations (Fotsis et al. 1997).

Results with different melanoma cell lines have demonstrated the antiproliferative effects of flavonoids against these cancerous cells and in the absence of cytotoxic effects (Martínez et al. 2003). These results point to a correlation between antiproliferative activity and flavonoid structure, and they have shown that the presence of the C_2-C_3 double bond on the C ring, conjugated with the 4-oxo function, is critical for this biological activity (Rodriguez et al. 2002). Sanchez et al. (2001) demonstrated the low cytotoxicity of four polymethoxyflavones that had an antiproliferative effect on two animal tumor cells (LLC-MK2 and C6 cells). When the effects of phenolic compounds were assayed in terms of inhibition of cell growth, tangeretin was the most effective flavonoid on B16F10 cells, followed by gallic acid, baicalein, myricetin, 7,3′-dimethylhesperetin, quercetin, and luteolin.

It can be concluded that the antiproliferative effects of flavonoids are mediated by the inhibition on several kinases and kinase inhibitors and involved in cell-cycle arrest and apoptosis and that this inhibition depends upon the particular structure of each flavonoid.

7.3.2.3 Anti-Invasive

Metastasis occurs when cancer cells invade beyond the boundaries of the primary tumor site and establish new tumors in distant organs. Because metastases are responsible for most cancer deaths, attention has focused on the mechanisms by which cancer cells acquire metastatic properties (Liu et al. 2006). The invasion of surrounding tissues by cancer cells involves several steps, including matrix metalloproteinase (MMP) secretion, migration, invasion, and adhesion. Citrus flavonoids have shown effects on all of these steps. The citrus flavonoids quercetin and apigenin have been reported to possess the ability to inhibit lung colonization *in vivo* by the melanoma B16-BL6 cell line in a dose-dependent manner (Caltagirone et al. 2000). The polymethoxylated flavones tangeretin and nobiletin caused a downregulation of secretion of MMPs in several cancerous lines, both *in vitro* and *in vivo* (Rooprai et al. 2001). Quercetin also inhibited the expression of MMP-2 and MMP-9 in prostate cancer PC-3 cells (Vijayababu et al. 2006). Apigenin has inhibitory effects on the *in vitro* motility and invasiveness of MO4 mouse cells into embryonic chick heart fragments (Bracke et al. 1989) and Hela Cx43 carcinoma cells (Czyz et al. 2004).

A study related to the antimetastatic capacity of flavonoids described how tangeretin inhibits the mobility of sarcoma cells (Bracke et al. 1989) and platelet aggregation through inhibition of the 12-lipoxygenase activity of the platelets. Similarly, it has been observed that tangeretin diminishes the expression of metaloproteases (MMP-2 and MMP-9) in several cell lines (Rooprai et al. 2001), which would also be related to possible antimetastatic activity.

Recent studies have shown that orally administered rutin has a chemoprotective effect on the colon, where it significantly reduces the number of neoplastic foci (Yang et al. 2000). Rutin reduced the area of invasion, both with regard to the percentage of implantation and the growth index, reductions which appear to confirm its antiproliferative effect. The greatest reduction in the number of metastatic nodules was obtained with diosmin; reductions also occurred with respect to the percentage of implantation, growth index, and invasion index in comparson to an ethanol control group (Yang et al. 1997).

These findings suggest the existence of certain structural factors that regulate the level of antimetastasic activity and that are basically related with the greater inhibition capacity of certain enzymatic activities or the ability to block the receptors responsible for the release of mediators of inflammation (Korthui et al. 2002) or platelet aggregation (Guerrero et al. 2005).

7.3.2.4 Antiangiogenic

Tumor growth is strongly dependent upon the formation of new blood vessels, which infiltrate the growing mass of tumor cells and provide oxygen and nutrients and removing metabolites. Decreasing the oxygen pressure in a growing tumor leads to hypoxia, one of the strongest stimuli for the expression of mediators of neovascularization (Dulak and Jozkowicz 2005). Blood vessels are formed in three different ways: vasculogenesis, angiogenesis, and arteriogenesis (Carmeliet 2003).

In vitro and *in vivo* investigations have indicated that some citrus flavonoids are able to inhibit several key events of the angiogenic process, such as the proliferation and migration of endothelial cells and vascular smooth muscle cells and the expression of two major proangiogenic factors, vascular endothelial growth factor and MMP-2 (Albini et al. 2005).

Several citrus flavonoids, such as naringin, rutin, and apigenin, showed significant inhibitory activity at micromolar concentrations in MDA human breast cancer cells, and naringin seemed to be the most active of the compounds studied. Flavonols have been proposed to act as chemopreventive agents in numerous epidemiological studies and have recently been shown to inhibit angiogenesis and the proliferation of tumor and endothelial cells. Quercetin was found to inhibit several steps of angiogenesis, including proliferation, migration, and tube formation of human microvascular dermal endothelial cells, in a dose-dependent manner (Tan et al. 2003). These findings suggest that quercetin has an antiangiogenic potential and that this effect may be related to a decrease in the expression and activity of MMP-2. Apigenin is a potent inhibitor of cell proliferation and angiogenesis, as it markedly inhibits the proliferation and, to a lesser degree, the migration of endothelial cells and capillary formation, independent of its inhibition of hyaluronidase activity (Fang et al. 2005).

7.3.3 Cardiovascular Properties

Several established risk factors for cardiovascular disease have been linked to excessive generation of ROS, known as a state of oxidative stress. For instance, in animal models of hyperlipidemia (Miller et al. 1998), hypertension (Morawietz et al. 2001), and diabetes, elevated levels of vascular superoxide anion production have been found. Moreover, clinical studies have demonstrated that hypercholesterolemia and diabetes in humans are also associated with increased vascular superoxide anion generation (Guzik et al. 2000). All these data strongly suggest that increased oxidative stress is involved in the pathophysiology of cardiovascular disease.

Several mechanisms have been proposed to explain how excessive production of ROS lead to vascular pathology. First, ROS are able to promote the oxidation of

LDL (Hiramatsu et al. 1987). Uptake of oxidatively modified lipoproteins by macrophages transforms these cells into foam cells, which are a key component of atherosclerotic plaques (Steinberg 1997). Second, superoxide anions rapidly inactivate endothelium-derived nitric oxide (NO), a molecule with intrinsic antiatherogenic properties, leading to endothelial dysfunction, which is a hallmark of early atherosclerosis (Darley-Usmar et al. 1995). Moreover, the reaction between superoxide anion and NO generates peroxynitrite ($ONOO^-$), which has been found to be cytotoxic to endothelial and vascular smooth muscle cells through a broad range of biological actions, such as lipid oxidation and mitochondrial DNA damage (Gutierrez et al. 2006). Third, ROS have been shown to be involved in increased expression of certain vascular proinflammatory genes that are pertinent to atherogenesis (Griendling et al. 2000).

Many epidemiological studies have shown that regular flavonoid intake is associated with a reduced risk of cardiovascular diseases (Middleton et al. 2000). In coronary heart disease, the protective effects of flavonoids include mainly antithrombotic, anti-ischemic, antioxidant, and vasorelaxant activities (Jendekova et al. 2006). It has been suggested that flavonoids decrease the risk of coronary heart disease by three major actions: improving coronary vasodilatation, decreasing the ability of platelets in the blood to clot, and preventing oxidation of LDLs.

7.3.3.1 Vasoprotective and Vasorelaxant

Endothelial cells synthesize and release a number of factors, including prostacyclin, NO, endothelium-derived hyperpolarizing factor, and endothelin, which are important in the regulation of vascular tone and the control of platelet and leukocyte adhesion, aggregation, and migration. NO appears to be the critical factor in the preservation of normal coronary vascular function, and there is a well-established correlation between coronary artery disease and impairment of NO activity.

Flavonoids prevent endothelial dysfunction *in vivo* and reduce blood pressure, oxidative stress, and end-organ damage in hypertensive animals. Moreover, some clinical studies have shown that flavonoid-rich foods can improve endothelial function in patients with hypertension and ischemic heart disease. Different studies have shown that flavonoids can cause vasorelaxation at physiological concentrations. The relaxation observed is largely endothelium and NO dependent, although other mechanisms also appear to be involved (Woodman and Chan 2004).

A diet rich in quercetin resuled in increases in NO production and cyclic GMP content in rat aorta in a resting state, and these changes were endothelium dependent (Benito et al. 2002). In other assays, flavones, such as apigenin, luteolin, or quercetin, inhibited NO production. In contrast, flavanones, such as naringenin, did not demonstrate significant inhibition. Other studies on the potential vasorelaxant, antioxidant, and cyclic nucleotide phosphodiesterase inhibitory effects of the citrus fruit flavonoids naringenin and hesperetin in intact rat aortic rings showed that their vasorelaxant effects are basically related to the inhibition of different phosphodiesterase isoenzymes (Orallo et al. 2004).

7.3.3.2 Antithrombotic

Platelet aggregation is the critical event during the initiation of coronary thrombosis, and flavonoids have been reported to modulate platelet function, thus reducing the risk of clot formation (Guerrero et al. 2005). Upon vascular damage, platelets adhere to exposed subendothelium, become activated, and secrete biologically active ligands, including adenosine diphosphate (ADP), serotonin, and thromboxane A2 (TxA2), a potent vasoconstrictor that leads to the deposition of circulating cells at the site of injury and the expansion of the thrombus.

Certain dietary flavonoids have been shown to inhibit platelet function and to impair enzymes involved in cellular signaling, such as cyclo-oxygenases and lypoxygenases, phosphodiesterases, tyrosine kinases, and phospholipases. They have also been postulated to have anticoagulant activity via inhibition of NAD(P)H:quinine acceptor oxidoreductase (Chen et al. 1993), an enzyme inhibited by oral anticoagulants, or by interfering with phosphatidylserine exposure (Bucki et al. 2003). Other mechanisms reportedly related to antiplatelet effects rely on antioxidant properties (Pignatelli et al. 2000). Quercetin has been shown to inhibit collagen-induced responses of platelets through selective blockade of the glycoprotein VI signaling pathway (Hubbard et al. 2003).

Different citrus flavonoids, including apigenin and luteolin, greatly affect the affinity for the thromboxane receptor. Thus, these flavonoids abrogate aggregation and dense granule secretion elicited by arachidonic acid, resulting in inhibition of reactivity, similar to that caused by aspirin (Guerrero et al. 2005).

7.3.3.3 Effect on Coronary Heart Disease

Flavonoids may be protective against CHD by influencing several processes, such as decreased LDL oxidation, increased HDL levels, reduction of cardiac mast cell mediator release, and decreased cardiovascular inflammation. One of the mechanisms to trigger CHD is a greater production of LDL cholesterol. A number of studies have shown that consumption of fruit and vegetables is associated with a reduction of elevated LDL cholesterol levels (Djoussé et al. 2004). Dauchet et al. (2004) conducted a prospective study to assess the relationship between the frequency of fruit (including citrus fruits) and/or vegetable consumption and CHD risk in France and Northern Ireland. They found a favorable relationship between the frequency of citrus fruit consumption and lower rates of acute coronary events in both France and Northern Ireland. Other epidemiological studies have shown a reduction of 12% in the risk of myocardial infarction events in association with citrus fruit consumption (Joshipura et al. 2001).

Increased LDL and especially oxidized LDL are recognized as risk factors in CHD. De Whalley et al. (1990) showed that certain flavonoids, such as quercetin, are potent inhibitors of the modification of LDL by mouse macrophages. Flavonoids also inhibited the cell-free oxidation of LDL mediated by $CuSO_4$. The flavonoids appeared to act by protecting LDL against oxidation caused by the macrophages, because they inhibited the generation of lipid hydroperoxides and protected R-tocopherol, a major lipophilic antioxidant carried in lipoproteins, from being consumed by oxidation in the LDL. Thus, the flavonoids protected R-tocopherol in LDL from oxidation, maintained their levels for longer periods of time, and delayed the onset of lipid peroxidation.

Citrus flavonoids, such as hesperetin, and naringenin, can also act as regulators of apolipoprotein B (apoB), the principal protein of the cholesterol-carrying LDL and the determinant for cellular recognition and uptake of LDL by the high-affinity LDL receptor (Kurowska et al. 2004). Other dietary experimental studies on rats have determined that hesperidin, the main citrus flavanone, increase HDL cholesterol, while it lowers LDL, plasma triglycerides, and total lipids (Monforte et al. 1995). The protective role of flavonoids in cardiac ischemia may also relate to their ability to inhibit mast cell secretion (Middleton et al. 2000).

7.3.4 Anti-Inflammatory and Antiallergic Activities

Plants rich in certain flavonoids have been traditionally used for their anti-inflammatory properties, and recently, attention has been given to isolated flavonoids, including those in citrus, as potential anti-inflammatory agents (Manthey et al. 2001). Inflammation is typically characterized by increased permeability of endothelial tissue and an influx of blood leukocytes into the interstitium, resulting in edema.

The approach to use bioflavonoids as a general anti-inflammatory and antiaging therapies for humans is controversial. Low bioavailability and loss of function because of metabolic processing are the two main arguments against the efficacy of dietary supplementation with plant flavonoids (Duthie et al. 2003). However, there is now sufficient evidence from *in vivo* research and clinical trials to support their use. Several key studies have shown that the anti-inflammatory properties of diosmin and hesperidin are due to its inhibition of the synthesis and biological activities of different proinflammatory mediators, mainly the arachidonic acid derivatives, prostaglandins E2 and F2 and thromboxane A2 (Duthie et al. 2003).

The possible activity of flavonoids in anti-inflammatory and antiallergic responses was well documented by Gabor (1986). Studies on citrus flavonoids by Galati et al. (1994) showed the anti-inflammatory dose-dependent activity of hesperidin, diosmin, and other flavonoids and their influence on the metabolism of arachidonic acid and histamine release. These flavonoids significantly inhibited lysosomal enzyme secretion and arachidonic acid release from membranes by inhibiting lipoxygenase, cycloxygenase, and phospholipase A2.

The inhibition of arachidonic acid release in the inflamed cells would provide less arachidonic substrate for the lipooxygenase and cycloxygenase pathways, leading to a lower levels of endoperoxides, prostaglandins, prostacycline, and thromboxanes on the one hand and hydroperoxy- and hydroxyeicosatrienoic acids and leukotrienes on the other (Gabor 1986). Such an effect confirms the decrease in histamine release, and in fact, histamine is known to act in the first stage of the inflammatory process (Middleton 1981).

Diosmin behaves as a powerful protective agent against inflammatory disorders. Diosmin reduced edema formation and inhibited the synthesis for prostaglandin E2 (78.5%), prostaglandin F2 (45.2%), and thromboxane B2 (59.5%). Intravenous injection of diosmin reduced hyperglycemia induced by injection of alloxan in rats. This effect was linked to its ability to scavenge active oxygen radicals, as demonstrated *in vitro* using human neutrophils or mouse peritoneal macrophages (Jean and Bodinier 1994). Middleton et al. (1981), who studied the effect of quercetin on ragweed antigen-induced basophil histamine release in subjects with hay fever, reported a dose-dependent inhibitory effect and an antagonistic effect of calcium in the inhibition of histamine release by quercetin. Other citrus flavonoids (hesperidin, tangeretin, and nobiletin) exhibited slight to moderate activity (Middleton et al. 2000).

7.3.5 Antimicrobial Activity

One of the properties of flavonoids most related to their physiological action in plants is their antifungal and antiviral activities. Some flavonoids have been considered phytoalexins or phytoanticipins. They are formed as antimicrobial barriers in plants in response to microbial infection. It should therefore not be surprising that they have been found to be effective antimicrobial compounds against a wide array of microorganisms. Because of their widespread ability to inhibit spore germination in plant pathogens, they have also been proposed for use against fungal pathogens in humans (Harborne and Williams 2000).

Fungal infections, especially ringworm infections and dermatophytosis, can affect various parts of the body, such as the skin, hair, and nails (Zheng et al. 2012). The increasing resistance of microorganisms against available antimicrobial agents is a major concern among scientists and clinicians worldwide (Cushinie and Lamb 2005). Pathways of resistance are complex, including activation of ATP binding cassette (ABC) transporters, activation of cytochrome P450 enzymes, and conjunction with glutathione (Wink et al. 2012). The World Health Organization cited that antimicrobial resistance is one of the three most serious problems for public health (Bassetti et al. 2011), and the development of new antifungal drugs is an urgent requirement (Zheng et al. 2012).

Increasingly, flavonoids and other natural products are becoming the subjects of anti-infective research, and many groups have isolated and identified the structures of flavonoids possessing antifungal, antiviral and antibacterial activities.

A survey of current available chemical data suggested that methylated flavones and, to a lesser extent, flavonols, frequently aglycones, are the main classes of antimicrobial and antiviral flavonoids found in medicinal plants. Some of these compounds have been isolated by bioassay-guided fractionation, after previously detecting activity on the part of the plant. Four flavones, 5,6,7,8-tetramethoxyflavone, 5,6,7-trimethoxyflavone, 4′-hydroxy-5,6,7,8-tetramethoxyflavone, and 4′-hydroxy-5,6,7-trimethoxyflavone, were isolated through bioassay-guided fractionation and evaluated for *in vitro* antimicrobial activity. These results showed that isolated flavones may be particularly useful against two pathogenic microorganisms, *Staphylococcus aureus* and *Candida albicans* (Bastos et al. 2009). Antimicrobial tests of 5-hydroxy-3,7,4′-trimethoxyflavone and 5,4′-dihydroxy-3,7-dimethoxyflavone indicated low antifungal activity against the fungi *Candida albicans* and *Trichophyton mentagrophytes* and low antibacterial activity against *Escherichia coli* and *Pseudomonas aeruginosa* (Ragasa et al. 2008).

The flavonoid quercetin-3-methylether showed antibacterial activity against the Gram-positive bacterium *Staphylococcus epidermidis* (Pistelli et al. 2009). Flavonoid sulfates such as quercetin 3,7-di-*O*-methyl 3-sulfate and kaempferol 7-*O*-methyl 3-sulfate inhibited the growth of *Mycobacterium tuberculosis* and *Klebsiella pneumoniae* (Habbu et al. 2009), while flavonoids such as quercetin 3-methyl ether have anti-*Helicobacter pylori* activity (Ustün et al. 2006).

Rutin, quercetin, and naringenin showed significant antimicrobial activity against Gram-positive bacteria and yeasts. Several quercetin glycosides have also been reported as antiviral components of some traditional medicines, such as quercetin 7-rhamnoside and quercetin 3-rhamnoside, which are active against the porcine epidemic diarrhea virus and influenza A virus replication (Choi et al. 2009). Similar results were also obtained for luteolin and luteolin 3′-glucuronyl acid methylester. Quercetin and hesperetin actively inhibited the infectivity and/or replication of herpes simplex viruses, poliovirus, parainfluenza virus, and syncytial viruses, although the quantitatively important grapefruit flavonoid naringin totally lacked this ability.

The flavones 5-hydroxy-7-methoxyflavone and 5,7-dimethoxyflavone inhibit HIV-1 protease. 5-Hydroxy-3,7-dimethoxyflavone inhibits hepatitis C virus and human cytomegalovirus protease, respectively. Other flavones with antiviral activity include luteolin-7-*O*-glucoside and 5-carboxymethyl-4′,7-dihydroxyflavone, which displayed inhibitory activity *in vitro* against hepatitis B virus (Cao et al. 2010).

Nobiletin, hesperidin, and naringin acted as antifungal agents against *Penicillium digitatum*, and the polymethoxyflavone nobiletin was the most effective compound, followed by the flavanones hesperidin and naringin (Ortuño et al. 2006). Also, another paper (Salas et al. 2011) reported the antifungal activity of isolated flavonoids from immature aborted fruits of the *Citrus* species, such as naringin, hesperidin, and neohesperidin, on four fungi often found as food contaminants: *Aspergillus parasiticus*, *Aspergillus flavus*, *Fusarium semitectum*, and *Penicillium expansum*. All the flavonoids showed the capacity to alter the growth of fungi, although the intensity of this activity depended on the type of fungus and compound used.

7.3.6 Ecological Function

Flavonoids constitute a diverse array of plant secondary compounds that perform a wide variety of physiological and ecological functions. The role of anthocyanin pigments as visual signals in angiosperms for attracting pollinators and fruit dispersal agents is well known, but these functions were acquired late in the evolutionary diversification of flavonoids. Less well known and probably more ancient functions of flavonoids include protection against the detrimental effects of UV radiation; mediation of interactions between pollen and stigma; defense against bacteria, pathogenic fungi, and herbivores; mediation of interactions between plants and mutualistic mycorhizzal fungi; and regulators of hormonal activity.

The flavonoid plays an important role in the relationship of plants with the environment and in particular with those animals that eat them. Monophagic insects are chemically attracted to certain plants by some compound or compounds, among which volatiles are important alongside others which determined palatability. For example, in the case of the silk worm, it is isoquercetin which is principally responsible for the worming interaction with mulberry.

With regard to the effectiveness of one type of compound or another in stimulating the feeding animal, it seems that flavonoid glycosides are better than aglycones. Quercetin 3-rutinoside (rutin) stimulated the feeding behavior of *Manduco sexta* in tobacco to a greater extent than the aglycon, quercetin (De Boer and Hanson 1987).

In addition to being an attractant (animals), flavonoids can also repel or even be toxic to others; rutin and isoquercitrin, for example, inhibit larval worm growth in the fruit (Duffey and Isman 1981) and heart of the tobacco plant (Elliger et al. 1980). Studies of the effect of flavonoids on aphids (Dreyer and Jones 1981) have shown that eriodictyol, homoeriodictyol, luteolin and, to a lesser extent, vitexin and naringenin have a repellent effect, while flavanones and flavone-*O*-glucosides do not. Curiously, neohesperidin and neohesperidin dihydrochalcone have a repellent effect that is 10 times less than that of the natural phloridzina.

Since certain products may be toxic to insects but not mammals, which they can attract or repel, growing interest has been expressed for the possibility of controlling pests via environmentally friendly means. The flavanone glycosides in inmature citrus fruits have a role in protecting young citrus tissue against herbivory or disease (Del Río et al. 2004), because they accumulate to very high concentrations in young tissue (mainly leaves and fruit) and are gradually diluted during continued development.

7.4 USES OF CITRUS FLAVONOIDS

7.4.1 Pharmaceutical Compounds

Diosmin is a flavonoid that can be isolated from various plant sources or derived from the flavonoid Hesperidin. Chemically, Diosmin is a flavone derivative, which is defined as the 7-rhamnoglucoside 5,7,3′-trihydroxy-4′-methoxyflavone. Diosmin can be manufactured by extracting hesperidin from citrus rinds and converting the hesperidin to diosmin, and was introduced in Europe as both a diuretic and urinary antiseptic. For over 30 years, diosmin has also been used as a phlebotonic and vascularprotecting agent, and in 1969, diosmin was considered as a therapeutic agent. A diosmin-hesperidin formulation was first launched in European countries as a vegetal extract drug product in 1971 to treat chronic venous insufficiency (CVI) functional symptoms.

Diosmin-hesperidin formulations are used worldwide, not only to neat CVI, but also for a wide range of other venocapillary disorders, including varicose veins, venous stasis ulcers, subconjunctival and retinal hemorrhage, and gingival bleeding (Ramelet 2001). Mechanisms of action include improvement of venous tone, increased lymphatic drainage, protection of capillary bed microcirculation, inhibition of inflammatory reactions, and reduced capillary permeability. Diosmin causes a significant decrease in plasma levels of endothelial adhesion molecules and reduces neutrophil activation, thus providing protection against microcirculatory damage (Lyseng-Williamson and Perry 2003).

Hesperidin methyl chalcone (HMC) is a methylated derivative of the flavonoid hesperidin present in different citrus fruits. The methylation of hesperidin under alkaline conditions produces HMC (Figure 7.8). Hesperidin methyl chalcone is widely used not only in pharmaceutical but also in dietary supplement and cosmetic products Hesperidin methylchalcone has been shown to help strengthen capillaries by increasing capillary resistance and decreasing capillary permeability. Increased venous motility, tone, and the ability of vessels to dilate are also enhanced by HMC,

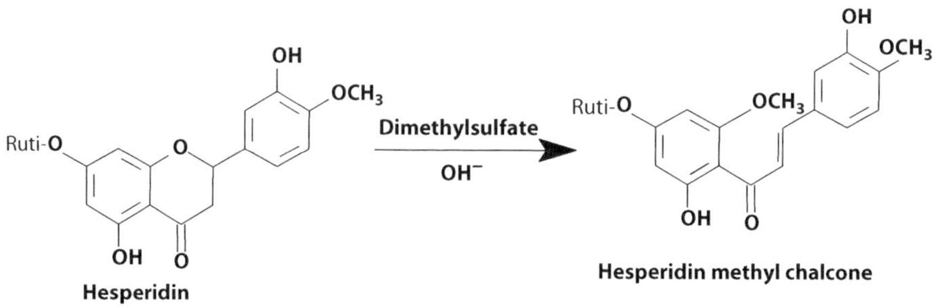

FIGURE 7.8 Synthesis and structure of hesperidin methyl chalcone (HMC).

and is often used to reduce dark circles under the eyes. Also, it is also known to reduce melanin synthesis, thus lightening the skin, and it also protects skin from UV-caused damage and cell ageing. Furthermore, HMC restores the barrier function of the skin, improving hydration, regulating pH and promoting skin cell proliferation (Domange et al. 1994).

7.4.2 Nutraceuticals

"Nutraceutical" is the term used to describe a medicinal or nutritional component that includes a food, plant, or naturally occurring material which may have been purified or concentrated and that is used for the improvement of health by preventing or treating a disease. It is often thought that nutraceuticals have to occupy a narrow strip of legislative ground between pharmaceuticals and food, but in reality their position is much more complex. Nutraceuticals may range from isolated nutrients, dietary supplements, and diets to herbal products and processed products, such as cereals, soups, and beverages. A nutraceutical is any nontoxic food extract supplement that has scientifically proven health benefits for both the treatment and prevention of disease. There is no universal definition of nutraceuticals and/or functional foods, as the definitions vary across countries and markets. All foods are generally functional, because they provide nutrients and energy to sustain growth and support vital cellular processes. Functional foods, however, are generally considered to go beyond the provision of basic nutrients to potentially offer additional benefits, such as reducing the risk of disease and/or promoting optimal health to the consumer (Wildman and Kelley 2007).

Citrus fruits, due to the phenolic compounds in their composion, have the property to modulate the body function that contributes to the prevention of a disease, and they can be considered "functional foods." The modern concept of functional food for the general population was proposed by the Japanese Academic Society in early 1986, and legislation for functional foods was first implemented as FOSHU ("Foods for Specified Health Use").

Phenolic compounds are some of the main active ingredients contained in functional foods and nutraceutical formulations, especially the flavonoids, due to their action as potent antioxidants, metal chelators, and anti-inflammatory, antiallergic, hepatoprotective, antithrombotic, antiviral, and anticarcinogenic activities. Bioavailability, which refers to the body's ability to fully or partially absorb ingested bioactives, is crucial to the ability to exert beneficial effects. The bioavailability and efficacy of active ingredients in nutraceuticals and functional foods are important considerations in their formulation (Havrlentová et al. 2011).

7.4.3 Cosmeceuticals

Cosmeceuticals are cosmetics that contain ingredients that are known to be beneficial to humans in some way. This term was popularized by Albert Kligman. Flavonoids are multiactive components used in common cosmetics, primarily for antioxidant and soothing actions. Flavonoids are powerful antioxidants, and in virtually all cases they outperform artificial antioxidants and poor antioxidants, such as tocopherol and ascorbic acid. They enable control of oxidative stress and peroxidation of membrane lipids. Flavonoids are powerful enzyme modulators; in many cases, it has been shown that they are positive or negative cofactors for various enzymes, such as phosphodiesterases, hyaluronidases, elastases, aldose reductases, lipoxydases, and lipoxygenases (Arct and Pytkowska 2008).

The application of flavonoids in pharmacy/medical technology and food supplements has been well developed. The application of flavonoids in personal care and cosmetic products is less developed, although the use of botanical products is common practice. Major applications of flavonoids are in antiaging, anticellulite, anticouperosis, and skin-lightening products, and application categories include soothing, astringent, or bactericidal properties, and many others (Epstein 2009).

Despite this, their multiactive properties are far from being fully used. It is well known that many flavonoids provide protection from telangiectasias and petechias caused by ruptured blood vessels. Thus, the notion of a strengthening effect of these compounds on blood vessel walls is common. The activity of flavonoids on skin blood vessels is complex. Three main components of their activity can be distinguished: blood vessel protection, platelet aggregation prevention, and deceased capillary permeability (Brand-Garnys et al. 2007).

Free radicals are often brought connected to the occurrence of skin defects and premature skin aging, particularly in conjunction with overexposure to UV-A and UV-B radiation (photoaging; dermatoheliosis) (Matsumura and Ananthaswamy 2004). In cosmetic practice, it is common to compensate these effects by using vitamin C and vitamin E. Topical application of flavonoids, such as quercetin, has been shown to be at least 60 times more active than ascorbic acid and 350 times more active than tocopherol. In actual fact, tocopherol is only a weak antioxidant. It was also found that many flavonoids are able to protect human dermal fibroplasts significantly better than vitamin C and E (Saija and Tomaino 1998; Brand-Garnys et al. 2007).

In addition, flavonoids are also able to protect products vulnerable to oxidation, such as ascorbic acid and tocopherol, and it is therefore not surprising that, for example, the combination of vitamin C plus rutine is much more effective than the individual ingredients (Brand-Garnys 2007). Many flavonoids exhibit antiallergic properties, and some are claimed to have antiviral properties (Lin and Du 2012). Hesperidin, rutin, quercetin, and quercitrin were examined for their inhibition of the actions of vesicular stomatitis virus (VSV) on mouse fibroblasts, and hesperidin was also tested against influenza virus. The results showed the compoudn was preventive rather than curative, but this application seems to be most suitable indeed for skin care products for the inhibition of topically active viruses, such as herpes simplex virus. It goes without saying that cosmetic products, especially for facial care, are usually prophylactic rather than reparative (Brand-Garnys 2007).

Also a large number of references have been made to the anti-inflammatory, anti-thrombotic, diuretic, fungicide and bactericide activity of flavonoids important co-factor for various enzymes; skin lightning is a beautiful exponent thereof. Rutin is well known inhibitor for the enzymes aldose reductase, lipoxygenase and phosphodiesterase. It dramatically strengthens the capillaries, has bactericide and antiviral properties, and a variety of properties that are pharmaceutical by nature. Rutin is also highly appreciated for its ability to avoid erythraemia. There is also evidence that quercetin and quercitrin inhibit elastase, collagenase, hyaluronidase and tyrosinase. All these enzymes are most significant to the cosmetic chemist as they enable to control and manipulate the quality of the skin (Hosseinzadeh and Nassiri-Asl 2014).

7.4.4 Food Additives

Other uses of flavonoids mainly involve their capacity to modify the flavor and/or taste of different compounds and preparations used in the food industry. Among the flavonoids and derivatives that occur in the *Citrus* genus, compounds with widely differing properties exist; some have a very bitter taste, others are tasteless, while others, such as the dihydrochalcones (Horowitz and Gentili 1969), are extremely sweet.

One of the most important structural differences related to taste is the type of glycosidic chain, for example, rutinoside (rhamnopyranose-*R*-1,6-glucopyranose) and neohesperidoside (rhamnopyranose-*R*-1,2-glucopyranose). The first produces tasteless and the second is a bitter compound (Horowitz and Gentili 1969), so that the neohesperidoside flavanones naringin, poncirin, and neohesperidin are bitter while the rutinoside flavanones isonaringin, neoponcirin, and hesperidin are tasteless. Another very important structural factor is the C-ring of the flavonoid structure. When the flavanones naringin and neohesperidin are converted into their corresponding flavones, rhoifolin and neodiosmin, this leads to loss of bitterness. This fact suggests that the planar structure of flavones, with higher conjugation levels, suppresses the taste responses, while the less-conjugated, nonplanar flavanones favor the taste responses (Horowitz and Gentili 1969).

Another type of reaction which alters the C-ring of flavanones is their conversion into chalcones and dihydrochalcones (Horowitz and Gentili 1963). These compounds are extremely sweet when derived from neohesperidoside flavanones (naringin dihydrochalcone and neohesperidin dihydrochalcone) but tasteless when they come from rutinoside flavanones (isonaringin dihydrochalcone and hesperidin dihydrochalcone). Even among the neohesperidoside dihydrochalcones, there is a gradation of taste; the presence of a free hydroxyl in the B-ring is necessary for a greater degree of sweetness (Horowitz and Gentili 1969). Yet other citrus flavonoids

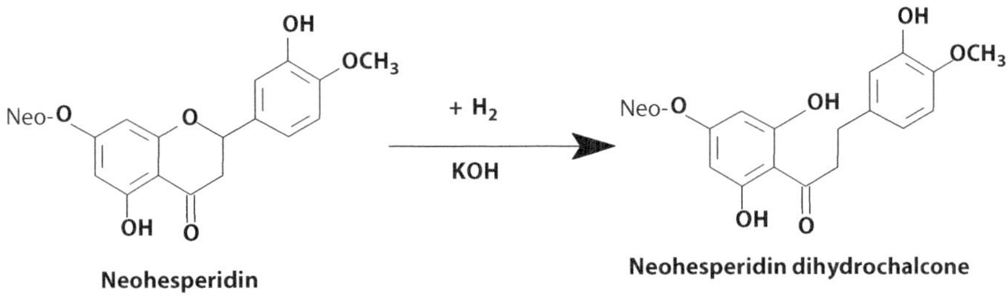

FIGURE 7.9 Synthesis and structure of neohesperidin dihydrochalcone (Neo DHC).

exist which, although tasteless, can alter the taste of fruit juice and other food products. For example, the addition of the flavones neodiosmin to citric juices can significantly reduce the perception threshold of the bitterness produced by limonin (Guadagni et al. 1976).

The dihydrochalcone glucosides of naringin, neohespiridin, and hespiridin are 300, 1100, and 300 times as sweet as sucrose, respectively, on a weight basis and are potential sweeteners without calories (Horowitz and Gentili 1963, 1969). These sweeteners have potential for industrial production because the synthetic sweetener saccharin has some side effects. Neohesperidin dihydrochalcone, also known as neohesperidin DHC, Neo-DHC, and NHDC, is a nonnutritive sweetener and artificial sweetener approved for use in the European Union. NHDC was discovered in 1963, and it is synthesized from naringin and neohesperidin (Figure 7.9), found in grapefruit and bitter orange, and was developed in a research program to find methods for minimizing the taste of bitter flavorants in citrus juices.

Bitterness is a common flavor characteristic in the fruits of some species of the *Citrus* genus, and it is determined by the quantity (concentration) and composition of branched-chain flavanone glycosides, the prevailing flavonoids in citrus (Gattuso et al. 2007). The bitter flavanone 7-*O*-neohesperidosides (e.g., neohesperidin and naringin) are the dominant and, in some cases, the only flavanone glycosides in bitter *Citrus* species (i.e., pomelo, grapefruit, and bitter orange) and comprise the branched-chain disaccharide neohesperidose (rhamnose-2-*O*-glucose) O-linked to position 7 of the flavanone.

7.5 SUMMARY

Citrus fruits have high contents of many bioactive compounds, such as vitamin C, carotenoids, and flavonoids. Four types of flavonoids, flavanones, flavones, flavonols, and anthocyanins, occur in *Citrus* species. Flavanones are the most abundant, and the highly methoxylated flavones occur in much lower concentrations; many of these compounds are found in glycosylated and aglycone states. Naringin is one of the phenolic flavonoids present in different citrus fruits and is the flavonoid most abundant in grapefruit and pomelos. Hesperidin is present in higher levels in lemons, limes, sweet oranges, tangerines, and tangor citrus fruits. Neohesperidin (hesperetin-7-neohesperidoside) is the principal flavonoid of *C. aurantium*, and significant amounts also occur in grapefruit. Several cultivars of red oranges, such as Tarocco, Moro, and Sanguinello, are characterized by the presence of anthocyanins in both the rind and fruit juice vesicles, and cyanidin 3-glucoside and cyanidin 3-(6′-malonyl)-β-glucoside are the main components. The best-described property of almost every group of flavonoids is their capacity to act as antioxidants, as they can block free radicals, and this may have an additive effect to the endogenous scavenging compounds. Many citrus flavonoids act as quenchers of singlet oxygen, superoxide anion scavenging, hydroxyl scavengers, reduce the nitric oxide injury by interfering with inducible nitric oxide synthase activity, are strong scavengers of lipoid radicals, and form complexes with transition metals. In relation with anticarcenogenic properties, citrus flavonoids inhibit carcinogenesis, and they may act in the different developmental stages of malignant tumors by protecting DNA against oxidative damage, inactivating carcinogens, inhibiting the expression of the mutagenic genes and enzymes responsible for activating procarcinogenic substances, and activating the systems responsible for xenobiotic detoxification. Many epidemiological studies have shown that regular flavonoid intake is associated with a reduced risk of cardiovascular disease, and the protective effects of flavonoids include mainly antithrombotic, anti-ischemic, anti-oxidant, and vasorelaxant activities. On the other hand, some citrus flavonoids have antifungal, antiviral, and antibacterial activities. Citrus fruits, due to their phenolic compounds, have the property to modulate the body function that contributes to the prevention of a disease and thus can be considered "functional foods". Other uses of flavonoids mainly involve their capacity to modify the flavor and/or taste of different compounds and preparations used in the food industry. The dihydrochalcone glucosides of naringin, neohespiridin, and hespiridin are potential sweeteners without calories.

REFERENCES

Afanas'ev, I. B., A. I. Dorozhko, A. V. Brodskii, V. A. Kostyuk, and A. I. Potapovitch. 1989. Chelating and free radical scavenging mechanisms of inhibitory action of rutin and quercetin in lipid peroxidation. *Biochemical Pharmacology* 38:1763–1769.

Albini, A., F. Tosetti, R. Benelli, and D. M. Noonan. 2005. Tumor inflammatory angiogenesis and its chemoprevention. *Cancer Research* 65:10637–10641.

Arct, J., and K. Pytkowska. 2008. Flavonoids as components of biologically active cosmeceuticals. *Clinics in Dermatology* 26:347–357.

Barreca, D., C. Bisignano, G. Ginestra, and G. Gattuso. 2013. Polymethoxylated, *C*- and *O*-glycosil flavonoids in tangelo (*Citrus reticulata* × *Citrus paradisi*) juice and their influence on antioxidant properties. *Food Chemistry* 141:1481–1488.

Bassetti, M., F. Ginocchio, and M. Mikulska. 2011. New treatment option against gram-negative organisms. *Critical Care* 15:1–9.

Bastos, M. L., M. R. Lima, L. M. Conserva, V. S. Andrade, E. M. Rocha, and R. P. Lemos. 2009. Studies on the antimicrobial activity and brine shrimp toxicity of *Zeyheria tuberculosa* (Vell.) Bur. (Bignoniaceae) extracts and their main constituents. *Annals of Clinical Microbiology and Antimicrobials* 8:16–25.

Benavente-García, O., J. Castillo, and J. A. Del Río. 1993. Changes in neodiosmin levels during the development of *Citrus aurantium* leaves and fruits. Postulation of a neodiosmin biosynthetic pathway. *Journal of Agricultural and Food Chemistry* 41:1916–1919.

Benavente-García, O., J. Castillo, F. Sabater, and J. A. Del Río. 1995. Characterisation of S-adenosyl-L-methionine:eriodictyol 4′-O-methyltransferase from *Citrus aurantium*. *Plant Physiology and Biochemistry* 33:263–271.

Benavente-García, O., J. Castillo, F. R. Marín, A. Ortuño, and J. A. Del Río. 1997. Uses and properties of citrus flavonoids. *Journal of Agricultural and Food Chemistry* 45:4505–4515.

Benito, S., D. López, M. P. Sáiz, S. Buxaderas, J. Sánchez, P. Puig-Paradella, and M. T. Mitjavila. 2002. A flavonoid-rich diet increases nitric oxide production in rat aorta. *British Journal of Pharmacology* 135:910–916.

Bors, W., W. Hellers, C. Michel, and M. Saran. 1990. Radical chemistry of flavonoids antioxidant. In Emerit, I., L. Packer, and C. Auclair (eds.), *Antioxidants in Therapy and Preventive Medicine,* pp. 165–170. New York: Plenum Press.

Bracke, M. E., B. M. Vyncke, N. A. Van Larebeke, E. A. Bruyneel, G. K. De Bruyne, G. H. De Pestel, W. J. De Coster, M. F. Espeel, and M. M. Mareel. 1989. The flavonoid tangeretin inhibits invasion of MO4 mouse cells into embryonic chick heart *in vitro*. *Clinical and Experimental Metastasis* 7:283–300.

Brand-Garnys, E. E., H. Denzer, H. Meijer, and H. M. Brand. 2007. Flavonoids: A review for cosmetic application, part two. *Journal of Applied Cosmetology* 25:145–159.

Bravo, L. 1998. Polyphenols: Chemistry, dietary sources, metabolism, and nutritional significance. *Nutrition Review* 56:317–333.

Bucki, R., J. J. Pastore, F. Giraud, J. C. Sulpice, and P. A. Janmey. 2003. Flavonoid inhibition of platelet procoagulant activity and phosphoinositide synthesis. *Thrombosis and Haemostasis* 1:1820–1828.

Caltagirone, S., C. Rossi, A. Poggi, F. O. Ranelletti, P. G. Natali, M. Brunetti, F. B. Aiello, and M. Piantelli. 2000. Flavonoids apigenin and quercetin inhibit melanoma growth and metastatic potential. *Journal of Cancer* 87:595–600.

Cao, Y., D. H. Tan, J. J. Chen, G. Z. Zeng, Y. B. Ma, Y. P. Wu, H. Yan, J. Yang, L. F. Lu, and Q. Wang. 2010. Bioactive flavones and biflavones from *Selaginella moellendorfii* Hieron. *Fitoterapia* 81:253–258.

Carmeliet, P. 2003. Angiogenesis in health and disease. *Nature Medicine* 9:653–660.

Castillo, J., O. Benavente-García, and J. A. Del Río. 1992. Naringin and neohesperidin levels during development of leaves, flowerbuds, and fruits of *Citrus aurantium. Plant Physiology* 99:67–73.

Chang, W. S., Y. J. Lee, F. J. Lu, and H. C. Chiang. 1993. Inhibitory effects of flavonoids on xanthine oxidase. *Anticancer Research* 13:2165–2170.

Chen, S., J. Hwang, and P. S. K. Deng. 1993. Inhibition of NAD(P)H:quinine acceptor oxidoreductase by flavones: A structureactivity study. *Archives of Biochemistry and Biophysics* 302:72–77.

Choi, S. H., H. C. Ko, S. Y. Ko, Y. H. Hwang, J. G. Park, S. H. Kang, S. H. Han, S. H. Yun, and S. J. Kim. 2007. Correlation between flavonoid content and the NO production inhibitory activity of peel extracts from various citrus fruits. *Biological and Pharmaceutical Bulletin* 30:772–778.

Choi, H. J., J. H. Song, K. S. Park, and D. H. Kwon. 2009. Inhibitory effects of quercetin 3-rhamnoside on influenza A virus replication. *European Journal of Pharmaceutical Sciences* 37:329–333.

Cushinie, T. T. P., and A. J. Lamb. 2005. Antimicrobial activity of flavonoids. *International Journal of Antimicrobial Agents* 26:343–356.

Czyz, J., Z. Madeja, U. Irmer, W. Korohoda, and D. F. Hulser. 2004. Flavonoid apigenin inhibits motility and invasiveness of carcinoma cells *in vitro. Cancer Cell Biology* 114:12–18.

Darley-Usmar, V., H. Wiseman, and B. Halliwell. 1995. Nitric oxide and oxygen radicals: A question of balance. *FEBS Letters* 369:131–135.

Darmon, N., V. Ferrandiz, M. T. Canal, and S. Mitjavilla. 1990. Activite antiradicallaire de flavonoides vis-a-vis de l'anion superoxide et du radical hydroxyle. *Liaison-Grupe Polyphenols Bulletin* 15:158–162.

Dauchet, L., J. Ferrières, D. Arveiler, J. W. Yarnell, F. Gey, P. Ducimetiére, J. B. Ruidavets, B. Haas, A. Evans, A. Bingham, P. Amouyel, and J. Dallongeville. 2004. Frequency of fruit and vegetable consumption and coronary heart disease in France and Northern Ireland: The PRIME study. *British Journal of Nutrition* 92:963–972.

De Boer, G., and F. E. Hanson. 1987. Feeding responses to solanaceous allelochemicals by larvae of the tobacco hornworm *Munduca sexta*. *Entomologia Experimentalis et Applicata* 45:123–131.

De Whalley, C. V., S. M. Rankin, R. S. Hoult, W. Jessup, and D. S. Leake. 1990. Flavonoids inhibit the oxidative modification of low density lipoproteins by macrophages. *Biochemical Pharmacology* 39:1743–1750.

Del Río, J. A., P. Gomez, A. G. Baidez, M. C. Arcas, J. M. Botia, and A. Ortuno. 2004. Changes in the levels of polymethoxyflavones and flavanones as part of the defense mechanism of *Citrus sinensis* (cv. Valencia Late) fruits against *Phytophthora citrophthora*. *Journal of Agricultural and Food Chemistry* 52:1913–1917.

Djoussé, L., D. K. Arnett, H. Coon, M. A. Province, L. L. Moore, and R. C. Ellison. 2004. Fruit and vegetable consumption and LDL cholesterol: The National Heart, Lung, and Blood Institute Family Heart Study. *American Journal of Clinical Nutritrion* 79:213–217.

Domange, J. R., S. Bougaret, and L. Yubero. 1994. Pharmacological studies of Cyclo 3 Fort and its constituents (*Ruscus* extract, hesperidin methyl chalcone, ascorbic acid): An up to date review. *Clinical Hemorheology* 14:7–13.

Dreyer, D. L., and K. C. Jones. 1981. Feeding deterrency of flavonoides and related phenolics toward *Schizophis graminum* and *Myzuz persice*: Aphid feeding deterrents in wheat. *Phytochemistry* 20:2489–2493.

Duffey, S. S., and M. B. Isman. 1981. Inbibition of insect larval growth by phenolics in glandular trichomes of tomato leaves. *Experientia* 37:574–576.

Dugo, P., L. Mondello, D. Morabito, and G. Dugo. 2003. Characterization of the anthocyanins fraction of Sicilian blood orange juice by micro-HPLC-ESI/MS. *Journal of Agricultural and Food Chemistry* 51:1173–1176.

Dulak, J., and A. Jozkowicz. 2005. Anti-angiogenic and anti-inflammatory effects of statins: Relevance to anti-cancer therapy. *Current Cancer Drug Targets* 5:579–594.

Duthie, G. G., P. T. Gardner, and J. A. Kyle. 2003. Plant polyphenols: Are they the new magic bullet? *Proceedings of the Nutrition Society* 62:599–603.

Elliger, C. A., B. C. Chan, and A. C. Waiss. 1980. Flavonoids as larval growth inhibitors. Structural factors governing toxicity. *Naturwissenchafften* 67:358–360.

Epstein, H. 2009. Cosmeceuticals and polyphenols. *Clinics in Dermatology* 27:475–478.

Fang, J., C. Xia, Z. Cao, J. Z. Zheng, E. Reed, and B.-H. Jiang. 2005. Apigenin inhibits VEGF and HIF-1 expression via PI3K/AKT/p70S6K1 and HDM2/p53 pathways. *FASEB Journal* 19:342–353.

Finkel, T., and N. J. Holbrook. 2000. Oxidants, oxidative stress and the biology of ageing. *Nature* 408:239–247.

Fotsis, T., M. S. Pepper, E. Aktas, S. Breit, S. Rasku, H. Adlercreutz, K. Wahala, R. Montesano, and L. Schweigerer. 1997. Flavonoids, dietary-derived inhibitors of cell proliferation and *in vitro* angiogenesis. *Cancer Research* 57:2916–2921.

Gabor, M. 1986. Anti-inflammatory and anti-allergic properties of flavonoids. In Cody, V., E. Middleton, J. B. Harborne, and A. Beretz (eds.), *Plant Flavonoids in Biology and Medicine: Biochemical, Pharmacological, and Structure-Activity Relationships*, pp. 471–480. New York: Alan R. Liss.

Galati, E. M., M. T. Monforte, S. Kirjavainen, A. M. Forestieri, A. Trovato, and M. M. Tripodo. 1994. Biological effects of hesperidin, a citrus flavonoid (note I): Antiinflammatory andanalgesic activity. *Farmaco* 40:709–712.

Gattuso, G., D. Barreca, C. Gargiulli, U. Leuzzi, and C. Caristi. 2007. Flavonoid composition of *Citrus* juice. *Molecules* 12:1641–1673.

Griendling, K. K., D. Sorescu, B. Lassegue, and M. Ushio-Fukai. 2000. Modulation of protein kinase activity and gene expression by reactive oxygen species and their role in vascular physiology and pathophysiology. *Arteriosclerosis Thrombosis and Vascular Biology* 20:2175–2183.

Guadagni, D. G., V. P. Maier, and J. G. Turnbaugh. 1976. Effect of neodiosmin on threshold and bitterness of limonin in water and orange juice. *Journal of Food Science* 41:681–684.

Guerrero, J. A., M. A. Lozano, J. Castillo, O. Benavente-García, V. Vicente, and J. Rivera. 2005. Flavonoids inhibit platelet function through binding to the thromboxane A2 receptor. *Thrombosis and Haemostasis* 2:1–8.

Gutierrez, J., S. W. Ballinger, V. M. Darley Usmar, and A. Landar. 2006. Free radicals, mitochondria, and oxidized lipids: The emerging role in signal transduction in vascular cells. *Circulation Research* 99:924–932.

Gutteridge, J. M. C., and B. Halliwell. 1994. *Antioxidants in Nutrition, Health, and Disease.* New York: Oxford University Press.

Guzik, T. J., N. E. West, E. Black, D. McDonald, C. Ratnatunga, R. Pillai, and K. M. Channon. 2000. Vascular superoxide production by NAD(P)H oxidase: Association with endothelial dysfunction and clinical risk factors. *Circulation Research* 86:85–90.

Habbu, P. V., K. M. Mahadevan, R. A. Shastry, and H. Manjunatha. 2009. Antimicrobial activity of flavonoid sulphates and other fractions of *Argyreia speciosa* (Burm. f.) Boj. *Indian Journal of Experimental Biology* 47:121–128.

Halliwell, B. 1994. Free radicals and antioxidants: A personal view. *Nutrition Reviews* 52:253–265.

Harborne, J. B., and C. A. Williams. 2000. Advances in flavonoid research since 1992. *Phytochemistry* 55:481–504.

Hashimoto, N., T. Noda, S. J. Kim, H. Yamauchi, S. Takigawa, C. Matsuura-Endo, T. Suzuki, K. H. Han, and M. Fukushima. 2010. Colored potato extracts induce superoxide dismutase-2 mRNA via ERK1/2 pathway in HepG2 cells. *Plant Foods for Human Nutritrion* 65:266–270.

Havrlentová, M., Z. Petruláková, A. Burgárová, F. Gago, A. Hlinková, and E. Šturdík. 2011. Cereal β-glucans and their significance for the preparation of functional foods: A review. *Czech Journal of Food Sciences* 29:1–14.

Heo, H. Y., S. J. Lee, C. H. Kwon, S. W. Kin, D. H. Sohn, and W. W. Au. 1994. Anticlastogenic effects of galangin against bleomycin-induced chromosomal aberrations in mouse spleen lymphocytes. *Mutation Research* 311:225–229.

Hertog, M. G. L., E. J. M. Feskeens, C. H. Hollman, M. B. Katan, and D. Kromhout. 1993. Dietary antioxidant flavonoids and risk of coronary heart disease: De Zutphen Elderly Study. *Lancet* 342:1007–1011.

Hillebrand, S., M. Schwarz, and P. Winterhalter. 2004. Characterization of anthocyanins and pyranoanthocyanins from blood orange [*Citrus sinensis* (L.) Osbeck] juice. *Journal of Agricultural and Food Chemistry* 52(24):7331–7338.

Hiramatsu, K., H. Rosen, J. W. Heinecke, G. Wolfbauer, and A. Chait. 1987. Superoxide initiates oxidation of low density lipoprotein by human monocytes. *Arteriosclerosis* 7:55–60.

Hirano, T., M. Gotoh, and K. Oka. 1994. Natural flavonoids and lignans are potent cytostatic agents against human leukaemic HL-60 cells. *Life Sciences* 55:1061–1069.

Honzel, D., S. G. Carter, K. A. Redman, A. G. Schauss, J. R. Endres, and G. S. Jensen. 2008. Comparison of chemical and cell-based antioxidant methods for evaluation of foods and natural products: Generating multifaceted data by parallel testing using erythrocytes and polymorphonuclear cells. *Journal of Agricultural and Food Chemistry* 56:8319–8325.

Horowitz, R., and B. Gentili. 1963. Dihydrochalcones derivatives and their use as sweetening agents. U.S. Patent 3,087,821.

Horowitz, R., and B. Gentili. 1969. Taste and structure in phenolic glycosides. *Journal of Agricultural and Food Chemistry* 17:696–700.

Horowitz, R., and B. Gentili. 1977. Flavonoids constituents of *Citrus*. In Nagy, S., P. E. Shaw, and M. K. Veldhuis (eds.), *Citrus Science and Technology*, vol. 1, pp. 397–426. Westport, CT: AVI Publising.

Hosseinzadeh, H., and M. Nassiri-Asl. 2014. Review of the protective effects of rutin on the metabolic function as an important dietary flavonoid. *Journal of Endocrinological Investigation* 37:783–788.

Hubbard, G. P., J. M. Stevens, M. Cicmil, T. Sage, P. A. Jordan, C. M. Williams, J. A. Lovegrove, and J. M. Gibbins. 2003. Quercetin inhibits collagen-stimulated platelet activation through inhibition of multiple components of the glycoprotein VI signaling pathway. *Thrombosis and Haemostasis* 1:1079–1088.

Hudson, B. J. F., and J. I. Lewis. 1983. Polyhydroxy flavonoid antioxidants for edible oils, structure-activity relationships. *Food Chemistry* 10:47–55.

Huk, I., V. Brovkovych, and V. J. Nanobash. 1998. Bioflavonoid quercetin scavenges superoxide and increases nitric oxide concentration in ischaemia-reperfusion injury: An experimental study. *British Journal of Surgery* 85:1080–1085.

Jean, T., and M. C. Bodinier. 1994. Mediators involved in inflammation: Effects of Daflon 500 mg on their release. *Angiology* 45:554–559.

Jendekova, L., S. Kojsova, R. Andriantsitohaina, and O. Pechanova. 2006. The time-dependent effects of provinols on brain NO synthase activity in L-NAME-induced hypertension. *Physiological Research* 55:S31–S37.

Joshipura, K. J., F. B. Hu, J. E. Manson, M. J. Stampfer, E. B. Rimm, F. E. Speizer, G. Colditz, A. Ascherio, B. Rosner, D. Spiegelman, and W. C. Willett. 2001. The effect of fruit and vegetable intake on risk for coronary heart disease. *Annals of Internal Medicine* 134:1106–1114.

Jung, K. A., and M. K. Kwak. 2010. The Nrf2 system as a potential target for the development of indirect antioxidants. *Molecules* 15:7266–7291.

Kandaswami, C., E. Perkins, D. S. Soloniuk, G. Drzewiecki, and E. Middleton, Jr. 1991. Antiproliferative effects of citrus flavonoids on a human squamous cell carcinoma in vitro. *Cancer Letters* 56:147–152.

Kawaii, S., Y. Tomono, E. Katase, K. Ogawa, and M. Yano. 1999. Quantitation of flavonoid constituents in *Citrus* fruits. *Journal of Agricultural and Food Chemistry* 47:3565–3571.

Khan, M., Z. Huma, and O. Dangles. 2014. A comprehensive review on flavanones, the major citrus polyphenols. *Journal of Food Composition and Analysis* 33:85–104.

Kooststra, M. 1994. Protection from UV-B induced DNA damage by flavonoids. *Plant Molecular Biology* 26:771–774.

Korthui, R. J., and D. C. Gute. 2002. Anti-inflammatory actions of a micronized, purified flavonoid fraction in ischemia/reperfusion. *Advances in Experimental Medicine and Biology* 505:181–190.

Kuhnau, J. 1976. The flavonoids. A class of semi-essential food components: Their role in human nutrition. *World Review of Nutrition and Dietetics* 24:117–191.

Kuntz, S., U. Wenzel, and H. Daniel. 1999. Comparative analysis of the effects of flavonoids on proliferation, cytotoxicity, and apoptosis in human colon cancer cell lines. *European Journal of Nutritrion* 38:133–142.

Kuo, S. M. 1996. Antiproliferative potency of structurally distinct dietary flavonoids on human colon cancer cells. *Cancer Letters* 110:41–48.

Kurowska, E. M., J. A. Manthey, A. Casaschi, and A. G. Theriault. 2004. Modulation of HepG2 cell net apolipoprotein B secretion by the Citrus polymethoxyflavone tangeretin. Lipids 39:143–151.

Lin, A. L., and G. H. Du. 2012. Antiviral properties of phytochemicals. In Patra, A. K. (ed.), *Dietary Phytochemicals and Microbes*, pp. 93–126. Dordrecht: Springer Science.

Liu, H., D. C. Radisky, C. M. Nelson, H. Zhang, J. E. Fata, R. A. Roth, and M. Bissell. 2006. Mechanism of Akt1 inhibition of breast cancer cell invasion reveals a protumorigenic role for TSC2. *Proceedings of the National Academy of Sciences USA* 103:4134–4139.

Lyseng-Williamson, K. A., and C. M. Perry. 2003. Micronised purified flavonoid fraction: A review of its use in chronic venous insufficiency, venous ulcers and haemorrhoids. *Drugs* 63:71–100.

Maccarone, E., P. Rapisarda, F. Fanella, E. Arena, and L. Mondello. 1998. Cyanidin 3-(6′-malonyl)-β-glucoside. One of the major anthocyanins in blood orange juice. *Italian Journal of Food Safety* 10:367–372.

Manthey, J.A., N. Guthrie, and K. Grohmann. 2001. Biological properties of citrus flavonoids pertaining to cancer and inflammation. *Current Medicinal Chemistry* 8(2):135–153.

Manthey, J. A., and N. Guthrie. 2002. Antiproliferative activities of *Citrus* flavonoids against six human cancer cell lines. *Journal of Agricultural and Food Chemistry* 50:5837–5843.

Martínez, C., J. Yañez, M. Alcaraz, O. Bernavente-Garcia, J. Castillo, J. Lorente, and V. Vicente. 2003. Effects of several polyhydroxylated flavonoids on the growth of B16F10 melanoma and Melan-a cell lines. Influence of sequential oxidation state on the flavonoid skeleton. *Melanoma Research* 13:3–9.

Matsumura, Y., and H. N. Ananthaswamy. 2004. Toxic effects of ultraviolet radiation on the skin. *Toxicology and Applied Pharmacology* 195:298–308.

Middleton, E., G. Drzewiwcki, and D. Krishnarao. 1981. Quercetin: An inhibitor of antigen-induced human basophil histamine. *Journal of Immunology* 127:546–550.

Middleton, E., Jr., C. Kandaswami, and T. C. Theoharides. 2000. The effects of plant flavonoids on mammalian cells: Implications for inflammation, heart disease, and cancer. *Pharmacological Reviews* 52:673–751.

Miller, F. J., Jr., D. D. Gutterman, C. D. Rios, D. D. Heistad, and B. L. Davidson. 1998. Superoxide production in vascular smooth muscle contributes to oxidative stress and impaired relaxation in atherosclerosis. *Circulation Research* 82:1298–1305.

Monforte, M. T., A. Trovato, S. Kirjavainen, A. M. Forestier, E. M. Galati, and R. B. Lo Curto. 1995. Biological effects of hesperidin, a *Citrus* flavonoid (note II): Hypolipidemic activity on experimental hypercholesterolemia in rat. *Pharmacology* 50:595–599.

Morawietz, H., M. Weber, U. Rueckschloss, N. Lauer, A. Hacker, and G. Kojda. 2001. Upregulation of vascular NAD(P)H oxidase subunit gp91phox and impairment of the nitric oxide signal transduction pathway in hypertension. *Biochemical and Biophysical Research Communications* 85:1130–1135.

Morel, I., G. Lescoat, P. Cogrel, O. Sergent, N. Pasdecoup, P. Brissot, P. Cillard, and J. Cillard. 1993. Antioxidant and iron chelating activities of the flavonoids catechin, quercetin and diosmetin on iron-loaded rat hepatocyte cultures. *Biochemical Pharmacology* 1:13–19.

Ooghe, W. C., S. J. Ooghe, C. M. Detavernier, and A. Huyghebaert. 1994. Characterization of orange juice (*Citrus sinensis*) by polymethoxylated flavones. *Journal of Agricultural and Food Chemistry* 42:2191–2195.

Orallo, F., E. Alvarez, H. Basaran, and C. Lugnier. 2004. Comparative study of the vasorelaxant activity, superoxide-scavenging ability and cyclic nucleotide phosphodiesterase-inhibitory effects of hesperetin and hesperidin. *Naunyn-Schmiedeberg's Archives of Pharmacology* 370:452–463.

Ortuño, A., A. Báidez, P. Gómez, M. C. Arcas, I. Porras, A. García-Lidón, and J. A. Del Río. 2006. *Citrus paradisi* and *Citrus sinensis* flavonoids: Their influence in the defence mechanism against *Penicillium digitatum*. *Food Chemistry* 98:351–358.

Peterson, J., J. Dwyer, G. Beecher, S. Bhagwat, S. Gebhardt, D. Haytowitz, and J. Holden. 2006. Flavanones in oranges, tangerines (mandarins), tangors and tangelos: A compilation and review of the data from the analytical literature. *Journal of Food Composition and Analysis* 19:S66–S73.

Pham-Huy, L. A., H. He, and C. Pham-Huy. 2008. Free radicals, antioxidants in disease and health. *International Journal of Biomedical Science* 4:89–96.

Piantelli, M., A. Rinelli, E. Macri, N. Maggiano, L. M. Larocca, A. Scerrati, R. Roselli, M. Iacoangeli, G. Scambia, A. Capella, and F. O. Ranelletti. 1993. Type II estrogen binding sites and antiproliferative activity of quercetin in human meningiomas. *Cancer* 63:193–198.

Pignatelli, P., F. M. Pulcinelli, A. Celestini, L. Lenti, A. Ghiselli, P. P. Gazzaniga, and F. Violi. 2000. The flavonoids quercetin and catechin synergistically inhibit platelet function by antagonizing the intracellular production of hydrogen peroxide. *American Journal of Clinical Nutrition* 72:1150–1155.

Pincemail, J., C. Deby, Y. Lion, P. Braguet, P. Hans, K. Drien, and R. Goutier. 1986. Role of flavonoids in lipoperoxidation and radicalar reactions. In Farkas, L., M. Gabor, and F. Kallay (eds.), *Flavonoids and Bioflavonoids, Proceedings of the 7th Hungarian Bioflavonoid Symposium, Szeged, Hungary, 1985*, pp. 423–436. New York: Elsevier Science.

Pistelli, L., A. Bertoli, C. Noccioli, J. Mendez, R. A. Musmanno, T. Maggio, and G. Coratza. 2009. Antimicrobial activity of *Inga fendleriana* extracts and isolated flavonoids. *Natural Product Communications* 4:1679–1683.

Ragasa, C. Y., M. C. Lapina, J. J. Lee, E. H. Mandia, and J. A. Rideout. 2008. Secondary metabolites from *Tectona philippinensis*. *Natural Product Research* 22:820–824.

Ramelet, A. A. 2001. Clinical benefits of Daflon 500 mg in the most severe stages of chronic venous insufficiency. *Angiology* 52:S49–S56.

Rice-Evans, C. A., N. J. Miller, P. G. Bolwell, J. B. Bramley, and J. B. Pridham. 1995. The relatives activities of plant-derived polyphenolic flavonoids. *Free Radical Research* 22:375–383.

Rodriguez, J., J. Yañez, V. Vicente, M. Alcaraz, O. Benavente-García, J. Castillo, J. Lorente, and J. A. Lozano. 2002. Effects of several flavonoids on the growth of B16F10 and SK-MEL-1 melanoma cell lines: Relationship between structure and activity. *Melanoma Research* 12:99–107.

Rooprai, H. K., A. Kandanearatchi, S. L. Maidment, M. Christidou, G. Trillo-Pazos, D. T. Dexter, G. J. Rucklidge, W. Widmer, G. J. Pilkington. 2001. Evaluation of the effects of swainsonine, captopril, tangeretin and nobiletin on the biological behavior of brain tumour cells *in vitro*. *Neuropathology and Applied. Neurobiology* 27:29–39.

Saija, A., and A. Tomaino. 1998. Influence of different penetration enhancers on in vitro skin permeation and in vivo photoprotective effect of flavonoids. *International Journal of Pharmaceutics* 175:85–94.

Salah, N., N. J. Miller, G. Paganga, L. Tijburg, G. P. Bolwell, and C. Rice-Evans. 1995. Polyphenolic flavonols as scavenger of aqueous phase radicals and as chain-breaking antioxidants. *Archives of Biochemistry and Biophysics* 2:339–346.

Salas, M. P., G. Céliz, and H. Geronazzo. 2011. Antifungal activity of natural and enzymatically modified flavonoids isolated from *Citrus* species. *Food Chemistry* 124:1411–1415.

Sanchez, I., J. Calderon, B. Ruiz, J. Tellez, L. Calzada, and J. Taboada. 2001. *In vitro* cytotoxicity of flavonoids against MK2 and C6 tomour cells. *Phytotherapy Research* 15:290–293.

Shimoi, K., S. Masuda, M. Furogori, S. Esaki, and N. Kinae. 1994. Radioprotective affect of antioxidative flavonoids in γ-ray irradiated mice. *Carcinogenesis* 15:2669–2672.

Shoskes, D. A. 1998. Effect of bioflavonoids quercetin and curcumin on ischemic renal injury: A new class of renoprotective agents. *Transplantation* 66:147–152.

Sorata, Y., U. Takahama, and M. Kimura. 1984. Protective effect of quercetin and rutin on photosensittized lysis of human erythrocytes in the presence of hemato porphyrin. *Biochimica et Biophyica Acta* 799:313–317.

Speciale, A., J. Chirafisi, A. Saija, and F. Cimino. 2011. Nutritional antioxidants and adaptive cell responses: An update. *Current Molecular Medicine* 11:770–789.

Stapleton, A. E., and V. Walbot. 1994. Flavonoids can protect maize DNA from the induction of ultraviolet radiation damage. *Plant Physiology* 105:881–889.

Steinberg, D. 1997. Low density lipoprotein oxidation and its pathobiological significance. *Journal of Biological Chemistry* 272:20963–20966.

Sun, Y., S. Gu, Z. Liu, Y. Zhang, and X. Zhang. 2010. Simultaneous determination of flavonoids in differents parts of *Citrus reticulate* "Chachi" fruit by high-performance liquid chromatography–photodiode array detection. *Molecules* 15:5378–5388.

Takahama, U., R. J. Youngman, and E. F. Elsnert. 1974. Transformation of quercetin by singlet oxygen generated by a photosensitized reaction. *Photobiochemistry and Photobiology* 7:175–181.

Tan, W. F., L. P. Lin, M. H. Li, Y. X. Zhang, Y. G. Tong, D. Xiao, J. Ding. 2003. Quercetin, a dietary-derived flavonoid, possesses antiangiogenic potential. *European Journal of Pharmacology* 459:255–262.

Torel, J., J. Cillard, and P. Cillard. 1986. Antioxidant activity of flavonoids and reactivity with peroxy radical. *Phytochemistry* 2:383–385.

Uckoo, R. M., G. K. Jayaprakasha, V. M. Balasubramaniam, and B. S. Patil. 2012. Grapefruit (*Citrus paradise* Macfad) phytochemicals composition is modulated by household processing techniques. *Journal of Food Science* 77:921–926.

Ustün, O., B. Ozcelik, Y. Akyon, V. Abbasoglu, and E. Yesilada. 2006. Flavonoids with anti-*Helicobacter pylori* activity from *Cistus laurifolius* leaves. *Journal of Ethnopharmacology* 108:457–461.

Valko, M., D. Leibfritz, J. Moncol, M. T. Cronin, M. Mazur, and J. Telser. 2007. Free radicals and antioxidants in normal physiological functions and human disease. *International Journal of Biochemistry and Cell Biology* 39:44–84.

Venturini, N., T. Barboni, F. Curk, J. Costa, and J. Paolini. 2014. Volatile and flavonoid composition of the peel of *Citrus medica* L. var. Corsican fruit for quality assesment of its liqueur. *Food Technology and Biotechnology* 52:403–410.

Vijayababu, M. R., A. Arunkumar, P. Kanagaraj, P. Venkataraman, G. Krishnamoorthy, and J. Arunakaran. 2006. Quercetin downregulates matrix metalloproteinases 2 and 9 proteins expression in prostate cancer cells (PC-3). *Molecular and Cellular Biochemistry* 287:109–116.

Wildman, R. E. C., and M. Kelley. 2007. Neutraceuticals and functional foods. In Wildman, R. E. C. (ed.), *Handbook of Nutraceuticals and Functional Foods*, pp. 1–22. Boca Raton, FL: CRC Press.

Wink, M., M. L. Ashour, and L. El-Readi. 2012. Secondary metabolites from plants inhibiting ABC transporters and microbes to cytotoxic and antimicrobial agents. *Frontiers in Microbiology* 3:1–15.

Woodman, O. L., and E. Chan. 2004. Vascular and anti-oxidant actions of flavonols and flavones. *Clinical and Experimental Pharmacology and Physiology* 31:786–790.

Yañez, J., V. Vicente, M. Alcaraz, J. Castillo, O. Benavente-García, M. Canteras, and J. A. Teruel. 2004. Cytotoxicity and antiproliferative activities of several phenolic compounds against three melanocytes cell lines: Relationship between structure and activity. *Nutrition and Cancer* 49:191–199.

Yang, M., T. Tanaka, Y. Hirose, T. Deguchi, H. Mori, and Y. Kawada. 1997. Chemopreventive effects of diosmin and hesperidin on *N*-butyl-*N*-(4-hydroxybutyl) nitrosamine-induced urinary bladder carcinogenesis in male ICR mice. *International Journal of Cancer* 73:719–724.

Yang, K., S. A. Lamprecht, Y. Liu, H. Shinozaki, KFan, D. Leung, H. Newmark, V. E. Steele, G. J. Kelloff, and M. Lipkin. 2000. Chemoprevention studies of the flavonoids quercetin and rutin in normal and azoxymethane-treated mouse colon. *Carcinogenesis* 21:1655–1660.

Zhang, M., C. Duan, Y. Zang, Z. Huang, and G. Liu. 2011. The flavonoid composition of flavedo and juice from pummelo cultivar (*Citrus grandis* (L.) Osbeck) and the grapefruit cultivar (*Citrus paradisi*) from China. *Food Chemistry* 129:1530–1536.

Zheng, N., Z. Wang, Y. Shi, and J. Lin. 2012. Evaluation of the antifungal activity of total flavonoids extract from *Patrinia villosa* Juss and optimization by response surface methodology. *African Journal of Microbiology Research* 6:586–593.

8 Neuroprotection by Dietary and Citrus Flavonoids

Sam-Long Hwang, Ping-Hsiao Shih, and Gow-Chin Yen

CONTENTS

8.1 INTRODUCTION

Neurodegenerative disorders have been reported to be related to aging, inflammation, and oxidative stress. These risk factors contribute to the degeneration of neural cells, leading to neuronal dysfunction and disorders. Neural cells in aging brains are more vulnerable to oxidative stress because of their high metabolic activity and low antioxidant defense capacity (Aruoma 2002; Floyd et al. 2001; Halliwell 1992). Functional deterioration of neural cells and neurotoxic substances cause the vicious cycles of oxidative stress, inflammation, and disruption of calcium homeostasis during brain aging, which can result in neurotoxicity, memory and cognition declines, and neurodegenerative diseases (Annunziato et al. 2002; Butterfield et al. 1999; Yuan and Yankner 2000). Additionally, hydrogen peroxide is produced in β-amyloid (Aβ) aggregation, dopamine oxidation, and brain ischemia/reperfusion, leading to protein and lipid oxidation and DNA damage in neurons (Mattson 2004; Oikawa et al. 2006; Warner et al. 2004).

Many studies have reported that dietary factors such as flavonoids improve neuronal function and synaptic plasticity. Mechanisms underlying such actions have primarily been characterized through antioxidant and anti-inflammatory bioactivities and signaling regulation at the molecular level (Gomez-Pinilla 2008). Flavonoids exhibit characteristics of both antioxidant and signaling molecules. As signaling molecules, flavonoids interact with key cellular receptors or proteins (kinases and enzymes) that are involved in signaling regulation of physiological responses or gene expression (Kong et al. 2000; Mattson 2004; Williams et al. 2004). Furthermore, it has been suggested that flavonoids are more likely to exert neuroprotective actions through modulation of intracellular signaling associated with neuronal survival, death, and differentiation, as well as through mitochondrial maintenance (Spencer 2008).

Citrus fruits and products are consumed globally. Citrus fruits are rich in flavonoids, mainly flavanones, flavones, and polymethoxyflavones (Gonzalez-Molina et al. 2010; Manthey et al. 2001). Citrus flavonoids exhibit antioxidant, anticarcinogenic, anti-inflammatory, and signaling bioactivity and can traverse the blood–brain barrier (Dimpfel 2006; Manthey et al. 2001; Youdim et al. 2003). Therefore, citrus flavonoids have the potential to intervene in neurodegeneration and promote brain functions.

8.2 DIETARY AND CITRUS FLAVONOIDS

Polyphenols are the most abundant natural antioxidants in people's diets (Tapiero et al. 2002). The main class of polyphenols is flavonoids, of which the basic chemical structure contains a heterocyclic C_6-C_3-C_6 skeleton. Flavonoids can be grouped into flavanols, flavones, isoflavones, flavanols, and flavanones, based on the oxidization of the heterocyclic (C_3) ring. Flavonoids also include anthocyanins, proanthocyanidins, stilbenes (resveratrols), and lignans (Rice-Evans et al. 1995; Tapiero et al. 2002). Dietary sources of flavonoids are fruits, vegetables, and plant-derived beverages, including fruit juices, tea, coffee, and red wine. Foods high in flavonoid content include apples, carrots, cauliflower, citrus fruits, soybeans, and tomatoes. Among these foods, citrus plants are the most important economical fruit trees, with an annual production of nearly 1.02 hundred million tons worldwide (Gonzalez-Molina et al. 2010). Citrus flavonoids include three major subgroups: flavanones (mainly di- and tri-*O*-glycosides), flavone glycosides (mainly di- and tri-*O*-glycosides and *C*-glycosides), and polymethoxyflavones (Manthey et al. 2001). The citrus *O*-glycosides are mainly in the form of rutinosides or neohesperidosides. The main citrus rutinosides are hesperidin, narirutin, eriocitrin, isorhoifolin, and diosmin. These compounds are tasteless and are found primarily in oranges (*Citrus sinensis* L.), tangerines (*C. reticulate* L.), and lemons (*C. limon* L.). The neohesperidosides, such as naringin, neoeriocitrin, neodiosmin, and neohesperidin, have a bitter taste and are mostly found in hybrids of grapefruit (*C. paradise* Macf.) and pomelo (*C. grandis* L.). Diosmetin and luteolin, the main citrus flavonoe aglycones, are found in general in citrus plants and lemons, respectively. Citrus flavonols are only found in lemons (*C. limon* L.). Citrus polymethoxyflavones, including tangeretin, nobiletin, natsudaidain, and heptamethoxyflavone, are abundant in oranges, lemons, and tangerines (Gonzalez-Molina et al. 2010; Manthey et al. 2001). Additionally, new compounds of polymethoxyflavones and hydroxylated polymethoxychalcones have also been found in sweet oranges (*Citrus sinesis*) (Li et al. 2006). New flavonoid compounds (flavones, flavonol derivatives, and flavanone) with *C*-glycosides and *O*-glycosides have recently been identified in blood orange juice (*Citrus sinensis* L. Osbeck) and these compounds have been validated for their antioxidant properties and inhibition of acetylcholinesterase activity (Barreca et al. 2016).

8.3 ANTIOXIDANT PROPERTIES OF DIETARY AND CITRUS FLAVONOIDS

It is known that the antioxidant effects of flavonoids are mainly attributed to their scavenging of oxygen-derived free radicals (Lin et al. 2002; Miyake et al. 2006). By actions of hydrogen donation and metal ion binding, as well as the resonance effect of phenoxyl radical stabilization, flavonoids can exert antioxidant activity (Bors et al. 2001; Rice-Evans et al. 1996). They act as reactive oxygen species (ROS) scavengers, quenchers of singlet oxygen formation, reducing agents and metal chelators, chain-breaking antioxidants, and protectors of ascorbic acid (Kandaswami and Middleton 1997). Compared to other dietary flavonoids, citrus flavonoids exhibit moderate to strong oxygen radical and ROS scavenging activities. Nevertheless, citrus flavonoids possess higher activity for inhibition of lipid peroxidation and scavenging benzoylperoxide radicals ($PhCOO^{\bullet}$), methyl methacrylate radicals ($R^{\bullet}$), 2,2-diphenyl-1-picrylhydrazyl (DPPH), alkyperoxyl radicals ($ROO^{\bullet}$), and peroxynitrite ($ONOO^{-}$) (Hirata et al. 2005; Kim et al. 2004; Sawa et al. 1999; Yokozawa et al. 2002b; Yu et al. 2005). The most abundant and widespread citrus flavonoids are hesperidin, hesperetin, naringin, naringenin, and neohesperidin, and these exhibit moderate antioxidant activities. In contrast, kaempferol, luteolin, neoeriocitrin, and scutellarein exert strong antioxidant activities. Generally, antioxidant activities of citrus glycosides are weaker than those of the aglycones; however, the inhibition of hesperidin on the formation of $R^{\bullet}$ and polyunsaturated fatty acid-derived free radicals is stronger than that of hesperetin (Hirata et al. 2005; Jung et al. 2003). Hesperetin also exerts good intracellular $ONOO^{-}$ scavenging activity (Kim et al. 2004; Pollard et al. 2006). An *in vivo* study showed that isorhamnetin effectively suppressed the peroxidation of lipids in the blood, liver, and kidneys of rats with diabetes (Yokozawa et al. 2002a).

8.4 FLAVONOIDS: ANTIOXIDANTS OR SIGNALING MOLECULES?

Studies on the bioactivities of flavonoids have focused on the transfer of their antioxidant activities to signaling regulation (Spencer 2008; Williams et al. 2004).

Antioxidant activities of flavonoids are insufficient to explain their cellular bioactivity. Concentrations of flavonoids (in the high nanomolar range) and their metabolites (in the low micromolar range) are lower than those of small molecules of nutrients (present in high micromolar concentrations), such as ascorbic acid and α-tocopherol, in the blood, plasma, and brain. Their antioxidant activities are not comparable with those of such nutrients *in vivo* (Williams et al. 2004). Increasing evidence suggests that flavonoids protect neural cells against oxidative stress by actions other than antioxidant activity. It has been shown that the neuroprotection of flavonoids against oxidative stress-induced apoptosis is better than that of ascorbate, with 10 times above the dose of flavonoids in neurons (Schroeter et al. 2001). Like dose effects of genistein against Aβ-induced apoptosis in primary hippocampal neurons, a low dose (0.8 μM) of citrus flavonoids was more effective than a high dose (50 μM) in inhibiting caspase-3 activity and DNA damage in hydrogen peroxide-induced PC12 cells (Zeng et al. 2004; Hwang and Yen 2008). In hydrogen peroxide-treated cortical neurons, neuroprotective effects of citrus flavanones (hesperetin and 5-nitro-hesperetin) were not reflected by their antioxidant property but through the activation of prosurvival Akt and extracellular signal-regulated kinase 1/2 (ERK1/2) signaling regulation (Vauzour et al. 2007). Furthermore, high levels of ROS and intracellular calcium elevate caspase-3 and JNK activities, which can be suppressed by Akt signaling pathways (Crossthwaite et al. 2002; McGinnis et al. 1999; Vauzour et al. 2007). Hesperetin, isorhamnetin, and isosakuranetin have been shown to activate Akt signaling (Hwang and Yen 2009; Vauzour et al. 2007). Based on the higher ROS or intracellular calcium level but lower caspase-3 and JNK activities of hydrogen peroxide-treated PC12 cells after treatment with a high concentration (50 μM) of isorhamnetin or isosakuranetin rather than treatment with a low concentration (0.8 μM) of the flavonoid, it has also been suggested that the flavonoids act more as signaling molecules than as antioxidants to protect cells from oxidative stress (Hwang and Yen 2009).

8.5 BENEFICIAL BIOACTIVITIES OF FLAVONOIDS ON NEUROLOGICAL PROCESSES VIA SIGNALING REGULATIONS

Flavonoids exert many beneficial bioactivities, including antioxidant action, anti-inflammation, antitumorigenesis, reduction of plasma cholesterol and blood sugar levels, reduction in blood pressure, and neuroprotection (Williams et al. 2004). As neuroprotective agents against oxidative stress, flavonoids can directly scavenge free radicals, activate prosurvival regulatory pathways, or indirectly increase endogenously defensive gene expression (Williams et al. 2004). *In vitro* and animal studies have shown that neuroprotective effects of flavonoids against oxidative damage and amyloid-derived neurotoxicity are attributed to their antioxidant properties, anti-inflammation, and signaling regulation (Hwang et al. 2012a).

Apparent beneficial bioactivities of flavonoids on neurological processes include: (a) interacting with neural signaling pathways for pro-cellular survival, antineurodegeneration, and memory and cognition enhancements, via the upregulation of antioxidant enzymes and proteins involved in synaptic plasticity and neuronal repair; (b) protecting against neurotoxins and neuroinflammation via the modulation of neuroprotective and anti-inflammatory signaling pathways; (c) eliciting changes in blood flow to the brain and facilitating more-efficient cerebral blood flow for optimal brain function and adult neurogenesis in the hippocampus; (d) inhibiting neuropathological processes in specific brain regions via the suppression of Aβ aggregation and proamyloidogenic process and the promotion of Aβ clearance and nonamyloidogenic process (Williams and Spencer 2012). These flavonoid-modulated neurological processes can potentially prevent age-related neurodegenerative disorders and dementia, as well as enhancing cognition.

It is known that memory enhancement is controlled at the molecular level (Kandel 2001). Short-term memory involves preexisting protein modifications, and long-term memory requires new mRNA and protein syntheses; the cyclic AMP (cAMP) element binding protein (CREB) binds to the promoter regions of genes associated with memory and synaptic plasticity and plays a critical role in the process of long-term memory (Spencer 2009). Hence, it is plausible to identify effective agents that trigger CREB activity, leading to memory consolidation. Five flavonoid-associated signaling pathways can modulate such a process, including (a) cAMP-dependent protein kinase (protein kinase A [PKA]), (b) protein kinase B (PKB/Akt), (c) protein kinase C (PKC), (d) calcium-calmodulin kinase (CaMK), and (e) ERK pathways. All these signaling pathways can converge to CREB and lead to synaptic plasticity by formation of stable long-term potentiation (LTP) at synapses (Spencer 2009). These processes involve interactions with key cellular receptors and proteins, CREB phosphorylation, brain-derived neurotrophic factor expression and release, neurotransmitter (glutamate) release, phosphatidylinositol 3-kinase (PI3K)/mammalian target of rapamycin (mTOR) signaling, and sustained synthesis of Arc (the cytoskeletal immediate-early gene *Arc/Arg3.1*), which is essential for memory consolidation synthesis and synapse growth (Bramham et al. 2010; Spencer 2009).

Brain-derived neurotrophic factor (BDNF) is essential for neural cell survival, differentiation, and synaptic activity and is broadly expressed in the brain (Einat and Manji 2006). It elevates the survival of basal forebrain cholinergic neurons, hippocampus, and parietal cortex, areas that show very serious injury in Alzheimer's disease brains (Michalski and Fahnestock 2003). Further, BDNF is robustly expressed in hippocampal neurons after synaptic stimulation. It interacts with pre- or postsynaptic tropomyosin receptor kinase B (TrkB) and triggers processes that lead to LTP consolidation for learning and memory (Spencer 2009). LTP requires changes in the molecular composition and structure of neurons which are dependent on mRNA and protein synthesis (Waltereit et al. 2001). *Arc* mRNA is robustly upregulated following LTP stimulation and is distributed to dendritic processes for local translation and facilitation of synapse-specific modifications for dendritic spinal growth. The Arc protein regulates α-amino-3-hydroxy-5-methyl-4-isoxazolepropionic acid (AMPA) receptor trafficking at synapses and F-actin polymerization in the spine. It thus plays an important role in regulation of protein synthesis-dependent forms of synaptic plasticity (Bramham et al. 2010; Pintchovski et al. 2009). Arc expression is upregulated by signaling cascades involving BDNF, CREB, ERK, PKA, PKC, and serum response factor (Bramham et al. 2010). All of these factors are potential targets of flavonoids, and such flavonoid-induced events in brain neurons will help to enhance memory and cognition functions.

8.6 NEUROPROTECTION VIA CITRUS FLAVONOIDS AND EFFECTS ON MEMORY AND COGNITION

An increasing number of studies have shown that citrus flavonoids exhibit positive effects on neuroprotection, memory, and cognition. It has been suggested that hesperetin (at 0.1 μM) protects cortical neurons from oxidative damage by activating receptor-mediated prosurvival Akt and ERK1/2 signaling pathways, which inhibit the activation of proapoptotic proteins, including apoptosis signal-regulating kinase-1 (ASK1), caspase-9, and caspase-3 (Vauzour et al. 2007). A high dose of hesperetin protected cortical neurons against oxidative stress, Aβ-associated neurotoxicity, and glutamate-induced excitotoxicity (Cho 2006). With long-term ingestion of hesperetin (by mice at 50 mg/kg of body weight for 5 weeks), the oxidation of lipids and proteins was obviously decreased; the expression of antioxidant enzymes, including catalase, Mn-superoxide dismutase (Mn-SOD), Cu,Zn-SOD, glutathione (GSH) peroxidase, and glutathione reductase, as well as the GSH/GSSG ratio, were obviously increased in the brains of mice (Choi and Ahn 2008). Autophagy regulation plays a critical role in the progression and/or development of neurodegenerative diseases. A recent study showed that Aβ-stimulated autophagy promotes the impairment of neuronal energy metabolism by mechanisms of glucose uptake, glucose transporters (GLUTs), and insulin signaling cascades; it was suggested that hesperetin and hesperidin treatments improve Aβ-impaired glucose utilization through downregulation of neuronal autophagy (Huang et al. 2012).

Naringin, a well-known citrus flavanone glycoside, exerts neuroprotection in a neurotoxin model of Parkinson's disease (PD) *in vivo*. It induces glia-derived neurotrophic factor (GDNF), an important neurotrophic factor for adult dopaminergic (DA) neurons, and attenuates the level of tumor necrosis factor-α (TNF-α) in microglia (Leem et al. 2014).

Naringenin, a aglycone of naringin, protects cells against Aβ-associated oxidative damage and enhances the memory performance of scopolamine-treated mice with amnesia (Heo et al. 2004a). It also exerts an inhibitory effect on acetylcholinesterase and hence may have potential in dementia treatment (Heo et al. 2004b). Accumulating evidence suggests that diabetes is strongly associated with the risk of Alzheimer's disease (AD) and memory deficits. In studies using diabetes-associated AD-type neurodegeneration and cognitive impairment animal models, naringnenin pretreatment reversed streptozotocin-induced alterations in cognitive behavioral, biochemical, antioxidant enzyme, histopathological, and choline acetyltransferase expression in rat hippocampus via its antioxidant potential and cholinesterase inhibitory action (Khan et al. 2012; Rahigude et al. 2012). In addition, a recent study revealed that naringenin protects against 6-hydroxydopamine (6-OHDA)-induced neurotoxicity via the E2-related factor 2 (Nrf2)/antioxidant response element (ARE) signaling pathway in models of PD, both *in vitro* and *in vivo*. Unlike in the control group, naringenin failed to block 6-OHDA neurotoxicity or induce Nrf2-dependent cytoprotective genes against 6-OHDA-induced oxidative damage in Nrf2 small interfering RNA (siRNA)-treated SH-SY5Y cells. In the animal study, oral administration of naringenin protected mice against 6-OHDA-induced nigrostriatal dopaminergic neurodegeneration and oxidative injury (Lou et al. 2014).

Nobiletin exerts antineuroinflammatory activity on suppression of microglial activation via inhibition of the release of nitric oxide (NO) and expression of inducible NO synthase, which is similar to that of minocycline, a well-known microglial inactivator. It also inhibits the release of proinflammatory cytokines (tumor necrosis factor and interleukin-1β), JNK, and p38 MAPK activation, and translocation (from the cytoplasm) and expression of nuclear factor κB in the nucleus (Cui et al. 2010). Furthermore, nobiletin exhibits neurotrophic action to induce neurite outgrowth via a non-tyrosine kinase receptor A (TrkA)-mediated PKA/MEK/ERK signaling pathway to increase CREB phosphorylation in cultural hippocampal neurons (Nagase et al. 2005a). It can promote the neurological processes of memory and cognition. The administration of nobiletin can rescue impaired memory of olfactory-bulbectomized mice with cholinergic neurodegeneration in the hippocampus (Nagase et al. 2005b; Nakajima et al. 2007). Studies have shown that nobiletin and its metabolite, 4′-demethylnobiletin, rescue learning impairment associated with *N*-methyl-D-aspartate (NMDA) receptor antagonism via cAMP/PKA/ERK/CREB signaling (Al Rahim et al. 2009; Nakajima et al. 2007).

Onozuka et al. (2008) indicated that nobiletin administration improved memory deficits in mice by decreasing the Aβ burden and plaques and activating ERK in the hippocampus. Nobiletin also enhanced PKA-mediated phosphorylation of the GluR1 receptor, the subunit of the AMPA receptor, at Ser845 and the postsynaptic AMPA receptor sensitivity to glutamate in the murine hippocampus. These neurological processes are crucial for maintaining the strength of basal synaptic transmission and regulating the long-term potential (Matsuzaki et al. 2008). Furthermore, nobiletin ameliorates cognitive impairment, oxidative burden, and hyperphosphorylation of tau in senescence-accelerated mouse prone 8 (SAMP8), a model of aging characterized by the early onset of learning and memory impairments and pathological features of AD. It restores the decrease in the GSH/GSSG ratio and increases activities of glutathione peroxidase and manganese-SOD in brains of SAMP8 mice (Nakajima et al. 2013). In a triple-transgenic mouse model of AD (3XTg-AD) in which the mice progressively develop amyloid plaques, neurofibrillary tangles, and cognitive impairments, a 3-month period of nobiletin treatment reversed short-term memory and recognition memory impairments and reduced soluble $A\beta_{1\text{-}40}$ levels in the brain in 3XTg-AD mice. It also reduced ROS levels in the hippocampus of both 3XTg-AD and wild-type mice. Nobiletin thus has potential uses in treatment and prevention of AD (Nakajima et al. 2015). Recently, a remarkable clinical study performed in Japan showed that long-term administration of dried orange peel (the Chinese traditional herb with multiple physiological benefits), which is rich in nobiletin, to patients with AD may improve their cognition and behavior performance (Seki et al. 2013).

8.7 ANTI-INFLAMMATORY ACTIONS OF CITRUS FLAVONOIDS CONFERRING NEUROPROTECTIVE EFFECTS

Inflammation is a normal response to cellular stress, oxidative damage, tissue trauma, bleeding, and infection. However, uncontrolled inflammation has been associated with many diseases, such as diabetes, atherosclerosis, cancer, and neurodegenerative disorders (Benavente-García and Castillo 2008; Wang and Smart 1999).

Distinct biological mediators influence different steps of the inflammation cascade. Anti-inflammatory agents can exhibit therapeutic properties by suppressing actions or syntheses of the mediators (Manthey et al. 2001). In inflammation, phospholipase A2, cyclooxygenase, and lipoxygenase catalyze key reactions involved in syntheses of proinflammatory arachidonic acid derivatives (AAD). It is known that flavonoids can inhibit such reactions (Gil et al. 1994; Landolfi et al. 1984). The AAD are essential for activating neutrophils and stimulate ROS formation in inflammatory tissues (Dana et al. 1994; Rossi et al. 1976; Zallen et al. 1998). ROS-mediated inflammation and its mediators play central roles in many neurodegenerative disorders. ROS induces chronic inflammation, and proinflammatory mediators such as NO and TNF in microglia. A high level of NO is neurotoxic because of the formation of peroxynitrite (Benavente-García and Castillo 2008). Citrus flavonoids can exhibit neuroprotection by their anti-inflammatory effects. Hesperidin and diosmin exert anti-inflammatory properties by blocking the synthesis and actions of ADD (Manthey et al. 2001). Apigenin and diosmin can effectively inhibit NO formation and TNF-α release induced by lipopolysaccharide or *advanced glycation end products* in microglia (Shanmugam et al. 2008). Citrus flavonoids are known to suppress phosphodiesterase and kinase activation in the initiation stage of inflammation. The activation of these enzymes induces the expression of proinflammatory TNF-α and protein kinase activity. Citrus flavonoids can also inhibit the induction of endothelial cell adhesion molecules triggered by cytokines via blocking the adhesion of neutrophils, monocytes, and other leukocytes from injured regions (Benavente-García and Castillo 200; Manthey et al. 2001).

8.8 MOLECULAR MECHANISMS UNDERLYING THE NEUROPROTECTION OF DIETARY AND CITRUS FLAVONOIDS

8.8.1 Demand for Flavonoids as Pharmacological and Neuroprotective Agents

Trends of major strategy for age-related neurodegenerative diseases and dementia include dietary prevention and pharmaceutical therapies in modern clinical neurological practice. The action of such neuroprotective compounds can be by their interactions with signaling-related transporters, receptors, or key enzymes in neural cells. For examples, estrogen and estrogen receptor (ER)-triggered signaling pathways play pivotal roles in controlling neuronal differentiation and survival in brain tissues (Habauzit et al. 2011). Hence, estrogen can exert neuritogenic effects and neuroprotection against physiological stress and degenerative pathologies such as Alzheimer's and Parkinson's diseases. Additionally, glutamate acts as a necessary excitatory neurotransmitter for regulating brain functions. Ceftriaxone, a β-lactam antibiotic, can exhibit neuroprotective effects by activating the major glutamate transporter for glutamate uptake in astroglial cells (Lee et al. 2008; Su et al. 2003). Further, rapamycin exhibits neuroprotection by inactivating mammalian target of rapamycin (mTOR), which leads to the suppression of neuronal apoptosis and autophagy in several models of neurodegenerative diseases (Carloni et al. 2012; Wu et al. 2009). Pharmacological compounds exhibit sound action as signaling molecules in neurobiological processes. However, they usually exert side effects in humans. Accumulating evidence has indicated that flavonoids are relatively safe and exert signaling properties in neuroprotection and cognition improvement. The demand of flavonoids for intervention in neurodegenerative disorders are increasing.

8.8.2 Neuroprotective Signaling Pathways of Flavonoids

Williams et al. (2004) indicated that flavonoids may have a chemically structural affinity to bioactive proteins, including receptors and kinases involved in cellular physiology regulation. They interact with proteins associated with kinase signaling cascades for cellular survival or apoptosis, such as those in Akt/PKB, MAPK, PI-3K, PKC, and tyrosine kinase pathways. Flavonoids may also alter bioactivity of the proteins via their ATP binding sites (e.g., mitochondria ATPase, calcium membrane ATPase, PKA, PKC, and topoisomerase) or benzodiazepine binding sites (e.g., GABA-A, adenosine receptors). However, flavonoids can modulate cellular signaling by regulating calcium homeostasis and mitochondrial function. As second messengers, calcium ions can trigger kinase (Akt, ERK, or JNK) signaling; ATP is required for mitochondrion–mediated regulatory pathways (Williams and Spencer 2004). Studies have shown that hesperetin, naringenin, quercetin, and resveratrol inhibit the activity of kinases by interacting with ATP binding sites (Huang et al. 1999; So et al. 1996; Spencer et al. 2003; Walker et al. 2000). Nevertheless, flavonoids can exert neuroprotective effects against apoptosis by interacting with the upstream kinase (MAPKKK) of JNK, mitochondria permeability transition pores, which control the release of Ca^{2+} and cytochrome C via benzodiazepine binding sites, and mitochondrion-related proapoptotic factors, such as Smac/Diablo (Hwang et al. 2012a; Srinivasula et al. 2001; Williams et al. 2004).

It is important to know that different cellular bioactivities may exist for a flavonoid between its high and low doses (Williams et al. 2004). Studies have shown that low doses (in the nanomolar to low micromolar range) of flavanols (epicatechin, EGC [epicatechin gallate], or EPCG [epigallocatechin gallate]) exert neuroprotective actions through suppression of JNK and caspase 3 and activation of antiapoptotic MAPKs and PKC (Levites et al. 2002; Spencer et al. 2001a,b). However, high doses of EGC and EGCG induce apoptosis by triggering sustained activation of MAPK/JNK in cancer cells. In neurons, a high dose of quercetin (>30 μM) inhibited the pro-cellular survival of the PI-3K/Akt signaling pathway (Matter et al. 1992; Spencer et al. 2003); a low dose of quercetin (<20 μM) activates MAPK pathways to induce pro-cellular survival gene expression and defensive responses (Kong et al. 2000). For intervention in neurodegenerative disorders and cognition enhancement, the effective dose and cytotoxicity of flavonoids are important issues. Citrus flavonoids are promising candidates because of their low cytotoxicity for normal cells.

8.8.3 Mechanisms Underlying the Neuroprotection of Citrus Flavonoids

It is known that citrus flavonoids such as hesperidin, hesperetin, isorhamnetin, isosakuranetin, and neohesperidin, at physiological and high doses, exhibit multiple mechanisms of neuroprotection, including inhibition of ROS formation and caspase-3 activity, decreases in membrane and DNA damage, enhancement of antioxidant enzyme activity, maintenance of calcium homeostasis and mitochondrial potential, antineuroinflammination, and pro-cellular survival signaling regulation (Cui et al. 2010; Hwang and Yen 2008, 2009). Under neurotoxic insults, citrus flavonoids most likely act as signaling molecules associated with endogenous defense and pro-survival responses in neural cells (Hwang and Yen 2009; Vauzour et al. 2007). Both such mechanisms may also involve the control of *de novo* protein synthesis in neurons. The induced proteins can be antioxidant enzymes, defensive proteins, or crucial for controlling neuronal differentiation, long-term potentiation, and memory consolidation, such as Arc and CREB. Thus, citrus flavonoids have the potential to influence cell survival, memory, and cognition at the molecular and cellular levels in the brain (Spencer 2009).

Hesperetin and its chemical structure counterparts, isorhamnetin and isosakuranetin, differentially activate prosurvival signaling kinases, such as Akt/PKB and MAPK/p38, based on their chemical structures and doses used in PC12 cells under oxidative stress (Hwang and Yen 2009). These flavonoids also suppress the activation of the proapoptotic kinase JNK. Studies have shown that hesperetin can activate the ER, and physiological levels of estrogen exhibit neuroprotective effects via both ER- and tyrosine kinase receptor (Trks)-mediated signaling (Bourque et al. 2009; Lee and McEwen 2001; Liu et al. 2008). However, only physiologically relevant concentrations (0.1 and/or 1.0 μM) of hesperetin exhibit neuroprotective effects against oxidative damage via both ER- and TrkA-mediated actions to inhibit the decrease in cell viability, scavenge ROS, maintain calcium homeostasis, and suppress caspase-3 activity in cells. These receptor-mediated signaling pathways, triggered by hesperetin (1.0 μM), also induce protein expressions of BDNF, PPARγ coactivator 1α (PGC-1α), and the selective Alzheimer's disease indicator-1 (seladin-1) (Hwang and Yen 2011). They are known to be protective against apoptosis, oxidative damage, and Aβ-related neurotoxicity (Greeve et al. 2000; Han et al. 2000; St.-Pierre et al. 2006). PGC-1α can be highly inducible in most tissues. Hence, it is an almost ideal protector against the mitochondrial dysfunction-associated damage seen in Parkinson's and Alzheimer's diseases (St.-Pierre et al. 2006). Additionally, ER and TrkA are known to be expressed in most AD-vulnerable brain regions. Hesperetin thus has the potential for intervening in neurodegenerative disorders, particularly for AD. However, molecular mechanisms underlying the neuroprotection of a high dose (50 μM) of hesperetin are suggested to be different from such receptor-mediated actions (Hwang and Yen 2011).

Membrane-localized ER (mER) triggers parallel MAPK/ERK, PKA, Akt/PKB, and PKC pathways; TrkA can mediate MAPK/ERK, PI3K/Akt, phospholipase Cγ/PKC, and cAMP/PKA signaling pathways (Vasudevan and Pfaff 2007). It has been shown that hesperetin rapidly induces PGC-1α and seladin-1 and simultaneously triggers the activation of PI3K, PKA, PKC, MAPK/ERK, and CREB signaling pathways via both ER and TrkA (Hwang et al. 2012b). Such receptor-mediated parallel pathways have cross-talk effects and converge on different transcriptional factors, collaborating to speed, promote, and sustain the expression of PGC-1α and seladin-1. These findings support the proposed principles of mER-mediated parallel signaling pathways (Vasudevan and Pfaff 2007) and explain why physiologically relevant concentrations of hesperetin, with known moderate antioxidant and low estrogen activities, can exert great neuroprotective effects against oxidative stress. Thus, a more complete intervention in oxidative damage-related neurodegeneration using a dietary flavonoid is probable. Further, aging brains are responsive to the action of estrogens, which confer memory and cognition enhancements. Estrogens affect the basal forebrain and regulate the cholinergic neurons projecting into the cerebral cortex and hippocampus, where they play an important role in memory and cognitive functions (McEwen 2001; Okada et al. 2015). ER is associated with the induction of calmodulin, CaMK II, and NMDA receptor expressions. They are important for long-term potentiation, synaptic differentiation, synapse formation, or localization of receptors in synapses (Manthey and Behl 2006; McEwen 2001). Thus, hesperetin may potentially induce such proteins via the novel mER- and TrkA-mediated pathways to collaborate and enhance CREB and ER transcriptional actions. Furthermore, the pathways may also converge on the sites of serum response element to enhance Arc expression, strengthening memory formation (Bramham et al. 2010; McEwen 2001). Taken together, an approach to more completely intervene in age-related neurodegenerative disorders and for memory and cognition enhancements using citrus flavonoids is favorable.

8.9 APPLICATIONS OF CITRUS FLAVONOIDS TO OTHER NEURODEGENERATIVE DISORDERS

In addition to aging-related neurodegeneration, other neurological disorders, such as stroke (cerebrovascular attack/ischemia-reperfusion), epilepsy, and Huntington's disease (HD), may also lead to memory and cognitive deficits. HD, an inherited neurodegenerative disease, is characterized by the progressive loss of neurons in the striatum (Gopinath et al. 2011). Vicious cycles of oxidative stress and mitochondrial dysfunction, excitotoxicity, or inflammation are implicated in these neurodegenerative disorders (Guar et al. 2009; Gopinath et al. 2011). To prevent or delay such oxidative stress induced neurotoxicity by using antioxidants or pharmacological agents that augment endogenous defense is reasonable for therapeutic treatments. Oral administration of hesperidin can limit the extent of rat brain damage following stroke via the suppression of free radicals and its associated inflammation (Raza et al. 2011). Pretreatment of dietary hesperidin also attenuates nitric oxide (NO)-mediated cerebral ischemic injury and associated memory dysfunction by improving oxidative defense and mitochondrial complex enzyme activities and attenuating histopathological changes in rat hippocampus (Guar and Kumar 2010). Naringin pretreatment exerts neuroprotective effects in an animal model of HD through a NO mechanism and antioxidant and antiapoptotic effects (Kumar and Kumar 2010; Gopinath et al. 2011). Naringin also has therapeutic potential for temporal lobe epilepsy and ischemia-reperfusion cerebral injury via its antioxidant and anti-inflammatory activities and reversion in histopathological alterations in cortex, striatum, and hippocampus areas (Guar et al. 2009; Golechha et al. 2011). Animal studies have shown that nobiletin exerts neuroprotection against cerebral ischemia-reperfusion injury by multiple mechanisms, including Akt/CREB/BDNF and the Bcl-2 signaling pathway, ameliorating blood–brain barrier permeability, anti-inflammatory effects, and anti-apoptotic effects. It also improves brain ischemia-induced learning and memory deficits through stimulation of CaMKII and CREB phosphorylation (Yamamoto et al. 2009; Zhang et al. 2013; Yasuda et al. 2014). Furthermore, *in vivo* studies have shown that 3,5,6,7,8,3′,4′-heptamethoxyflavone (HMF), a citrus flavonoid, exerts neuroprotection against memory impairment and neuronal cell death in a global cerebral ischemia mouse model, by mechanisms of BDNF production, neurogenesis, and anti-inflammatory effects. HMF, administrated for 3 days immediately after ischemic surgery, protects mice against ischemia-induced memory dysfunction, rescues neuronal cell death in the CA1 cell layer, increases the production of BDNF, stimulates the autophosphorylation of CaMK II, and suppresses microglial activation in the hippocampus. It has been suggested that HMF-induced BDNF and neurogenesis may be mediated by ERK1/2 and CREB signaling (Okuyama et al. 2012, 2014). The *in vitro/in vivo* effects of neuroprotection and proposed mechanisms of citrus flavonoids are summarized in Table 8.1.

8.10 APPLIED ISSUES OF CITRUS FLAVONOIDS AS NEUROPROTECTIVE AGENTS

Citrus flavonoids are known to exert neuroprotective and pharmacological activities, such as antioxidative damage, anti-inflammation, and amelioration of memory impairment and pathology of neurodegenerative disorders. The methoxylated and polymethoxylated flavonoids are primarily found in peels of *Citrus* species. Issues of their poor solubility and bioavailability might require high doses of oral administration to reach therapeutic plasma concentrations in the brain, and this may limit dietary and clinical uses of citrus flavonoids (Scalbert and Williamson 2000). The solubility, stability, and bioavailability of phytochemicals are main concerns for their practical applications. However, encapsulation processes of flavonoids with protein, cyclodextrin, hydrophobically modified starch, or soluble dietary fiber can improve their solubility, stability, absorption, and bioavailability after oral administration. Nanoformulation developments are alternative approaches to enhancing the solubility and bioavailability of the flavonoids. It has been shown that solubility, oral absorption, anti-inflammatory, and neuroprotective effects of curcumin are dramatically enhanced by the preparation of curcumin nanoformulations in aluminum-treated mice with memory impairment (Kakkar and Kaur 2011). An amorphous solid dispersion of nano-sized citrus nobiletin for improving oral bioavailability, hepatoprotection, and distribution in the brain has been developed (Onoue et al. 2011, 2013). Furthermore, the preparation of a quercetin nanoformulation with increased solubility, bioavailability, and improved sustained release via nanoencaspulation has also been developed (Wang et al. 2014). Nevertheless, studies on safe toxicological profiles and efficiency of the flavonoid nanoformulations and human trials need to be conducted to establish their safety and effectiveness in clinical applications as improved therapeutic agents or for the prevention of neurodegenerative disorders.

Unintended reactions of phytochemicals are another concern for their uses in health foods and therapeutic applications. Flavonoids can interact with ATP binding cassette (ABC) efflux transporters, such as P-glycoprotein (P-gp). They also

TABLE 8.1
***In Vitro* and *In Vivo* Effects of Proposed Neuroprotective Mechanisms of Citrus Flavonoids**

Flavonoid	Subgroup	Effective Dose	Model(s)	Nature of Damage	Neuroprotective Molecular Effect(s)	Reference
Hesperetin	Flavanone	100 nmol/L (IVT)	Primary cortical neurons	Oxidative stress	↑ Akt, ERK1/2 ↓ ASK1, caspases	Vauzour et al. 2007
		50 mg/kg (IVV)	Brain	Oxidative stress	↑ SOD, GSH/GSSG ↓ Lipid and protein peroxidation	Choi and Ahn 2008
		1 μM (IVT)	Aβ-induced cytotoxicity in Neuro-2A cells	Glucose uptake, insulin signaling	↑ Insulin-stimulated neuronal glucose uptake ↓ Autophagy	Huang et al. 2012
		0.8/50 μM (IVT)	H_2O_2-induced cytotoxicity in PC12 neuronal cells	Oxidative stress	↑ Catalase, p-Akt, ↓: JNK, p38	Hwang et al. 2009
		0.1/50 μM (IVT)	H_2O_2-induced cytotoxicity in PC12 neuronal cells	Oxidative stress	↑ Akt, ERK, CREB, PGC-1α, seladin-1 ↓ ROS, calcium, caspase-3	Hwang et al. 2011
Hesperidin	Flavanone	50 mg/kg (IVV)	Middle cerebral artery occlusion	Oxidative stress, neuroinflammation	↑ Antioxidant enzymes, GSH ↓ Proinflammatory cytokines, iNOS, GFAP	Raza et al. 2011
		50 mg/kg (IVV)	Bilateral common carotid artery occlusion and reperfusion	Oxidative stress; nitric oxide signals; mitochondrial activity	↑ Antioxidant enzymes, respiratory enzymes, memory performance	Gaur and Kumar 2010
		50 mg/kg (IVV)	3-NP-induced Huntington's syndrome	Nitric oxide signaling	↑ Locomotor activity, oxidative defense, mitochondrial ETC activities	Kumar and Kumar 2010

(Continued)

TABLE 8.1 (CONTINUED)
***In Vitro* and *In Vivo* Effects of Proposed Neuroprotective Mechanisms of Citrus Flavonoids**

Flavonoid	Subgroup	Effective Dose	Model(s)	Nature of Damage	Neuroprotective Molecular Effect(s)	Reference
Naringin	Flavanone	80 mg/kg (IVV)	Neurotoxin-induced PD	Inflammation effects	↑ Glia-derived neurotrophic factor ↓ TNF-α, microglial activation	Leem et al. 2014
		50 mg/kg (IVV)	3-NP-induced Huntington's syndrome	Nitric oxide signaling	↑ Locomotor activity, oxidative defense, mitochondrial ETC activities	Kumar and Kumar 2010
Naringenin	Flavanone	4.5 mg/kg (IVV)	Scopolamine-induced amnesia	Oxidative stress	↑ Passive avoidance memory ↓ Acetylcholinesterase	Heo et al. 2004a
Nobiletin	Flavone	50 mg/kg (IVV)	Streptozotocin-induced hippocampus damage	Oxidative stress	↑ GPx, GR, GST, SOD, CAT, ChAT ↓: MDA, spatial learning and memory	Khan et al. 2012
		50 μM (IVT)	LPS-induced neuroinflammation	Inflammatory effects	↓ Proinflammatory cytokines, NF-κB	Cui et al. 2010
		10/50 mg/kg (IVV)	SAMP8 mouse	Oxidative stress	↑ Recognition memory, context-dependent fear memory, GSH/GSSG, GPx, Mn-SOD ↓ Carbonyl protein, tau phosphorylation	Nakajima et al. 2013
		50 mg/kg (IVV)	Alzheimer's disease (3XTg-AD) mouse model	NMDA-ERK signaling	↓ Soluble Aβ, amyloid plaques, neurofibrillary tangles, cognitive impairments, ROS	Nakajima et al. 2015

Abbreviations: 3-NP, 3-nitropropionic acid; ChAT, choline acetyltrasnferase; ETC, electron transport chain; GFAP, glial fibrillary acidic protein; IVT, *in vitro*; IVV, *in vivo*; PD, Parkinson's disease; SAMP8, senenscence-accelerated mouse prone 8 model.

interact with metabolizing enzymes of xenobiotics, such as cytochrome P450 (CYP) enzymes. As functional foods or pharmacological agents, citrus flavonoids might interact with conventional drugs in therapies (Pal and Mitra 2006; Manthey et al. 2001). Prescription drugs can be substrates of efflux transporters or metabolizing enzymes. Thus, interactions of flavonoids with metabolizing enzyme or drug efflux proteins can lead to higher or subtherapeutic plasma drug concentrations or improvement of drug absorption. Hence, interactions between citrus flavonoids and medicinal agents whose metabolism is by P-gp mediated efflux or CYP-mediated metabolism need to be clarified. Many studies have shown that citrus flavonids, such as diosmetin, hesperetin, naringenin, naringin, nobiletin, and polymethoxylated flavones, have interactions with efflux transporters or drug-metabolizing enzymes (Bharti et al. 2014; Pingili et al. 2016; Quintieri et al. 2010; Surichan et al. 2012; Takanaga et al. 2000).

8.11 CONCLUSIONS

Citrus flavonoids exert antioxidant, anti-inflammatory, and neuroporotective effects against age-related and other neurodegenerative disorders. Bioactivities of citrus flavonoids on neurological processes include the upregulation of signaling pathways for endogenous defense, pro-cellular survival and synaptic plasticity, and the suppression of neuropathological processes in specific brain regions. As neuroprotecive agents, citrus flavonoids benefit users with relative safety, traversing the blood–brain barrier, exerting multiple protective mechanisms and enhancements of memory and cognition. Hesperidin, hesperetin, naringenin, and nobiletin are potential therapeutic agents for neurodegenerative disorders, such as Alzheimer's and Parkinson's diseases. Hesperidin and naringin could be possibly applied to dementias caused by stroke, epilepsy, and HD. Practical issues of citrus flavonoids as neuroprotective agents cover solubility, stability and bioavailability enhancements, and unintended flavonoid–drug interactions, as well as safety and human trials.

ACKNOWLEDGMENT

This research was supported in part by the Ministry of Education, Taiwan, under the ATU Plan.

REFERENCES

Al Rahim, M., A. Nakajima, D. Saigusa, N. Tetsu, Y. Maruyama, M. Shibuya, H. Yamakoshi, Y. Tomioka, Y. Iwabuchi, Y. Ohizumi, and T. Yamakuni. 2009. 4′-Demethylnobiletin, a bioactive metabolite of nobiletin enhancing PKA/ERK/CREB signaling, rescues learning impairment associated with NMDA receptor antagonism via stimulation of the ERK cascade. *Biochemistry* 48:7713–7721.

Annunziato, L., A. Pannaccione, M. Cataldi, A. Secondo, P. Castaldo, G. Di Renzo et al. 2002. Modulation of ion channels by reactive oxygen and nitrogen species: A pathophysiological role in brain aging? *Neurobiology of Aging* 23:819–834.

Aruoma, O. I. 2002. Neuroprotection by dietary antioxidants: New age of research. *Die Nahrung* 46:381–382.

Barreca, D., G. Gattuso, G. Laganà, and U. Leuzzi. 2016. *C*- and *O*-glycosyl flavonoids in Sanguinello and Tarocco blood orange (*Citrus sinensis* L. Osbeck) juice: Identification and influence on antioxidant properties and acetylcholinesterase activity. *Food Chemistry* 196:619–627.

Benavente-García, O., and J. Castillo. 2008. Update on uses and properties of citrus flavonoids: New findings in anticancer, cardiovascular, and anti-inflammatory activity. *Journal of Agricultural and Food Chemistry* 56:6185–6205.

Bharti, S., N. Rani, B. Krishnamurthy, and D. S. Arya. 2014. Preclinical evidence for the pharmacological actions of naringin: A review. Planta Medica 80:437–451.

Bors, W., C. Michel, and K. Stettmaier. 2001. Structure–activity relationships governing antioxidant capacities of plant polyphenols. *Methods in Enzymology* 335:166–180.

Bourque, M., D. E. Dluzen, and T. Di Paolo. 2009. Neuroprotective actions of sex steroids in Parkinson's disease. *Frontiers in Neuroendocrinology* 30:142–157.

Bramham, C. R., M. N. Alme, M. Bittins, S. D. Kuipers, R. R. Nair, B. Pai et al. 2010. The arc of synaptic memory. *Experimantal Brain Research* 200:125–140.

Butterfield, D. A., B. Howard, S. Yatin, T. Koppal, J. Drake, K. Hensley et al. 1999. Elevated oxidative stress in models of normal brain aging and Alzheimer's disease. *Life Sciences* 65:1883–1892.

Carloni, S., G. Buonocore, M. Longini, F. Proietti, and W. Balduini. 2012. Inhibition of rapamycin-induced autophagy causes necrotic cell death associated with Bax/Bad mitochondrial translocation. *Neuroscience* 203:160–169.

Cho, J. 2006. Antioxidant and neuroprotective effects of hesperidin and its aglycone hesperetin. *Archives of Pharmaceutical Research* 29:699–706.

Choi, E. J., and W. S. Ahn. 2008. Neuroprotective effects of chronic hesperetin administration in mice. *Archives of Pharmaceutical Research* 31:1457–1462.

Crossthwaite, A. J., S. Hasan, and R. J. Williams. 2002. Hydrogen peroxide-mediated phosphorylation of ERK1/2, Akt/PKB and JNK in cortical neurones: Dependence on Ca^{2+} and PI3-kinase. *Journal of Neurochemistry* 80:24–35.

Cui, Y., J. Wu, S. C. Jung, D. B. Park, Y. H. Maeng, J. Y. Hong et al. 2010. Anti-neuroinflammatory activity of nobiletin on suppression of microglial activation. *Biological and Pharmaceutical Bulletin* 33:1814–1821.

Dana, R., H. L. Malech, and R. Levy. 1994. The requirement for phospholipase A2 for activation of the assembled NADPH oxidase in human neutrophils. *Biochemistry Journal* 297(Pt 1):217–223.

Dimpfel, W. 2006. Different anticonvulsive effects of hesperidin and its aglycone hesperetin on electrical activity in the rat hippocampus in-vitro. *Journal of Pharmacy and Pharmacology* 58:375–379.

Einat, H., and H. K. Manji. 2006. Cellular plasticity cascades: Genes-to-behavior pathways in animal models of bipolar disorder. *Biological Psychiatry* 59:1160–1171.

Floyd, R. A., M. West, and K. Hensley. 2001. Oxidative biochemical markers: Clues to understanding aging in long-lived species. *Experimental Gerontology* 36:619–640.

Gil, B., M. J. Sanz, M. C. Terencio, M. L. Ferrandiz, G. Bustos, M. Paya et al. 1994. Effects of flavonoids on Naja naja and human recombinant synovial phospholipases A2 and inflammatory responses in mice. *Life Sciences* 54:PL333–PL338.

Golechha, M., U. Chaudhry, J. Bhatia, D. Saluja, and D. Singh Arya. 2011. Naringin protects against kainic acid-induced status epilepticus in rats: Evidence for an antioxidant, anti-inflammatory and neuroprotective intervention. *Biological and Pharmaceutical Bulletin* 34:360–365.

Gomez-Pinilla, F. 2008. Brain foods: The effects of nutrients on brain function. *Nature Reviews Neuroscience* 9:568–578.

Gonzalez-Molina, E., R. Dominguez-Perles, D. A. Moreno, and C. Garcia-Viguera. 2010. Natural bioactive compounds of Citrus limon for food and health. *Journal of Pharmaceutical and Biomedical Analysis* 51:327–345.

Gopinath, K., D. Prakash, and G. Sudhandiran. 2011. Neuroprotective effect of naringin, a dietary flavonoid against 3-nitropropionic acid induced neuronal apoptosis. *Neurochemistry International* 59:1066–1073.

Greeve, I., I. Hermans-Borgmeyer, C. Brellinger, D. Kasper, T. Gomez-Isla, C. Behl et al. 2000. The human DIMINUTO/DWARF1 homolog seladin-1 confers resistance to Alzheimer's disease-associated neurodegeneration and oxidative stress. *Journal of Neuroscience* 20:7345–7352.

Guar, V., A. Aggarwal, and A. Kumar. 2009. Protective effect of naringin against ischemic reperfusion cerebral injury: Possible neurobehavioral, biochemical and cellular alterations in rat brain. *European Journal of Pharmacology* 616:147–154.

Guar, V., and A. Kumar. 2010. Hesperidin pre-treatment attenuates NO-mediated cerebral ischemic reperfusion injury and memory dysfunction. *Pharmacological Reports* 62:635–648.

Habauzit, D., G. Flouriot, F. Pakdel, and C. Saligaut. 1992. Effects of estrogens and endocrine-disrupting chemicals on cell differentiation-survival-proliferation in brain: Contributions of neuronal cell lines. *Journal of Toxicology and Environmental Health B Critical Reviews* 14:300–327.

Halliwell, B. 1992. Reactive oxygen species and the central nervous system. *Journal of Neurochemistry* 59:1609–1623.

Han, B. H., A. D'Costa, S. A. Back, M. Parsadanian, S. Patel, A. R. Shah et al. 2000. BDNF blocks caspase-3 activation in neonatal hypoxia-ischemia. *Neurobiology of Disease* 7:38–53.

Heo, H. J., D. O. Kim, S. C. Shin, M. J. Kim, B. G. Kim, and D. H. Shin. 2004a. Effect of antioxidant flavanone, naringenin, from *Citrus junoson* neuroprotection. *Journal of Agricultural and Food Chemistry* 52:1520–1525.

Heo, H. J., M. J. Kim, J. M. Lee, S. J. Choi, H. Y. Cho, B. Hong et al. 2004b. Naringenin from Citrus junos has an inhibitory effect on acetylcholinesterase and a mitigating effect on amnesia. *Dementia and Geriatric Cognitive Disorders* 17:151–157.

Hirata, A., Y. Murakami, M. Shoji, Y. Kadoma, and S. Fujisawa. 2005. Kinetics of radical-scavenging activity of hesperetin and hesperidin and their inhibitory activity on COX-2 expression. *Anticancer Research* 25:3367–3374.

Huang, S. M., S. Y Tsai, J. A. Lin, C. H. Wu, and G. C. Yen. 2012. Cytoprotective effects of hesperetin and hesperidin against amyloid beta-induced impairment of glucose transport through downregulation of neuronal autophagy. *Molecular Nutrition and Food Research* 56:601–609.

Huang, Y. T., J. J. Hwang, P. P. Lee, F. C. Ke, J. H. Huang, C. J. Huang et al. 1999. Effects of luteolin and quercetin, inhibitors of tyrosine kinase, on cell growth and metastasis-associated properties in A431 cells overexpressing epidermal growth factor receptor. *British Journal of Pharmacology* 128:999–1010.

Hwang, S. L., P. H. Shih, and G. C. Yen. 2012a. Neuroprotective effects of citrus flavonoids. *Journal of Agricultural and Food Chemistry* 60:877–885.

Hwang, S. L., J. A. Lin, P. H. Shih, C. T. Yeh, and G. C. Yen. 2012b. Pro-cellular survival and neuroprotection of citrus flavonoid: The actions of hesperetin in PC12 cells. *Food and Function* 3:1082–1090.

Hwang, S. L., and G. C. Yen. 2008. Neuroprotective effects of the citrus flavanones against H_2O_2-induced cytotoxicity in PC12 cells. *Journal of Agricultural and Food Chemistry* 56:859–864.

Hwang, S. L., and G. C. Yen. 2009. Modulation of Akt, JNK, and p38 activation is involved in citrus flavonoid-mediated cytoprotection of PC12 cells challenged by hydrogen peroxide. *Journal of Agricultural and Food Chemistry* 57:2576–2582.

Hwang, S. L., and G. C. Yen. 2011. Effect of hesperetin against oxidative stress via ER- and TrkA-mediated actions in PC12 cells. *Journal of Agricultural and Food Chemistry* 59:5779–5785.

Jung, H. A., M. J. Jung, J. Y. Kim, H. Y. Chung, and J. S. Choi. 2003. Inhibitory activity of flavonoids from Prunus davidiana and other flavonoids on total ROS and hydroxyl radical generation. *Archives of Pharmaceutical Research* 26:809–815.

Kakkar, V., and I. P. Kaur. 2011. Evaluating potential of curcumin loaded solid lipid nanoparticles in aluminium induced behavioural, biochemical and histopathological alterations in mice brain. *Food and Chemical Toxicology* 49:2906–2913.

Kandaswami, K., and E. Middleton, Jr. 1997. Flavonoids as antioxidants, in *Natural Antioxidants: Chemistry, Health Effects, and Application*, ed. F. Shahidi, pp. 174–203. Champaign, IL: AOCS Press.

Kandel, E. R. 2001. The molecular biology of memory storage: A dialogue between genes and synapses. *Science* 294:1030–1038.

Khan, M. B., M. M. Khan, A. Khan, M. E. Ahmed, T. Ishrat, R. Tabassum et al. 2012. Naringenin ameliorates Alzheimer's disease (AD)-type neurodegeneration with cognitive impairment (AD-TNDCI) caused by the intracerebroventricular-streptozotocin in rat model. *Neurochemistry International* 61:1081–1093.

Kim, J. Y., K. J. Jung, J. S. Choi, and H. Y. Chung. 2004. Hesperetin: A potent antioxidant against peroxynitrite. *Free Radical Research* 38:761–769.

Kong, A. N., R. Yu, C. Chen, S. Mandlekar, and T. Primiano. 2000. Signal transduction events elicited by natural products: Role of MAPK and caspase pathways in homeostatic response and induction of apoptosis. *Archives of Pharmacal Research* 23:1–16.

Kumar, P., and A. Kumar. 2010. Protective effect of hesperidin and naringin against 3-nitropropionic acid induced Huntington's like symptoms in rats: Possible role of nitric oxide. *Behavioral Brain Research* 206:36–48.

Landolfi, R., R. L. Mower, and M. Steiner. 1984. Modification of platelet function and arachidonic acid metabolism by bioflavonoids. Structure–activity relations. *Biochemical Pharmacology* 33:1525–1530.

Lee, S. G., Z. Z. Su, L. Emdad, P. Gupta, D, Sarkar, A. Borjabad et al. 2008. Mechanism of ceftriaxone induction of excitatory amino acid transporter-2 expression and glutamate uptake in primary human astrocytes. *Journal of Biological Chemistry* 283:13116–13123.

Lee, S. J., B. S. McEwen. 2001. Neurotrophic and neuroprotective actions of estrogens and their therapeutic implications. *Annual Review of Pharmacology and Toxicology* 41:569–591.

Leem, E., J. H. Nam, M. T. Jeon, W. H. Shin, S. Y. Won, S. J. Park, M. S. Choi, B. K. Jin, U. J. Jung, and S. R. Kim. 2014. Naringin protects the nigrostriatal dopaminergic projection through induction of GDNF in a neurotoxin model of Parkinson's disease. *Journal of Nutritional Biochemistry* 25:801–806.

Levites, Y., T. Amit, M. B. Youdim, and S. Mandel. 2002. Involvement of protein kinase C activation and cell survival/cell cycle genes in green tea polyphenol (–)-epigallocatechin 3-gallate neuroprotective action. *Journal of Biological Chemistry* 277:30574–30580.

Li, S., C. Y. Lo, and C. T. Ho. 2006. Hydroxylated polymethoxyflavones and methylated flavonoids in sweet orange (*Citrus sinensis*) peel. *Journal of Agricultural and Food Chemistry* 54:4176–4185.

Lin, C. M., C. T. Chen, H. H. Lee, and J. K. Lin. 2002. Prevention of cellular ROS damage by isovitexin and related flavonoids. *Planta Medica* 68:365–367.

Liu, L., D. M. Xu, and Y. Y. Cheng. 2008. Distinct effects of naringenin and hesperetin on nitric oxide production from endothelial cells. *Journal of Agricultural and Food Chemistry* 56:824–829.

Lou, H., X. Jing, X. Wei, H. Shi, D. Ren, and X. Zhang. 2014. Naringenin protects against 6-OHDA-induced neurotoxicity via activation of the Nrf2/ARE signaling pathway. *Neuropharmacology* 79:380–388.

Manthey, D., and C. Behl. 2006. From structural biochemistry to expression profiling: Neuroprotective activities of estrogen. *Neuroscience* 138:845–850.

Manthey, J. A., K. Grohmann, and N. Guthrie. 2001. Biological properties of citrus flavonoids pertaining to cancer and inflammation. *Current Medicinal Chemistry* 8:135–153.

Matsuzaki, K., K. Miyazaki, S. Sakai, H. Yawo, N. Nakata, S. Moriguchi et al. 2008. Nobiletin, a citrus flavonoid with neurotrophic action, augments protein kinase A-mediated phosphorylation of the AMPA receptor subunit, GluR1, and the postsynaptic receptor response to glutamate in murine hippocampus. *European Journal of Pharmacology* 578:194–200.

Matter, W. F., R. F. Brown, and C. J. Vlahos. 1992. The inhibition of phosphatidylinositol 3-kinase by quercetin and analogs. *Biochemical and Biophysical Research Communications* 186:624–631.

Mattson, M. P. 2004. Pathways towards and away from Alzheimer's disease. *Nature* 430:631–639.

McEwen, B. S. 2001. Estrogens effects on the brain: Multiple sites and molecular mechanisms. *Journal of Applied Physiology* 91:2785–2801.

McGinnis, K. M., K. K. Wang, and M. E. Gnegy. 1999. Alterations of extracellular calcium elicit selective modes of cell death and protease activation in SH-SY5Y human neuroblastoma cells. *Journal of Neurochemistry* 72:1853–1863.

Michalski, B., and M. Fahnestock. 2003. Pro-brain-derived neurotrophic factor is decreased in parietal cortex in Alzheimer's disease. *Brain Research Molecular Brain Research* 111:148–154.

Miyake, Y., C. Sakurai, M. Usuda, S. Fukumoto, M. Hiramitsu, K. Sakaida et al. 2006. Difference in plasma metabolite concentration after ingestion of lemon flavonoids and their aglycones in humans. *Journal of Nutritional Science and Vitaminology* 52:54–60.

Nagase, H., N. Omae, A. Omori, O. Nakagawasai, T. Tadano, A. Yokosuka et al. 2005a. Nobiletin and its related flavonoids with CRE-dependent transcription-stimulating and neuritegenic activities. *Biochemical and Biophysical Research Communications* 337:1330–1336.

Nagase, H., T. Yamakuni, K. Matsuzaki, Y. Maruyama, I. Kasahara, Y. Hinohara et al. 2005b. Mechanism of neurotrophic action of nobiletin in PC12D cells. *Biochemistry* 44:13683–13691.

Nakajima, A., Y. Aoyama, T. T. Nguyen, E. J. Shin, H. C. Kim, S. Yamada, T. Nakai, T. Nagai, A. Yokosuka, Y. Mimaki, Y. Ohizumi, and K. Yamada. 2013. Nobiletin, a citrus flavonoid, ameliorates cognitive impairment, oxidative burden, and hyperphosphorylation of tau in senescence-accelerated mouse. *Behavioural Brain Research* 250:351–360.

Nakajima, A., Y. Aoyama, E. J. Shin, Y. Nam, H. C. Kim, T. Nagai, A. Yokosuka, Y. Mimaki, T. Yokoi, Y. Ohizumi, and K. Yamada. 2015. Nobiletin, a citrus flavonoid, improves cognitive impairment and reduces soluble Aβ levels in a triple transgenic mouse model of Alzheimer's disease (3XTg-AD). *Behavioural Brain Research* 289:69–77.

Nakajima, A., T. Yamakuni, K. Matsuzaki, N. Nakata, H. Onozuka, A. Yokosuka et al. 2007. Nobiletin, a citrus flavonoid, reverses learning impairment associated with *N*-methyl-D-aspartate receptor antagonism by activation of extracellular signal-regulated kinase signaling. *Journal of Pharmacology and Experimental Therapeutics* 321:784–790.

Oikawa, S., I. Hirosawa, S. Tada-Oikawa, A. Furukawa, K. Nishiura, and S. Kawanishi. 2006. Mechanism for manganese enhancement of dopamine-induced oxidative DNA damage and neuronal cell death. *Free Radical Biology and Medicine* 41:748–756.

Okada, K., K. Nishizawa, T. Kobayashi, S. Sakata, K. Kobayashi. 2015. Distinct roles of basal forebrain cholinergic neurons in spatial and object recognition memory. *Scientific Reports* 5:13158.

Okuyama, S., M. Morita, K. Miyoshi, Y. Nishigawa, M. Kaji, A. Sawamoto, T. Terugo, N. Toyoda, N. Makihata, Y. Amakura, M. Yoshimura, M. Nakajima, and Y. Furukawa. 2014. 3,5,6,7,8,3′,4′-Heptamethoxyflavone, a citrus flavonoid, on protection against memory impairment and neuronal cell death in a global cerebral ischemia mouse model. *Neurochemistry International* 70:30–38.

Okuyama, S., N. Shimada, M. Kaji, M. Morita, K. Miyoshi, S. Minami, Y. Amakura, M. Yoshimura, T. Yoshida, S. Watanabe, M. Nakajima, and Y. Furukawa. 2012. Heptamethoxyflavone, a citrus flavonoid, enhances brain-derived neurotrophic factor production and neurogenesis in the hippocampus following cerebral global ischemia in mice. *Neuroscience Letters* 528:190–195.

Onoue, S., T. Nakamura, A. Uchida, K. Ogawa, K. Yuminoki, N. Hashimoto, A. Hiza, Y. Tsukaguchi, T. Asakawa, T. Kan, and S. Yamada. 2013. Physicochemical and biopharmaceutical characterization of amorphous solid dispersion of nobiletin, a citrus-polymethoxylated flavone, with improved hepatoprotective effects. *European Journal of Pharmaceutal Sciences* 49:453–460.

Onoue, S., A. Uchida, H. Takahashi, Y. Seto, Y. Kawabata, K. Ogawa, and K. Yuminoki. 2011. Development of high-energy amorphous solid dispersion of nanosized nobiletin, a citrus polymethoxylated flavone, with improved oral bioavailability. *Journal of Pharmaceutical Sciences* 100:3793–3801.

Onozuka, H., A. Nakajima, K. Matsuzaki, R. W. Shin, K. Ogino, S. Saigusa et al. 2008. Nobiletin, a citrus flavonoid, improves memory impairment and Aβ pathology in a transgenic mouse model of Alzheimer's disease. *Journal of Pharmacology and Experimental Therapeutics* 326:739–744.

Pal, D., and A. K. Mitra. 2006. MDR- and CYP3A4-mediated drug-herbal interactions. *Life Sciences* 78:2131–2145.

Pingili, R., S. Vemulapalli, S. S. Mullapudi, S. Nuthakki, S. Pendyala, and N. Kilaru. 2016. Pharmaccokinetic interaction study between flavanones (hesperetin, naringenin) and rasagiline mesylate in winstar rats. *Drug Development and Industrial Pharmacy* 42:1110–1117.

Pintchovski, S. A., C. L. Peebles, H. J. Kim, E. Verdin, and S. Finkbeiner. 2009. The serum response factor and a putative novel transcription factor regulate expression of the immediate-early gene Arc/Arg3.1 in neurons. *Journal of Neuroscience* 29:1525–1537.

Pollard, S. E., M. Whiteman, and J. P. Spencer. 2006. Modulation of peroxynitrite-induced fibroblast injury by hesperetin: A role for intracellular scavenging and modulation of ERK signalling. *Biochemical and Biophysical Research Communications* 347: 916–923.

Quintieri, L., S. Bortoiozzo, S. Strgliotto, S. Moro, M. Pavanetto, A. Nasi, P. Palatini, and M. Floreani. 2010. Flavonoids diosmetin and hesperetin are potent inhibitors of cytochrome P450 2C9-mediated drug metabolism in vitro. *Drug Metabolism and Pharmacokinetics* 25:466–476.

Rahigude, A., P. Bhutada, S. Kaulaskar, M. Aswar, and K. Otari. 2012. Participation of antioxidant and cholinergic system in protective effect of naringenin against type-2 diabetes-induced memory dysfunction in rats. *Neuroscience* 226:62–72.

Raza, S. S., M. M. Khan, A. Ahmad, M. Ashafaq, G. Khuwaja, R. Tabassum et al. 2011. Hesperidin ameliorates functional and histological outcome and reduces neuroinflammation in experimental stroke. *Brain Research* 1420:93–105.

Rice-Evans, C. A., N,. J. Miller, P. G. Bolwell, P. M. Bramley, and J. B. Pridham. 1995. The relative antioxidant activities of plant-derived polyphenolic flavonoids. *Free Radical Research* 22:375–383.

Rice-Evans, C. A., N. J. Miller, and G. Paganga. 1996. Structure–antioxidant activity relationships of flavonoids and phenolic acids. *Free Radical Biology and Medicine* 20:933–956.

Rossi, F., D. Romeo, and P. Patriarca. 1976. Metabolic perturbation of the inflammatory cells. *Agents and Actions* 6:50–59.

Sawa, T., M. Nakao, T. Akaike, K. Ono, and H. Maeda. 1999. Alkylperoxyl radical-scavenging activity of various flavonoids and other phenolic compounds: Implications for the anti-tumor-promoter effect of vegetables. *Journal of Agricultural and Food Chemistry* 47:397–402.

Scalbert, A., and G. Williamson. 2000. Dietary intake and bioavailability of polyphenols. *Journal of Nutrition* 130:2073S–2085S.

Schroeter, H., J. P. Spencer, C. Rice-Evans, and R. J. Williams. 2001. Flavonoids protect neurons from oxidized low-density-lipoprotein-induced apoptosis involving c-Jun N-terminal kinase (JNK), c-Jun and caspase-3. *Biochemistry Journal* 358:547–557.

Seki, T., T. Kamiya, K. Furukawa, M. Azumi, S. Ishizuka, S. Takayama et al. 2013. Nobiletin-rich *Citrus reticulata* peels, a kampo medicine for Alzheimer's disease: A case series. *Geriatrics and Gerontology International* 13:236–238.

Shanmugam, K., L. Holmquist, M. Steele, G. Stuchbury, K. Berbaum, O. Schulz et al. 2008. Plant-derived polyphenols attenuate lipopolysaccharide-induced nitric oxide and tumour necrosis factor production in murine microglia and macrophages. *Molecular Nutrition and Food Research* 52:427–438.

So, F. V., N. Guthrie, A. F. Chambers, M. Moussa, and K. K. Carroll. 1996. Inhibition of human breast cancer cell proliferation and delay of mammary tumorigenesis by flavonoids and citrus juices. *Nutrition and Cancer* 26:167–181.

Spencer, J. P. 2008. Flavonoids: Modulators of brain function? *British Journal of Nutrition* 99(E Suppl 1):ES60–ES77.

Spencer, J. P. 2009. The impact of flavonoids on memory: Physiological and molecular considerations. *Chemical Society Reviews* 38:1152–1161.

Spencer, J. P., C. Rice-Evans, and R. J. Williams. 2003. Modulation of pro-survival Akt/protein kinase B and ERK1/2 signaling cascades by quercetin and its in vivo metabolites underlie their action on neuronal viability. *Journal of Biological Chemistry* 278:34783–34793.

Spencer, J. P., H. Schroeter, A. R. Rechner, and C. Rice Evans. 2001a. Bioavailability of flavan-3-ols and procyanidins: Gastrointestinal tract influences and their relevance to bioactive forms in vivo. *Antioxidants and Redox Signaling* 3:1023–1039.

Spencer, J. P., H. Schroeter, B. Shenoy, S. K. Srai, E. S. Debnam, and C. Rice-Evans. 2001b. Epicatechin is the primary bioavailable form of the procyanidin dimers B2 and B5 after transfer across the small intestine. *Biochemical and Biophysical Research Communications* 285:588–593.

Srinivasula, S. M., R. Hegde, A. Saleh, P. Datta, E. Chiozaki, J. Chai, R. A. Lee, P. D. Robbins, T. Fernandes-Alnemri, Y. Shi, and E. S. A. Alnemri. 2001. A conserved XIAP-interaction motif in caspase-9 and Smac/DIABLO regulates caspase activity and apoptosis. *Nature* 410:112–116.

St.-Pierre, J., S. Drori, M. Uldry, J. M. Silvaggi, J. Rhee, S. Jager et al. 2006. Suppression of reactive oxygen species and neurodegeneration by the PGC-1 transcriptional coactivators. *Cell* 127:397–408.

Su, Z. Z., M. Leszczyniecka, D. C. Kang, D. Sarkar, W. Chao, D. J. Volsky et al. 2003. Insights into glutamate transport regulation in human astrocytes: Cloning of the promoter for excitatory amino acid transporter 2 (EAAT2). *Proceedings of the National Academy of Sciences USA* 100:1955–1960.

Surichan, S., V. P. Androutsopoulos, S. Sifakis, E. Koutala, A. Tsatsakis, R. R. Arroo, and M. R. Boarder. 2012. Bioactivation of the citrus flavonoid nobiletin by CYP1 enzymes in MCF7 breast adenocarcinoma cells. *Food and Chemical Toxicology* 50:3320–3328.

Takanaga, H., A. Ohnishi, S. Yamada, H. Matsuo, S. Morimoto, Y. Shoyama, H. Ohtani, and Y. Sawada. 2000. Polymethoxylated flavones in orange juice are inhibitors of P-glycoprotein but not cytochrome P450 3A4. *Journal of Pharmacology and Experimental Therapeutics* 293:230–236.

Tapiero, H., K. D. Tew, G. N. Ba, and G. Mathe. 2002. Polyphenols: Do they play a role in the prevention of human pathologies? *Biomedicine and Pharmacotherapy* 56:200–207.

Vasudevan, N., and D. W. Pfaff. 2007. Membrane-initiated actions of estrogens in neuroendocrinology: Emerging principles. *Endocrine Reviews* 28:1–19.

Vauzour, D., K. Vafeiadou, C. Rice-Evans, R. J. Williams, and J. P. Spencer. 2007. Activation of pro-survival Akt and ERK1/2 signalling pathways underlie the anti-apoptotic effects of flavanones in cortical neurons. *Journal of Neurochemistry* 103:1355–1367.

Walker, E. H., M. E. Pacold, O. Perisic, L. Stephens, P. T. Hawkins, M. P. Wymann et al. 2000. Structural determinants of phosphoinositide 3-kinase inhibition by wortmannin, LY294002, quercetin, myricetin, and staurosporine. *Molecular Cell* 6:909–919.

Waltereit, R., B. Dammermann, P. Wulff, J. Scafidi, U. Staubli, G. Kauselmann et al. 2001. Arg3.1/Arc mRNA induction by Ca^{2+} and cAMP requires protein kinase A and mitogen-activated protein kinase/extracellular regulated kinase activation. *Journal of Neuroscience* 21:5484–5493.

Wang, H. Q., and R. C. Smart. 1999. Overexpression of protein kinase C-alpha in the epidermis of transgenic mice results in striking alterations in phorbol ester-induced inflammation and COX-2, MIP-2 and TNF-alpha expression but not tumor promotion. *Journal of Cell Science* 112(Pt 20):3497–3506.

Wang, S., R. Su, S. Nie, M. Sun, J. Zhang, D. Wu, and N. Moustaid-Moussa. 2014. Application of nanotechnology in improving bioavailability and bioactivity of diet-derived phytochemicals. *Journal of Nutritional Biochemistry* 25:363–376.

Warner, D. S., H. Sheng, and I. Batinic-Haberle. 2004. Oxidants, antioxidants and the ischemic brain. *Journal of Experimental Biology* 207:3221–3231.

Williams, R. J., and J. P. Spencer. 2012. Flavonoids, cognition, and dementia: Actions, mechanisms, and potential therapeutic utility for Alzheimer disease. *Free Radical Biology and Medicine* 52:35–45.

Williams, R. J., J. P. Spencer, and C. Rice-Evans. 2004. Flavonoids: Antioxidants or signalling molecules? *Free Radical Biology and Medicine* 36:838–849.

Wu, P., C. Jiang, Q. Shen, and Y. Hu. 2009. Systematic gene expression profile of hypothalamus in calorie-restricted mice implicates the involvement of mTOR signaling in neuroprotective activity. *Mechanisms of Ageing and Development* 130:602–610.

Yamamoto, Y., N. Shioda, F. Han, S. Moriguchi, A. Nakajima, A. Yokosuka, Y. Mimaki, Y. Sashida, T. Yamakuni, Y. Ohizumi, and K. Fukunaga. 2009. Nobiletin improves brain ischemia-induced learning and memory deficits through stimulation of CaMKII and CREB phosphorylation. *Brain Research* 1295:218–229.

Yasuda, N., T. Ishii, D. Oyama, T. Fukuta, Y. Agato, A. Sato, K. Shimizu, T. Asai, T. Asakawa, T. Kan, S. Yamada, Y. Ohizumi, and N. Oku. 2014. Neuroprotective effect of nobiletin on cerebral ischemia-reperfusion injury in transient middle cerebral artery-occluded rats. *Brain Research* 1559:46–54.

Yokozawa, T., H. Y. Kim, E. J. Cho, J. S. Choi, and H. Y. Chung. 2002a. Antioxidant effects of isorhamnetin 3,7-di-*O*-beta-D-glucopyranoside isolated from mustard leaf (*Brassica juncea*) in rats with streptozotocin-induced diabetes. *Journal of Agricultural and Food Chemistry* 50:5490–5495.

Yokozawa, T., T. Nakagawa, and K. Kitani. 2002b. Antioxidative activity of green tea polyphenol in cholesterol-fed rats. *Journal of Agricultural and Food Chemistry* 50:3549–3552.

Youdim, K. A., M. S. Dobbie, G. Kuhnle, A. R. Proteggente, N. J. Abbott, and C. Rice-Evans. 2003. Interaction between flavonoids and the blood-brain barrier: In vitro studies. *Journal of Neurochemistry* 85:180–192.

Yu, J., L. Wang, R. L. Walzem, E. G. Miller, L. M. Pike, and B. S. Patil. 2005. Antioxidant activity of citrus limonoids, flavonoids, and coumarins. *Journal of Agricultural and Food Chemistry* 53:2009–2014.

Yuan, J., and B. A. Yankner. 2000. Apoptosis in the nervous system. *Nature* 407:802–809.

Zallen, G., E. E. Moore, J. L. Johnson, D. Y. Tamura, M. Barkin, H. Stockinger et al. 1998. New mechanisms by which secretory phospholipase A2 stimulates neutrophils to provoke the release of cytotoxic agents. *Arch Surg.* 133:1229–1233.

Zeng, H., Q. Chen, and B. Zhao. 2004. Genistein ameliorates beta-amyloid peptide (25-35)-induced hippocampal neuronal apoptosis. *Free Radical Biology and Medicine* 36:180–188.

Zhang, L., H. Zhao, X. Zhang, L. Chen, X. X. Bai, and J. Zhang. 2013. Nobiletin protects against cerebral ischemia via activating the p-Akt, p-CREB, BDNF and Bcl-2 pathway and ameliorating BBB permeability in rat. *Brain Research Bulletin* 96:45–53.

9 Health Benefits of Orange Juice and Citrus Flavonoids

Danielle R. Gonçalves, Paula Souza Ferreira, Elizabeth A. Baldwin, and Thais Borges Cesar

CONTENTS

9.1 INTRODUCTION: ORANGE JUICE AND HEALTH IMPLICATIONS

Previous studies have associated orange juice consumption with prevention of cardiovascular disease, and this effect is mainly attributed to its nutrients and bioactive compounds, such as vitamin C and citrus flavonoids. The main citrus flavonoids found in orange juice are hesperidin and naringenin, which can affect several metabolic routes that improve blood serum antioxidant capacity and anti-inflammatory performance, while decreasing insulin resistance protecting against diabetes and metabolic syndrome. In addition, orange juice is linked with better nutrition and satiation contributing to body weight maintenance (Ribeiro et al. 2017).

Primarily, orange juice is a natural source of vitamin C, folic acid, calcium, potassium, and magnesium, and blood levels of vitamin C were substantially increased after regular consumption of orange juice. In addition, it was verified a positive dose-response effect between the consumption of orange juice and levels of vitamin C and folic acid in the blood stream (Franke et al. 2005). Orange juice is also a source of energy and can be easily incorporated in a healthy diet plan, because orange juice has moderate amount of calories per gram of juice (0.45 kcal/g). Besides the criticism of the media and some health professionals about the "high calorie content" of orange juice and other fruit juices, epidemiological studies have shown that 100% fruit juices, as orange juice, are associated with a lower body mass index and a healthier diet in children and adults (O'Neil et al. 2011).

Many compounds of citrus fruits have been associated with health properties, and therefore they have been used to ameliorate various diseases. The evidence comes from the beginning of civilization, passing through the discovery of new worlds during the 16th and 17th centuries, when finally, in the 18th century, it was discovered that citrus fruits and some vegetables in the diet could cure and prevent scurvy. Currently, citrus fruits are a well-known source of essential nutrients, including vitamin C, and other bioactive compounds that can be used to ameliorate acute respiratory infections, psoriasis, allergies, asthma, arthritis, and other conditions. They are also recognized as a widely available important source of antioxidants.

Besides the traditional nutrients, citrus juices also contain flavonoids. Hesperidin and naringin are the most prevalent flavonoids in oranges and grapefruit; and eriocitrin is the main flavonoid in lemons and limes. Inside the body, eriocitrin is generated as a metabolite from hesperidin. On the other hand, there is another set of citrus compounds extracted from fruit tissue and peel, named polymethoxyflavones, for example, tangeritin, nobiletin, and heptamethoxyflavone. Previous studies have shown large potential of these compounds as anti-inflammatories or hypolipidemic drugs. The chemical structures of flavonoids and polymethoxyflavones are shown in Figure 9.1.

Citrus fruits in general are excellent sources of phytonutrients, vitamins, and minerals. For example, citrus contains polyphenols, flavonoids (naringin, hesperidin, neohesperidin, citronin, narirutin, and others), polymethoxylated flavones (tangeritin, nobilitin, sinensetin, and others) (Bai et al. 2013; Baldwin et al. 2014; Murata 1997; U.S. Department of Agriculture [USDA] 2016), and carotenoids (β-cryptoxanthin, α- and β-carotene, lutein, and zeaxanthin in general, and lycopene in blood oranges and red grapefruit) (Bai et al. 2013; Riso et al. 2005). Citrus fruits also contain important minerals and vitamins like potassium, ascorbic acid, and folate. Flavonoids are common in fruits and vegetables, and have antioxidant and anti-inflammatory properties that protect low-density lipoprotein (LDL) from oxidation, preventing the development of atherosclerosis (Aptekmann and Cesar 2013).

Carotenoids are considered to impart health benefits by reducing certain cancers and eye disease, including β-carotene, lycopene, lutein, and zeaxanthin in part because they are antioxidants and pro-vitamin A compounds (Johnson 2002). Potassium can lower blood pressure (Cappuccio and MacGregor 1991), while ascorbic acid, or vitamin C, is important to the immune system and protects endothelial cells and LDL from oxidant stress (Sabharwal and May 2008). Vitamin C also acts as an antioxidant and reduces risk of atherosclerosis (Boekholdt et al. 2006). Folic acid, or vitamin B9, helps the body convert carbohydrates into glucose for energy and is part of the vitamin B complex. It is essential for proper brain function, mental and emotional health, and is important for production of DNA and RNA. This is especially important for fast-growing tissue, such as in pregnancy, and helps to

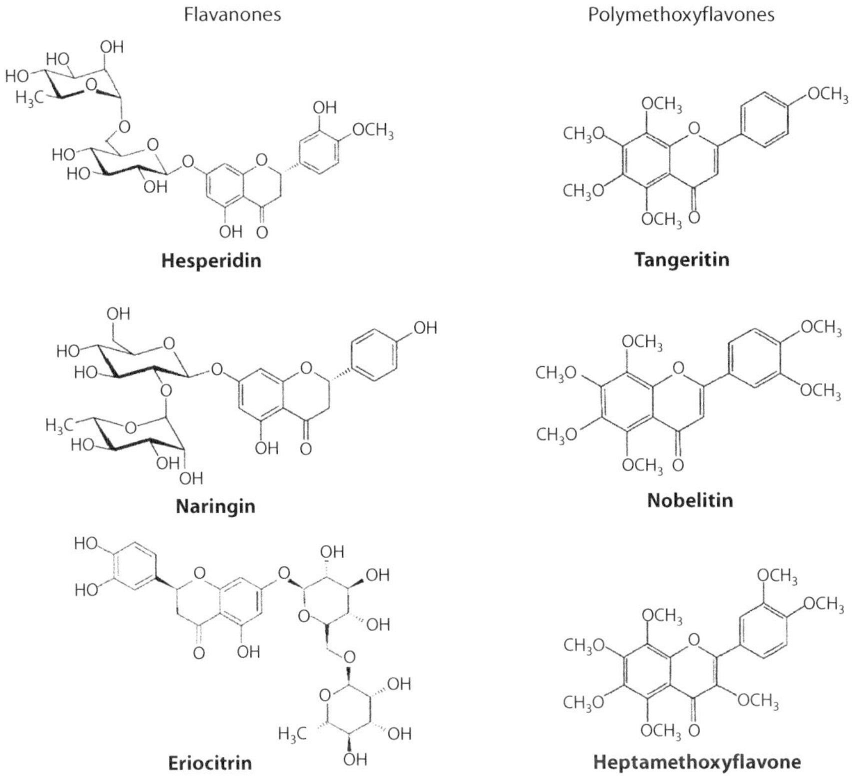

FIGURE 9.1 Chemical structures of the bioactive compounds flavanones and polymethoxyflavones found in citrus fruits.

prevent neural tube birth defects including cleft palate, spina bifida, and brain damage. Folic acid may also reduce risk of heart disease by working with vitamins B_6 and B_{12} to lower homocysteine levels (Baldwin et al. 2014; Boushey et al. 1995), although supplements were not found to be preventative (Lonn et al. 2006). The focus of this chapter, however, will be mostly on phytonutrients, and more specifically flavonoids and their role in human health, but the benefits of vitamins and minerals, will also be discussed.

9.2 EFFECT OF ORANGE JUICE AND CITRUS FLAVONOIDS ON METABOLIC SYNDROME AND CARDIOVASCULAR DISEASE

Cardiovascular diseases (CVDs) are considered the leading cause of death globally, killing more people annually than any other cause (World Health Organization [WHO] 2016). CVDs include coronary heart disease (CHD), cerebrovascular disease, peripheral vascular disease, and rheumatic heart disease (WHO 2016). CHD, specifically, is a common term for the buildup of plaque in the heart's arteries that can lead to a heart attack. The most important behavioral risk factors for CVD include

TABLE 9.1
Ethnic-Specific Waist Circumference Thresholds for Abdominal Obesity

Ethnicity	Male	Female
Europid	≥102 cm; ≥94 cm[a]	≥88 cm; ≥80 cm[a]
South Asian, Central and South American	≥90 cm	≥80 cm
Chinese and Japanese	≥90 cm	≥80 cm
Middle Eastern; Sub-Saharan African	≥94 cm	≥80 cm

Source: Adapted from O'Neill S, O'Driscoll L, *Obes Rev* 2015; 16(1): 1–12.

[a] For Europid, the literature typically reports that ≥94 cm for men and ≥80 cm for women are suitable waist circumference (WC) cutoffs, however in practice a WC of ≥102 for men and ≥88 for women are more typically used.

an unhealthy diet and lack of physical inactivity. These behavioral risk factors may be exhibited in individuals as increased blood pressure, increased blood glucose, increased blood lipids, and an overweight body mass or obesity (i.e., risk factors for CHD) (WHO 2016). Moreover, the major modifiable, treatable, or controllable risk factors for CHD are high blood total cholesterol levels, high blood LDL cholesterol, low blood HDL cholesterol, high blood triglycerides, high blood pressure, diabetes, obesity or being overweight, and tobacco use. The risk factors that cannot be controlled include age, gender, and family history of CHD (American Heart Association [AHA] 2016).

In this sense, metabolic syndrome, a complex disorder represented by a clustering of reversible metabolic alterations, is caused mainly by unhealthy behavior risk factors and can increase the risk factors for CVD (CHD and cerebrovascular disease) and diabetes mellitus type 2. There are five metabolic parameters that are related to metabolic syndrome risk factors: high triglyceride level (≥150 mg/dL); low HDL cholesterol level (<40 mg/dL in men or <50 mg/dL in women); high blood pressure (PAS ≥130 mm Hg and PAD ≥85 mm Hg); high fasting blood glucose (≥100 mg/dL); and abdominal obesity (increased waist circumference, where the thresholds for measuring require ethnic and nation specificity, as shown in Table 9.1). The presence of at least three of these risk factors is considered in the diagnosis of metabolic syndrome (Alberti et al. 2009).

9.2.1 Weight, Body Mass Index, and Abdominal Obesity

Excess body weight and associated comorbidities are a public health concern in many developed countries, and increased waist circumference is a focal component of metabolic syndrome, adopted by the National Cholesterol Education Program (Grundy et al. 2004). Obesity has a multifactorial and complex etiology, likely resulting from the presence of one or more factors: lifestyle, diet, genetics, and physiological or behavioral factors that ultimately result in energy imbalance (Rampersaud and Valim 2017). Values of body mass index (BMI) and waist circumference exceeding the recommended upper limits are known to be predictive of future risk factors that predispose individuals to CVD (Flint et al. 2010). According to the International Diabetes Federation (IDF), abdominal obesity is a prerequisite risk factor for the diagnosis of metabolic syndrome, because it is easily assessed using waist circumference and independently associated with each of the other metabolic syndrome components, including insulin resistance (Alberti et al. 2006).

The importance of abdominal obesity appears to be primarily due to high-turnover distribution of free fatty acids (FFAs) to other body organs. The large amounts of FFAs released by the metabolically active intra-abdominal adipose tissue to the portal system of the liver may interfere with hepatic insulin clearance. The secretion of cytokines by the intra-abdominal adipose tissue, including leptin, adiponectin, resistin, interleukins (IL) such as IL-1 and IL-6, and tumor necrosis factor alpha (TNF-α), is involved in the control of energy balance. The imbalanced release of these factors is associated with increased metabolic disorder (Han and Lean 2016).

High consumption of fruits, juices, and nonstarchy vegetables has been associated with changes in body composition, reflected in BMI and abdominal obesity. The consumption of pure fruit juices, without added sweeteners, has not been related to weight gain (O'Neil et al. 2011; Wang et al. 2011), but to the protection against chronic diseases such as CHD and some cancers because of the contribution of bioactive compounds, including flavonoids and vitamins (Aptekmann and Cesar 2013). Hence, consumption of fresh food may not help to reduce body weight and waist circumference if it is not accompanied by a caloric restriction diet, but it may help to delay or prevent their increase (Deopurkar et al. 2010; Morand et al. 2011). It has been reported that orange juice consumption contributes to the development of weight gain and obesity (Popkin et al. 2006; Rivera et al. 2008). However, emerging studies demonstrate that the intrinsic sugars found in orange juice do not seem to manifest the negative health effects associated with the excess intake of added sugars, including obesity (Rampersaud and Valim 2017), and this has been confirmed in studies of chronic orange juice intake. These studies found that daily orange juice intake does not promote weight gain nor increase body measures of metabolic syndrome (Cesar, Rodrigues et al. 2010; Morand et al. 2010; O'Neil et al. 2011). In addition, orange juice is considered a nutrient-dense beverage and has been associated with a high-quality diet in children and adults. A number of clinical studies report no deleterious effects on anthropometric measurements when 100% orange juice is included as an intervention to the usual or study diet (Rampersaud and Valim 2017).

In fact, there is a significant increase in the percentage of dietary energy derived from daily orange juice intake, and the energy and carbohydrate provided by orange juice (112 calories and 26 grams of carbohydrates in each 250 milliliters), but it was thought that orange juice may negatively influence the energy balance (energy consumed > energy expended), and contribute to weight gain if it is not adjusted to the dietary requirements or consumed in place of other drinks and fruits in the diet (USDA, 2016). In our studies on chronic orange juice consumption, we have not found changes in body weight, BMI, body fat, and waist circumference of healthy adults, even in those who already presented elevated BMI and abdominal obesity (Cesar, Aptekmann et al. 2010; Cesar, Rodrigues et al. 2010; Dourado and Cesar 2015; Lima et al. 2012; Silveira et al. 2015).

Furthermore, even the long-term daily consumption (12 months) or the large amounts of carbohydrates (52 g in 500 mL and 78 g in 750 mL) and energy (224 calories in 500 mL and 336 calories in 750 mL) found in the orange juice offered in our studies did not contributed to weight gain or obesity, in opposition to the idea that 100% fruit juice contributes to weight gain (Aptekmann and Cesar 2013; Basile et al. 2010; Cesar, Rodrigues et al. 2010; USDA 2016). In a 12-month cross-sectional study, orange juice consumers' body weight, waist circumference, and body fat were not different from that of nonconsumers (Aptekmann and Cesar 2013). In another study, healthy women, with excess of body weight drank 500 mL of orange juice daily for 8 weeks. At the end of the study, it was found lower waist circumference in those women, suggesting a reduction of abdominal fat, even with increased energy and carbohydrate intake during the study period. In the same study, the men were instructed to drink even more juice (750 mL/day), and their anthropometric measures remained unchanged (Basile et al. 2010). Similarly, an observational study reported a more favorable body mass index in orange juice consumers compared to non-consumers, as well as a decreased risk for obesity and metabolic syndrome (O'Neil et al. 2012).

In addition, the chronic intake of red-fleshed sweet orange juice, which is a variety with a reddish color due to greater concentrations of the carotenoids β-carotene and lycopene, was not implicated in body anthropometric measures and body composition changes of normal weight and overweight individuals of both genders (Lima et al. 2012; Silveira et al. 2015). Furthermore, the 8-week period of red-fleshed orange juice intake (750 mL/day) showed a significant increase in the energy intake and carbohydrates of men, and a mean increase of vitamin C and folate of all subjects. Although the carbohydrate and energy intake increased for the men, women did not exhibit a change in carbohydrate and energy intake (Lima et al. 2012; Silveira et al. 2015).

Another study aimed to show the effect of orange juice on metabolic syndrome parameters of normal weight and overweight adults with increased waist circumference. The study found no changes in anthropometric measures and body composition in either group. However, they found that there were improvements in the lipid profile (reduction in total cholesterol and LDL cholesterol), the immune response (increase in systemic IL-12), systemic inflammation (reduction in high-sensitivity C-reactive protein [hs-CRP]), and in serum oxidative stress or antioxidant capacity (reduction of lipid peroxidation and enhancement of total antioxidant capacity) for both normal and overweight subjects (Dourado et al. 2015).

It is likely that other factors were responsible for this effect, such as bioactive compounds and the substantial amounts of vitamins and minerals regularly present in the diet of the volunteers. It has been shown that the intake of orange juice after a high-fat and high-carbohydrate meal neutralizes its proinflammatory effect and prevents endotoxin increase and Toll-like receptor expression (Ghanim et al. 2010). This effect was attributed to the antioxidant and anti-inflammatory actions of orange juice flavanones. Hence, moderate consumption of orange juice may provide meaningful nutritional and dietary benefits and do not appear to negatively impact body weight and body composition (Rampersaud and Valim 2017). The increased intake of vitamin C, folate, and flavonoids, as well as the fiber content of both yellow- and red-fleshed orange juice, are indicated as contributors to the beneficial effects of these juices for metabolic syndrome parameters, and they do not appear to adversely affect weight gain and body composition.

9.2.2 Glucose Metabolism

Due to the significant amount of sugar (fructose, glucose, and sucrose) in orange juice, there is a concern about the elevation of blood glucose and blood triglycerides after acute or chronic intake. However, orange juice has a moderate glycemic index, with 22 grams of sugar in 250 mL (11 g of sucrose and 5.5 g of fructose and glucose) (USDA 2016). It was shown to maintain blood glucose levels within normal limits even when ingested right after meals rich in carbohydrates and lipids (Ghanim et al. 2010). In addition, orange juice consumption enhances circulating concentrations of hydrophilic and lipophilic phytochemicals, and is among one of the primary contributors of total flavonoids in the diet, specifically of flavanones. It is estimated that each 100 g of orange juice has 11.95 mg of hesperidin, 2.14 mg of naringenin, and 0.17 mg of eriodictyol (Franke et al. 2005; Rampersaud and Valim 2017; USDA 2016).

Two interventional studies, which evaluated regular consumption of orange juice, did not find changes in plasma levels of glucose, insulin, HOMA-IR (homeostasis model assessment-insulin resistance) values, or glycated hemoglobin. Both male and female normal and overweight subjects were instructed to drink three cups of yellow- and red-fleshed orange juices (250 mL each) for 8 consecutive weeks. After orange juice intervention, no effect on body weight and anthropometric measures were observed, but several markers of metabolic syndrome were improved (Dourado et al. 2015; Silveira et al. 2015). The overweight volunteers of both studies had serum lipids at the borderline high classification of lipoprotein levels, and showed higher levels of total cholesterol, LDL cholesterol, triglycerides, and TNF-α than the normal-weight subjects. Despite the elevated levels of cholesterol and TNF-α of the volunteers, the yellow and red orange juices decreased systemic inflammation and improved serum lipid profiles. TNF-α level is commonly elevated in overweight individuals, and is associated with increased waist circumference and metabolic disorders. Of major importance was the improved insulin sensitivity promoted by red orange juice. A 28% decrease in the HOMA-IR index was shown, as well as a 25% decrease of fasting insulin in normal-weight individuals, suggesting a decrease in insulin resistance (Dourado et al. 2015; Silveira et al. 2015). Although the overweight individuals did not show this improvement, 16% of the individuals who had insulin resistance were no longer insulin resistant after the red-fleshed orange juice intervention (Silveira et al. 2015). In another study, normolipidemic and normoglycemic men with elevated BMI had also benefited from orange juice consumption. They showed a significant decrease in fasting glucose levels after drinking orange juice for 8 consecutive weeks (Basile et al. 2010).

9.2.3 Cholesterol and Lipid Metabolism

Dyslipidemias are a major risk factor for the development of atherosclerosis, followed by visceral fat and obesity, which are related to higher rates of morbidity and mortality from CHD. According to the "Third Report of the Expert Panel on Detection, Evaluation, and Treatment of High Blood Cholesterol in Adults (Adult Treatment Panel III)," reductions in LDL cholesterol decrease CHD risk (Stone et al. 2013). Moreover, reducing elevated levels of atherogenic cholesterol (total and LDL cholesterol) is recommended to decrease the risk of CHD and related conditions (Jacobson et al. 2015).

The hypolipidemic effect of orange juice has been attributed to its flavanones and polymethoxylated flavones (PMFs). The hypocholesterolemic effect of citrus juices was first postulated to be due to their principal citrus flavanones, hesperetin and naringenin, aglycones from hesperidin and naringin, respectively, found as glycosides in both oranges and grapefruits (Bok et al. 1999; Chiba et al. 2003). One study found a significantly lower LDL cholesterol level after 3 weeks of orange juice supplementation to rabbits with diet-induced hypercholesterolemia. The observed changes were associated with significant decreases in liver cholesterol esters, but not with increases in the fecal excretion of cholesterol, which suggests a direct effect on liver LDL metabolism (Kurowska et al. 2000). In another study, Wistar rats were fed a richly saturated fatty acid diet along with orange juice, hesperidin, or orange juice with added hesperidin, for 30 days. The results showed a significant decrease in total cholesterol and HDL cholesterol after hesperidin supplementation, but not after supplementation with orange juice or orange juice with added hesperidin, highlighting the hypocholesterolemic role of this citrus flavanone (Vinueza et al. 2008). In rats, HDL is the lipoprotein that transports cholesterol to the body organs, rather than the LDL, resembling the role of the LDL cholesterol in humans (Vinueza et al. 2008).

An *in vitro* model of cells was utilized to study the effect of hesperetin and naringenin in the net secretion of LDL-associated apoB. HepG2 cells are human hepatoma cells that are commonly used to study the regulation of hepatic production and catabolism of VLDL (very low density lipoprotein) and LDL, which are atherogenic apoB-containing lipoproteins. The cells were exposed to hesperetin and naringenin, and it was found that there was a reduction in LDL-associated apoB, confirming the effect of citrus flavonoids on hepatic lipoprotein metabolism (Borradaile et al. 1999; Kurowska and Manthey 2002). Furthermore, mechanistic studies showed that hesperetin and naringenin exert their cholesterol-lowering action by interfering with the availability of neutral lipids, especially cholesteryl esters, required for the assembly and secretion of lipoproteins by reduced incorporation of ^{14}C acetate into cellular cholesteryl esters (Kurowska and Manthey 2002). In addition, hesperidin and naringin (in their glycoside form) were shown to reduce the activity of acyl-coenzyme A: cholesterolacyltransferase (ACAT) in the liver tissue of rats (Bok et al. 1999).

Citrus PMFs have also shown cholesterol and triacylglycerol lowering potential *in vivo*, by its direct effects on liver lipids. Similarly to flavanones, citrus PMFs induced a substantial reduction of medium apoB in HepG2 cells. Five PMFs were examined in this study, and tangeretin was found to be the most active, followed by nobiletin and 3,5,6,7,8,3′,4′ heptamethoxyflavone, which showed slightly lower activities (86% and 81%–83% reductions, respectively). Sinensetin and tetra-*O*-methylscutellarein had similar effects, although they were less pronounced (Kurowska and Manthey et al. 2002). In addition, hamsters that were fed a high cholesterol diet, supplemented with 1% polymethoxyflavone (mainly tangeritin) or 3% flavanones (hesperidin and naringin) for 35 days, showed comparable reductions in the levels of total cholesterol, VLDL + LDL cholesterol, and triglycerides in the blood. This suggests that the PMFs have a greater hypolipidemic potential, implying higher hypolipidemic potency of the PMFs versus flavanones, as PMFs yielded comparable results with a lower dose (Kurowska et al. 2004). The hypocholesterolemic effect of citrus PMFs was also shown by the inhibition of 3-hydroxy-3-methyl-glutaryl coenzyme A reductase, and by the increased expression of LDL receptors in the liver of rats fed citrus peel extracts rich in PMFs (Bok et al. 1999).

Clinical studies have also found changes in blood cholesterol of subjects drinking orange juice regularly. The intake of 750 mL orange juice for 60 days was shown to reduce total cholesterol and LDL cholesterol in normocholesterolemic individuals by 11% and 15%, respectively (Cesar, Rodrigues et al. 2010). Significant inverse correlations were found between regular orange juice intake and total cholesterol, LDL cholesterol, and apoB in men and women with normal and high cholesterol levels. The prolonged intake of orange juice (480 mL for 12 months) by normo- and hypercholesterolemic individuals was associated with the reduction of CVD risk factors, as shown by significantly lower total cholesterol, LDL cholesterol, apoB, and LDL/HDL ratio in orange juice consumers with moderate hypercholesterolemia as well the normocholesterolemic consumers in comparison to nonconsumers (Aptekmann and Cesar 2013). Moreover, the blood lipid profile of normal-weight and overweight subjects was improved by regular intake of orange juice. In each group, the participants were instructed to consume 750 mL of orange juice without added sugars daily, divided into at least two intakes a day for 8 weeks, without any other dietetic intervention. It was found that the subjects had decreased total cholesterol and LDL cholesterol, suggesting that participants of both weight groups had benefited from orange juice intake. However, the normal weight subjects seemed to have a more pronounced reduction than overweight subjects (–11% and –8%, respectively) (Dourado et al. 2015).

Regular intake of orange juice also benefited individuals who were not following the dietary recommendations to prevent CHD (fat intake >30% of total diet energy and cholesterol >200 mg/day) (AHA 2016). Despite a high fat intake and an elevated intake of saturated fats during the entire study period, healthy women and men who volunteered to drink 500 mL and 750 mL orange juice for 8 weeks, respectively, had a significant decrease in total cholesterol and LDL cholesterol. Additionally, the serum triglyceride level decreased significantly in men, and although women's HDL cholesterol levels were below that recommended at baseline, their HDL cholesterol levels significantly increased after the orange juice intervention (Basile et al. 2010).

The red-fleshed orange juice has also been shown to improve the lipid profile of normal and overweight adults. Both male and female normal and overweight volunteers who drank 750 mL of red-fleshed orange juice during a period of 8 weeks exhibited reduced total cholesterol and LDL cholesterol levels. The normal-weight volunteers also had a decrease in HDL cholesterol and apolipoprotein A1 (apoA1) levels (Silveira et al. 2015). Normocholesterolemic and hypercholesterolemic subjects, both males and females, were evaluated on the first and last day of 60 consecutive days of their daily 750 mL orange juice consumption. After orange juice supplementation, LDL cholesterol levels were reduced in those with hypercholesterolemia, while remaining unchanged for those with normocholesterolemia. As has been ound in many studies, orange juice did not alter the HDL cholesterol of either group. Despite the unaltered HDL cholesterol level, orange juice consumption promoted changes in HDL metabolism by increasing the transfer of free cholesterol and decreasing the transfer of triglycerides to the HDL particle in both groups. As is well known, the free form of cholesterol is unstable and is stored in its esterified form in cells. Hence, this tendency is suggested to favor HDL cholesterol function by the reverse transport of cholesterol and cholesterol esterification, which may also protect the lipoprotein from rapid degradation in the plasma thus preventing a fall in HDL cholesterol level (Cesar, Aptekmann et al. 2010). The increased transfer of triglycerides to HDL occurs in hypertriglyceridemia and is suggested to cause a greater instability in lipoprotein particles. Hepatic lipase hydrolyzes the triglyceride-enriched HDL to form small dense HDL particles of abnormal composition (high triglyceride/cholesteryl ester ratio). Such small dense HDL particles have a shorter half-life in plasma than larger HDL particles, and as a result the HDL-cholesterol level falls, affecting its antiatherogenic properties (Kontush and Chapman 2006). These effects were observed regardless of whether the subject had hypercholesterolemia. Additionally, the activity of the antioxidative enzyme paraoxanase 1 (PON1) was decreased in the normocholesterolemic subjects (Cesar, Aptekmann et al. 2010). Considering the results of this study, the effects of orange juice consumption can be beneficial to both normo- and hypercholesterolemic subjects.

The reduction in LDL cholesterol elicited by orange juice consumption cannot be ascribed to changes in dietary cholesterol during the supplementation periods. As observed in our studies, changes in fat intake were not detected in the dietary questionnaires, and some had elevated fat intake during orange juice supplementation (Basile et al. 2010). Moreover, slight changes in cholesterol intake are unlikely to influence LDL cholesterol levels (Cesar, Aptekmann et al. 2010). Taken altogether, orange juice has an undoubtedly antiatherogenic effect, being a powerful food for the prevention of CHD, since its flavonoid composition can make important changes in the blood lipid profile and may help to fight atherogenic cholesterol.

9.2.4 Blood Pressure

The consumption of flavonoids from several citrus fruits and citrus juices in the diet has been associated with improvement of blood pressure, contributing to the potential for orange juice to prevent CHD (Kay et al. 2012; Mennen et al. 2004; Rangel-Huerta et al. 2015). Observational studies and clinical trials have pointed to the direct association between body weight and blood pressure. The importance of this relationship is reinforced by the high and increasing prevalence of excess body weight for individuals throughout the world (Appel et al. 2006).

It has been found that the chronic consumption of orange juice has been beneficial for blood pressure of overweight subjects. A group of overweight men, who consumed 500 mL of orange juice daily during a 35-day period, had a significantly decreased diastolic blood pressure (Basile et al. 2010). Similarly, 8 weeks of red-fleshed orange juice intake (750 mL/day) lowered the diastolic blood pressure of overweight subjects and the systolic blood pressure in normal-weight subjects

(Lima et al. 2012). Orange juice components, such as potassium, hesperidin, vitamin C, and folate, favor returning high blood pressure to normal and healthy levels, and prevention of the onset of CHD. The endothelium of blood vessels in healthy individuals expresses constitutive forms of nitric oxide synthase (NOS) and cyclooxygenase (COX-1), which produces the vasoactive hormones nitric oxide (NO) and prostacyclin, respectively. NO and prostacyclin are the major cardioprotective hormones, and both are released by endothelial cells and act in synergy to relax blood vessels and inhibit platelet activation, thereby limiting thrombosis (Mitchell et al. 2008). The ability of flavonoids to activate endothelial NO synthase is likely the mechanism underlying improved endothelial function. Other mechanisms underlying flavanones' blood pressure-lowering effects could be attributed to their action in the cyclooxygenase pathway, and to their inhibitory action of angiotensin-converting enzyme (Actis-Goretta et al. 2006; Morand et al. 2011).

Hesperidin, the main flavanone found in orange juice, has been shown to promote vascular relaxation and improve blood flow in the vascular endothelium through inducing NO production (Grassi et al. 2009). Flavanones have the highest bioavailability among flavonoid compounds, while hesperidin represents 90% of total flavanones in oranges. The remaining flavanones are comprised of naringin or narirutin. In particular, orange juice and pure hesperidin consumption is of interest with regard to their potential health benefits and their effect on diastolic blood pressure, because diastolic blood pressure is an indicator of peripheral vessel resistance. One study evaluated one group of subjects consuming orange juice and the other group consuming pure hesperidin at 500 mL/day for 4 weeks. The group that consumed orange juice had a decrease in diastolic blood pressure, while systolic blood pressure remained unaltered. The observed changes in microvascular endothelial reactivity were positively correlated with plasma concentrations of hesperetin after the intake of pure hesperidin or orange juice (Actis-Goretta et al. 2006; Morand et al. 2011).

Other components of orange juice may also affect blood pressure. Available data suggests that potassium has beneficial effects on blood pressure. Its protective effect against hypertension has been attributed to enhanced blood potassium levels, which increases the excretion of sodium (Appel et al. 2006). According to the "American Heart Society Scientific Statement of Dietary Approaches to Prevent and Treat Hypertension," the recommended potassium intake is set at 4.7 g per day, which corresponds to the average total potassium intake observed in clinical trials and the potassium content of the DASH diet (Appel et al. 2006). Orange juice is a good source of potassium, as every 250 mL supplies almost 5 g, which is more than the recommended amount (USDA 2016). Most interventional studies on orange juice have implemented 500 and 750 mL per day, reaching 9.9 and 14.9 g of potassium per day. Generally, these amounts do not pose risks for healthy populations with normal kidney function, because excess potassium is readily excreted in the urine (Appel et al. 2006). For patients with renal impairment and for those taking medications that increase renal potassium retention, orange juice should be ingested with caution.

Table 9.2 shows a summary of some studies about the effect of orange juice or hesperidin on cardiovascular risk factors, in both humans and animals.

TABLE 9.2
Briefing of Studies on Orange Juice and Hesperidin with Main Findings

Study	Orange Juice (OJ) or Hesperidin (HSP)	Amount and Study Duration	Model (*n*, gender, age)	Preview Factors	Results
Vinueza et al. 2008	HSP OJ HSP + OJ	25 mg/L 1:6 (OJ: H_2O) 25 mg/L + 1:6 (OJ:H_2O) 30 days	Rats 8 males	Control	↓ Total cholesterol and HDL-C
Cesar, Aptekmann et al. 2010	OJ	750 mL 60 days	Human 14 men and 15 women 25–55 y	>30% overweight	↓ Total cholesterol and LDL-C ↓ HDL-C and Apo A-1
Cesar, Rodrigues et al. 2010	OJ	750 mL 60 days	Human 12 men and 19 women 25–55 y	Hypercholesterolemia 6 men and 8 women	↓ Total cholesterol and LDL-C ↓ Transference of Phospholipids and TG ↑ Free cholesterol transfer ↓ Cholesteryl ester transfer
Basile et al. 2010	OJ	750 mL (men) 500 mL (women) 8 weeks	Human 20 men and 21 women 20–53 y	70% overweight (men) 33% overweight (women) 30% men high WC 23% high WC (women)	↓ Total cholesterol, ↓ LDL-C (men, women) ↑ HDL (women), ↓ WC (women) ↓ TG, ↓ Glucose, ↓ Diastolic blood pressure (men)
Nasser et al. 2011	OJ	750 mL 8 weeks	Human 26 men and 22 women 23–59 y	Normal	↓ Serum total antioxidant capacity

(Continued)

TABLE 9.2 (CONTINUED)
Briefing of Studies on Orange Juice and Hesperidin with Main Findings

Study	Orange Juice (OJ) or Hesperidin (HSP)	Amount and Study Duration	Model (*n*, gender, age)	Preview Factors	Results
Lima et al. 2012	Red-fleshed OJ	750 mL 3 weeks	Human 19 men and 16 women ≈34 y	35% overweight 15% obese 54% high WC	↓ Systolic blood pressure (normal weight) ↓ Diastolic Blood Pressure (overweight)
Aptekmann et al. 2013	OJ	480 mL 12 months	Human 103 men and 26 women 18–65 y	38% high cholesterol 58% overweight (men) 32% overweight (women)	↓Total cholesterol and LDL-C ↓ LDL/HDL ratio and Apo B (Normo- and hypercholesterolemic)
Silveira et al. 2015	Red-fleshed OJ	750 mL 8 weeks	Human 19 men and 16 women 23–59 y	67% overweight/high WC 33% obese	↓Total cholesterol, ↓LDL-C (normal and overweight) ↓ HDL, ↓ Apo A1 (normal weight) ↓ Fasting insulin, ↓HOMA-IR (normal weight) ↓SBP (normal weight) ↓ Serum CRP (normal and overweight) ↑ Serum total antioxidant capacity (normal and overweight)
Dourado et al. 2015	Orange juice	750 mL 8 weeks	Human 46 from both genders 23–59 y	54.3% overweight	↓ Total cholesterol, ↓ LDL-C (normal and overweight) ↓ Serum CRP and Lipid peroxidation (normal and overweight) ↑ Serum IL-12 ↑ Serum total antioxidant capacity (normal and overweight) (Normal and overweight)

9.3 EFFECT OF ORANGE JUICE AND CITRUS FLAVONOIDS ON INFLAMMATION AND OXIDATIVE STRESS OF CORONARY HEART DISEASE

The pathogenesis of metabolic syndrome has independent cardiometabolic risk factors, including molecules of hepatic, vascular, and immunologic origin contributing to the proinflammatory state that is commonly observed in individuals with CHD (Grundy et al. 2004). Pathogenic elements related to CHD are commonly associated with oxidative stress and are implicated in the development of vascular complications. Reactive oxygen species (ROS) and oxidative stress are strongly linked to endothelial dysfunction, inflammation, LDL oxidation, metabolic syndrome, and insulin resistance (Santilli et al. 2015). Specifically, the process LDL oxidation is directly involved in the development of atherosclerosis. Oxidative alteration of native LDL cholesterol is a key step in atherosclerosis genesis, and the protection of this lipoprotein against oxidation may be an efficient strategy to prevent or reduce the risk of atheroma progression events (Cesar, Rodrigues et al. 2010). In addition, individuals with CHD have a number of etiologically linked metabolic risk factors, which may or may not coexist with inflammatory markers such as C-reactive protein (CRP), uric acid, and cytokines (Han and Lean 2016).

For this reason, there is growing interest in the development of drugs to reduce the proinflammatory state as a way to reduce metabolic risk factors for CHD. Antioxidant substances may restrain the development of atherosclerosis by preventing the oxidative alteration of LDL cholesterol. The consumption of foods rich in flavonoids has been linked to the reduction of several risk factors for CHD (Hooper et al. 2008; Kim et al. 2016). In this sense, lipid-lowering drugs are indicated as reducers of CRP levels, suggesting an anti-inflammatory action (Grundy et al. 2004). The anti-inflammatory and antioxidant effect of orange juice was observed in either normal-weight and overweight individuals who consumed yellow orange juice (*Citrus sinensis*) or red-fleshed orange juice (navel orange, Cara Cara). Eight weeks of daily orange juice intake reduced CRP levels and lipid peroxidation in both groups, although this reduction was more pronounced in the overweight subjects (Dourado et al. 2015; Silveira et al. 2015). Overweight subjects are suggested to be more susceptible to the antioxidant effects of orange juice because of the accumulation of abdominal adiposity, which is implicated in raising oxidative stress. The relevance of this result lies in the ability of orange juice to influence the concentrations of CRP and LDL cholesterol, which are important predictive markers of inflammation and cardiovascular events. In addition, yellow orange juice and red-fleshed orange juice increased the total antioxidant capacity for both groups (Dourado et al. 2015; Silveira et al. 2015).

Corroborating the preceding study, adult healthy subjects of both genders who received daily doses of orange juice, 750 mL/day (3 cups a day) for 8 weeks, showed increased blood serum antioxidant capacity, but no effect was noted for lipid peroxidation (TBARS). Studies have shown that TBARS levels are elevated accordingly with cardiovascular risk factors, hypertension, hyperlipidemia, and diabetes. Therefore, the health status of the subjects did not showed an action of orange juice constituents. However, their blood serum antioxidant capacity improved due to the increase in vitamin C and flavonoid reserves, mainly flavanones, which were supplied by the orange juice (Nasser et al. 2011).

At some point, all of the components of orange juice are linked to a reduction in both CVD risk factors and metabolic syndrome by its valuable contribution to increase hydrophilic and lipophilic phytochemicals in the blood, as well to help consumers meet their nutritional requirements of folate and potassium (Carter et al. 2010; Franke et al. 2005; Kurowska et al. 2010; O'Neil et al. 2012). Even small amounts of folate have been associated with lower homocysteine levels (Moat et al. 2004). An inverse correlation was found between men with regular orange juice intake and their homocysteine levels (Aptekmann and Cesar 2013). A daily intake of 750 mL of orange juice increased folate concentrations by 18%, but did not affect homocysteine concentrations of adult individuals with elevated plasma total cholesterol and LDL cholesterol after 4 weeks of orange juice consumption (Kurowska, Spence et al. 2000).

The effect of folate on homocysteine helps explain its effects on CHD, because folic acid is essential for homocysteine remethylation. Population studies have shown that homocysteine levels are inversely related to serum folate, both in healthy individuals and individuals with CHD (Hatzis et al. 2006; Moat et al. 2004). Elevated homocysteine levels have been previously linked to altered DNA methylation levels in various diseases; however, folate-based methods of lowering homocysteine have had limited effects in reducing cardiovascular events. One possible reason for the limited efficacy of such therapy is that they have failed to reverse epigenetic changes induced by homocysteine. It is possible that individuals with high homocysteine levels have a homocysteine memory effect, suggested by the deleterious effects of prior and extended exposure to elevated homocysteine concentrations having long-lasting effects on target organs and genes, underestimating the benefit of homocysteinelowering therapies. Therefore, there is a requirement to begin homocysteine-modifying therapies at an early stage of atherosclerosis, as evidence suggests that they do not appear to be effective in reducing cardiovascular events in patients with advanced CHD (Krishna et al. 2013).

It is found that blood levels of vitamin C are significantly increased after the intake of at least two cups of orange juice for at least 2 weeks (Rampersaud and Valim 2017). It was suggested that vitamin C status is preferably improved by eating foods rich in vitamin C, in addition to not smoking and refraining from other dietary habits that may prevent the depletion of ascorbic acid. The dietetic intake of vitamin C or of foods with high vitamin C content was inversely correlated with hypertension, whereas supplemental vitamin C did not show this correlation (Buijsse et al. 2015).

9.4 EFFECT OF ORANGE JUICE AND CITRUS FLAVONOIDS ON INFLAMMATION MARKERS

Inflammation is a critical component for the development of human diseases. Under normal conditions, inflammation is a defensive and protective mechanism that responds to tissue injury; however, exacerbated regulation and prolonged inflammation has been found to contribute to chronic inflammatory conditions, causing increases in free radicals and inflammatory cytokine mobilization and growth factors.

Although chronic inflammation is not the major cause of disease, it contributes to their pathogenesis through the alteration of biomolecules, the induction of genetic damage, and through cellular dysfunction and enhancement of cellular proliferation. The major human diseases linked to chronic inflammation include neurological and cardiovascular diseases, metabolic disorders, obesity, and cancer (Ho et al. 2012). A large number of studies have shown that nutritional choices, such as fruit or unsweetened fruit juices, could minimize oxidative stress and decrease the exacerbated inflammatory response present in chronic diseases. In various experimental models, the regular consumption of orange juice has been related to a reduction in the number of inflammatory biomarkers following the stimulation of an inflammatory response (Coelho et al. 2013).

It has been discovered that orange juice or hesperidin, its major flavonoid, have led to modulation of gene expression profiles of chemokines associated with inflammatory or atherogenic processes, including lipid efflux, chemotaxis, and molecule and platelet adhesion. These chemokines are regulated by the NF-κB transcription factor, and their modulation by orange juice and hesperidin comes from an upregulation of the expression of NF-κB inhibitor (IκB) (Milenkovic et al. 2011). Foreign molecules in the organism can activate NF-κB (i.e., microorganisms and their products, which enter into the cell core and bind to promoter molecules, activating genes that contribute to the adaptive immune and proinflammatory cytokine secretion). During this course of action, IκB is phosphorylated and creates a complex with NF-κB in the cytosol, where this inhibitor is able to arrest NF-κB and block its action. When this complex is broken, IκB is quickly degraded (Janeway et al. 2002).

Some studies have shown that the consumption of meals with pro-inflammatory macronutrients like those that are high fat, high cholesterol (HFHC) can also increase NF-κB binding, which stimulates the expression of tumor necrosis factor-α (TNF-α), interleukin-1β (IL-1β), and the suppressor of cytokine signaling 3 (SOCS3). SOCS3 is a protein that is elevated in obese individuals and can interfere in the signal transduction of insulin, a hormone that promotes the distribution and storage of nutrients. SOCS3 also affects leptin signal transduction, an adipokine that regulates the satiety and reduction of food intake. Despite the fact that many health professionals consider orange juice as a high caloric drink, its regular consumption is not shown to stimulate the alterations associated with a high caloric intake. This effect is most likely attributed to the flavonoids naringenin and hesperidin, which exert antioxidant and anti-inflammatory effects (Deopurkar et al. 2010).

In addition to these effects, the regular intake of orange juice is able to suppress TLR4 expression, a receptor for endotoxin lipopolysaccharides (LPS), after a proinflammatory induction (Ghanim et al. 2010). Toll receptors are a signaling pathway of innate immunity on multicellular organisms, and LPS is a component of the cell wall of gram-negative bacteria that can activate TLR4 and trigger immune responses (Janeway et al. 2002). *In vitro* analyses show that treatment with orange juice on LPS-stimulated macrophages increased NO (nitric oxide) secretion, while the treatment only with hesperidin suppressed around 50% of this secretion. These results suggest that orange juice has an antimicrobial function, probably due to its high vitamin C content, which is capable of elevating iNOS (nitric oxide synthase), the enzyme responsible for catalyzing the production of NO, at the mRNA protein level. On the other hand, the suppression of NO secretion observed after hesperidin treatment would be important in situations involving excessive NO production (Dourado et al. 2013), such as septic shock or atherosclerosis. An overproduction of LPS initiates an inflammation cascade because it increases the secretion of proinflammatory cytokines (Manthey and Cesar 2013). However, it was observed that in this case, orange juice can suppress TNF-α level and contribute to neutralizing persistent inflammation. This effect could be attributed to the hesperidin present in orange juice, because this citrus flavanone has been shown to reduce levels of TNF-α (63%), and IL-10 and IL-12 by 47% and 29%, respectively. Thus, the ingestion of orange juice supports the immune response, enhancing the effector functions of macrophages related to antimicrobial activity. Additionally, hesperidin attenuates the immune response and therefore may prevent the inflammatory process (Dourado et al. 2013). Therefore, orange juice can be characterized as a noninflammatory macronutrient food with the capacity to neutralize oxidative and inflammatory effects (Ghanim et al. 2010).

In addition to the flavanone hesperidin, oranges are a source of citrus polymethoxylated flavones (PMFs), which can also influence cytokine production at the level of gene transcription. It was already observed that citrus PMFs were also able to block the LPS-induced production of TNF-α, IL-10, and MIP-1α in an *in vitro* model with human monocytes. Studies have shown PMFs, such as 3,5,6,7,8,3′,4′-heptamethoxyflavone (HMF), could modulate inflammation *in vitro* through the inhibition of human phosphodiesterase-4 activity, and that this inhibition leads to substantial increases in cAMP levels in human monocytes. The inhibition of phosphodiesterase activity, along with increased cAMP levels, is a known link with decreased cytokine production. In an animal study, rodents were LPS-stimulated to develop carrageenan/paw edema, an inflammatory process that involves the formation antigen/antibody complex (Arthus reaction) and ultimately results in necrosis. The HMF administered by intraperitoneal injection was nearly as effective in inhibiting the edema as hydrocortisone, a known drug to control this kind of inflammation (Manthey and Cesar 2013).

To analyze HMF effects on brain damage, mice were subjected to LPS-induced sickness and the microglial cells were removed and immunostained with IBA1 antibody. Microglia, which is responsible for protecting the CNS (central nervous system) against various types of pathogenic factors under physiological conditions, is readily activated in response to injuries or immunological challenges. HMF treatment showed the suppression of LPS-induced activation of microglia, with a reduction in IL-1β mRNA expression, and a tendency to reduce COX-2 mRNA expression (Okuyama et al. 2015). Brain damage can induce sickness, anorexia, and body weight loss, however, HMF administration has also been shown to suppress LPS-induced body weight loss. The feeding mechanisms by which HMF interferes with body weight balance could be through the leptin (adipocyte-derived hormone)-PI3K-PDE pathway and the ghrelin (appetite-related hormone)-adenylate cyclase-cAMP-PKA system in the hypothalamus. Therefore, it is very likely that, HMF inhibits inflammation in the brain and recovers food intake and body weight loss; thus, HMF might be beneficial as a therapeutic or preventative compound for various neurological diseases in humans (Okuyama et al. 2015).

In another study to evaluate HMF treatment on brain tissue, mice were induced to cerebral ischemia by surgical intervention. It was observed that HMF-active

CREB (cAMP response element-binding protein) and ERK1/2 in the hippocampus of the ischemic brain, in addition to inducing BDNF (brain-derived neurotrophic factor), which is not observed with surgery treatment itself. These results suggest that HMF treatment accelerated neurogenesis in specific areas of the brain (dentate gyrus subgranular zone [SGZ] or subventricular zone [SVZ]) after ischemia (Okuyama et al. 2012).

9.5 EFFECT OF ORANGE JUICE AND CITRUS FLAVONOIDS ON CANCER

One of the most serious complications of inflammation is its evolution into the tumorigenesis of various cancers. Chronic inflammation is involved in every stage of cancer development and has been related to cytokines that regulate the inflammation, which can promote growth, attenuating apoptosis, and facilitating the invasion and metastasis. Thus, proinflammatory markers have become targets for cancer chemoprevention (Dourado et al. 2013; Ho et al. 2012). In one animal study, mice were subjected to a topical application of TPA (tetradecanoylphorbol-13-acetate), a classical skin tumor promoter with leucocyte infiltration and epidermal proliferation. It was found that the treatment with PMFs decreased inflammatory parameters, and reduced iNOS and cyclooxygenase-2 (COX-2) protein expression at the molecular level. The groups treated with PMFs showed that the diameters of the tumors tended to be smaller, which may have contributed to decreased iNOS, COX-2 (cyclooxygenase-2). It was also observed that PMF treatment interfered with MAPKs and PI3K/Akt and PKC signaling, leading to blocking the activation of transcription factors such as NF-κB and STAT3 (Ho et al. 2012). PI3K/Akt signal transduction plays a critical role in the control of cell growth and proliferation. The increased Akt activation or indirect changes in its regulation results in stronger cell survival signaling, which is a common feature in various forms of human cancers. Many downstream substrates of Akt kinase have been identified, including those related to chemotherapeutic resistance in cancer cells, and they directly or indirectly regulate apoptosis. Akt also regulates the NF-κB pathway via phosphorylation and activation of inhibitory κ-B kinase and RelA. The NF-κB transcription factors themselves regulate immune responses, cell growth, and apoptosis. Thus, the inhibition of NF-κB activation offers a potential strategy for the treatment of different malignancies (Arafa et al. 2009), and PMFs have shown a great potential as chemopreventive agents to be used in the treatment of inflammation associated with tumorigenesis (Ho et al. 2012).

Recent studies have shown that orange juice, in addition to modulating cytokines, has a potential cytotoxic and proapoptotic effect on the Loucy cell line, a T acute lymphoblastic leukemia (T-ALL) model. It was found that the treatment with orange juice from both yellow and red pulp oranges was able to reduce the viability of the Loucy cell line; the red orange juice induced early and late apoptosis while the yellow orange juice induced only late apoptosis. Consumption of orange juice was also capable of promoting cell cycle arrest in G0/G1, decreasing the accumulation of cells in the G2/M phase. In this study, no inhibition was observed in cell growth with hesperidin treatment. Therefore, the results with both yellow and red orange juices were mainly due to the interaction between nutrients and other nonnutritive compounds of juices, such as ascorbic acid and carotenoids (β-carotene and lycopene, respectively), and the differences in their effects during the apoptosis process suggest that orange juice composition could determine the mechanisms that initiate programmed cell death (Dourado et al. 2015).

A recurrent problem in cancer treatment is the chemotherapy resistance that can occur in some patients and the increased cisplatin dosage (chemotherapeutic drug), which causes severe cytotoxicity. Thus, some researchers have looked for compounds, such as citrus flavonoids, that would act in synergy with these drugs and would not be cytotoxic to the patients. In a previous study, the pentamethoxyflavone tangeretin was examined for its effect on cisplatin-resistant ovarian cancer cells. It was found that tangeretin alone, as well as tangeretin associated with cisplatin, caused an accumulation of G2-M cells. Additionally, the preexposure with TAN improved the sensitization of cancer cells to cisplatin leading to cell death through apoptosis. These findings suggest that tangeretin has a pharmacological effect as a valuable therapeutic adjuvant (Arafa et al. 2009).

9.6 EFFECT OF ORANGE JUICE AND CITRUS FLAVONOIDS ON PHYSICAL ACTIVITY

It is suggested that individuals who regularly engage in moderate exercise also tend to consume a greater amount of flavonoid-rich fruits and vegetables. On the other hand, those who eat healthy foods would have more energy to perform physical activity, so the engagement in physical activity may help to foster changes in dietary behavior (Loprinzi 2015). For many decades, daily physical exercise has been designed as an adjuvant treatment for chronic illnesses. It is a low-risk activity, and the combination of regular exercise with bioactive compounds in reducing chronic disease risk factors has been a recent approach for new studies (Oliveira et al. 2013).

In a recent study, rats that underwent continuous or interval swimming training were supplemented with hesperidin and compared to control groups submitted only to the continuous or interval swimming training. Both groups of rats showed improvement in biochemical and oxidative biomarkers 4 weeks after the intervention. The supplementation with hesperidin did not affect the weight gain of rats during the 4-week period, but swim training was a significant factor in reducing the weight gain of the rats. This suggests that energy expenditure during exercise was the key factor in maintaining body weight. Continuous exercise increased the oxidative stress in animals that performed the continuous swimming exercise; however, the hesperidin supplementation increased the antioxidant capacity in this group markedly (over 100%). It was also observed that the training by itself, as well as training in association with hesperidin, decreased total cholesterol, LDL cholesterol, and triglycerides, and increased HDL cholesterol, besides the reduction on blood glucose. This suggests that isolated hesperidin in rats can increase the number of GLUT-2 and GLUT-4 carriers, enhancing cellular signaling glucose and consequently reducing insulin resistance (Oliveira et al. 2013).

It is known that physical activity increases reactive oxygen species (ROS) production. This comes from the human endogenous antioxidant system in response to a regular physical training and can be involved in the process of training adaptation, in addition to prevent chronic diseases (Gomez-Cabrera et al. 2008). However, some researchers have recommended the use of antioxidants to decrease these oxidative damages. Rats undergoing acute exercise were administrated eriocitrin, and it was observed that these rats had lower levels of oxidative stress markers caused by exhaustive exercise, including DT (o-dityrosine) and NT (nitrotyrosine). Eriocitrin was also very effective for the suppression of redox effects in the rat liver. There was a depletion in GSSG (oxidized glutathione) and a preservation of GSH (reduced glutathione), while the level of total glutathione tended to increase in the rat liver during acute exercise. In this study, the authors suggest that eriocitrin metabolites could scavenge the free radicals generated by lipid peroxidation and prevent their formation. These effects could be the result of the absorbed eriocitrin metabolites, which have shown a strong capability of DPPH (1, 1-diphenyl-2-picrylhydrazyl) radical scavenging (Minato et al. 2003).

Another study observed that there was a decreased blood lactate concentration in sedentary overweight/obese pre- and postmenopausal middle-aged women upon the introduction of aerobic physical activity and the consumption of 500 mL/day of orange juice. This indicates an improvement in physical performance with less fatigue. It was verified that energy intake from orange juice was the same at the end of the study in the experimental group versus a control group (no orange juice intake), and physical activity promoted a reduction of total body fat and adipose tissue in both groups. It is possible that the intake of orange juice influenced this result by providing extra energy as well as essential nutrients, such as β-carotene and vitamin C (Aptekmann and Cesar 2013).

Vitamin C is a natural water-soluble free radical scavenger with the ability to donate two electrons from the double bond of the six-carbon molecule. During this reaction, vitamin C is oxidized and generates a stable intermediate product, dehydroascorbic acid (DHA), which can be taken up by erythrocytes and reduced to vitamin C again via endogenous glutathione reductase (GSH) by enzymatic (NADH/NADPH) or nonenzymatic systems. This process is called ascorbate (vitamin C) recycling and is responsible for maintaining vitamin C plasma concentrations (Mendiratta et al. 1998; Nimse et al. 2015; Padayatty et al. 2003). Studies have shown that the complete bioavailability of vitamins occurs at 200 mg (Levine et al. 1996), with an excretion half-life of around 30 minutes. This data suggests that vitamin C can redistribute from the blood plasma into other organs after its absorption; thus, the regular consumption of a high dose of vitamin C could be stored in the tissues (Hickey et al. 2008). However, it is important to point out that in previous studies, the supplementation of vitamin C capsules rendered a small difference in plasma concentrations of vitamin C compared to the plasma concentrations of vitamin C obtained after the consumption of vitamin C-rich foods. Thus, it would be more beneficial to recommend the consumption of fruits and vegetables rich in vitamin C rather than vitamin C capsules to obtain potential benefits (Padayatty et al. 2004).

However, the common practice of taking vitamin C supplements during training should be carefully recommended. Training efficiency and endurance capacity is directly related to the mitochondrial content and depends on VO_2 max (maximal capacity to take up, transport, and utilize oxygen during exercise), and the adaptations of the cardiovascular system during the exercises are also dependent on the VO_2 max. Then, it was explored if vitamin C administered to both rats and humans who underwent physical training could increase VO_2 max, endurance capacity, and skeletal muscle mitochondrial biogenesis. Vitamin C supplementation resulted in high vitamin C plasma concentrations in both experimental models, but this intervention only partially decreased the improvement in VO_2 max associated with exercise training and significantly hampered endurance capacity in animals. On the other hand, the animals undergoing exercise training without vitamin C supplementation had an increased maximal exercise time and increased mRNA concentrations of Mn-SOD and GPx in their skeletal muscle as a result of the increase in mitochondrial synthesis. Therefore, despite the fact that free radicals are considered to be harmful to the organism, during physical activity, their low concentrations may be considered beneficial because they act as signals to enhance defenses rather than being deleterious, as they can be at higher concentrations (Gomez-Cabrera et al. 2008).

REFERENCES

Actis-Goretta L, Ottaviani JI, Fraga CG. Inhibition of angiotensin converting enzyme activity by flavanol-rich foods. *J Agric Food Chem* 2006; 54: 229–34.

Alberti KG, Eckel RH, Grundy SM, Zimmet PZ, Cleeman JI, Donato KA, Fruchart JC et al. Harmonizing the metabolic syndrome: A joint interim statement of the International Diabetes Federation Task Force on Epidemiology and Prevention; National Heart, Lung, and Blood Institute; American Heart Association; World Heart Federation; International Atherosclerosis Society; and International Association for the Study of Obesity. *Circulation* 2009; 120(16): 1640–5.

American Heart Association. Coronary artery disease—Coronary heart disease. 2016. http://www.heart.org/HEARTORG/Conditions/More/MyHeartandStrokeNews/Coronary-Artery-Disease-Coronary-Heart-Disease_UCM_436416_Article.jsp#.V18QYfkrLIU.

Appel LJ, Brands MW, Daniels SR, Karanja N, Elmer PJ, Sacks FM. Dietary approaches to prevent and treat hypertension: A scientific statement from the American Heart Association. *Hypertension* 2006; 47(2): 296–308.

Aptekmann NP, Cesar TB. Long-term orange juice consumption is associated with low LDL-cholesterol and apolipoprotein B in normal and moderately hypercholesterolemic subjects. *Lipids Health Dis* 2013; 12: 119.

Arafa el-SA, Zhu Q, Barakat BM, Wani G, Zhao Q, El-Mahdy MA, Wani AA. Tangeretin sensitizes cisplatin-resistant human ovarian cancer cells through downregulation of phosphoinositide 3-kinase/Akt signaling pathway. *Cancer Res.* 2009; 69(23): 8910–7. doi:10.1158/0008-5472.CAN-09-1543.

Bai J, Manthey JA, Ford BL, Luzio G, Cameron RG, Narciso J, Baldwin EA. Effect of extraction, pasteurization and cold storage on flavonoids and other secondary metabolites in fresh orange juice. *J Sci Food Agri* 2013; 93: 2771–81.

Baldwin, EA, Bai, J, Plotto, A, Ritenour, MA. Citrus fruit quality assessment; producer and consumer perspectives. *Stewart Postharvest Rev* 2014; 23: 1–7.

Basile LG, Lima, CG, Cesar, TB. Daily intake of pasteurized orange juice decreases serum cholesterol, fasting glucose, and diastolic blood pressure in adults. *Proc Fla State Hort Soc* 2010; 123: 228–33.

Boekholdt SM, Meuwese MC, Day NE, Luben R, Welch A, Wareham NJ, Khaw K-T. Plasma concentrations of ascorbic acid and C-reactive protein, and risk of future coronary artery disease, in apparently healthy men and women: The EPIC-Norfolk prospective population study. *Br J Nutr* 2006; 96: 516–22.

Bok S-H, Lee S-H, Park Y-B, Bae K-H, Son K-H, Jeong T-S, Choi M-S. Plasma and hepatic cholesterol and hepatic activities of 3-hydroxy-3-methyl-glutaryl-CoA reductase and acyl CoA: Cholesterol transferase are lower in rats fed citrus peel extract or a mixture of citrus bioflavonoids. *J Nutr* 1999; 129: 1182–5.

Borradaile NM, Carroll KK, Kurowska EM. Regulation of HepG2 cell apolipoprotein B metabolism by the citrus flavanones hesperetin and naringenin. *Lipids* 1999; 34(6): 591–8.

Boushey CJ, Beresford SA, Omenn GS, Motulsky AG. A quantitative assessment of plasma homocysteine as a risk factor for vascular disease: Probable benefits of increasing folic acid intakes. *JAMA* 1995; 274: 1049–57.

Buijsse B, Jacobs DR Jr, Steffen LM, Kromhout D, Gross MD. Plasma ascorbic acid, a priori diet quality score, and incident hypertension: A prospective cohort study. *PLoS One* 2015; 10(12): e0144920.

Cappuccio FP, MacGregor GA. Does potassium supplementation lower blood pressure? A meta-analysis of published trials. *J Hypertension* 1991; 9: 465–73.

Carter B, Monsivais P, Drewnowski A. Absorption of folic acid and ascorbic acid from nutrient comparable beverages. *J Food Sci* 2010; 75(9): H289–93. doi:10.1111/j.1750-3841.2010.01844.x.

Cesar TB, Aptekmann NP, Araujo MP, Vinagre CC, Maranhão RC. Orange juice decreases low-density lipoprotein cholesterol in hypercholesterolemic subjects and improves lipid transfer to high-density lipoprotein in normal and hypercholesterolemic subjects. *Nutr Res* 2010; 30(10): 689–94.

Cesar TB, Rodrigues LU, Araujo MSP, Aptekmann NP. Suco de laranja reduz o colesterol em indivíduos normolipidêmicos. *Rev Nutr* 2010; 23(5): 779–89.

Chiba H, Uehara M, Wu J, Wang X, Masuyama R, Suzuki K, Kanazawa K, Ishimi Y. Hesperidin, a citrus flavonoid, inhibits bone loss and decreases serum and hepatic lipids in ovariectomized mice. *J Nutr* 2003; 133: 1892–7.

Coelho RC, Hermsdorff HH, Bressan J. Anti-inflammatory properties of orange juice: Possible favorable molecular and metabolic effects. *Plant Foods Hum Nutr* 2013; 68(1): 1–10. doi:10.1007/s11130-013-0343-3.

Deopurkar R, Ghanim H, Friedman J, Abuaysheh S, Sia CL, Mohanty P, Viswanathan P, Chaudhuri A, Dandona P. Differential effects of cream, glucose, and orange juice on inflammation, endotoxin, and the expression of Toll-like receptor-4 and suppressor of cytokine signaling-3. *Diabetes Care*. 2010 May 1; 33(5): 991–7.

Dourado GK, Cesar TB. Investigation of cytokines, oxidative stress, metabolic, and inflammatory biomarkers after orange juice consumption by normal and overweight subjects. *Food Nutr Res* 2015; 59: 28147.

Flint AJ, Rexrode KM, Hu FB, Glynn RJ, Caspard H, Manson JE, Willett WC, Rimm EB. Body mass index, waist circumference, and risk of coronary heart disease: A prospective study among men and women. *Obes Res Clin Pract* 2010 Jul; 4(3): e171–e181.

Franke AA, Cooney RV, Henning SM, Custer LJ. Bioavailability and antioxidant effects of orange juice components in humans. *J Agric Food Chem* 2005; 53(13): 5170–8.

Ghanim H, Sia CL, Upadhyay M, Korzeniewski K, Viswanathan P, Abuaysheh S, Mohanty P, Dandona P. Orange juice neutralizes the proinflammatory effect of a high-fat, high-carbohydrate meal and prevents endotoxin increase and Toll-like receptor expression. *Am J Clin Nutr* 2010 Apr; 91(4): 940–9. doi: 10.3945/ajcn.2009.28584.

Gomez-Cabrera MC, Domenech E, Romagnoli M, Arduini A, Borras C, Pallardo FV, Sastre J, Viña J. Oral administration of vitamin C decreases muscle mitochondrial biogenesis and hampers training-induced adaptations in endurance performance. *Am J Clin Nutr* 2008; 87(1): 142–9.

Grassi D, Desideri G, Groce G, Tiberti S, Aggio A, Ferri C. Flavonoides, vascular function and cardiovascular protection. *Curr Pharm Des* 2009; 15: 1072–84.

Grundy SM, Brewer HB Jr, Cleeman JI, Smith SC Jr, Lenfant C, National Heart, Lung, and Blood Institute, American Heart Association. Definition of metabolic syndrome: Report of the National Heart, Lung, and Blood Institute/American Heart Association conference on scientific issues related to definition. *Arterioscler Thromb Vasc Biol* 2004; 24(2): e13–18.

Han TS, Lean ME. A clinical perspective of obesity, metabolic syndrome and cardiovascular disease. *JRSM Cardiovasc Dis* 2016 Feb 25; 5: 2048004016633371. doi:10.1177/2048004016633371. eCollection 2016 Jan–Dec.

Hatzis CM, Bertsias GK, Linardakis M, Scott JM, Kafatos AG. Dietary and other lifestyle correlates of serum folate concentrations in a healthy adult population in Crete, Greece: A cross-sectional study. *Nutr J* 2006; 5: 1–10.

Hickey S, Roberts HJ, Miller NJ. Pharmacokinetics of oral vitamin C. *J Nutr Environ Med* 2008; 0(0): 000–000.

Ho CT, Pan MH, Lai CS, Li S. Polymethoxyflavones as food factors for the management of inflammatory diseases. *J Food Drug Anal* 2012; 20(suppl. 1): 337–41.

Hooper L, Kroon PA, Rimm EB, Cohn JS, Harvey I, Le Cornu KA, Ryder JJ, Hall WL, Cassidy A. Flavonoids, flavonoid-rich foods, and cardiovascular risk: A meta-analysis of randomized controlled trials. *Am J Clin Nutr* 2008; 88(1): 38–50.

Jacobson TA, Ito MK, Maki KC, Orringer CE, Bays HE, Jones PH, McKenney JM et al. National Lipid Association recommendations for patient-centered management of dyslipidemia: Part 1—Full report. *J Clin Lipidol* 2015; 9(2): 129–6.

Janeway C, Travers P, Walport M, Shlomchik; trad. Cristina Bonorino et al. *Imunologia: Sistema imune na saúde e na doença*. 5a edicao. Porto Alegre: *Artmed*, 2002.

Johnson EJ. The role of carotenoids in human health. *Nutr Clin Care* 2002, 5: 56–65.

Kay CD, Hooper L, Kroon PA, Rimm EB, Cassidy A. Relative impact of flavonoid composition, dose and structure on vascular function: A systematic review of randomised controlled trials of flavonoid-rich food products. *Mol Nutr Food Res* 2012; 56(11): 1605–16.

Kim K, Vance TM, Chun OK. Greater total antioxidant capacity from diet and supplements is associated with a less atherogenic blood profile in U.S. adults. *Nutrients* 2016; 8(1): pii: E15.

Kontush A, Chapman MJ. Antiatherogenic small, dense HDL-guardian angel of the arterial wall? *Nat Clin Pract Cardiovasc Med* 2006; 3: 144–53.

Krishna SM, Dear A, Craig JM, Norman PE, Golledge J. The potential role of homocysteine mediated DNA methylation and associated epigenetic changes in abdominal aortic aneurysm formation. *Atherosclerosis* 2013; 228(2): 295–305. doi: 10.1016/j.atherosclerosis.2013.02.019.

Kurowska EM, Borradaile NM, Spence JD, Carroll KK. Hypocholesterolemic effects of dietary citrus juices in rabbits. *Nutr Res* 2000; 20: 121–9.

Kurowska EM, Manthey JA. Regulation of lipoprotein metabolism in HepG2 cells by citrus flavonoids. *Adv Exp Med Biol* 2002; 505: 173–9.

Kurowska EM, Manthey JA, Casaschi A, Theriault AG. Modulation of HepG2 cell net apolipoprotein B secretion by the citrus polymethoxyflavone, tangeretin. *Lipids* 2004; 39(2): 143–51.

Kurowska EM, Spence JD, Jordan J, Wetmore S, Freeman DJ, Piché LA, Serratore P. HDL-cholesterol-raising effect of orange juice in subjects with hypercholesterolemia. *Am J Clin Nutr* 2000; 72(5): 1095–100.

Levine M, Conry-Cantilena C, Wang Y, Welch RW, Washko PW, Dhariwal KR, Park JB, Lazarev A, Graumlich JF, King J, Cantilena LR. Vitamin C pharmacokinetics in healthy volunteers: Evidence for a recommended dietary allowance. *Proceedings of the National Academy of Sciences* 1996 Apr 16; 93(8): 3704–9.

Lima CG, Basile LG, Silveira JQ, César TB. Ingestão regular do suco de laranja vermelha reduz pressão arterial de adultos. *J Health Sci Inst* 2012; 30(1): 59–63.

Lonn E, Yusuf S, Arnold M, Sheridan P, Pogue J, Micks M, McQueen M, Probstfield J, Fodor G, Held C. Homocysteine lowering with folic acid and B vitamins in vascular disease. *N Engl J Med* 2006; 354: 1567–77.

Loprinzi PD. Association between accelerometer-determined physical activity and flavonoid-rich fruit and vegetable consumption among a national sample of U.S. adults. *Prev Med Rep* 2015; 3: 58–61. doi: 10.1016/j.pmedr.2015.12.001.

Manthey JA, Cesar TB. Issues surrounding the anti-inflammatory actions of the citrus polymethoxylated flavones. In Tropical and Subtropical Fruits: Flavors, Color, and Health Benefits. *ACS Symposium Series* 2013; 1129(Chapter 10): 159–169. American Chemical Society.

Mendiratta S, Qu ZC, May JM. Erythrocyte ascorbate recycling: Antioxidant effects in blood. *Free Radic Biol Med* 1998; 24(5): 789–97.

Mennen LI, Sapinho D, de Bree A, Arnault N, Bertrais S, Galan P, Hercberg S. Consumption of foods rich in flavonoids is related to a decreased cardiovascular risk in apparently healthy French women. *J Nutr* 2004; 134(4): 923–6.

Milenkovic D, Deval C, Dubray C, Mazur A, Morand C. Hesperidin displays relevant role in the nutrigenomic effect of orange juice on blood leukocytes in human volunteers: A randomized controlled cross-over study. *PLoS ONE* 2004; 6(11): e26669. doi:10.1371/journal.pone.0026669.

Minato K, Miyake Y, Fukumoto S, Yamamoto K, Kato Y, Shimomura Y, Osawa T. Lemon flavonoid, eriocitrin, suppresses exercise-induced oxidative damage in rat liver. *Life Sci* 2003; 72(14): 1609–16.

Mitchell JA, Ali F, Bailey L, Moreno L, Harrington LS. Role of nitric oxide and prostacyclin as vasoactive hormones released by the endothelium. *Exp Physiol* 2008; 93(1): 141–7. Epub 2007.

Moat SJ, Lang D, McDowell IF, Clarke ZL, Madhavan AK, Lewis MJ, Goodfellow J. Folate, homocysteine, endothelial function and cardiovascular disease. *J Nutr Biochem* 2004; 15: 64–79.

Morand C, Dubray C, Milenkovic D, Lioger D, Martin JF, Scalbert A, Mazur A. Hesperidin contributes to the vascular protective effects of orange juice: A randomized crossover study in healthy volunteers. *Am J Clin Nutr* 2011 Jan; 93(1): 73–80.

Murata T. Citrus. In: *Postharvest Physiology and Storage of Tropical and Subtropical Fruits*, Mitra S (ed.). Wallingford, UK: CAB International, 1997, pp. 21–47.

Nasser ALM, Dourado GK, Manjate DA, Carlos IZ, Cesar TB. Avaliação do estresse oxidativo no sangue de consumidores habituais de suco de laranja. *Rev Ciênc Farm Básica Apl* 2011; 32(2): 00–00.

Nimse SB, Pal D. Free radicals, natural antioxidants, and their reaction mechanisms. *RSC Adv* 2015; 5: 27986–28006. doi: 10.1039/c4ra13315c.

Okuyama S, Miyoshi K, Tsumura Y, Amakura Y, Yoshimura M, Yoshida T, Nakajima M, Furukawa Y. 3,5,6,7,8,3',4'-heptamethoxyflavone, a citrus polymethoxylated flavone, attenuates inflammation in the mouse hippocampus. *Brain Sci* 2015; 5(2): 118–29. doi:10.3390/brainsci5020118.

Okuyama S, Shimada N, Kaji M, Morita M, Miyoshi K, Minami S, Amakura Y, Yoshimura M, Yoshida T, Watanabe S, Nakajima M, Furukawa Y. Heptamethoxyflavone, a citrus flavonoid, enhances brain-derived neurotrophic factor production and neurogenesis in the hippocampus following cerebral global ischemia in mice. *Neurosci Lett* 2012; 528(2): 190–5. doi: 10.1016/j.neulet.2012.08.079.

Oliveira DM, Dourado GK, Cesar TB. Hesperidin associated with continuous and interval swimming improved biochemical and oxidative biomarkers in rats. *J Int Soc Sports Nutr* 2013; 10(27): 1–7. doi: 10.1186/1550-2783-10-27.

O'Neil CE, Nicklas TA, Rampersaud GC, Fulgoni VL 3rd. One hundred percent orange juice consumption is associated with better diet quality, improved nutrient adequacy, and no increased risk for overweight/obesity in children. *Nutr Res* 2011; 31(9): 673–82.

O'Neil CE, Nicklas TA, Rampersaud GC, Fulgoni VL 3rd. 100% Orange juice consumption is associated with better diet quality, improved nutrient adequacy, decreased risk for obesity, and improved biomarkers of health in adults: National Health and Nutrition Examination Survey, 2003–2006. *Nutr J* 2012; 11: 107.

O'Neill S, O'Driscoll L. Metabolic syndrome: A closer look at the growing epidemic and its associated pathologies. *Obes Rev* 2015; 16(1): 1–12.

Padayatty SJ, Katz A, Wang Y, Eck P, Kwon O, Lee JH, Chen S, Corpe C, Dutta A, Dutta SK, Levine M. Vitamin C as an antioxidant: Evaluation of its role in disease prevention. *J Am Coll Nutr* 2003; 22(1): 18–35.

Popkin BM, Armstrong LE, Bray GM, Caballero B, Frei B, Willett WC. A new proposed guidance system for beverage consumption in the United States. *Am J Clin Nutr* 2006; 83: 529–42.

Rampersaud GC, Valim MF. 100% citrus juice: Nutritional contribution, dietary benefits, and association with anthropometric measures. *Crit Rev Food Sci Nutr* 2017; 57(1): 129–140.

Rangel-Huerta OD, Aguilera CM, Martin MV, Soto MJ, Rico MC, Vallejo F, Tomas-Barberan F, Perez-de-la-Cruz AJ, Gil A, Mesa MD. Normal or high polyphenol concentration in orange juice affects antioxidant activity, blood pressure, and body weight in obese or overweight adults. *J Nutr* 2015; 145(8): 1808–16.

Ribeiro C, Dourado G, Cesar T. Orange juice allied to a reduced-calorie diet results in weight loss and ameliorates obesity-related biomarkers: A randomized controlled trial. *Nutrition* 2017; 38: 13–19. http://dx.doi.org/10.1016/j.nut.2016.12.020.

Riso P, Visioli F, Gardana C, Grande S, Brusamolino A, Galvano F, Galvano G, Porrini M. Effects of blood orange juice intake on antioxidant bioavailability and on different markers related to oxidative stress. *J Agri Food Chem* 2005; 53: 941–7.

Rivera JA, Muñoz-Hernández O, Rosas-Peralta M, Aguilar-Salinas CA, Popkin BM, Willett WC. Comité de Expertos para las Recomendaciones [Beverage consumption for a healthy life: Recommendations for the Mexican population]. *Salud Publica Mex* 2008; 50(2): 173–95.

Sabharwal AK and May JM. α-Lipoic acid and ascorbate prevent LDL oxidation and oxidant stress in endothelial cells. *Mol Cell Biochem* 2008; 309: 125–32.

Santilli F, D'Ardes D, Davì G. Oxidative stress in chronic vascular disease: From prediction to prevention. *Vascul Pharmacol* 2015; 74: 23–37. doi: 10.1016/j.vph.2015.09.003.

Silveira JQ, Dourado GK, Cesar TB. Red-fleshed sweet orange juice improves the risk factors for metabolic syndrome. *Int J Food Sci Nutr* 2015; 66(7): 830–6.

Stone NJ, Robinson J, Lichtenstein AH, Bairey Merz CN, Blum CB, Eckel RH, Goldberg AC et al. 2013 ACC/AHA guideline on the treatment of blood cholesterol to reduce atherosclerotic cardiovascular risk in adults: A report of the American College of Cardiology/American Heart Association Task Force on Practice Guidelines. *Circulation* 2013; 129(25 Suppl 2): S1–45.

U.S. Department of Agriculture, Agricultural Research Service. USDA National Nutrient Database for Standard Reference, Release 28 slightly revised May, 2016. Nutrient Data Laboratory home page, 2016. http://ndb.nal.usda.gov/ndb/foods.

Vinueza JC, Faria JB, César TB. Hesperidina diminui o colesterol sanguíneo de ratos alimentados com gordura saturada. *Alim Nutr Araraquara* 2008; 19(4): 473–79.

Wang Y, Lloyd B, Yang M, Davis CG, Lee SG, Lee W, Chung SJ, Chun OK. Impact of orange juice consumption on macronutrient and energy intakes and body composition in the US population. *Public Health Nutr* 2012; 15(12): 2220–7. doi: 10.1017/S1368980012000742.

World Health Organization. Cardiovascular diseases (CVDs). 2016. http://www.who.int/mediacentre/factsheets/fs317/en/.

10 Citrus Flavonoids and Their Effect on Obesity

Shi Feng, Laura Reuss, Joon Hyuk Suh, and Yu Wang

CONTENTS

10.1 INTRODUCTION

Obesity is a chronic disease that increases the risk of suffering from a variety of life-threatening diseases including diabetes, hypertension, cardiovascular diseases, and also certain cancers. It is the result of a chronic imbalance between energy intake and energy expenditure, and it is characterized by an increase in both the size and number of adipocytes. The only obesity treatment medication currently available, Orlistat, has serious side effects that include increased blood pressure, dry mouth, constipation, headache, and insomnia (Yun 2010). Sibutramine, a previously prescribed weight loss medication, was withdrawn by the U.S. Food and Drug Administration in 2010 due to patients suffering an increased risk of cardiovascular adverse events. Therefore, in recent years, natural alternatives that exhibit antiobesity potential have been widely investigated. For instance, natural products such as crude extracts or isolated compounds found in edible plants, more specifically, fruits and vegetables, have been shown to reduce body weight and subsequently prevent diet-induced obesity (Han et al. 2005; Lopes et al. 2005; Moro and Basile 2000). The possible benefits found in dietary plants are due largely to the functional ingredients and antioxidant nutraceuticals, which are also known as phytochemical substances. The potential for using edible plants and their derivatives as antiobesity agents is of great interest, and numerous investigations into the important contributions of phytochemicals are currently underway.

Among these phytochemicals are flavonoids, which provide significant human health benefits, primarily because of their powerful pharmacological abilities to act as radical scavengers (Cook and Samman 1996). Ubiquitous in fruits and vegetables, flavonoids are an important group of polyphenolic compounds, with citrus fruits being the richest sources of them (Yao et al. 2004). The chemical properties of citrus flavonoids will be discussed in this chapter, and the ability of citrus flavonoids to act upon different therapeutic targets and inhibit obesity will also be reviewed.

10.2 CITRUS FLAVONOIDS AND THEIR CHEMICAL PROPERTIES

10.2.1 Introduction to Flavonoids

Ubiquitous to plants, flavonoids comprise a group of polyphenolic compounds with various chemical structures. More than 4,000 flavonoid types have been identified. Many of these are involved in protecting plants from fungal parasites, herbivores, pathogens, and oxidative cell injury (Cook and Samman 1996). In addition, flavonoids are also plant pigments and assist in pollination (Cody et al. 1986). Flavonoids are considered an integral component of the human diet (Clifford 2000; Harborne et al. 1984), as they are responsible for the color and taste of plant-based foods, as well as the prevention of fat oxidation and the integrity of vitamins and enzymes. Although difficult to estimate accurately, the proposed average human dietary intake of flavonoids is approximately 100 to 1,000 mg/day (Aherne and O'Brien 2002). Flavonoids are organized into various classes based on their molecular structure. The seven major classes of flavonoids, which are all common in the human diet, include flavones, flavonols, flavanones, flavanonols, flavanols, anthocyanins, and isoflavones (Aherne and O'Brien 2002; Bravo 1998) (Table 10.1). Flavones are mainly found in herbs, flavonols are in vegetables and fruits, flavanones and flavanonols are

TABLE 10.1
Major Classes of Flavonoids, Representative Individual Compounds, and Typical Food Sources

Class	Representative Flavonoids	Food Sources
Flavones	Apigenin, chrysin, luteolin	Parsley, berries, celery
Flavonols	Quercetin, kaempferol, myricetin	Onions, apples, chia seeds
Flavanones	Hesperidin, naringenin, fisetin,	Citrus fruits
Flavanonols	Taxifolin	Lemon, aurantium
Flavanols	Catechins, epigallocatechin gallate	Cocoa, tea
Anthocyanins	Cyanidin, delphinidin, malvidin	Berries, grapes
Isoflavones	Daizein, genistein, glycitein	Legumes

in citrus fruits, flavanols are in tea, anthocyanins are in fruits, and isoflavones are in legumes (Peterson and Dwyer 1998).

10.2.2 Biosynthesis of Flavonoids

Flavonoids are derived from phenylalanine and malonyl-coenzyme A (CoA) through a general phenylpropanoid pathway (Dixon et al. 2002) which serves as a rich source of metabolites in plants. Synthesis starts with the catalyzation of malonyl-CoA and 4-coumaroyl CoA via chalcone synthase (CHS), which then leads to the nine flavonoid subclasses that include chalcones, aurones, isoflavonoids, flavones, flavonols, flavandiols, anthocyanis, condensed tannins, and phlobaphene pigments (Winkel-Shirley 2001, 2002) (Figure 10.1).

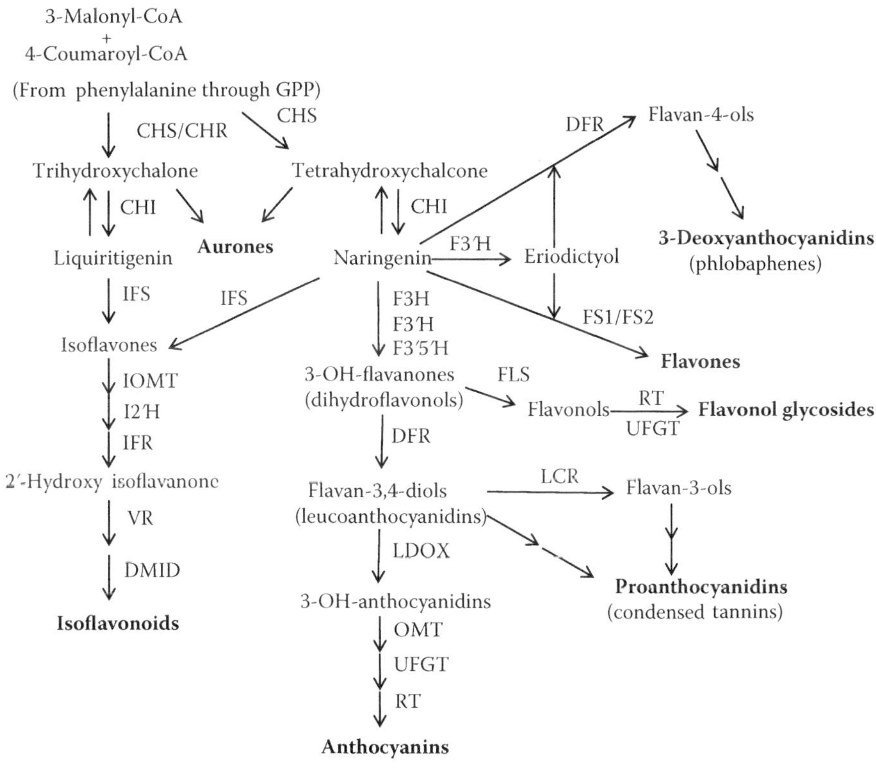

FIGURE 10.1 Schematic of the flavonoid biosynthetic pathway. CHR, chalcone reductase; CHS, chalcone synthase; DFR, dihydroflavonol 4-reductase; DMID, 7,2′-dihydroxy, 4′-methoxyisoflavanol dehydratase; F3H, flavanone 3-hydroxylase; F3′H, flavonoid 3′-hydroxylase; F3′5′H, flavonoid 3′5′-hydroxylase; FLS, flavonol synthase; FS1/FS2, flavone synthase; GPP, general phenylpropanoid pathway; I2′H, isoflavone 2′-hydroxylase; IFR, isoflavone reductase; IFS, isoflavone synthase; IOMT, isoflavone O-methyltransferase; LCR, leucoanthocyanidin reductase; LDOX, leucoanthocyanidin dioxygenase; OMT, O-methyltransferase; RT, rhamnosyl transferase; UFGT, UDP flavonoid glucosyl transferase; VR, vestitone reductase. (From Winkel-Shirley, B., *Plant Physiology* 126:485–493, 2002.)

10.2.3 Major Flavonoids in Citrus and Their Structural and Chemical Properties

Flavonoids are widely present in numerous plant-based foods, and especially so in *Citrus* species. Specifically, hesperidin, naringin, and polymethoxylated flavones (PMFs) are characteristic citrus flavonoids (Nogata et al. 2006). The citrus peel, as well as the seeds, are rich in these flavonoids, yet the flavonoid compositions found in the peel and seeds are not consistent in citrus fruits. Additionally, the flavonoid compositions found in the peel and seeds are also quite different from those found in the juices. In general, flavonoids make contributions to citrus fruit and product quality through multiple aspects, such as fruit appearance, taste, and nutritional value. For example, hesperidin can lead to the formation of sediments in juice products that produce an undesirable cloudiness (Mizrahi and Berk 1970). It is also worth mentioning that food preparation and the processing of fresh fruits may greatly decrease flavonoid content due to leaching of compounds into or out of storage containers and/or the removal of the peel and seeds, some of the richest flavonoid sources located in the fruit (Polydera et al. 2005; Sánchez-Moreno et al. 2005). To date, over 60 types of citrus flavonoids have been identified. Some of these citrus flavonoid compositions have been used for both classification and discrimination of *Citrus* species. For instance, a factorial discrimination analysis of the flavanone glycoside composition in citrus juice was successfully used in the differentiation of lemon, lime, grapefruit, and sweet orange (Mouly et al. 1994). Six flavanone glycosides, eriocitrin, neoeriocitrin, narirutin, naringin, hesperidin, and neohesperidin, found in the citrus juices of 124 samples, were separated and quantified by liquid chromatography (LC)-UV detection. The major flavanone glycosides in each citrus group were determined as follows: in lemon and lime, eriocitrin and hesperidin; in sweet orange, narirutin and hesperidin; in grapefruit, narirutin and naringin.

The major groups of flavonoids found in citrus are flavanones, flavones, and flavonols (Benavente-García et al. 1997). The structures of these important citrus flavonoids are shown in Figure 10.2. Anthocyanins have also been identified as citrus flavonoids, but only in blood oranges (Horowitz and Gentili 1977). Of these three major flavonoid types, flavanones are the most abundant in citrus. The flavonone skeleton consists of two aromatic rings connected through a dihydropyrone ring. In addition, it is believed that the 7-*O*-glycosylflavanone is the most plentiful of all citrus fruit flavonoids (Benavente-García et al. 1995), whereas flavones and flavonols have been identified in low concentrations in citrus. However, the highly methoxylated form of flavones indicates a higher biological activity compared to flavanones. Citrus flavanones mainly exist as one of two forms, either aglycone or glycoside (Tripoli et al. 2007). Naringenin and hesperetin are considered the most important flavanone among the aglycones, whereas two classified types of the glycoside form include neohesperidosides and rutinosides (Maxcheix et al. 1990). Together, either naringin, neohesperidin, or neoeriocitrin, which all carry a bitter taste, combine with neohesperidose to form a glycoside flavanone. In a similar manner, either hesperidin, narirutin, or didymin, each of which has no perceivable taste, can combine with rutinose (a disaccharide residue) and also form a glycoside flavanone (Tripoli et al. 2007). Flavanones are commonly present

Flavanone aglycone forms

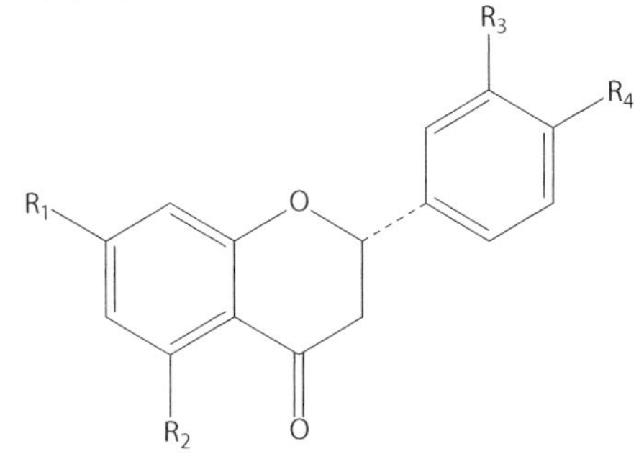

Naringenin	R_1=OH; R_2=OH; R_3=H; R_4=OH
Hesperetin	R_1=OH; R_2=OH; R_3=OH; R_4=OCH_3
Isosakuranetin	R_1=OH; R_2=OH; R_3=H; R_4=OCH_3

Flavanone neohesperidoside forms

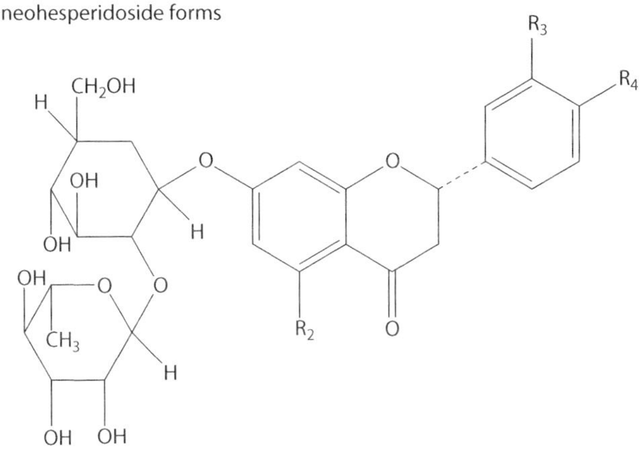

Naringin	R_2=OH; R_3=H; R_4=OH
Neohesperidin	R_2=OH; R_3=OH; R_4=OCH_3
Neoeriocitrin	R_2=OH; R_3=OH; R_4=OH

FIGURE 10.2 Structures of the characteristic citrus flavonoids. *(Continued)*

in disaccharide form, which contributes to the typical citrus fruit taste (Maxcheix et al. 1990). According to an early review, the main components identified in citrus juices include flavanone-*O*-glycosides and flavone-*O*-glycosides (Gattuso et al. 2007). In addition, currently identified flavonoid types found in citrus juice include hesperetin, naringenin, taxifolin, isosakuranetin, and eriodictyol. Conversely, the aglycone forms of flavanones are not very common in citrus juices due to their low solubility in water. However, PMFs do occur frequently in the essential oils of citrus peels (Miyake et al. 1997).

Flavanone rutinside forms

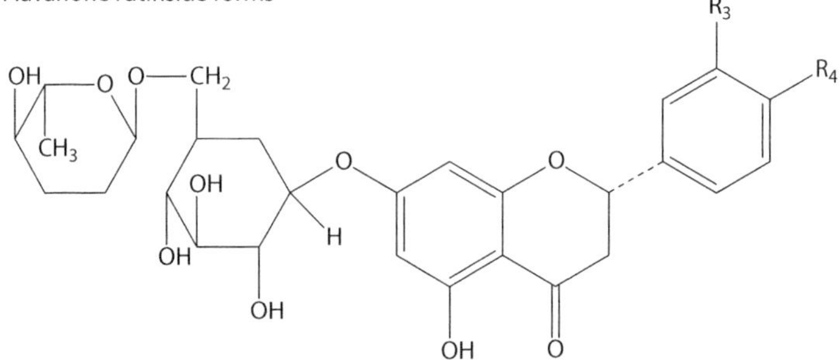

Narirutin	R_3=H; R_4=OH
Hesperidin	R_3=OH; R_4=OCH_3
Didymin	R_3=H; R_4=OCH_3
Eriocitrin	R_3=OH; R_4=OH

Flavone and flavonol aglyvone forms

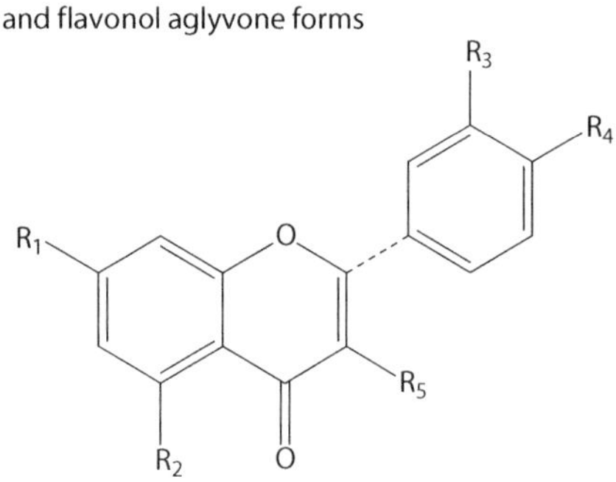

Apigenin	R_1=OH; R_2=OH; R_3=H; R_4=OH; R_5=H
Luteolin	R_1=OH; R_2=OH; R_3=OH; R_4=OH; R_5=H
Quercetin	R_1=OH; R_2=OH; R_3=OH; R_4=OH; R_5=OH
Kampferol	R_1=OH; R_2=OH; R_3=H; R_4=OH; R_5=OH

Polymethoxylated flavones

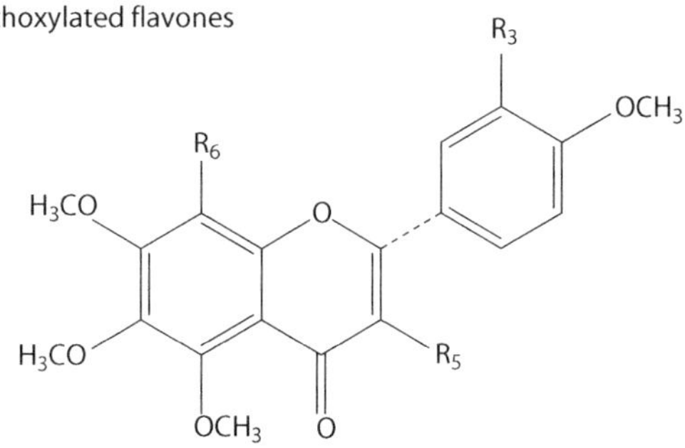

Tangeretin	R_3=H; R_5=H; R_6=OCH_3
Nobiletin	R_3=OCH_3; R_5=H; R_6=OCH_3

FIGURE 10.2 (CONTINUED) Structures of the characteristic citrus flavonoids.

10.2.4 General Benefits of Flavonoids and Their Possible Applications

Flavonoids are linked to a range of health benefits based on their antioxidant properties. Not only do they exhibit antioxidant activity in several ways, but they also demonstrate antiradical activities, antilipoperoxidation activities, and activities of metal chelation (Bombardelli and Morazzoni 1993). Flavonoids are known as powerful radical scavengers, which is attributed to their hydrogen-donating ability (Burda and Oleszek 2001). It is well accepted that flavonoids may act as antioxidants through the inhibition of free radical-mediated cytotoxicy and lipid peroxidation (Lyons-Wall and Samman 1997). Research has also shown that flavonoids inhibit many aging and degenerative events involving reactive oxygen species, due to their role as potent radical scavengers (Benavente-García et al. 1997). It has also been demonstrated that the antioxidant activities of flavonoids may reduce the risk of cardiovascular disease (Duthie et al. 2000). Additionally, epidemiological and animal studies of flavonoids have revealed possible protective effects against cardiovascular diseases, as well as some types of cancer (Manthey et al. 2001; Middleton et al. 2000). In general, flavonoids have been reported to exhibit a wide range of biological benefits, including anti-inflammatory activity, antimicrobial activity, and the prevention of both atherosclerosis and cancer.

Traditionally, flavonoid-rich plants have been used to treat inflammation. Even with low bioavailability and the risk of degradation due to metabolism, the use of bioflavonoids is well supported through the results of numerous *in vivo* research and clinical trials. Additional advantages of flavonoid use include lower costs and fewer side effects compared to current anti-inflammatory drugs. In a similar manner, plant flavonoids are also believed to be potent anticancer agents (Elangovan et al. 1994; Hertog et al. 1993; Hirano et al. 1994). Concurrently, flavonoids have shown antifungal and antiviral activities as well (Huet 1982). The protective activity of flavonoids against viruses is influenced through a structure–activity relationship (Kaul et al. 1985). Although much remains to be explored through investigation into the antiviral activities of flavonoids, it is evident that there lies strong potential in replacing synthetic drugs with a natural flavonoid product acting as an antiviral agent.

10.3 CITRUS FLAVONOIDS AND THEIR ANTIOBESITY ACTIVITIES

Obesity, defined as a body mass index (BMI) of ≥30 kg/m^2, is a chronic disease that increases risk factors leading to a variety of life-threatening diseases. Obesity is the result of a chronic imbalance of energy intake and energy expenditure. It is characterized by an increase in both the size and number of adipocytes, which are the primary site for energy storage in the body and will accumulate triglycerides when exposed to an excess nutritional intake. The only current medication available for the treatment of obesity, Orlistat, has been shown to cause serious side effects that raise increasing concerns in overweight or obese patients. Therefore, recent research is focused on searching for safe natural alternatives that exhibit antiobesity potential. The potential use of plants, herbs, and their derivatives as antiobesity agents is of great interest, and numerous investigations into the importance of phytochemicals

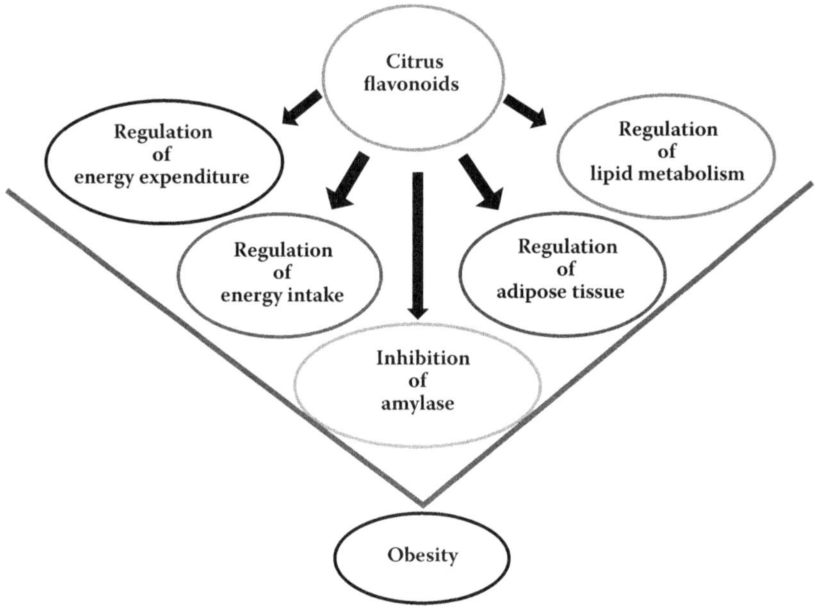

FIGURE 10.3 Citrus flavonoids and their potential antiobesity targets.

are currently underway. According to the recent epidemiological studies, possible mechanisms of antiobesity and those relationships with citrus flavonoids can be summarized by the following: regulation of energy intake and expenditure, regulation of lipid metabolism, regulation of adipose tissue, and inhibition of amylase (Figure 10.3). In this chapter, potential antiobesity components of citrus flavonoids belonging to the family of botanical polyphenols that inhibit obesity through various therapeutic targets are discussed.

10.3.1 Citrus Flavonoids as Regulators of Energy Intake

Energy intake is regulated by an appetite-mediated neural network centrally located in the hypothalamus of the human brain (Schwartz et al. 2000). Appetite suppression occurs when compounds, such as dietary constituents, act on the various neurotransmitter pathways regulating psychological and behavioral expression of appetite, metabolism, and peripheral physiology, as well as the neural pathway functions of the central nervous system (CNS) (Halford and Blundell 2000). Additionally, signaling hormones, such as leptin, adiponectin, resistin, ghrelin, glucagon-like peptide-1 (GLP-1), cholecystokinin (CCK), and peptide YY (PYY), which are produced in peripheral tissues such as adipose tissue, the pancreas, and gut, function to link peripheral adiposity and energy homeostasis to CNS-regulated food intake (Kim and Park 2011) (Table 10.2). Therefore, certain citrus flavonoids can be used as appetite suppressants by targeting these specific signaling hormones that control appetite.

TABLE 10.2
Targets Involved in Regulation of Energy Intake

Target	Produced by	Function in Energy Balance
Leptin	Adipose tissue	Lowering food intake
Adiponectin	Adipose tissue	Stimulating food intake during fasting
Resistin	Adipose tissue	Lowering food intake
Ghrelin	Stomach	Stimulating appetite and food intake
GLP-1	Pancreas, intestine, brainstem	Lowering food intake
CCK	Intestine	Lowering food intake
PYY	Intestine	Lowering food intake

The main signaling hormones secreted by adipose tissue include leptin, adiponectin, and resistin, and they are referred to as adipocytokines. Adipocytokines are associated with energy regulation, and it is believed that control of adipocytokine secretion is key in targeting obesity prevention. Leptin suppresses food intake, subsequently inducing weight loss through mediation of the long-term regulation of energy balance (Klok et al. 2007). The urge to eat is reduced because adipocyte energy storage is linked via negative feedback pathways in the brain, thus regulating food intake (Zhang et al. 1994). Animal studies have demonstrated that the primary center regulating food intake and body weight is located in the hypothalamus (Satoh et al. 1997; Schwartz et al. 1996). After leptin is released from adipose tissue, it eventually binds to the hypothalamic leptin receptors, thus relaying information about the status of the body's energy stores (Golden et al. 1997; Meister 2000; Sahu 2003). Adiponectin has been proven to regulate energy homeostasis as well as glucose and lipid metabolism (Yamauchi et al. 2002). Impaired glucose and lipid metabolism is indicative of obesity (Kahn and Flier 2000); consequently, obesity has been associated with decreased plasma adiponectin levels (Maeda et al. 1996). Additionally, adiponectin suppresses hepatic glucose production (Berg et al. 2001; Combs et al. 2001). However, the central effect of adiponectin on energy balance remains both unclear and controversial. Originally named for its resistance to insulin, the peptide hormone resistin has been shown to promote short-term satiety in rats when centrally administered. This suggests the possibility that resistin serves as a neuropeptide involved in regulating energy homeostasis (Tovar et al. 2005). Investigation into the antidiabetic properties of tangeretin in mice fed a high-fat diet (HFD) indicated that administration of HFD plus 200 mg/kg of tangeretin significantly altered the secretion of adipocytokines, such as adiponectin, leptin, and resistin (Kim et al. 2012b). Additionally, isolated rat adipocytes treated with anthocyanins (cyanidin or cyanidin 3-glucoside), flavonoids identified in the blood orange (Rapisarda et al. 2000) have demonstrated enhanced secretion of leptin and adiponectin (Tsuda et al. 2004). The dried, immature fruit of *Citrus aurantium* L. ("zhiqiao" in Chinese) is a traditional Chinese medicine. Naringenin and hesperetin, both flavonoids derived from this herb, have been shown to upregulate transcription of adiponectin in differentiated 3T3-L1 cells (Liu et al. 2008). In a study of insulin resistant hamsters, supplementation with citrus PMFs increased adiponectin through the regulation of adipocytokines (Li et al. 2006). PMFs are specific to *Citrus*, and the two most common PMFs, tangeretin and nobiletin, occur largely in the peels of both tangerines and oranges (Horowitz and Gentili 1977). Numerous studies have specifically focused on nobiletin, as it has been found to enhance the production of adiponectin proteins at a concentration of 10 μM (Kunimasa et al. 2009) and also to increase the expression of adipokines in the white adipose tissue of obese diabetic ob/ob mice (Lee et al. 2010). Similarly, a very recent study using the murine preadipocyte cell line 3T3-L1 concluded that nobiletin and tangeretin modulated adipocytokine secretion as well as increased the secretion of adiponectin (Miyata et al. 2011). *Citrus grandis* (L.) Osbeck (red wendun) leaves have long been used in traditional Chinese medicine. Two flavone glycosides, rhoifolin and cosmosiin from red wendun leaves, were observed to enhance adiponectin secretion in differentiated 3T3-L1 adipocytes (Rao et al. 2011). In the following years, hesperidin, and naringin supplementation was demonstrated to significantly enhance adiponectin expression (Mahmoud et al. 2013).

Ghrelin, a hormone primarily secreted in the stomach, is another peripheral signal with centrally localized effects. The hypothalamus is largely responsible for mediating the effects of ghrelin on energy balance (Klok et al. 2007). Ghrelin secretion in mammals is increased through fasting (Kim and Park 2011); this, then stimulates appetite. Therefore, ghrelin might be a useful target in obesity prevention, and ghrelin antagonists could decrease food intake. The neuropeptide glucagon-like peptide-1 (GLP-1) is largely released from the pancreas, intestine, and brainstem (Kim and Park 2011). It enhances satiety, which in turn reduces energy intake and is therefore a potential physiological regulator for controlling appetite and energy intake in humans (Flint et al. 1998). Secretion of GLP-1, followed by its pancreatic receptor binding, is responsible for both short-term appetite control as well as long-term maintenance of body-weight (Drucker 2006). Cholecystokinin (CCK) is a satiety peptide distributed widely throughout the gastrointestinal tract and the central nervous system. It suppresses food intake by binding the G-protein-coupled CCK receptor, which then signals the hypothalamus to control appetite (Moran 2000). CCK also indicates physiological effects by delaying gastric emptying (Little et al. 2005). Another satiety gut hormone, PYY, also known as peptide tyrosine tyrosine, is secreted in the gut and influences gastrointestinal responses in a manner similar to CCK. Low-dose intravenous infusion of PYY has been demonstrated to reduce food intake in both animals and humans (Batterham et al. 2002; Chelikani et al. 2006). It has been shown that naringin and naringenin, both components in the aqueous extract of *Poncirus fructus*, are responsible for the activation of ghrelin receptors through calcium imaging and whole-cell patch clamp methods (Jang et al. 2013). Consequently, hesperetin and its glycoside hesperidin were demonstrated to stimulate the release of cholecystokinin (CCK) from enteroendocrine STC-1 cells. Additionally, hesperetin significantly and dose dependently stimulated CCK secretion, with a 50% effecctive concentration (EC_{50}) of 0.050 mM compared to the untreated control (Kim et al. 2013). Following this, a similar investigation using STC-1 cells showed that naringenin significantly ($p < 0.05$) stimulated CCK secretion, whereas naringin did not, suggesting the importance naringenin plays in the role of appetite regulation and satiety (Park et al. 2014).

10.3.2 Citrus Flavonoids as Regulators of Energy Expenditure

There are several ways to maintain a body's energy balance, such as expending excess energy that enters the body as food via the basal metabolic rate (BMR), physical exercise, or adaptive thermogenesis. Among these, adaptive thermogenesis regulates energy expenditure in response to environmental temperature and food intake (Westerterp 2004; Yun 2010). Brown adipose tissue (BAT), one of the three types of adipocytes, is critical for controlling energy balance. Energy expenditure is regulated in mammalian BAT through nonshivering thermogenesis that dissipates excess energy as heat (Cannon and Nedergaard 2004), during which uncoupling protein-1 (UCP1) plays a key role.

UCP1 is a mitochondrial protein uniquely expressed in BAT and functions to create a fatty acid-activated uncoupling of respiration (Rousset et al. 2004). UCP1 uncouples substrate oxidation by converting ADP to ATP, thereby leading to the generation of heat and increased energy expenditure. Unlike UCP1, UCP2 is highly expressed in a wide range of adult human tissues, such as cells in the lymphatic system, macrophages, and pancreatic islet cells. It is not involved in adaptive thermogenesis but may, however, contribute to resting energy expenditure in adult humans (Bouchard et al. 1997). Therefore, it is believed that substances able to increase UCP gene expression may effectively prevent obesity through increased energy expenditure (Kumar et al. 1999). Large amounts of natural products and plant extracts have been studied as antiobesity therapies, and some plant-derived flavonoids have been linked to the upregulation of UCPs (Goto et al. 2012; Hasani-Ranjbar et al. 2013; Sergent et al. 2012; Vermaak et al. 2011). However, investigations into the effects of citrus flavonoids on UCPs are limited. The grapefruit flavonoid naringenin has been shown to contribute to the induction of peroxisome proliferator-activated receptor (PPAR)-regulated fatty acid oxidation genes, such as UCP-1 in human hepatocytes (Goldwasser et al. 2010). Additionally, an animal study of dietary naringenin showed decreased plasma triglycerides and adiposity in treated rats, as well as markedly enhanced expression of UCP-2 (Cho et al. 2011). Similarly, nobiletin has been shown to increase mRNA expression of UCP-2 in high-fat diet-induced obese mice (Lee et al. 2013). Collectively, this indicates investigations into citrus flavonoids are extremely meaningful due to the huge potential of flavonoids in antiobesity therapy.

10.3.3 Citrus Flavonoids as Regulators of Lipid Metabolism

Lipid metabolism disorder is one key characteristic of obesity. In general, enhanced lipolysis and reduced lipogenesis are two major approaches in the regulation of fat deposits. Therefore, the enzymes involved in these processes, including acetyl coenzyme A (CoA) carboxylase (ACC), fatty acid synthase (FAS), and hormone-sensitive lipase (HSL) could be selectively targeted in the development of antiobesity drugs. ACC is known to regulate the metabolism of fatty acids. It indirectly inhibits the oxidation of fatty acids through the synthesis of malonyl-CoA, an inhibitor of fatty acid oxidation. FAS, as it is named, functions to catalyze fatty acid synthesis. Additionally, it is well accepted that inhibition of FAS can reduce food intake and body weight. HSL, which is highly expressed in adipose tissue, is an intracellular neutral lipase that hydrolyzes triacylglycerols, diacylglycerols, monoacylglycerols, and cholesteryl esters, as well as other lipid and water-soluble substrates (Kraemer and Shen 2002). It is largely responsible for the mobilization of free fatty acids from adipose triacylglycerol (TAG) stores. It has been demonstrated that HSL is one of the major enzymes that contribute to TAG breakdown in *in vitro* assays and in organ cultures of murine white adipose tissue (WAT) (Schweiger et al. 2006). 5′-AMP-activated protein kinase (AMPK) signaling plays a critical role in sensing energy availability at the cellular level and regulating lipid metabolism (Ix and Sharma 2010). Activation of the AMPK pathway inhibits fatty acid synthesis and increases fatty acid oxidation and lipolysis through regulation of FAS, ACC, and HSL (Figure 10.4).

An early study investigated the lipid-lowering effect of luteolin by using a cell model of steatosis induced by palmitate (Liu et al. 2011). Incubation of HepG2 cells with palmitate markedly increased lipid accumulation, and it was observed that luteolin enhanced AMPKα. In addition, intracellular triglyceride (TG) measurements indicated that the luteolin-mediated reduction of enhanced TG was blocked by pretreatment with the AMPK inhibitor. This study suggested that activation of

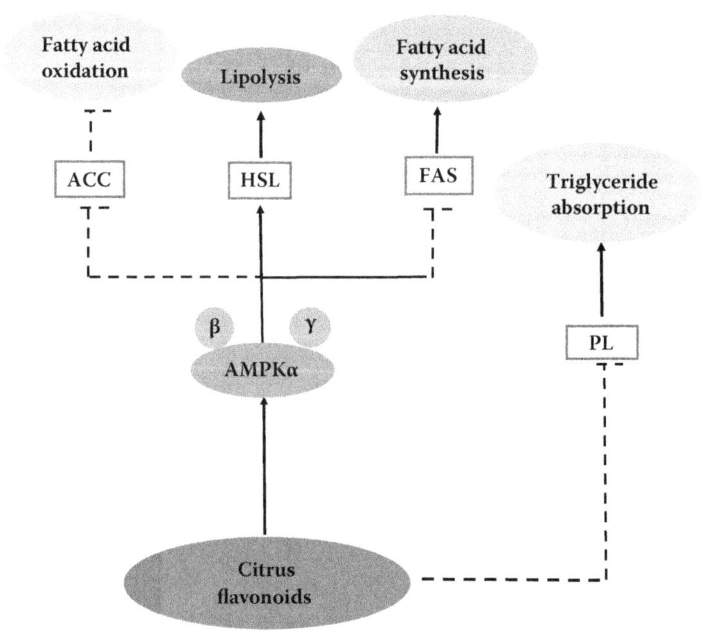

FIGURE 10.4 Possible antiobesity mechanisms of citrus flavonoids as regulators of lipid metabolism.

the AMPK signaling pathway was one possible mechanism of the lipid-lowering effect of luteolin. A subsequent study explored the effects of naringin on metabolic syndrome in mice (Pu et al. 2012). In this study, the high-fat diet-induced mice developed obesity, which was then attenuated by naringin. This lipid-lowering effect was attributed to inhibiting fatty acid synthesis pathways, as well as increasing fatty acid oxidation, both of which depend on AMPK activation. Therefore, it was suggested that naringin fought against metabolic syndrome in high-fat diet-induced mice through an AMPK-dependent mechanism which involved multiple types of intracellular signaling and the reduction of oxidative damage. Additionally, through the AMPK signaling pathway, enzymes such as FAS can also be targeted directly, as a potential treatment for obesity. Citrus flavonoids such as luteolin, quercetin, kaempferol, taxifolin, and hesperetin have previously been shown to have inhibitory effects on animal fatty acid synthase (Li and Tian 2004).

Concurrently, inhibition of dietary triglyceride absorption via inhibition of pancreatic lipase (PL) has also been applied as a potential new approach for the treatment of obesity (Birari and Bhutani 2007) (Figure 10.4). Commonly occurring citrus flavonoids, such as luteolin (Yamamoto et al. 2000) and hesperidin (Han et al. 2000), as well as neohesperidin (Kawaguchi et al. 1997), have all been shown to inhibit PL to different extents. These inhibitory abilities suggest strong potentials as obesity treatments.

10.3.4 Citrus Flavonoids as Regulators of Adipose Tissue

Adipose tissue, which is composed of adipocytes, plays a vital role in energy balance and lipid homeostasis in the human body. Based on the energy balance demand of the body, adipocytes primarily store triglycerides and then, when necessary, release them in the form of monoglycerides and free fatty acids (Yun 2010). As caloric intake increases, adipose tissue grows by both cell size and cell number via adipogenesis. Adipogenesis is a complex process during which preadipocytes and fibroblast-like preadipocytes evolve into mature adipocytes (Ali et al. 2013). This generation of new adipocytes is controlled by several transcription factors regulating preadipocyte proliferation and adipogenesis, such as the CCAAT enhancer binding proteins (C/EBPs) family, PPARγ, and cyclic AMP (cAMP)-responsive element binding protein (CREB). All are responsible for the transactivation of adipocyte genes involved in morphological changes of the cell, lipid metabolism, and synthesis of adipocyte-specific peptides and cytokines during terminal differentiation (Kim and Park 2011).

PPARγ is considered a master regulator of adipocyte differentiation. PPARγ2 activates a lipogenic transcription factor named sterol regulatory element binding protein (SREBP1) (Fajas et al. 1999). SREBP1 subsequently induces PPARγ by stimulating the PPARγ promoter activity and production of an endogenous ligand (Kim et al. 1998b). It can also initiate lipogenesis, lead to an increase of fatty acid uptake, and promote lipid accumulation (Kim et al. 1998a). Increased expression of PPARγ results in terminal differentiation by transcription of downstream genes. In the C/EBP family, C/EBPβ and C/EBPδ are expressed early in adipocyte differentiation to promote C/EBPα and PPARγ activation (Farmer 2006). This gives rise to the induction of adipocyte-specific genes, such as adipocyte protein 2 (aP2), FAS, and lipoprotein lipase (LPL), responsible for adipocyte differentiation (Farmer 2005; Gregoire et al. 1998). CREB also plays a critical role in molecular control of the preadipocyte-adipocyte transition. It is constitutively expressed prior to, as well as during, adipogenesis, and overexpression of a constitutively active CREB in 3T3-L1 preadipocytes is both necessary and sufficient to initiate adipogenesis (Reusch et al. 2000). Therefore, it is believed that modulation of cellular and molecular events in adipogenesis could serve as an effective means to control body weight gain and obesity.

Recently, the effects of citrus aurantium flavonoids (CAF) on the inhibition of adipogenesis and adipocyte differentiation in 3T3-L1 cells was studied (Kim et al. 2012a). The insulin-induced expression of C/EBPβ and PPARγ mRNA and protein was significantly downregulated in a dose-dependent manner following CAF treatment. Meanwhile, CAF dramatically decreased the expression of C/EBPα. This study indicated that CAF downregulated the expression of C/EBPβ and subsequently inhibited the activation of PPARγ and C/EBPα, which led to the suppression of adipogenesis in 3T3-L1 adipocytes. Similarly, it was illustrated in another recent study that nobiletin significantly suppressed the differentiation of 3T3-L1 preadipocytes into adipocytes (Kanda et al. 2012) upon induction with insulin, together with a cAMP elevator such as 3-isobutyl-1-methylxanthine (IBMX), by downregulating the expression of the gene encoding PPARγ2. In addition, nobiletin decreased the phosphorylation of CREB. Interestingly, some contradictory conclusions have been made on nobiletin. In an earlier study, nobiletin was demonstrated to enhance differentiation of 3T3-L1 preadipocytes (Saito et al. 2007). It was also observed that nobiletin increased accumulation of lipid droplets in adipocytes in a dose-dependent manner and increased the expression of genes critical for acquisition of the adipocyte phenotype. The increased expression of C/EBPβ was also demonstrated by nobiletin, as was the activation of CREB. Therefore, further studies into citrus flavonoid effects on obesity prevention through the regulation of adipogenesis are imperative to elucidate any confusion.

In addition to the modulation of adipogenesis, reduction of adipocytes by apoptosis is another potential target for antiobesity treatment. Apoptosis is a normal phenomenon of cell death necessary to maintain homeostasis and can lead to overall fat loss achieved through the reduction of adipose tissue by decreasing the number of adipocytes. It has been shown that apoptosis occurs in mature adipocytes in *in vitro* cell culture models, as well as in *in vivo* studies in rodents and humans (Herold et al. 2013). Moreover, it has been shown that preadipocytes can also undergo apoptotic cell death. Natural products have been illustrated as potential agents in inducing adipocyte apoptosis. Therefore, it is not a surprise that citrus flavonoids have also been studied for their abilities to induce apoptosis and fight obesity (Zhang and Huang 2012). In an early study, the relationship between the influences of flavonoids on cell population growth was investigated (Hsu and Yen 2006). The results indicated that the inhibitory effects of flavonoids, such as naringenin, hesperidin, naringin, and quercetin, on 3T3-L1 preadipocytes reduced the total cell population by 28.3%, 11.1%, 5.6%, and 71.5%, respectively. All four of these flavonoids can be found in *Citrus* species. Furthermore, this cell apoptosis assay showed that quercetin increased the number of apoptotic cells. The study indicated that quercetin

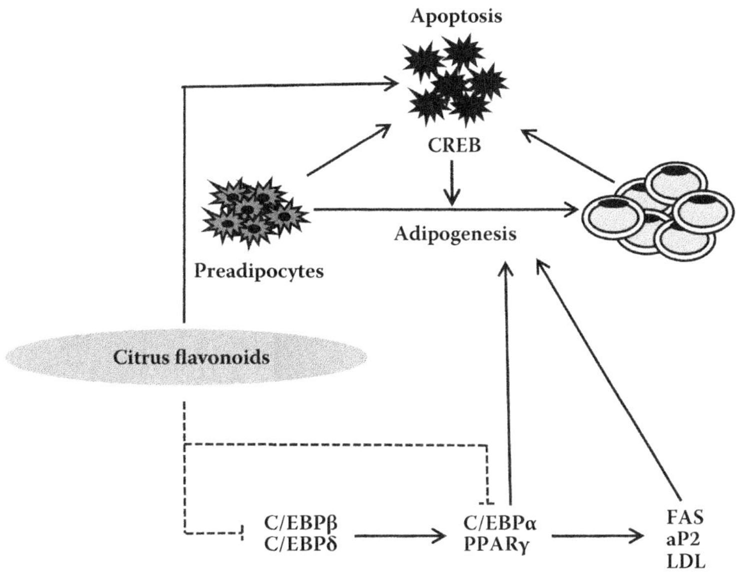

FIGURE 10.5 Possible antiobesity mechanisms of citrus flavonoids as regulators of adipose tissue.

efficiently inhibits cell population growth, as well as inducing apoptosis in 3T3-L1 preadipocytes. Another study investigated the effects of quercetin on adipogenesis and apoptosis in 3T3-L1 cells (Ahn et al. 2008). Here, the exposure of 3T3-L1 preadipocytes to quercetin resulted in attenuated adipogenesis and decreased expression of adipogenesis-related factors and enzymes. At the same time, treatment of 3T3-L1 adipocytes with quercetin resulted in the induction of apoptosis. Although investigations into the regulation of adipose tissue by citrus flavonoids are still in the initial stages, the antiobesity effects already demonstrated by citrus flavonoids suggest enormous potential for further investigation into citrus flavonoids for this purpose (Figure 10.5).

10.3.5 Citrus Flavonoids as Regulators in Other Mechanisms

α-Amylase, the major form of amylase found in humans, is involved in the digestion of starch. Salivary α-amylase is known to be mainly involved in the initiation of the digestion of starch in the oral cavity (Nater and Rohleder 2009), while pancreatic α-amylase is a key enzyme of dietary carbohydrate digestion in humans. Inhibition of α-amylase may be effective in retarding carbohydrate digestion and glucose absorption, which can be considered an additional strategy for the treatment of disorders in carbohydrate uptake, such as diabetes and obesity. In the study of the structural requirements for inhibition of human salivary α-amylase by flavonoids (Lo Piparo et al. 2008), the role of quercetin and luteolin as functional α-amylase inhibitors was confirmed. Similarly, the inhibitory activities of flavonoids quercetin and luteolin have been described to inhibit porcine pancreatic α-amylase (Tadera et al. 2006). Therefore, even with limited evidence and studies, the inhibition of α-amylase can still be considered a potential target for citrus flavonoids in obesity prevention.

10.4 SUMMARY

Over recent decades, the prevalence of obesity has greatly increased in the population of the Western world. Obesity is a chronic disease that should be considered a threat to one's general health, since it is associated with an increased risk for maladies such as diabetes, hypertension, cardiovascular diseases, and even cancers. Currently, limited antiobesity treatments are available on the market, since very few of them meet the desired expectations for both drug effectiveness as well as drug safety. Due to these circumstances, exploration of natural sources and their antiobesity effects has been actively pursued in recent years. Among these natural products, citrus fruits stand out as a strong potential source due to their popularity and abundance of flavonoids. Based on the growing understanding of the origins and mechanisms of obesity, more and more therapeutic targets for antiobesity have been identified, and the application of citrus flavonoids as antiobesity treatments has been largely investigated. Here, antiobesity activities of citrus flavonoids that act upon different therapeutic targets and inhibit obesity were discussed. According to numerous recent epidemiological studies, the possible mechanisms of antiobesity effects related to citrus flavonoids can be summarized as follows: regulation of energy intake and expenditure, regulation of lipid metabolism, and regulation of adipose tissue and inhibition of amylase. Since citrus flavonoids possess the potential to target multiple pathways that lead to obesity, there is a strong possibility that they may be developed into antiobesity agents through further efforts focused on additional research and clinical trials. The structure–activity relationship is also a future focus for such drug development.

REFERENCES

Aherne, S. A., and N. M. O'Brien. 2002. Dietary flavonols: Chemistry, food content, and metabolism. *Nutrition* 18:75–81.

Ahn, J., H. Lee, S. Kim, J. Park, and T. Ha. 2008. The anti-obesity effect of quercetin is mediated by the AMPK and MAPK signaling pathways. *Biochemical and Biophysical Research Communications* 373:545–549.

Ali, A. T., W. E. Hochfeld, R. Myburgh, and M. S. Pepper. 2013. Adipocyte and adipogenesis. *European Journal of Cell Biology* 92:229–236.

Batterham, R. L., M. A. Cowley, C. J. Small, H. Herzog, M. A. Cohen, C. L. Dakin, A. M. Wren, A. E. Brynes, M. J. Low, and M. A. Ghatei. 2002. Gut hormone PYY3-36 physiologically inhibits food intake. *Nature* 418:650–654.

Benavente-García, O., J. Castillo, F. R. Marin, A. Ortuño, and J. A. Del Río. 1997. Uses and properties of citrus flavonoids. *Journal of Agricultural and Food Chemistry* 45:4505–4515.

Benavente-García, O., J. Castillo, F. Sabater, and J. A. Del Río. 1995. Characterization of a *S*-adenosyl-L-methionine:eriodictyol 4′-*O*-methyltransferase from *Citrus aurantium*. *Plant Physiology and Biochemistry* 33:263–271.

Berg, A. H., T. P. Combs, X. Du, M. Brownlee, and P. E. Scherer. 2001. The adipocyte-secreted protein Acrp30 enhances hepatic insulin action. *Nature Medicine* 7:947–953.

Birari, R. B., and K. K. Bhutani. 2007. Pancreatic lipase inhibitors from natural sources: Unexplored potential. *Drug Discovery Today* 12:879–889.

Bombardelli, E., and P. Morazzoni. 1993. The flavonoids: New perspectives in biological activities and therapeutics. *Chimica Oggi* 11:25–28.

Bouchard, C., L. Pérusse, Y. C. Chagnon, C. Warden, and D. Ricquier. 1997. Linkage between markers in the vicinity of the uncoupling protein 2 gene and resting metabolic rate in humans. *Human Molecular Genetics* 6:1887–1889.

Bravo, L. 1998. Polyphenols: Chemistry, dietary sources, metabolism, and nutritional significance. *Nutrition Reviews* 56:317–333.

Burda, S., and W. Oleszek. 2001. Antioxidant and antiradical activities of flavonoids. *Journal of Agricultural and Food Chemistry* 49:2774–2779.

Cannon, B., and J. Nedergaard. 2004. Brown adipose tissue: Function and physiological significance. *Physiological Reviews* 84:277–359.

Chelikani, P. K., A. C. Haver, and R. D. Reidelberger. 2006. Dose-dependent effects of peptide YY (3-36) on conditioned taste aversion in rats. *Peptides* 27:3193–3201.

Cho, K. W., Y. O. Kim, J. E. Andrade, J. R. Burgess, and Y.-C. Kim. 2011. Dietary naringenin increases hepatic peroxisome proliferators-activated receptor α protein expression and decreases plasma triglyceride and adiposity in rats. *European Journal of Nutrition* 50:81–88.

Clifford, M. N. 2000. Anthocyanins: Nature, occurrence and dietary burden. *Journal of the Science of Food and Agriculture* 80:1063–1072.

Cody, V., E. Middleton, and J. B. Harborne. 1986. Plant flavonoids in biology and medicine: Biochemical, pharmacological, and structure-activity relationships. Proceedings of a symposium held in Buffalo, NY, July 22–26, 1985. New York: Liss.

Combs, T. P., A. H. Berg, S. Obici, P. E. Scherer, and L. Rossetti. 2001. Endogenous glucose production is inhibited by the adipose-derived protein Acrp30. *Journal of Clinical Investigation* 108:1875–1881.

Cook, N., and S. Samman. 1996. Flavonoids: Chemistry, metabolism, cardioprotective effects, and dietary sources. *Journal of Nutritional Biochemistry* 7:66–76.

Dixon, R. A., L. Achnine, P. Kota, C. J. Liu, M. Reddy, and L. Wang. 2002. The phenylpropanoid pathway and plant defence: A genomics perspective. *Molecular Plant Pathology* 3:371–390.

Drucker, D. J. 2006. The biology of incretin hormones. *Cell Metabolism* 3:153–165.

Duthie, G. G., S. J. Duthie, and J. A. Kyle. 2000. Plant polyphenols in cancer and heart disease: Implications as nutritional antioxidants. *Nutrition Research Reviews* 13:79–106.

Elangovan, V., N. Sekar, and S. Govindasamy. 1994. Chemopreventive potential of dietary bioflavonoids against 20-methylcholanthrene-induced tumorigenesis. *Cancer Letters* 87:107–113.

Fajas, L., K. Schoonjans, L. Gelman, J. B. Kim, J. Najib, G. Martin, J.-C. Fruchart, M. Briggs, B. M. Spiegelman, and J. Auwerx. 1999. Regulation of peroxisome proliferator-activated receptor γ expression by adipocyte differentiation and determination factor 1/sterol regulatory element binding protein 1: Implications for adipocyte differentiation and metabolism. *Molecular and Cellular Biology* 19:5495–5503.

Farmer, S. R. 2005. Regulation of PPARγ activity during adipogenesis. *International Journal of Obesity* 29(Suppl 1):S13–S16.

Farmer, S. R. 2006. Transcriptional control of adipocyte formation. *Cell Metabolism* 4:263–273.

Flint, A., A. Raben, A. Astrup, and J. J. Holst. 1998. Glucagon-like peptide 1 promotes satiety and suppresses energy intake in humans. *Journal of Clinical Investigation* 101:515–520.

Gattuso, G., D. Barreca, C. Gargiulli, U. Leuzzi, and C. Caristi. 2007. Flavonoid composition of citrus juices. *Molecules* 12:1641–1673.

Golden, P. L., T. J. Maccagnan, and W. M. Pardridge. 1997. Human blood–brain barrier leptin receptor. Binding and endocytosis in isolated human brain microvessels. *Journal of Clinical Investigation* 99:14–18.

Goldwasser, J., P. Y. Cohen, E. Yang, P. Balaguer, M. L. Yarmush, and Y. Nahmias. 2010. Transcriptional regulation of human and rat hepatic lipid metabolism by the grapefruit flavonoid naringenin: Role of PPARα, PPARγ and LXRα. *PLoS One* 5:e12399.

Goto, T., A. Teraminami, J.-Y. Lee, K. Ohyama, K. Funakoshi, Y.-I. Kim, S. Hirai, T. Uemura, R. Yu, and N. Takahashi. 2012. Tiliroside, a glycosidic flavonoid, ameliorates obesity-induced metabolic disorders via activation of adiponectin signaling followed by enhancement of fatty acid oxidation in liver and skeletal muscle in obese–diabetic mice. *Journal of Nutritional Biochemistry* 23:768–776.

Gregoire, F. M., C. M. Smas, and H. S. Sul. 1998. Understanding adipocyte differentiation. *Physiological Reviews* 78:783–809.

Halford, J. C., and J. E. Blundell. 2000. Pharmacology of appetite suppression, in *Progress in Drug Research*, ed. E. Jucker, pp. 25–58. Basel, Switzerland: Springer.

Han, L., Y. Kimura, and H. Okuda. 2005. Anti-obesity effects of natural products. *Studies in Natural Products Chemistry* 30:79–110.

Han, L.-K., B.-J. Xu, Y. Kimura, Y.-N. Zheng, and H. Okuda. 2000. Platycodi radix affects lipid metabolism in mice with high fat diet-induced obesity. *Journal of Nutrition* 130:2760–2764.

Harborne, J. B., B. L. Turner, and J. Harborne. 1984. *Plant Chemosystematics*. London: Academic Press.

Hasani-Ranjbar, S., Z. Jouyandeh, and M. Abdollahi. 2013. A systematic review of anti-obesity medicinal plants: An update. *Journal of Diabetes and Metabolic Disorders* 12:28.

Herold, C., H. O. Rennekampff, and S. Engeli. 2013. Apoptotic pathways in adipose tissue. *Apoptosis* 18:911–916.

Hertog, M. G., E. J. Feskens, D. Kromhout, P. Hollman, and M. Katan. 1993. Dietary antioxidant flavonoids and risk of coronary heart disease: The Zutphen Elderly Study. *Lancet* 342:1007–1011.

Hirano, T., M. Gotoh, and K. Oka. 1994. Natural flavonoids and lignans are potent cytostatic agents against human leukemic HL-60 cells. *Life Sciences* 55:1061–1069.

Horowitz, R., and B. Gentili. 1977. Flavonoid constituents of citrus. *Citrus Science and Technology* 1:397–426.

Hsu, C. L., and G. C. Yen. 2006. Induction of cell apoptosis in 3T3-L1 pre-adipocytes by flavonoids is associated with their antioxidant activity. *Molecular Nutrition and Food research* 50:1072–1079.

Huet, R. 1982. Constituents of citrus fruits with pharmacodnamic effect: Citroflavoncids. *Fruits* 39:267–271.

Ix, J. H., and K. Sharma. 2010. Mechanisms linking obesity, chronic kidney disease, and fatty liver disease: The roles of fetuin-A, adiponectin, and AMPK. *Journal of the American Society of Nephrology* 21:406–412.

Jang, Y., S.-W. Kim, J. Oh, G.-S. Hong, E.-K. Seo, U. Oh, and W.-S. Shim. 2013. Ghrelin receptor is activated by naringin and naringenin, constituents of a prokinetic agent Poncirus fructus. *Journal of Ethnopharmacology* 148:459–465.

Kahn, B. B., and J. S. Flier. 2000. Obesity and insulin resistance. *Journal of Clinical Investigation* 106:473–481.

Kanda, K., K. Nishi, A. Kadota, S. Nishimoto, M.-C. Liu, and T. Sugahara. 2012. Nobiletin suppresses adipocyte differentiation of 3T3-L1 cells by an insulin and IBMX mixture induction. *Biochimica et Biophysica Acta* 1820:461–468.

Kaul, T. N., E. Middleton, and P. L. Ogra. 1985. Antiviral effect of flavonoids on human viruses. *Journal of Medical Virology* 15:71–79.

Kawaguchi, K., T. Mizuno, K. Aida, and K. Uchino. 1997. Hesperidin as an inhibitor of lipases from porcine pancreas and Pseudomonas. *Bioscience Biotechnology and Biochemistry* 61:102–104.

Kim, G.-S., H. J. Park, J.-H. Woo, M.-K. Kim, P.-O. Koh, W. Min, Y.-G. Ko, C.-H. Kim, C.-K. Won, and J.-H. Cho. 2012a. Citrus aurantium flavonoids inhibit adipogenesis through the Akt signaling pathway in 3T3-L1 cells. *BMC Complementary and Alternative Medicine* 12:31.

Kim, H. Y., M. Park, K. Kim, Y. M. Lee, and M. R. Rhyu. 2013. Hesperetin stimulates cholecystokinin secretion in enteroendocrine STC-1 cells. *Biomolecules and Therapeutics* 21:121–125.

Kim, J. B., P. Sarraf, M. Wright, K. M. Yao, E. Mueller, G. Solanes, B. B. Lowell, and B. M. Spiegelman. 1998a. Nutritional and insulin regulation of fatty acid synthetase and leptin gene expression through ADD1/SREBP1. *Journal of Clinical Investigation* 101:1–9.

Kim, J. B., H. M. Wright, M. Wright, and B. M. Spiegelman. 1998b. ADD1/SREBP1 activates PPARγ through the production of endogenous ligand. *Proceedings of the National Academy of Sciences USA* 95:4333–4337.

Kim, K.-H., and Y. Park. 2011. Food components with anti-obesity effect. *Annual Review of Food Science and Technology* 2:237–257.

Kim, M. S., H. J. Hur, D. Y. Kwon, and J.-T. Hwang. 2012b. Tangeretin stimulates glucose uptake via regulation of AMPK signaling pathways in C2C12 myotubes and improves glucose tolerance in high-fat diet-induced obese mice. *Molecular and Cellular Endocrinology* 358:127–134.

Klok, M., S. Jakobsdottir, and M. Drent. 2007. The role of leptin and ghrelin in the regulation of food intake and body weight in humans: A review. *Obesity Reviews* 8:21–34.

Kraemer, F. B., and W.-J. Shen. 2002. Hormone-sensitive lipase control of intracellular tri-(di-) acylglycerol and cholesteryl ester hydrolysis. *Journal of Lipid Research* 43:1585–1594.

Kumar, M. V., G. D. Sunvold, and P. J. Scarpace. 1999. Dietary vitamin A supplementation in rats: Suppression of leptin and induction of UCP1 mRNA. *Journal of Lipid Research* 40:824–829.

Kunimasa, K., S. Kuranuki, N. Matsuura, N. Iwasaki, M. Ikeda, A. Ito, Y. Sashida, Y. Mimaki, M. Yano, and M. Sato. 2009. Identification of nobiletin, a polymethoxyflavonoid, as an enhancer of adiponectin secretion. *Bioorganic and Medicinal Chemistry Letters* 19:2062–2064.

Lee, Y.-S., B.-Y. Cha, S.-S. Choi, B. K. Choi, T. Yonezawa, T. Teruya, K. Nagai, and J.-T. Woo. 2013. Nobiletin improves obesity and insulin resistance in high-fat diet-induced obese mice. *Journal of Nutritional Biochemistry* 24:156–162.

Lee, Y.-S., B.-Y. Cha, K. Saito, H. Yamakawa, S.-S. Choi, K. Yamaguchi, T. Yonezawa, T. Teruya, K. Nagai, and J.-T. Woo. 2010. Nobiletin improves hyperglycemia and insulin resistance in obese diabetic ob/ob mice. *Biochemical Pharmacology* 79:1674–1683.

Li, B. H., and W. X. Tian. 2004. Inhibitory effects of flavonoids on animal fatty acid synthase. *Journal of Biochemistry* 135:85–91.

Li, R. W., A. G. Theriault, K. Au, T. D. Douglas, A. Casaschi, E. M. Kurowska, and R. Mukherjee. 2006. Citrus polymethoxylated flavones improve lipid and glucose homeostasis and modulate adipocytokines in fructose-induced insulin resistant hamsters. *Life Sciences* 79:365–373.

Little, T., M. Horowitz, C. Feinle-Bisset. 2005. Role of cholecystokinin in appetite control and body weight regulation. *Obesity Reviews* 6:297–306.

Liu, J. F., Y. Ma, Y. Wang, Z. Y. Du, J. K. Shen, and H. L. Peng. 2011. Reduction of lipid accumulation in HepG2 cells by luteolin is associated with activation of AMPK and mitigation of oxidative stress. *Phytotherapy Research* 25:588–596.

Liu, L., S. Shan, K. Zhang, Z. Q. Ning, X. P. Lu, and Y. Y. Cheng. 2008. Naringenin and hesperetin, two flavonoids derived from Citrus aurantium up-regulate transcription of adiponectin. *Phytotherapy Research* 22:1400–1403.

Lopes, S., E. Martins, and G. Frare. 2003. Detecção de Candidatus Liberibacter americanus em Murraya paniculata. *Summa Phytopathol* 31:48–49.

Lo Piparo, E., H. Scheib, N. Frei, G. Williamson, M. Grigorov, and C. J. Chou. 2008. Flavonoids for controlling starch digestion: Structural requirements for inhibiting human α-amylase. *Journal of Medicinal Chemistry* 51:3555–3561.

Lyons-Wall, P., and S. Samman. 1997. Flavonoids: Dietary perspectives and health benefits. *Nutrition Society of Australia* 21:106–114.

Maeda, K., K. Okubo, I. Shimomura, T. Funahashi, Y. Matsuzawa, and K. Matsubara. 1996. Paradoxical decrease of an adipose-specific protein, adiponectin, in obesity. *Biochemical and Biophysical Research Communications* 221:286–289.

Mahmoud, A. M., O. M. Ahmed, A. Abdel-Moneim, and M. B. Ashour. 2013. Upregulation of PPARγ mediates the antidiabetic effects of citrus flavonoids in type 2 diabetic rats. *International Journal of Bioassays* 2:756–761.

Manthey, J. A., N. Guthrie, and K. Grohmann. 2001. Biological properties of citrus flavonoids pertaining to cancer and inflammation. *Current Medicinal Chemistry* 8:135–153.

Maxcheix, J., A. Fleuriet, and J. Billot. 1990. The main phenolics of fruits, in *Fruit Phenolics*, pp. 1–98. Boca Raton, FL: CRC Press.

Meister, B. 2000. Control of food intake via leptin receptors in the hypothalamus. *Vitamins and Hormones* 59:265–304.

Middleton, E., C. Kandaswami, and T. C. Theoharides. 2000. The effects of plant flavonoids on mammalian cells: Implications for inflammation, heart disease, and cancer. *Pharmacological Reviews* 52:673–751.

Miyake, Y., K. Yamamoto, Y. Morimitsu, and T. Osawa. 1997. Isolation of C-glucosylflavone from lemon peel and antioxidative activity of flavonoid compounds in lemon fruit. *Journal of Agricultural and Food Chemistry* 45:4619–4623.

Miyata, Y., H. Tanaka, A. Shimada, T. Sato, A. Ito, T. Yamanouchi, and H. Kosano. 2011. Regulation of adipocytokine secretion and adipocyte hypertrophy by polymethoxyflavonoids, nobiletin and tangeretin. *Life Sciences* 88:613–618.

Mizrahi, S., and Z. Berk. 1970. Physico-chemical characteristics of orange juice cloud. *Journal of the Science of Food and Agriculture* 21:250–253.

Moran, T. H. 2000. Cholecystokinin and satiety: Current perspectives. *Nutrition* 16:858–865.

Moro, C., and G. Basile. 2000. Obesity and medicinal plants. *Fitoterapia* 71:S73–S82.

Mouly, P. P., C. R. Arzouyan, E. M. Gaydou, and J. M. Estienne. 1994. Differentiation of citrus juices by factorial discriminant analysis using liquid chromatography of flavanone glycosides. *Journal of Agricultural and Food Chemistry* 42:70–79.

Nater, U., and N. Rohleder. 2009. Salivary alpha-amylase as a non-invasive biomarker for the sympathetic nervous system: Current state of research. *Psychoneuroendocrinology* 34:486–496.

Nogata, Y., K. Sakamoto, H. Shiratsuchi, T. Ishii, M. Yano, and H. Ohta. 2006. Flavonoid composition of fruit tissues of *Citrus* species. *Bioscience Biotechnology and Biochemistry* 70:178–192.

Park, M., K. Kim, Y. M. Lee, M. R. Rhyu, and H. Y. Kim. 2014. Naringenin stimulates cholecystokinin secretion in STC-1 cells. *Nutrition Research and Practice* 8:146–150.

Peterson, J., and J. Dwyer. 1998. Flavonoids: Dietary occurrence and biochemical activity. *Nutrition Research* 18:1995–2018.

Polydera, A., N. Stoforos, and P. Taoukis. 2005. Effect of high hydrostatic pressure treatment on post processing antioxidant activity of fresh navel orange juice. *Food Chemistry* 91:495–503.

Pu, P., D.-M. Gao, S. Mohamed, J. Chen, J. Zhang, X.-Y. Zhou, N.-J. Zhou, J. Xie, and H. Jiang. 2012. Naringin ameliorates metabolic syndrome by activating AMP-activated protein kinase in mice fed a high-fat diet. *Archives of Biochemistry and Biophysics* 581:61–70.

Rao, Y. K., M.-J. Lee, K. Chen, Y.-C. Lee, W.-S. Wu, and Y.-M. Tzeng. 2011. Insulin-mimetic action of rhoifolin and cosmosiin isolated from Citrus grandis (L.) Osbeck leaves: Enhanced adiponectin secretion and insulin receptor phosphorylation in 3T3-L1 cells. *Evidence-Based Complementary and Alternative Medicine* doi:10.1093/ecam/nep204.

Rapisarda, P., F. Fanella, and E. Maccarone. 2000. Reliability of analytical methods for determining anthocyanins in blood orange juices. *Journal of Agricultural and Food Chemistry* 48:2249–2252.

Reusch, J. E., L. A. Colton, and D. J. Klemm. 2000. CREB activation induces adipogenesis in 3T3-L1 cells. *Molecular and Cellular Biology* 20:1008–1020.

Rousset, S., M.-C. Alves-Guerra, J. Mozo, B. Miroux, A.-M. Cassard-Doulcier, F. Bouillaud, and D. Ricquier. 2004. The biology of mitochondrial uncoupling proteins. *Diabetes* 53:S130–S135.

Sahu, A. 2003. Leptin signaling in the hypothalamus: Emphasis on energy homeostasis and leptin resistance. *Frontiers in Neuroendocrinology* 24:225–253.

Saito, T., D. Abe, and K. Sekiya. 2007. Nobiletin enhances differentiation and lipolysis of 3T3-L1 adipocytes. *Biochemical and Biophysical Research Communications* 357:371–376.

Sánchez-Moreno, C., L. Plaza, P. Elez-Martínez, B. De Ancos, O. Martín-Belloso, and M. P. Cano. 2005. Impact of high pressure and pulsed electric fields on bioactive compounds and antioxidant activity of orange juice in comparison with traditional thermal processing. *Journal of Agricultural and Food Chemistry* 563:4403–4409.

Satoh, N., Y. Ogawa, G. Katsuura, T. Tsuji, H. Masuzaki, J. Hiraoka, T. Okazaki, M. Tamaki, M. Hayase, and Y. Yoshimasa. 1997. Pathophysiological significance of the obese gene product, leptin, in ventromedial hypothalamus (VMH)-lesioned rats: Evidence for loss of Its satiety effect in VMH-lesioned rats. *Endocrinology* 138:947–954.

Schwartz, M. W., R. J. Seeley, L. A. Campfield, P. Burn, and D. G. Baskin. 1996. Identification of targets of leptin action in rat hypothalamus. *Journal of Clinical Investigation* 98:1101–1106.

Schwartz, M. W., S. C. Woods, D. Porte, R. J. Seeley, and D. G. Baskin. 2000. Central nervous system control of food intake. *Nature* 404:661–671.

Schweiger, M., R. Schreiber, G. Haemmerle, A. Lass, C. Fledelius, P. Jacobsen, H. Tornqvist, R. Zechner, and R. Zimmermann. 2006. Adipose triglyceride lipase and hormone-sensitive lipase are the major enzymes in adipose tissue triacylglycerol catabolism. *Journal of Biological Chemistry* 281:40236–40241.

Sergent, T., J. Vanderstraeten, J. Winand, P. Beguin, and Y.-J. Schneider. 2012. Phenolic compounds and plant extracts as potential natural anti-obesity substances. *Food Chemistry* 135:68–73.

Tadera, K., Y. Minami, K. Takamatsu, and T. Matsuoka. 2006. Inhibition of α-glucosidase and α-amylase by flavonoids. *Journal of Nutritional Science and Vitaminology* 52:149–153.

Tovar, S., R. Nogueiras, L. Y. Tung, T. R. Castañeda, M. J. Vázquez, A. Morris, L. M. Williams, S. L. Dickson, and C. Diéguez. 2005. Central administration of resistin promotes short-term satiety in rats. *European Journal of Endocrinology* 153:R1–R5.

Tripoli, E., M. La Guardia, S. Giammanco, D. Di Majo, and M. Giammanco. 2007. Citrus flavonoids: Molecular structure, biological activity and nutritional properties. A review. *Food Chemistry* 104:466–479.

Tsuda, T., Y. Ueno, H. Aoki, T. Koda, F. Horio, N. Takahashi, T. Kawada, and T. Osawa. 2004. Anthocyanin enhances adipocytokine secretion and adipocyte-specific gene expression in isolated rat adipocytes. *Biochemical and Biophysical Research Communications* 316:149–157.

Vermaak, I., A. M. Viljoen, and J. H. Hamman. 2011. Natural products in anti-obesity therapy. *Natural Product Reports* 28:1493–1533.

Westerterp, K. R. 2004. Diet induced thermogenesis. *Nutrition and Metabolism* 1:5.

Winkel-Shirley, B. 2001. Flavonoid biosynthesis. A colorful model for genetics, biochemistry, cell biology, and biotechnology. *Plant Physiology* 126:485–493.

Winkel-Shirley, B. 2002. Biosynthesis of flavonoids and effects of stress. *Current Opinion in Plant Biology* 5:218–223.

Yamamoto, M., S. Shimura, Y. Itoh, T. Ohsaka, M. Egawa, and S. Inoue. 2000. Anti-obesity effects of lipase inhibitor CT-II, an extract from edible herbs, Nomame Herba, on rats fed a high-fat diet. *International Journal of Obesity and Related Metabolic Disorders* 24:758–764.

Yamauchi, T., J. Kamon, Y. A. Minokoshi, Y. Ito, H. Waki, S. Uchida, S. Yamashita, M. Noda, S. Kita, and K. Ueki. 2002. Adiponectin stimulates glucose utilization and fatty-acid oxidation by activating AMP-activated protein kinase. *Nature Medicine* 8:1288–1295.

Yao, L. H., Y. Jiang, J. Shi, F. Tomas-Barberan, N. Datta, R. Singanusong, and S. Chen. 2004. Flavonoids in food and their health benefits. *Plant Foods for Human Nutrition* 59:113–122.

Yun, J. W. 2010. Possible anti-obesity therapeutics from nature: A review. *Phytochemistry* 71:1625–1641.

Zhang, Y., and C. Huang. 2012. Targeting adipocyte apoptosis: A novel strategy for obesity therapy. *Biochemical and Biophysical Research Communications* 417:1–4.

Zhang, Y., R. Proenca, M. Maffei, M. Barone, L. Leopold, and J. M. Friedman. 1994. Positional cloning of the mouse obese gene and its human homologue. *Nature* 372:425–432.

11 Dietary Fiber from Citrus and Its Antioxidant Activity

Ma. Ángeles Valdivia-López

CONTENTS

11.1 INTRODUCTION

Dietary fiber (DF) is a mixture of plant carbohydrate polymers: oligosaccharides and polysaccharides (e.g., cellulose, hemicelluloses, β-glucan, pectic substances, gums, resistant starch, inulin, and fructo-oligosaccharides), oligofructose, and the soy bean oligosaccharides raffinose and stachyose. These polymers may be accompanied by lignin and other noncarbohydrate components, for example, polyphenols, waxes, saponins, cutin, phytates, and resistant proteins. Chitosan is an example of a fiber of animal origin, derived from chitin and contained in the exoskeletons of invertebrates, such as the carapaces of crustaceans and insects and also in minor species, such as fungal mycelia. Chitosan is the second most abundant polysaccharide in nature after cellulose. Another definition of dietary fiber indicates focuses on it as a mixture of nondigestible carbohydrates and lignin not absorbed in the human intestine.

Each polysaccharide present in DF is characterized by its sugar residues and the nature of the glycosidic bond. Chemical quantitation of DF in foods is complex, since it primarily depends on the chemical nature of the fiber, degree of polymerization, and presence of oligosaccharides for an accurate determination of its composition.

In recent decades, the role of DF has become important because of the abundant and comprehensive evidence from published reports which have indicated numerous health benefits derived from an increased intake of DF, including reduced risks of coronary heart disease, diabetes, obesity, and some forms of cancer (Mann and Cummings 2009).

Consumption of fiber has additional health benefits due to the presence of compounds with different chemical natures, that is, polyphenols, phenolic acids, carotenoids, and other compounds with antioxidant capacities. Fiber-rich products can fortify foods, increase their DF content, and result in healthy items for consumption in fitness-promoting foods low in calories, cholesterol, and fat. They also may serve as functional ingredients to improve the physical and structural properties of hydration, oil-holding capacity, viscosity, texture, sensory characteristics, and shelf-life. According to current recommendations (Food and Nutrition Board, Institute of Medicine 2001), the average requirement of DF is 25 g/day for women younger than 50 years, 21 g/day for women older than 50 years, 38 g/day for men younger than 50 years, and 30 g/day for men older than 50 years. Most nutritionists and dietitians suggest that about 20%–30% of our DF intake should come from soluble fiber. Consumption of the appropriate fruit or fiber source is essential to achieve these recommendations.

11.2 CITRUS FIBER

Citrus fruits comprise one of the main types of fruit products; they are a food commodity in most of the world as a result of their appealing organoleptic characteristics and their nourishing value. The genus *Citrus*, a genus of flowering plants, belongs to the Rutacea subfamily Aurantaoideae, which originated in tropical and subtropical southeast regions of the world. Sweet orange (*Citrus sinensis* L. Osbeck), bitter orange (*Citrus aurantium* L.), lemon (*Citrus lemon* L. Burm.), mandarin (*Citrus reticulata* Blanco), bergamot (*Citrus bergamia* Risso), grapefruit (*Citrus paradisi* Mcfadyen), acid key lime (*Citrus aurantifolia* Swingle), and acid Persian lime (*Citrus latifolia* Tanaka) are the *Citrus* species used for industrial processing. The common orange (*Citrus sinensis*) and the sour orange (*Citrus aurantium*) are considered hybrids of the pomelo with some mandarin. Limes (*Citrus aurantifolia*) and lemons (*Citrus lemon*) originated through the hybridization of citron with some primitive papeda (a wild species from Asia of no commercial value). The kumquat (*Fortunella* spp.) is a relative of citrus, both belonging to the Rutaceae family.

The citrus industry is of paramount importance in North and South America (with Brazil, the United States, and Mexico the major producers), in the Mediterranean region (with Spain the main producer), and in China (Meléndez-Martínez et al. 2008). Brazil, the United States, and China are the main citrus-producing countries, followed by India, Mexico, Spain, and Italy (Figure 11.1) (Siddiq 2012).

According to Izquierdo and Sendra (2003), production of lemons and limes in 2002 in FAO countries was 11.2 million tons. They also reported that than one-third of harvested citrus fruit was processed to obtain several products, mainly juices.

Among the many sources of DF, citrus by-products have a high potential of use. After juice and essential oil extraction, the residue is mainly constituted of peels, albedo, and flavedo, which are almost one-fourth of the whole fruit mass, seeds, and fruit pulp.

The dry matter (DM) is fundamentally made up of soluble sugars (mainly glucose, fructose, and sucrose), the insoluble carbohydrates cellulose, hemicellulose,

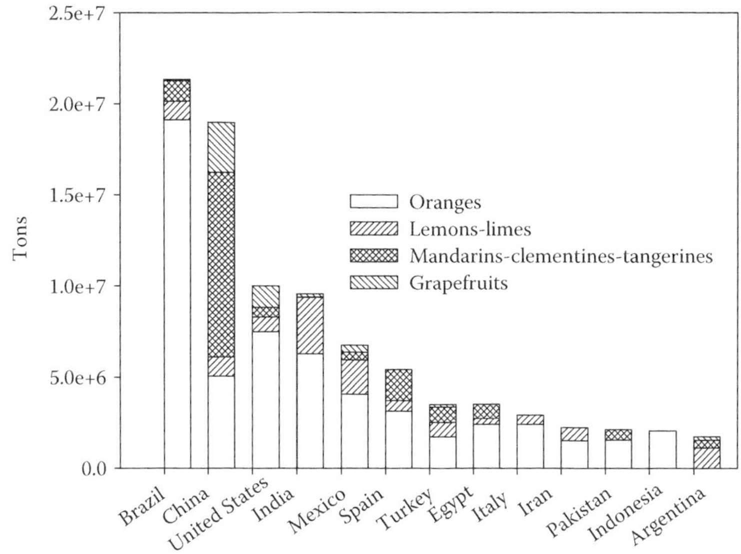

FIGURE 11.1 Citrus production in the world's 10 top citrus-producing countries in 2010. (From Siddiq, M. (ed.), *Tropical and Subtropical Fruits: Postharvest Physiology, Processing, and Packaging*, Wiley-Blackwell, 2012.)

and pectin, organic acids, and a considerable amount of flavonoids (including hesperidin in the majority of *Citrus* species, naringin in grapefruit, and bergamot in sour orange). During ripening of citrus, significant changes occur in the carbohydrates, especially in the pectin fraction. The molecular weight of pectins decreases, and there is a reduction in the degree of methylation. This drop is due to increased activities of polygalacturonases and pectin esterases. Insoluble protopectin is increasingly transformed into soluble forms of this molecule, and protopectin is tightly associated with cellulose in the cell wall matrix (Belitz et al. 2009).

Rezzadori et al. (2012) extensively documented what they referred to as systems for the recovery and use of waste from industrialized processing of oranges. In their proposals, they reviewed the production of ingredients for animal feed, bio-oil and charcoal production, essential oil extraction, pectin production, solid residues as adsorbents, and an integrated process for production of ethanol, biogas, and limonene. Additionally, they discussed the economic and environmental benefits of the proposed systems and the required investments estimated from data in the published literature.

Residues from orange juice extraction are potentially an excellent source of DF, because the residues are rich in pectin and may be available in large quantities (Grigelmo-Miguel and Martın-Belloso 1998). During orange juice production, only approximately half of an orange's fresh weight (FW) is transformed into juice, with the remainder consisting of large amounts of residue, peel, pulp, seeds, orange leaves, and whole orange fruits that do not reach the quality requirements to be processed. The washed pulp has been demonstrated to be a good source of soluble

DF (SDF) and insoluble DF (IDF), and the specific technological functions and functional properties allow it to be used as an ingredient in foods (Rezzadori et al. 2012).

The consumption of DF by humans plays an important role in the prevention of diseases and health problems, such as constipation, hemorrhoids, and hypercholesterolemia. Fruit fibers have good quality due to high total and soluble fiber contents, good functional properties like water and oil-holding capacities, good colonic fermentability, and low caloric content. Citrus fruits have better quality than other sources of DF because of the presence of associated bioactive compounds with antioxidant properties, which may exert greater health-promoting effects than DF itself (Benavente-García et al. 1997).

The presence of functional DF and antioxidants in citrus by-products allows their applications in food processing to obtain healthy products. DF powder produced from citrus residues also possesses good functional properties, including a high water retention capacity and swelling capacity, which are important properties for food formulation (Fernández-López et al. 2004).

In 1999, Larrauri published an article about the production of high-DF powders from fruit by-products and the potential preparation of those fibers in association with bioactive compounds. According to Larrauri, milling and screening have been the main steps in obtaining high-DF powders from cereals, and wet milling, washing, drying, and dry milling are very important in producing fibers from fruits. The author also concluded that the ideal DF should meet the following requirements: have no nutritionally objectionable components, be as concentrated as possible to provide the maximum physiological effect from minimum amounts, be bland in taste, color, texture, and odor, have a balanced composition of insoluble and soluble fractions and adequate amounts of associated bioactive compounds, have a good shelf life that does not adversely affect the food to which it is added, be compatible with food processing, have the right positive image in the eyes of the consumer with regard to source, wholesomeness, and other factors, have the expected physiological effects, and be reasonable in price. In Larrauri's review, he described the procedures to obtain high-DF powders from fruit by-products and the potential preparation of such fibers with associated bioactive compounds (Larrauri 1999).

Figuerola et al. (2005) conducted a study on some of the functional properties of fiber in concentrates of citrus fruit residues from extraction of grapefruit juice extraction (Ruby and Marsh cultivars), lemon (Eureka and Fino 49 cultivars), and orange (Valencia cultivar), in order to use them as potential fiber sources for the enrichment of foods. Most of the concentrates had more than 60% of the recommended total DF (TDF; measured on a dry matter basis). Among grapefruit varieties, Ruby has more TDF, with a higher IDF/SDF ratio; 92.7% of its TDF is insoluble fiber. In the case of lemon varieties, Fino 49 has the highest amount of TDF, and 90.8% of it is represented by insoluble fiber. The DF obtained from Valencia oranges was 64.3 g/100 g, with a 5.3:1 IDF/SDF ratio (Table 11.1). The water retention capacity (WRC) of the citrus concentrates was between 1.65 and 2.26 g/g; the swelling capacity (SWC) of concentrates was between 6.11 mL/g (DM) in Valencia oranges and 9.19 mL/g (DM) in Fino 49 lemons.

The values obtained could also be related to the amount of IDF found in the concentrates. It is known that the structural characteristics, the chemical composition of the fiber, and the water affinity of its components play important roles in the

TABLE 11.1
Dietary Fiber Composition and IDF/SDF Ratios for Various Fiber Sources

Fiber Concentrate	DF Composition (g/100 g DM)			
Source	**IDF**	**SDF**	**TDF**	**IDF/SDF Ratio**
Ruby grapefruit	56.0 ± 0.17	4.57 ± 0.36	62.6 ± 0.30	12.7:1
March grapefruit	37.8 ± 0.21	6.43 ± 0.45	44.2 ± 0.35	5.9:1
Eureka lemon	50.9 ± 0.20	9.20 ± 0.23	60.1 ± 0.22	5.5:1
Fino 49 lemon	62.0 ± 0.16	6.25 ± 0.16	68.3 ± 0.16	9.9:1
Valencia orange	54.0 ± 0.23	10.28 ± 0.30	64.3 ± 0.30	5.3:1

Note: Values are means ± standard deviations.

kinetics of water uptake. Water can be held in capillary structures of the fiber as a result of surface tension strength; in addition, water can interact with molecular components of DF through hydrogen bonding or dipole interactions. The fat adsorption capacity (FAC) depends on surface properties, overall charge density, thickness, and the hydrophobic nature of the fiber particle. Citrus concentrates have FAC values between 1.2 and 1.81 g of oil/g, as expected for fruit residues. The characteristics of the concentrates suggest many potential applications, for example, volume replacement, thickening, or texturizing for the development of foods with reduced calories and that are rich in DF (Figuerola et al. 2005). Also, WRC and FAC suggest some possibilities for the use of fibers as functional ingredients in food products. Similarly, some functional properties of the peel, pulp, and peel fiber from *Citrus hystrix* and *Citrus maxima* (red and white varieties) have been reported as potential DF sources in the enrichment of foods (Abirami et al. 2014).

Lemon (*Citrus limon* cv. Fino) juice industry by-products could be used to obtain high-DF powder with good microbial quality and favorable physicochemical characteristics to be used in food formulations. Processing conditions such as direct drying and washing previous to drying might affect fiber composition and properties. Water-holding capacity is enhanced by washing; it goes from 7 g of water/g of nonwashed fiber powder to 12.6 g of water/g of washed fiber powder. The oil-holding capacity, 6.7 g of oil/g of fiber powder, is not affected by these conditions. Water washing could prevent fiber browning during drying but might rinse away green components. Drying also causes a decrease in bacterial populations; Lario et al. (2004) reported an approximately 90% reduction in microbial counts as a result of drying.

Ubando-Rivera et al. (2005) showed that citrus by-products are a good source of DF as well as phytochemicals. They reported on a comparative study of the content of DF and associated antioxidant activities between Persian and Mexican lime peels. The TDF content of both varieties was high: 70.4% and 66.7%, respectively. The water-holding capacities (WHC) of DF concentrates were also high, 6.96–12.8 g/g. WHC is related to SDF, which is higher in the DF concentrate of Mexican limes. The components of SDF are uronic acids, at 13.2% for Persian limes and 14.3% for Mexican limes, and neutral sugars, at 7%–7.6%. IDF is the predominant fraction. IDF are neutral sugars which account for 26.8%–27.9% of TDF. The content of Klason lignin, 7.45%–7.67%, is similar to the values for other citrus fruits. The lime peel of both varieties has an appropriate ratio, 2.2:1 and 2.7:1, of SDF/IDF; these ratios represent a balanced proportion and indicate that DF in these varieties may confer benefits from the nutritional and health standpoints (Tables 11.2 and 11.3) (Ubando-Rivera et al. 2005).

On the other hand, DF concentrate from Mexican lime defatted seeds (*Citrus aurantifolia*) has 75% TDF. The SDF is 9.49% and IDS is 65.19%. The SDF/IDF ratio is 6.8, which is physiologically inappropriate (Herrera 2006).

Gorinstein et al. (2001) evaluated the contents of DF, total polyphenols, essential phenolics, ascorbic acid, and some trace elements of lemons, oranges, and grapefruits. They did not find significant differences in the contents of total, soluble, and insoluble DF for the studied peeled fruits or their peels. The contents of total, soluble, and insoluble DF in peels were significantly higher than in peeled fruits.

Recent studies have shown that peels of pomelo (*Citrus grandis* L. Osbeck cv. Kao Yai and cv. Kao Nampheung), residues obtained after juice extraction of tangerines (*Citrus reticulata* Blanco cv. Sainamphueng and cv. Bangmod), and peels and residues after juice extraction of kaffir lime (*Citrus hystrix* DC) are good sources of DF, with a TDF content of more than 60 g/100 g on (DW) and with a high proportion of SDF (Chinapongtitiwat et al. 2013).

In summary, the citrus-processing industry generates a significant amount of citrus by-products, mainly during the production of juices and essential oils. These residues can be a source of various products (Table 11.4). DF is not only desirable for its nutritional properties but also for its functional and technological properties.

TABLE 11.2
Dietary Fiber Composition of Powdered Lime Peel

	DF Composition (g/100 g DW)					
	Soluble DF		Insoluble DF			
Sample	Uronic Acid	Neutral Sugars	Uronic Acid	Neutral Sugars	Klason Lignin	Total DF
Mexican lime	14.3 ± 3.89[a]	7.59 ± 1.29[a]	13.1 ± 1.50[a]	27.9 ± 2.91[a]	7.67 ± 0.47[a]	70.4[a]
Persian lime	13.2 ± 3.95[a]	7.06 ± 1.68[a]	12.2 ± 0.91[a]	26.8 ± 1.25[a]	7.45 ± 0.92[a]	66.7[b]

Note: Values in a column with different letters are significantly different ($p \leq 0.05$).

TABLE 11.3
WHC and Swelling of Lime DF Concentrates

Sample	WHC (g H_2O/g DF)	Swelling (mL/g DF)
Mexican lime	12.84 ± 0.54[a]	13.64 ± 0.52[a]
Persian lime	6.96 ± 0.65[b]	11.34 ± 0.34[b]

Note: Values in a column with different letters are significantly different ($p \leq 0.05$).

TABLE 11.4
Some Known Uses and Health Benefits of Citrus Fiber

Form of Citrus	Use	Benefit	References
Citrus fiber	Ingredient in yogurt production and bakery products (muffins) for direct consumption	In bakery products, it can be an alternative for people requiring a low glycemic response	Sendra et al. 2010, Romero-Lopez et al. 2011
	Ingredient in bologna sausage and meat products, as a potential ingredient to reduce nitrite level	Helps to reduction residual nitrite levels and delay the oxidation process	Fernández-Ginés et al. 2003, Viuda-Martos et al. 2009
	Ingredient in preparation of low-calorie foods and for enrichment of foods	Peel, pulp, and peel fiber samples possess potential hypoglycemic effects	Abirami et al. 2014
	Citrus by-products and citrus pulp as feed ingredient (ruminants, ostrich, lamb)	High-energy feed for ruminant rations; in ostrich or lamb diets does not adversely affect meat quality; reduces feed costs	Bampidis et al. 2006, Lanza et al. 2004, Scerra et al. 2001
Citrus flavonoids	Used in formulated nutraceuticals; potential source of antioxidants; antimicrobial activity	Anticancer, cardiovascular, and anti-inflammatory activities; protective factor against free radicals and oxidative processes, antiradical activities, antilipoperoxidation activities, metal chelation; prevention of atherosclerosis; inhibition of tumor cell proliferation/tumor development; antimutagenic effects	Benavente-García et al. 2008, González-Molina et al. 2010

11.3 PHYSICOCHEMICAL PROPERTIES OF CITRUS DIETARY FIBER

The chemical nature of fibers is complex; DFs are constituted of a mixture of chemical compounds. Selection of analytical methods to investigate fibers depends on the composition of each particular fiber (Thebaudin et al. 1997). DF is composed of nondigestible carbohydrates, lignin, and other associated substances of vegetal origin, fibers of animal origin, and modified or synthetic nondigestible carbohydrate polymers. The latter are composed of (1) polysaccharides (cellulose, β-glucan, hemicelluloses, gums, mucilage, pectin, inulin, resistant starch); (2) oligosaccharides (fructo-oligosaccharides, oligofructose, polydextrose, galacto-oligosaccharides); and (3) soybean oligosaccharides (raffinose and stachyose).

DF obtained from different methods and sources, that is, different cultivars, genera, or species, might vary in their chemical composition and physicochemical properties, which subsequently will affect their use as additives in food applications and the physiological responses to them.

Chemical variability among bioactive compounds and its relationship with genetic and climatic factors has been studied by diverse authors (e.g., Dhuique-Mayer et al. 2005). These studies have focused principally on organic acids, sugars, and phenolic compounds of citrus fruit pulp, and it is known that their nature and concentration vary according to species, varieties, and also environmental and horticultural conditions, such as climate, rootstock, and irrigation (Albertini et al. 2006).

Bermejo et al. (2012) studied organic acids, vitamin C, and sugars for 20 different *Citrus* cultivars grown in the Mediterranean climate during the 2009–2010 seasons. They found the major differences in chemical composition could be attributed mainly to genetic factors; in addition, climatic and cultural factors affect fruit quality, and delays in fruit collection generally result in a loss of bioactive compounds. Lee and Kader (2000) highlighted that many pre- and postharvest factors influence the bioactive compound content of horticultural crops (preharvest factors include climatic conditions and cultural practices). Nagy (1980) reported that immature citrus fruits contained the highest concentration of vitamin C, whereas ripe fruits contained the lowest. Although the vitamin C concentration decreases during maturation of citrus fruits, the total vitamin C content per fruit tends to increase, because the total volume of juice and fruit size increase with advancing maturity. These data highlight the variations that can occur in citrus, which ultimately determine their physicochemical properties. However, there is scarce information regarding the changes in DF of citrus fruits in different stages of maturity and under diverse conditions of climate and field.

Viscous DFs include polysaccharides, such as gums, pectins, β-glucans, and other soluble compounds that thicken and swell when they are hydrated and dissolved with liquids, depending on the chemical composition and on the concentration. The technological functionality of fibers depends on their molecular characteristics or rheological properties, such as average particle diameter, storage modulus, loss modulus, and complex viscosity.

The rheological behavior and composition of citrus fiber play an important role in food product engineering, quality control, process control, and process equipment design. The citrus fibers are mostly composed of carbohydrates, which constitute up to 80% of the total composition. The most prevalent polysaccharides in citrus fibers are pectin, at 42.25%, cellulose, at 15.95%, and hemicellulose, at 10.06% (Lundberg et al. 2014). The acidic and therefore charged nature, for example, galacturonic acid, of pectin components is used in many applications due to its apparent viscosity or gelling capacity. Thus, pectin in citrus fiber is probably a contributing factor to its functional properties. Hemicellulose has high apparent viscosity when it is hydrated and has a high WHC because of its branched, generally amorphous, and noncrystalline structure. Although the chemical composition is different from pectin, hemicellulose could also contribute to the citrus fiber's apparent viscosity and WHC. Because citrus fiber solutions are generally non-Newtonian, their viscosity varies with shear rate. The physical size of the citrus fibers has an important impact on apparent viscosity, and the flow models have been used to describe its dependency with shear rate. For instance, a large particle size increases the apparent viscosity and yield stress.

In general, citrus cellulose, hemicelluloses, and lignin are nonviscous, and pectins and gums are viscous. Pectic substances and gums are easily fermented, whereas fermentability of hemicelluloses and celluloses also depends on solubility and crystallinity. This composition is impacted by soil and environmental conditions during growth, maturity at harvest, harvest date, plant parts included, and preparation of plants (Godoy et al. 2013).

Fiber-rich fractions, including SDF and IDF, alcohol-insoluble solids, and water-insoluble solids isolated from the peel of *Citrus sinensis* L. cv. Liucheng have good functional properties. The peel is high in insoluble fiber-rich fractions, 476–515 g/kg of peel, and also contains pectic polysaccharide-rich or SDF, at 94.1 g/kg of peel. Furthermore, insoluble fiber-rich fractions have WHCs of 15.5–16.7 mL/g, oil-holding capacities of 2.35–5.09 g/g, cation-exchange capacities of 454–997 meq/kg, and swelling properties of 14.6–21.1 mL/g. These values are considerably higher than those for cellulose. From these results, consumption of such peel insoluble fiber-rich fractions of desired physicochemical properties as sources of food fibers or low-calorie bulk ingredients can be recommended in food applications requiring oil and moisture retention.

The chemical composition and properties of different citrus fibers have been studied. Wang et al. (2015) reported on the physicochemical and *in vitro* biochemical properties of DF from five types of citrus fruit peels: orange, grapefruit, lemon, gonggan, and Ponkan. No significant difference in the total content of soluble and insoluble DFs in citrus fruit peels was observed. The binding capacities of SDFs of orange, lemon, gonggan, and Ponkan for sodium cholate and cholesterol were significantly lower than those of grapefruit SDF. The higher binding capacities of grapefruit SDF could be attributed to its porous network, which has a high degree of sponge-like structure. All the SDFs of citrus fruit peels exhibited a near-Newtonian fluid flow behavior, and the grapefruit SDF solution had higher viscosity than SDF solutions of other fruits. Regarding the water-holding, oil-holding, and swelling capacities of IDF, those of grapefruit and orange were the highest and lowest, respectively.

Chemical composition and physicochemical properties of DF can vary depending on storage conditions of citrus fruits. Dong et al. (2008) reported changes in DF, polygalacturonase, and cellulase of navel orange fruits (*Citrus sinensis* L. Osbeck "Cara Cara") under different storage conditions, that is, tree and room storage at

6°C and 75%–80% relative humidity. The contents of IDFs, hemicellulose, cellulose, and lignin of tree storage were lower than those with room storage, but a significant increase in SDF and water-soluble pectin occurred, compared with fruits in room storage. The results also indicated that tree storage is quite useful for regulating gene expression and controlling DF levels of oranges.

Residues of citrus juice production are a known source of fiber pectin, cold-pressed oils, essences, δ-limonene juice pulps and pulp wash, ethanol, seed oil, pectin, limonoids, and flavonoids. The main flavonoids found in *Citrus* species are hesperidin, narirutin, naringin, and eriocitrin. Peel and other solid residues of lemon waste mainly contain hesperidin and eriocitrin, with the latter predominating in liquid residues (Coll et al. 1998). Citrus seeds and peels were found to possess high antioxidant activity (Bocco et al. 1998). Although antioxidant compounds are clearly an important factor for health benefits, lemon seeds can be used for other industrial applications. Protein present in lemon seeds (*Citrus aurantifolia* Swingle) with molecular masses of 14–22 kDa has valuable functional properties. These proteins exhibit an ability to form plastic films when they interact with polyols such as sorbitol and glycerol. Such films exhibit superior mechanical properties and permeability to water vapor and oxygen compared to protein from other sources in terms of protein film properties (Alvarado-Suárez 2008).

11.4 ANTIOXIDANT COMPOUNDS OF FIBER

Currently, there is sufficient supported scientific evidence for the importance of antioxidants, as well as their properties and chemical structures. Similarly, there is an increasing interest to identify potential sources for the efficient production and extraction of various compounds possessing antioxidant activity.

Phenolic compounds constitute a large and ubiquitous group of phytochemicals. They are formed to protect plants from photosynthetic stress, ROS and herbivory. Phenolics are characterized as having at least one aromatic hydrocarbon ring attached with one or more hydroxyl groups. The simplest of the class are C_6-C_1 phenolic acids, such as gallic acid. In nature, phenolics are commonly found conjugated to sugars and organic acids and can be grouped as the flavonoids and the nonflavonoids (Crozier et al. 2006a). Among the well-known citrus bioactive compounds, flavonoids, especially the citrus-unique polymethoxyflavones and flavanone glycosides, have attracted considerable attention for their remarkable biological activities (Tripoli et al. 2007).

Among the phytochemicals that have been identified in citrus fruits, the most important are vitamin C, carotenoids, flavonoids, and others phytochemicals present in minor amounts. Citrus flavonoids have gained much interest due to their chemoprotective effects (Rapisarda et al. 1999).

Flavanones such as naringenin-7-*O*-rutinoside (narirutin) and hesperetin-7-*O*-rutinoside (hesperidin) are usually found in the citrus peel and to a lesser extent in the fleshy segments (Figure 11.2). Naringenin-7-*O*-neohesperidoside (naringin) can be found in grapefruit (*Citrus paradisi*) peel (Crozier et al. 2006b). This compound, together with hesperetin-7-*O*-neohesperidoside (neohesperidin) from bitter orange (*Citrus aurantium*), is intensely bitter (Crozier et al. 2006b).

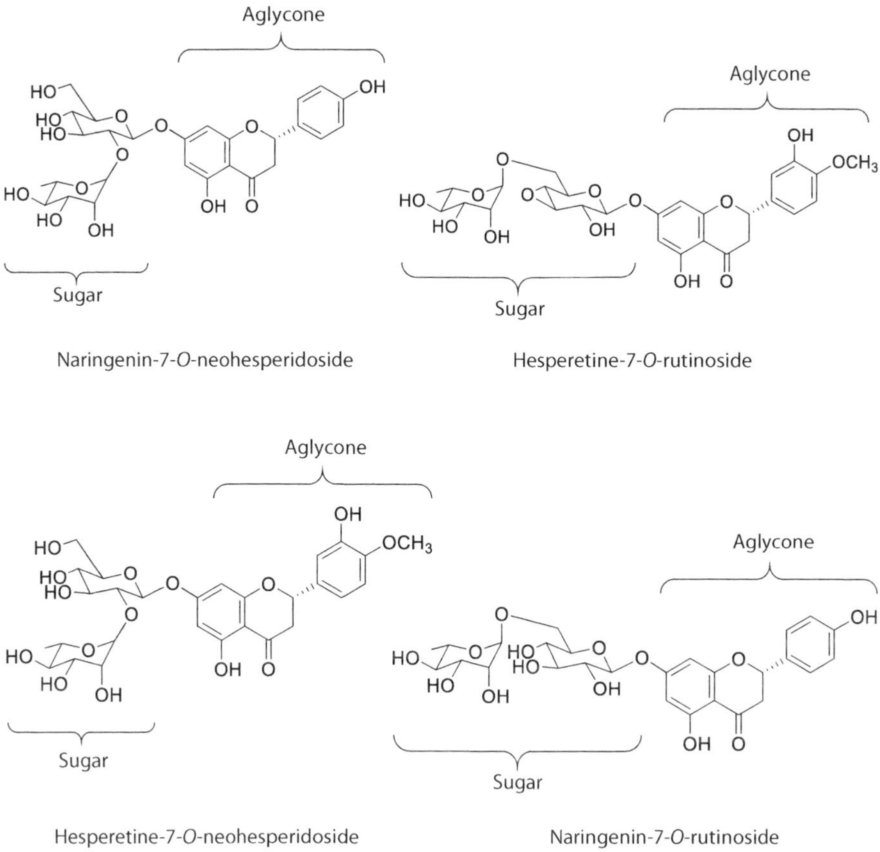

FIGURE 11.2 Structures of common citrus flavonoids.

Many studies have focused on the quantification of phenolic compounds and antioxidant capacities of citrus fruits, such as lime, grapefruit, sweet orange, lemon, and tangerine (Kelebek et al. 2008; Abad-Garcia et al. 2012; Goulas and Manganaris 2012).

Citrus and apple fibers are of higher quality than other DFs because of the presence of associated bioactive compounds, such as flavonoids, polyphenols, and carotenes (Fernández-Gines et al. 2003).

Gorinstein et al. (2001) conducted a comparative study on lemons, oranges, and grapefruits. The total radical-trapping antioxidative potential (TRAP) was assessed. The peeled lemons, oranges, and grapefruits contained 16,410, 15,410, and 13,510 mg/100 g of total phenols and their peels contained 19,011, 17,911, and 15,510 mg/100 g of total polyphenols, respectively. The content of total polyphenols in peeled lemons and their peels was significantly higher than in peeled oranges and grapefruits and their peels, respectively. The content of total polyphenols in the peels was significantly higher than in peeled fruits. The same results were obtained for the essential phenolics: ferulic, sinapic, *p*-coumaric, caffeic, and ascorbic acids. The TRAP was significantly higher in peeled lemons and their peels than in peeled oranges and grapefruits and their peels, respectively. In all three fruits, the TRAP was significantly higher in peels than in peeled fruits.

Ghasemi et al. (2009) reported the antioxidant activity determined by the α,α-diphenyl–picrylhydrazyl (DPPH) method for methanolic extracts of 13 commercially available citrus peels and tissues from Iran. Total phenolic content of the citrus samples, based on the Folin–Ciocalteu method, varied from 66.5 to 396.8 mg of gallic acid equivalent/g of extract, and the flavonoids content, based on the colorimetric $AlCl_3$ method, varied from 0.3 to 31.1 mg quercetin equivalent/g of extract. There was no correlation between the total phenolic and/or flavonoid contents and antioxidant activity in tissues and/or peels.

Ubando-Rivera et al. (2005) reported a comparative study on antioxidant activity in Persian lime (*Citrus latifolia*) and Mexican lime (*Citrus aurantifolia*) peels. The antioxidant activity of total extractable polyphenols (TEP) was studied, using three methods; azino-bis(3-ethylbenzothiazoline-6-sulfonic acid) (ABTS) radical scavenging activity, DPPH, and β-carotene–linoleic acid antioxidant assay. The TEP values in DF concentrates of Mexican and Persian limes were 10.55 and 19.90 mg/g, respectively. DF concentrates of Persian lime peel had greater polyphenol contents than peels of Mexican limes. In tests of ABTS and carotene, the polyphenols associated with the DF in both lime peel varieties showed good antioxidant capacity compared to Trolox, which was used as a control (Table 11.5).

Rodríguez (2004) studied the influence of ripeness of Persian and Mexican limes on the amount of flavanone glycosides. A higher amount was found in unripe lemons. Mexican lime had 3,850 mg/kg of peel of hesperidin and 366 mg/kg of peel of diosmin. The authors suggested that the peel of Mexican lemon is a potential source of functional ingredients, because it could reach up to 49.6 tons annually of hesperidin and 4.7 tons of diosmin from 12,900 tons obtained from Mexico lemon peel.

In an excellent review, M'hiri et al. (2014) discuss methods reported in the literature for the extraction of phenols from citrus peel. The review provided a critical comparison of the different methods for the extraction of phenolic compounds from citrus peel. The authors described methods such as conventional solvent extraction, advanced methods of extraction, for example, supercritical CO_2 extraction, pressurized fluid, ultrasound-assisted, microwave-assisted, and enzyme-assisted extractions. The review compiled valuable data that could be useful for the selection of an appropriate extraction method for bioactive compounds from vegetable sources. The main parameters influencing the extraction yield were also discussed.

TABLE 11.5
Antioxidant Activity, Determined via β-Carotene Bleaching

Sample	% Antioxidant Activity
Mexican lime	88.4 ± 0.32^a
Persian lime	89.1 ± 0.86^b
Trolox	89.5 ± 0.56^b

Note: Values in a column with different letters are significantly different at $p \leq 0.05$.

The same authors reported the effects of different operating conditions on four extraction techniques of antioxidants and their activities. The highest values for the total phenol and flavonoid contents were reached when ultrasound-assisted extraction (UAE) was carried out at 125 W during 30 min at 35°C, microwave-assisted extraction (MAE) at 200 W during 180 s, supercritical CO_2 extraction (SCE) at 80°C, 10 MPa, and high-pressure extraction at 50 MPa during 30 min at 35°C (M'hiri et al. 2015).

Increasing extraction yields of antioxidants from citrus peel is one of the challenges of this operation. Gamma-irradiation has been shown to be an acceptable technique for the retrieval of polyphenols. Kang et al. (2006) studied a process to produce polyphenols extracts from freeze-dried citrus peel powder prepared by extraction of citrus peels with 70% alcohol followed by gamma irradiation. Based on their results, the authors suggested that there may be opportunities to use citrus peel powder as a functional component in the food processing industry, with gamma irradiation to improve its color without a detrimental effect on its functional properties.

11.5 HEALTH BENEFITS OF BIOACTIVE COMPOUNDS OF FIBER

Fiber-rich by-products, abundant in DFs and bioactive compounds, are attractive to food processors, especially because consumers prefer natural supplements, as they are concerned that synthetic ingredients may present a health risk. These citrus components possess many nutritive and protective beneficial effects. Epidemiological studies have stressed that consumption of fruits has health benefits, for example, a reduced risk of coronary heart disease and stroke, as well as certain types of cancer. Apart from DF, these health benefits are mainly attributed to organic micronutrients, such as carotenoids, polyphenolics, tocopherols, vitamin C, and others (Schieber et al. 2001).

The major contribution of citrus antioxidant activity comes from the combination of phytochemicals and their synergistic action with vitamin C. The major phytochemicals in citrus fruits are terpenes and phenolic compounds, which possess anti-inflammatory and anticarcinogenic activities. Carotenoids and limonoids are terpenes released in the processing of juices. Citrus fruits are the main source of specific nutrients such as flavanones, for example, hesperetin and naringenin, usually present as glycosides, and the carotenoid cryptoxanthin, which are not present in other fruits in significant quantities (Codoñer-Franch and Valls-Bellés 2010).

The physiological effects are related to the physicochemical and functional properties of DF. It is well established that DFs obtained by different methods and from different sources, behave differently during their transit through the gastrointestinal tract, depending on their chemical composition and physicochemical characteristics, and also on the processing that a food has undergone (Chau and Huang 2003).

Consumption of citrus peel was associated with significant reduction total serum and high-density lipoprotein cholesterol levels. Citrus peel also exhibited a protective role against diabetic and peroxidative effects in a mouse model (Parmar and Kar 2007).

Huang and Ho (2010) studied seven citrus fruits, including grapefruit (*Citrus paradis*), hutoukan (*Citrus kotokan* Hayata), lemon (*Citrus limon* L. Bur), liucheng

(*C. sinensis* L. Osbeck), Ponkan (*C. reticulata* Blanco), murcott (*C. reticulata*, *C. sinensis*), and Tankan (*Citrus tankan* Hayata). They determined the inhibitory activities of citrus peel extracts on the production of the proinflammatory mediators PGE_2 and NO in LPS-activated RAW 264.7 cells. Among the tested citrus peels, Ponkan (*Citrus reticulata* Blanco) and Tankan (*Citrus tankan* Hayata) deserve special attention due to their outstanding inhibitory effect on PGE_2 and NO secretion. The authors also reported the fruit compositions in terms of flavanone glycosides and polymethoxy flavones. The polymethoxyflavone content, especially that of nobiletin, appeared to correlate well with the anti-inflammatory activities of certain citrus peel extracts.

Citrus fruit pulps represent an important source of phytochemicals with potent antioxidant capacity. Naringin, a flavanone glycoside, is one of the compounds present in significant amounts in citrus fruits (Ramful et al. 2011). A considerable number of published reports have associated therapeutic properties with naringin. Bharti et al. (2014) published a comprehensive review on preclinical evidence for the pharmacological actions of naringin. The authors documented and discussed effects of naringin on antioxidant, anti-inflammatory, atherosclerosis, cardiovascular disorders, diabetic complications, neuroprotection, hepatoprotection, cancer, bone diseases, infections, allergies, and others capacities. They concluded that available information suggests that naringin possesses therapeutic potential for various human disorders; however, further clinical investigation is evidently needed to provide significant insights into the mechanisms underlying the effects of naringin in humans (Bharti et al. 2014).

The citrus fruit yuzu (*Citrus ichangensis* × *C. reticulate*) is an important functional food that possesses several health-promoting properties. Nile and Park (2014) reviewed the composition, nutritional values, and functional properties of yuzu fruit, and their biological activity results were related to the potential impact on human health. Yuzu is a rich source of a wide variety of nonnutritive compounds, such as flavonoids, anthocyanins, phenolic acids, carotenoids, and tannins. Its bioactive compounds have been demonstrated to have numerous functional properties, such as antioxidant, anti-inflammatory, anticancer, antiplatelet, antiangiogenesis, and antimicrobial properties, both *in vitro* and *in vivo*. These diverse applications provided by the yuzu fruit, juice, peel, and seeds and its bioactive compounds are of great industrial importance.

The epicarp of citrus fruits, the flavedo, represents an interesting source of bioactive compounds that show a variety of activities, including as antioxidants, among others. Ramful et al. (2010) reported on the presence of bioactive compounds in various citrus fruits grown in Mauritius: oranges, Satsuma mandarins, clementines, mandarins, tangors, bergamots, lemons, tangelos, kumquats, calamondins, and grapefruits. The flavanone hesperidin was present in high concentrations in all flavedo extracts except those from grapefruit, where it was not detected. The content of hesperidin ranged from 83 ± 0.06 to 234 ± 1.73 mg/g FW. Poncirin, didymin, diosmin, isorhoifolin, and narirutin were present in all extracts, whereas naringin was present in the extract of only one mandarin variety. Most flavedo extracts have good DNA-protecting abilities according to cuphen assay results. Similarly, the total amount of phenolics shows a strong correlation with Trolox equivalent antioxidant capacity (TEAC), FRAP, and hypochlorous acid (HOCl) scavenging activity assays.

Lemons, like most citrus fruits, have been reported to have a high content of antioxidants and polyphenols. Lemon juice also contains high levels of ascorbic acid, which has been considered to facilitate the absorption of iron, hormones, and cell redox processes. Lemon by-products have high levels of flavonoids, particularly flavanones and flavones, and other phytochemicals, such as carotenoids. The flavonoid hesperidin, in particular, has shown beneficial effects for treatment of rheumatoid arthritis. This by-product has potential as a health-enhancing ingredient (González-Molina et al. 2010).

11.6 CITRUS FIBER AS A FUNCTIONAL INGREDIENT

The worldwide increasing concern for public health demands the development of functional foods with multiple health benefits. Currently, an increase in the use of functional ingredients for developing different food matrices has been observed. DF can also convey some functional properties to foods, for example, increased WHC, oil-holding capacity, emulsification, and/or gel formation. It has been demonstrated that DF incorporated into prepared foods, such as bakery products, dairy products, jams, meats, and soups, can modify texture properties, avoid syneresis, that is, the exudation of liquid from a gel caused by network contraction, stabilize high-fat foods and emulsions and improve shelf life.

The presence of fiber chemical components like pectin, lignin, cellulose, and hemicellulose, along with other compounds such as flavonoids, in a final product is more dependent on the industrial process than on the type of citrus. The chemical changes experienced by citrus fiber include loss of functional values, that is, SDF. Ascorbic acid content decreased when original waste products were transformed into fibers (Marin et al. 2007).

Fiber source and the type and degree of processing are the main factors that affect functional properties of high-fiber ingredients: solubility, viscosity, gelation, water-binding and oil-binding capacities, and mineral and organic molecule binding (Nelson 2001), thus affecting the functional quality of the intermediate and end products of manufacturing (Wang et al. 2002).

The use of citrus fiber has been investigated to assess its potential in various foods. Fiber-rich by-products can fortify foods, increase their DF content, and result in healthy products that are low in calories, cholesterol, and fat. An important issue for fiber utilization that provides health benefits and protection to foods is the content of antioxidant compounds present in fiber. They may also serve as functional ingredients to improve physical and structural properties of hydration, oil-holding capacity, viscosity, texture, sensory characteristics, and shelf life.

Meat products can be enriched with an adequate amount of DF by the appropriate selection of fiber sources and method of incorporation. Thus, it is expected that more acceptable novel meat products, with promising health benefits, will be available in the near future (Verma and Banerjee 2010).

Lemon fiber functionality has been studied in meat products. Fernández-Gines et al. (2004) reported that addition of lemon albedo to bologna sausages represented an improvement in their nutritional properties and may have beneficial effects, possibly as a result of the presence of active biocompounds which induced a decrease in residual nitrite levels. The formulations which provide products with sensory properties similar to conventional sausages were those with 2.5 and 5% raw albedo and 2.5, 5.0, and 7.5% cooked albedo. Lemon albedo showed potential as a good source of DF which can be used as a functional ingredient for meat products, like cooked sausages (Fernández-Gines et al. 2004).

Additionally, it has been shown that the use of orange fiber as an ingredient for dry-cured sausages has a protective effect in terms of oxidation and contributes to decrease residual nitrite levels, which could prevent nitrosamine formation. The most important phenolic compound in orange fiber is hesperidin, which has also been detected in sausages to which such fiber has been added. Results suggest that orange fiber is a potential functional ingredient, and further research on its use in other processed meats is required (Fernández-López et al. 2007).

Aleson-Carbonell et al. (2005) reported on the functionality of alternative health ingredients in fresh British-style sausages. Citrus fiber extracts and glucan-rich ingredients were used as extenders in addition to the conventional wheat rusk. Comparisons were established among samples without filler, that is, controls, and samples extended with the ingredients, alone or in combination at 7% levels. Addition of any of those ingredients, alone or in combination, reduced cooking losses and shrinkage and increased lightness in cooked sausages (Aleson-Carbonell et al. 2005).

11.7 CONCLUSIONS

The beneficial effects of food fibers have contributed to the development of a large and potential market for fiber-rich products and ingredients. In recent years, there has been a strong interest to find new sources of DF as a food ingredient for the food industry. Most of these materials obtained from citrus by-products could be used as functional ingredients for the development of healthy foods as a result of the properties of DF and bioactive compounds. A comprehensive study of the active compounds as well as their structures and physiological effects is essential for the design and development of new foods. Similarly, the design of processes or experimental strategies to extract and purify antioxidants in high yield to use them as nutraceuticals or in pharmaceutical formulations is a broad field of research.

Deeper knowledge on the relation between DF structure and functionality will help to understand complex fiber–food systems and to adequately exploit fiber potential applications to innovate and add valuable ingredients to foods. Knowing more about the rheological properties of citrus fiber will allow manufacturers to have more knowledge to design and manufacture food products that contain citrus fiber ingredients.

Improving this understanding will not only expand the ability to use fiber in more applications, but also it will help to create healthier foods.

REFERENCES

Abad-García, B., S. Garmón-Lobato, L. A. Berrueta, B. Gallon, and F. Vicente. 2012. On line characterization of 58 phenolic compounds in citrus fruit juices from Spanish cultivars by high-performance liquid chromatography with photodiode-array detection coupled to electrospray ionization triple quadrupole mass spectrometry. *Talanta* 99:213–224.

Abirami, A., G. Nagarani, and P. Siddhuraju. 2014. Measurement of functional properties and health promoting aspects-glucose retardation index of peel, pulp and peel fiber from *Citrus hystrix* and *Citrus maxima*. *Bioactive Carbohydrates and Dietary Fiber* 4(1):16–26.

Albertini, M. V., E. Carcouet, O. Pailly, C. Gambotti, F. Luro, and L. Berti. 2006. Changes in organic acids and sugars during early stages of development of acidic and acidless citrus fruit. *Journal of Agricultural and Food Chemistry* 54(21):8335–8339.

Aleson-Carbonell, L., J. Fernández-López, J. A. Pérez-Alvarez, and V. Kuri. 2005. Functional and sensory effects of fiber-rich ingredients on breakfast fresh sausages manufacture. *Food Science and Technology International* 11(2):89–97.

Alvarado-Suárez, L. A. 2008. Estudio para evaluar biopelículas proteínicas como barreras protectoras de alimentos. B.Sc. Thesis. Facultad de Química, UNAM, Mexico.

Bampidis, V., and P. H. Robinson. 2006. Citrus by-products as ruminant feeds: A review. *Animal Feed Science and Technology* 128(3–4):175–217.

Belitz, H. D., W. Grosch, and P. Schieberle. 2009. *Food Chemistry*, 4th rev., ext. ed., Berlin: Springer-Verlag.

Benavente-García, O., and J. Castillo. 2008. Update on uses and properties of citrus flavonoids: New findings in anticancer, cardiovascular, and anti-inflammatory activity. *Journal of Agricultural and Food Chemistry* 56(15):6185–6205.

Benavente-García, O., J. Castillo, and J. A. Del Río. 1997. Uses and properties of citrus flavonoids. *Journal of Agricultural and Food Chemistry* 45:4505–4515.

Bermejo, A., and A. Cano. 2012. Analysis of nutritional constituents in twenty *Citrus* cultivars from the Mediterranean area at different stages of ripening. *Food and Nutrition Sciences* 3:639–650.

Bharti, S., N. Rani, B. Krishnamurthy, and D. Arya. 2014. Preclinical evidence for the pharmacological actions of naringin: A review. *Planta Medica* 80:437–451.

Bocco, A., M. E. Cuvelier, H. Richard, and C. Berset. 1998. Antioxidant activity and phenolic composition of citrus peel and seed extracts. *Journal of Agricultural and Food Chemistry* 46:2123–2129.

Chau, C., and Y. Huang. 2003. Comparison of the chemical composition and physicochemical properties of different fibers prepared from the peel of *Citrus sinensis* L. cv. Liucheng. *Journal of Agricultural and Food Chemistry* 51(9):2615–2618.

Chinapongtitiwat, V., S. Jongaroontaprangsee, N. Chiewchan, and S. Devahastin. 2013. Important flavonoids and limonin in selected Thai citrus residues. *Journal of Functional Foods* 5:1151–1158.

Codoñer-Franch, P., and V. Valls-Bellés. 2010. Citrus as functional foods. *Current Topics in Nutraceutical Research* 8(4):173–184.

Coll, M. D., L. Coll, J. Laencine, and F. A. Tomas-Barberan. 1998. Recovery of flavanons from wastes of industrially processed lemons. *Zeitschrift für Lebensmittel-Untersuchung und-Forschung A* 206:404–407.

Crozier, A., I. B. Jaganath, and M. N. Clifford. 2006a. Phenols, polyphenols and tannins: An overview, in *Plant Secondary Metabolites and the Human Diet*, eds. A. Crozier, H. Ashihara, and M. N. Clifford, pp. 1–31. Oxford: Blackwell.

Crozier, A., T. Yokota, I. B. Jaganath, S. C. Marks, M. Saltmarsh, and M. N. Clifford. 2006b. Secondary metabolites in fruits, vegetables, beverages and other plant-based dietary components, in *Plant Secondary Metabolites and the Human Diet*, eds. A. Crozier, H. Ashihara, and M. N. Clifford, pp. 208–302. Oxford: Blackwell.

Dhuique-Mayer, C., C. Caris-Veyrat, P. Ollistrault, F. Curck, and M. J. Amiot. 2005. Varietal and interespecific influence on micronutrients contents in citrus from the Mediterranean area. *Journal of Agricultural and Food Chemistry* 53(6):2140–2145.

Dong, T., R. Xia, M. Wang, Z. Xiao, and P. Liu. 2008. Changes in dietary fiber, polygalacturonase, cellulase of navel orange (*Citrus sinensis* (L.) Osbeck "Cara Cara") fruits under different storage conditions. *Scientia Horticulturae* 116(4):414–420.

Fernández-Gines, J. M., J. Fernández-López, A. E. Sayas-Barber, and J. A. Pérez-Alvarez. 2003. Effects of storage conditions on quality characteristics of bologna sausages made with citrus fiber. *Journal of Food Science* 68(2):710–715.

Fernández-Gines, J. M., J. Fernández-López, E. Sayas-Barbera, E. Sendra, and J. A. Pérez-Alvarez. 2004. Lemon albedo as a new source of dietary fiber: Application to bologna sausages. *Meat Science* 67:7–13.

Fernández-López, J., J. M. Fernández-Ginés, L. Aleson-Carbonell, E. Sendra, E. Sayas-Barbera, and J. A. Pérez-Alvarez. 2004. Application of functional citrus by-products to meat products. *Trends in Food Science and Technology* 15:176–185.

Fernández-López, J., M. Viuda-Martos, E. Sendra, E. Sayas-Barberá, C. Navarro, and J. A. Pérez-Alvarez. 2007. Orange fiber as potential functional ingredient for dry-cured sausages. *European Food Research and Technology* 226:1–6.

Figuerola, F., M. L. Hurtado, A. M. Estevez, L. Chiffelle, and F. Asenjo. 2005. Fiber concentrates from apple pomace and citrus peel as potential fiber sources for food enrichment. *Food Chemistry* 91:395–401.

Food and Nutrition Board, Institute of Medicine. 2001. *Dietary Reference Intakes for Vitamin A, Vitamin K, Arsenic, Boron, Chromium, Copper, Iodine, Iron, Manganese, Molybdenum, Nickel, Silicon, Vanadium, and Zinc*. Washington, DC: National Academies Press.

Ghasemi, K., Y. Ghasemi, and M. A. Ebrahimzadeh. 2009. Antioxidant activity, phenol and flavonoid contents of 13 *Citrus* species peels and tissues. *Pakistan Journal of Pharmaceutical Sciences* 22(3):277–281.

Godoy, M. R. C, K. R. Kerr, and G. C. Fahey, Jr. 2013. Alternative fiber sources in companion animal nutrition. *Nutrients* 5(8):3099–3117.

González-Molina, E., R. Domínguez-Perles, D. A. Moreno, and C. García-Viguera. 2010. Natural bioactive compounds of citrus lemon for food and health. *Journal of Pharmaceutical and Biomedical Analysis* 51(2):327–345.

Gorinstein, S., O. Martın-Belloso, Y. Park, R. Haruenkit, A. Lojek, M. Ciz, A. Caspi, L. Libman, and S. Trakhtenberg. 2001. Comparison of some biochemical characteristics of different citrus fruits. *Food Chemistry* 74:309–315.

Goulas, V., and G. A. Manganaris. 2012. Exploring the phytochemical content and the antioxidant potential of citrus fruits grown in Cyprus. *Food Chemistry* 131(1):39–47.

Grigelmo-Miguel, N., and O. Martın-Belloso. 1998. Characterization of dietary fiber from orange juice extraction. *Food Research International* 31(5):355–361.

Herrera-Hernández, M. E. 2006. Semilla de limón mexicano (*Citrus aurantifolia*): Obtención de concentrado de fibra dietética con poder antioxidante. B.Sc. thesis, Facultad de Química, UNAM, Mexico.

Huang, Y., and S. Ho. 2010. Polymethoxy flavones are responsible for the anti-inflammatory activity. *Food Chemistry* 119:868–873.

Izquierdo, L., and J. M. Sendra. 2003. Citrus fruits composition and characterization. *In Encyclopedia of Food Sciences and Nutrition*, eds. B. Caballero, L. Trugo, and P. Finglas. Oxford: Academic Press.

Kang, H. J., S. P. Chawla, C. Jo, J. H. Kwon, and M. Byun. 2006. Studies on the development of functional powder from citrus peel. *Bioresource Technology* 97:614–620.

Kelebek, H., A. Canbas, and S. Selli. 2008. Determination of phenolic composition and antioxidant capacity of blood orange juices obtained from cvs. Moro and Sanguinello (*Citrus sinensis* (L.) Osbeck) grown in Turkey. *Food Chemistry* 107(4):1710–1716.

Lanza, M., V. Fasone, V. Galofaro, D. Barbagallo, M. Bella, and P. Pennisi. 2004. Citrus pulp as an ingredient in ostrich diet: Effects on meat quality. *Meat Science* 68(2):269–275.

Lario, Y., E. Sendra, J. Garcıa-Perez, C. Fuentes, E. Sayas-Barbera, J. Fernández-López, and J. A. Perez-Alvarez. 2004. Preparation of high dietary fiber powder from lemon juice by-products. *Innovative Food Science and Emerging Technologies* 5(1):113–117.

Larrauri, J. A. 1999. New approaches in the preparation of high dietary fibre powders from fruit by-products. *Trends in Food Science and Technology* 10:3–8.

Lee, S. K., and A. A. Kader. 2000. Preharvest and postharvest factors influencing vitamin C content of horticultural crops. *Postharvest Biology and Technology* 20(2000):207–220.

Lundberg, B., X. Pan, A. White, H. Chau, and A. Hotchkiss. 2014. Rheology and composition of citrus fiber. *Journal of Food Engineering* 125(1):97–104.

M'hiri, N., I. Ioannou, M. Ghoul, and N. Mihoubi Boudhrioua. 2014. Extraction methods of citrus peel phenolic compounds. *Food Reviews International* 30(4):265–290.

M'hiri, N., I. Ioannou, M. Ghoul, and N. Mihoubi Boudhrioua. 2015. Effect of different operating conditions on the extraction of phenolic compounds in orange peel. *Food and Bioproducts Processing* 96:161–170.

Mann, J. I., and J. H. Cummings. 2009. Possible implications for health of the different definitions of dietary fiber. *Nutrition Metabolism and Cardiovascular Diseases* 19:226–229.

Marín, F. R., C. Soler-Rivas, O. Benavente-García, J. Castillo, and J. Pérez-Alvarez. 2007. By-products from different citrus processes as a source of customized functional fibers. *Food Chemistry* 100:736–741.

Meléndez-Martínez, A. J., G. Britton, I. M. Vicario, and F. J. Heredia. 2008. The complex carotenoid pattern of orange juices from concentrate. *Food Chemistry* 109:546–553.

Nagy, S. 1980. Vitamin C contents of citrus fruit and their products: A review. *Journal of Agricultural and Food Chemistry* 28(1):8–18.

Nelson, A. L. 2001. Properties of high-fiber ingredients. *Cereal Foods World* 46(3):93–97.

Nile, S. H., and S. W. Park. 2014. Bioactive components and health-promoting properties of yuzu (*Citrus ichangensis* × *C. reticulata*). *Food Reviews International* 30(2):155–167.

Parmar, H. S., and A. Kar. 2007. Antidiabetic potential of *Citrus sinensis* and *Punica granatum* peel extracts in alloxan-treated male mice. *BioFactors* 31(1):17–24.

Ramful, D., T. Bahorun, E. Bourdon, E. Tarnus, and O. I. Aruoma. 2010. Bioactive phenolics and antioxidant propensity of flavedo extracts of Mauritian citrus fruits: Potential prophylactic ingredients for functional foods application. *Toxicology* 278:75–87.

Ramful, D., E. Tarnus, O. I. Aruoma, E. Bourdon, and T. Bahorun. 2011. Polyphenol composition, vitamin C content and antioxidant capacity of Mauritian citrus fruit pulps. *Food Research International* 44(7):2088–2099.

Rapisarda, P., A. Tomaino, R. Lo Cascio, F. Bonina, A. De Pasquale, and A. Saija. 1999. Antioxidant effectiveness as influenced by phenolic content of fresh orange juices. *Journal of Agricultural and Food Chemistry* 47(11):4718–4723.

Rezzadori, K., S. Benedetti, and E. R. Amante. 2012. Proposals for the residues recovery: Orange waste as raw material for new products. *Food and Bioproducts Processing* 90:606–614.

Rodríguez-Barba, C. E. 2004. Desarrollo de un esquema de extracción de Flavonas Glucosiladas a partir de Limón mexicano (*Citrus aurantifolia*). B.Sc. thesis, Facultad de Quimica, UNAM, Mexico.

Romero-Lopez, M. R., P. Osorio-Diaz, L. A. Bello-Perez, J. Tovar, and A. Bernardino-Nicanor. 2011. Fiber concentrate from orange (*Citrus sinensis* L.) bagase: Characterization and application as bakery product ingredient. *International Journal of Molecular Sciences* 12:2174–2186.

Scerra, V., P. Caparra, F. Foti, M. Lanza, and A. Priolo. 2001. Citrus pulp and wheat straw silage as an ingredient in lamb diets: Effects on growth and carcass and meat quality. *Small Ruminant Research* 40(1):51–56.

Schieber, A., F. C. Stintzing, and R. Carle. 2001. By-products of plant food processing as a source of functional compounds—Recent developments. *Trends in Food Science and Technology* 12:401–413.

Sendra, E., V. Kuri, J. Fernández-López, E. Sayas-Barberá, C. Navarro, J. Pérez-Alvarez. 2010. Viscoelastic properties of orange fiber enriched yogurt as a function of fiber dose, size and thermal treatment. *Food Science and Technology* 43(4):708–714.

Siddiq, M. (ed.). 2012. *Tropical and Subtropical Fruits: Postharvest Physiology, Processing and Packaging*. Hoboken, NJ: Wiley-Blackwell.

Thebaudin, J. Y., A. C. Lefebvre, M. Harrington, and C. M. Bourgeois. 1997. Dietary fibres: Nutritional and technological interest. *Trends in Food Science and Technology* 8(2):41–48.

Tripoli, E., M. La Guardia, S. Giammanco, D. Di Majo, and M. Giammanco. 2007. Citrus flavonoids: Molecular structure, biological activity and nutritional properties. A review. *Food Chemistry* 104:466–479.

Ubando-Rivera, J., A. Navarro-Ocaña, and M. A. Valdivia-López. 2005. Mexican lime peel: Comparative study on contents of dietary fiber and associated antioxidant activity. *Food Chemistry* 89:57–61.

Verma, A. K., and R. Banerjee. 2010. Dietary fiber as functional ingredient in meat products: A novel approach for healthy living. A review. *Journal of Food Science and Technology* 47(3):247–257.

Viuda-Martos, M., J. Fernández-López, E. Sayas-Barbera, E. Sendra, C. Navarro, and J. A. Pérez-Alvarez. 2009. Citrus co-products as technological strategy to reduce residual nitrite content in meat products. *Journal of Food Science* 74(8):R93–R100.

Wang, J., C. M. Rosell, and C. Benedito de Barber. 2002. Effect of the addition of different fibers on wheat dough performance and bread quality. *Food Chemistry* 79:221–226.

Wang, L., H. Xu, F. Yuan, Q. Pan, R. Fan, and Y. Gao. 2015. Physicochemical characterization of five types of citrus dietary fibers. *Biocatalysis and Agricultural Biotechnology* 4(2):250–258.

12 Antioxidative and Anti-Inflammatory Activities of Citrus Peel Extracts

Hyun-Dong Paik and Kee-Tae Kim

CONTENTS

12.1 INTRODUCTION

In the food industry, citrus fruits are mainly used in beverages such as juice and other citrus-based soft drinks. Citrus fruits constitute the largest sector of the world's fruit industry, with annual production estimated at greater than 100 million tons, a value that has increased continuously in the last decade (Cheigh et al. 2012; Li et al. 2006a).

During industrial processing of citrus fruit, large amounts of residues, including seeds, peels, rag, and pulp, are produced; the juice comprises only approximately 34% of total citrus fruit weight (Li et al. 2006b). However, these wastes are known to have economical uses, such as animal feed, dietary fiber, and fuel production (Bocco et al. 1998; Li et al. 2006b). Furthermore, citrus peels produced during citrus fruit processing have been used in folk medicine in Asia for the management of various diseases. For example, in the case of China, various *Citrus* species have been cultivated for over 1700 years and, recently, citrus ingredients are generally recognized as Traditional Chinese Medicines (TCMs) as well as food materials (Lu et al. 2006). For example, orange (Daidai) flower, which is used as a tea ingredient, the immature or mature whole fruit of *Fructus aurantii* are used as major ingredients to remedy flatulence and menopausal symptoms. The peel of mature *Citrus* (pericarpium citri reticulatae), which is a culinary seasoning, is used to reduce phlegm in the lungs. The peel of immature fruits of pericarpium citri reticulatae viride and dried fruit of *Citrus medica* L. var. Sarcodactylis (Noot.) Swingle are also used for gut and digestive disorders. In addition, chenpi, which is dried fruit peels of *Citrus reticulate*, has been widely used as a remedy to treat indigestion and to reduce respiratory inflammation condition, such as bronchitis and asthma, and improve cardiac and blood circulation as a traditional Chinese and Japanese medicine (Ho and Lin 2008). Zhi qiao and zhi shi are obtained from the mature and immature fruit peels of *Citrus aurantium* L., respectively, and also are used to remedy indigestion; furthermore, the fruit peel of *C. aurantium* has been used recently to cure dyspepsia and related conditions in Europe (Choi et al. 2007a).

Citrus peels contain various bioactive compounds, such as phenolic acids, flavonoids, limonoids, and fibers, as potential sources of biofunctional components (Schieber et al. 2001). Many researchers have reported that citrus peels contain more bioactive compounds than do fruit juices (de Carvalho et al. 2015).

Furthermore, citrus fruits have relatively higher concentrations of flavanones, flavanone glycosides, flavonol glycoside (such as rutin), and polymethoxylated flavones (such as tangeretin and nobiletin), which are very rarely found in other plant products, and the levels hydroxycinnamic acids of (a group of phenolic acids) are much higher in the peel than in the juice (Manthey and Grohmann 2001).

Citrus fruit peels contain large amounts of phenolic compounds, particularly phenolic acids and flavonoids (Brito et al. 2014). The major flavonoids in citrus are the flavanones hesperidin, naringin, narirutin, and neohesperidin (Benavente-García and Castillo 2008; Tripoli et al. 2007). Generally, the content and distribution of flavonoids in citrus vary, depending on the species and environmental factors (Nogata et al. 2006). Flavonoids have antioxidant, anticancer, antimicrobial, and anti-inflammatory effects (Arif et al. 2009; Benavente-García and Castillo 2008). Specifically, naringin and hesperidin have antioxidative and anti-inflammatory effects (Berkarda et al. 1998; Gabor 1998). Polymethoxylated flavones also have been studied for their pharmacological potential, including anti-inflammatory, anticanceric, and anitumor agents (Kawaii et al. 1999). Nobiletin is one of the polymethoxy flavones (5,6,7,8,3′,4′-hexamethoxy flavone), and it has anticancer, antiviral, and anti-inflammatory activities (Whitman et al. 2005). Recently, the demand for nobiletin in large quantities and high purity has increased because of its practical application as a food biofunctional or medicinal material. Rutin in citrus fruit peels has also been reported to have significant anti-inflammatory properties (Lee et al. 2015).

Figure 12.1 shows the structures of hesperidin, naringin, and nobiletin, which are the major flavonoids in citrus peels. Hesperidin and naringin are glucosides of hesperetin and naringenin, respectively.

For qualitative and quantitative analysis of these compounds, many researchers have used traditionally methods, including thin-layer chromatography and high-performance liquid chromatography, but recently newer techniques, such as the direct infusion electrospray ionization mass spectrometry method (Bagatela et al. 2015), the high-performance thin-layer chromatography method (Alam et al. 2014), and high-performance thin-layer chromatography coupled with desorption

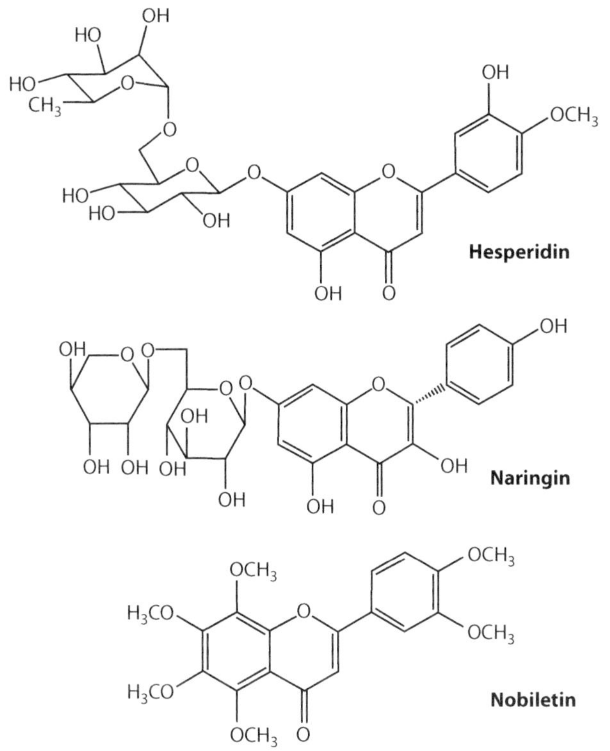

FIGURE 12.1 Structures of major flavonoids in extracts of citrus peels.

electrospray ionization mass spectrometry (Bagatela et al. 2015) have been developed and applied frequently.

A possible way to increase the industrial value of by-products (citrus peels and seeds) from fruit juice processing is to use them as functional food materials, such as natural antioxidative or anti-inflammatory agents. These compounds could have important physiological or ecological roles that would lead to increased interest in their commercialization by the food and pharmaceutical industries (Choi 2007b). Research on natural products and compounds derived from natural products has accelerated in recent years due to their importance in drug discovery.

12.2 EXTRACTION OF BIOACTIVE COMPOUNDS FROM CITRUS PEELS

12.2.1 Composition of Phytochemicals in Citrus Peels

Compositions of bioactive compounds vary according to the particular *Citrus* cultivar. Tables 12.1 and 12.2 show the compositions of major bioactive compounds (such as flavonoids, carotenoids, and polyphenolic acid) in peel extracts of representative

TABLE 12.1
Contents of Major Flavonoids in Peel Extracts of Representative *Citrus* Cultivars

	Flavonoid Content (mg/g Extract)				
Cultivar	**Narigin**	**Hesperidin**	**Rutin**	**Nobiletin**	**Reference**
C. reticulate Blanco	0.54	29.50	0.29	NT	Wang et al. 2008
C. tankan Hayata	0.58	23.40	0.26	NT	
C. grandis Osbeck (Taiwan)	23.90	0.32	0.18	NT	
C. grandis Osbeck cv.	29.80	0.34	0.17	NT	
C. sinensis L. Osbeck	0.36	20.70	0.23	NT	
C. limon L. Bur	1.51	9.42	0.29	NT	
C. unshi Marc	ND	4.47	2.27	1.21	Choi et al. 2007c
C. grandis Osbeck (Korea)	31.03	14.27	3.10	2.11	
C. iyo Hort.	ND	6.17	ND	2.53	
C. juno Sieb	ND	5.60	3.20	ND	

Abbreviation: ND, not detected; NT, not tested.

TABLE 12.2
Contents of Total Carotenoid, Phenolic Acid, and Pectin in Peel Extracts of Representative *Citrus* Cultivars

	Content in Peel Extract (mg/g of Extract)		
Cultivar	**Total Carotenoid**	**Total Phenolic Acid**	**Total Pectin**
C. reticulate Blanco	2.04	0.91	37.3
C. tankan Hayata	1.42	0.95	36.0
C. grandis Osbeck (Taiwan)	0.04	0.42	86.4
C. grandis Osbeck	0.02	0.50	81.9
C. sinensis (L.) Osbeck	0.45	0.63	43.7
C. limon (L.) Bur	0.11	0.64	65.2

Source: Wang, Y. C. et al., *Food Chemistry* 106:277–284, 2008.

Citrus cultivars. Furthermore, even in the same cultivar it has been shown that the compositions of these compounds are different according to the cultivating region (e.g., *C. grandis* Osbeck in Taiwan and Korea in Table 12.1).

Additionally, Wang et al. (2008) reported that in all of their samples tested, the total flavonoid content was greater than the total carotenoid content. Naringin was reported to be abundant in Peiyou peels (*C. grandis* Osbeck cv.; 29.8 ± 0.20 mg/g dry weight [DW], mean ± standard deviation), and Wendun peels (*C. grandis* Osbeck; 23.9 ± 0.32 mg/g DW), while hesperidin was abundant in Ponkan (*Citrus reticulata* Blanco; 23.4 ± 0.25 mg/g DW), Tankan (*C. tankan* Hayata; 29.5 ± 0.32 mg/g

DW), and Liucheng (*C. sinesis* L. Osbeck; 20.7 ± 0.38 mg/g DW) peels. Kumquat (*C. microcarpa*) peel contained the highest amount of diosmin (1.12 ± 0.03 mg/g DW) and quercetin (0.78 ± 0.003 mg/g DW). The caffeic acid content ranged from 3.06–80.8 μg/g (DW), lower than that of chlorogenic acid, ferulic acid, sinapic acid, and *q*-coumaric acid. For carotenoids, Ponkan, kumquat, and Liucheng peels contained the highest total amounts of lutein, zeaxanthin, β-cryptoxanthin, and β-carotene (total amounts of approximately 335 mg/g DW).

The composition of phytochemicals in citrus fruits can be different according to growing stages. Choi et al. (2007c) presented different contents of flavonoids in immature versus mature citrus fruit, and they concluded that the composition of most flavonoids in peels of citrus decreases with aging (Table 12.3).

12.2.2 Extraction of Phytochemicals from Citrus Peels

Flavonoids in citrus fruits have been traditionally extracted from by-product wastes such as peels by using ethanol solutions (de Rijke et al. 2006; Li et al. 2006a) and have been studied for high yields of bioactive extraction by many researches. The extraction yield for bioactive materials differs according to the concentration of ethanol. Jiang et al. (2014) evaluated the antioxidant-associated efficacy of flavonoids extracted with various concentrations of ethanol from a traditional Chinese medicine, Hua Ju Hong, which consists of the peels of *Citrus grandis* (L.) Osbeck. Water extract solution was subjected to an HPD-300 macroporous resin column and eluted with a gradient of 30%–90% ethanol. They found that the 80% ethanol extract had the greatest antioxidative activity *in vitro* and *in vivo*. For methanol extracts, flavones and glycosylated flavanones are most abundant, whereas hydrolyzed extracts contain mainly phenolic acids and flavonols (Bocco et al. 1998).

Ultrasonic extraction has been shown to be an effective technique for the isolation of bioactive compounds from plant sources. The ultrasonic method is more efficient in extracting hesperidin from Penggan peel than is the traditional solvent extraction method. At 40°C, the yield of hesperidin was higher with 60-kHz sonication for 60 min than with sonication at 20 or 100 kHz. Hesperidin was not degraded under these conditions. Ma et al. (2008a) also reported results of the extraction of hesperidin from Penggan (*Citrus reticulata*) peel using ultrasonic extraction.

It is important to note that the composition of phytochemicals differs with different extraction methods. For example, Ma et al. (2008b) reported that the contents of seven phenolic acids (caffeic acid, *p*-coumaric acid, ferulic acid, sinapic acid, protocatechuic acid, *p*-hydroxybenzoic acid, and vanillic acid) and two flavanone glycosides (narirutin and hesperidin) in extracts obtained from Satsuma mandarin (*Citrus unshiu* Marc.) by ultrasonic treatment were significantly higher than those extracted by the maceration method. Moreover, the phytochemical contents of extracts increased as both treatment time and temperature increased. Ultrasonic power had a positive effect on the contents of extracts. However, the phenolic acids may become degraded by ultrasonication at higher temperatures for a long time. Ma et al. (2008b) reported that the contents of caffeic acid, *p*-coumaric acid, ferulic acid, and *p*-hydroxybenzoic acid decreased by 35.33%–48.90% with sonication at 40°C for 20 min. They concluded that ultrasonic extraction techniques must be used

Table 12.3
Flavonoid Contents in Peel Extracts from Immature and Mature Citrus Fruits from the Same Tree

	Flavonoid Contents in Peel Extracts (mg/g of Extract)							
	Narigin		Hesperidin		Rutin		Nobiletin	
Cultivar	Immature	Mature	Immature	Mature	Immature	Mature	Immature	Mature
C. unshi Marc	ND	ND	17.23	4.47	3.07	2.27	1.21	0.75
C. grandis Osbeck (Korea)	55.41	31.03	27.59	14.27	1.96	3.10	2.11	1.19
C. iyo Hort.	3.73	ND	15.83	6.17	1.80	ND	2.53	1.27
C. juno Sieb	6.37	ND	23.20	5.60	ND	3.20	ND	ND

Source: Choi, S. Y. et al., *Pharmaceutical Bulletin* 30:772–778, 2007c.
Note: Immature fruits were harvested from July through September 2005, and mature fruits were harvested from November 2005 through January 2006.
Abbreviation: ND, not detected.

carefully to enhance the yields of phenolic acids from Satsuma mandarin peels. In contrast, Ma et al. (2008c) showed that the total phenolic content and antioxidative activity of extracts from Penggan peel increased with increasing ultrasonic treatment time and temperature, especially near the ultrasonic irradiation surface, showing that ultrasonic power had a positive effect. Therefore, both the optimization of ultrasonic variables and the ultrasonic device should be considered in practice.

Nayak et al. (2015) also studied the recovery of polyphenols from *Citrus sinensis* peels via microwave-assisted extraction with acetone and compared their yields to those from conventional acetone extraction, ultrasound, and accelerated-assisted solvent extraction, based on a response surface method. The highest total phenolic content (TPC) and total antioxidant activity were achieved with 5% acetone, 500 W of microwave power, 122-s treatment time, and 25 mL solvent/g of solids.

Xu et al. (2008) reported the efficiency of the infusion cooking method on extracting minerals and phenolic compounds (flavanone glycosides [FGs], polymethoxylated flavones [PMFs], and phenolic acids), and the antioxidant activities of hot water extracts of dried peels of two citrus varieties, Satsuma mandarin and Ponkan. They showed that narirutin, nobiletin, and tangeretin were more easily extracted than hesperidin and that the hot water extract of Ponkan had a much higher content of phenolic compounds acids, FGs, and PMFs and higher antioxidative activity than did Satsuma mandarin. Furthermore, they suggested that a second extraction was necessary because significant amounts of minerals and phenolic compounds were obtained at 100°C for 30 min.

Cheigh et al. (2012) extracted flavanones such as hesperidin and narirutin from *Citrus unshiu* peel by using subcritical water extraction (SWE), and they determined the optimum conditions by varying the extraction temperature and time under high pressure (100 ± 10 atm). The maximum yields of hesperidin and narirutin were obtained at 160°C for 10 min in amounts of 72 ± 5 mg/g of peel and 11.7 ± 0.8 mg/g peel, respectively, accounting for approximately 99% of the total amount of these flavanones in the raw material. In addition, they reported that the hesperidin yield by SWE was 1.9-, 3.2-, and 34.2-fold greater than the yield with ethanol, methanol, or hot water, respectively. The narirutin yield was 1.2-, 1.5-, and 3.7-fold higher with the same methods. Min et al. (2014) also prepared citrus peel extract using a combined process of SWE and acid hydrolysis. They reported that SWE extracted greater amounts of phenolic compounds than conventional hot water or ethanol extraction and that acid hydrolysates of SWE had greater 2,2-diphenyl-1-picrylhydrazyl (DPPH) radical scavenging activities and antioxidant activities.

The supercritical extraction method can also be useful for extraction of citrus peels, because the yield is relatively high and it is an ecofriendly method. Jerkovic et al. (2015) extracted various citrus varieties with supercritical CO_2 at 40°C and 10 MPa and profiled the extracts by using gas chromatography and mass spectrometry. They obtained high yields. They determined the composition of each essential oil and reported differences in the compositions of the oils among citrus varieties. The major compound found was limonene (up to 54.3%), along with oxygenated monoterpenes (such as linalool, α-terpineol, linalyl acetate, geranyl acetate, and citral), sesquiterpenes, and coumarin derivatives.

Li et al. (2006a) developed a practical and efficient isolation method from sweet orange (*Citrus sinensis*) peel to obtain nobiletin in gram quantities with only one purification cycle. They used a silica gel flash column and eluted peel extract by a mixed solvent system of ethyl acetate and hexanes. After the eluted fractions, mainly containing nobiletin and 5,6,7,4′-tetramethoxyflavone, were vacuum evaporated, the concentrate was loaded onto a Regis chiral column, and gram quantities of nobiletin and 5,6,7,4′-tetramethoxyflavone were eluted when they used ethanol and hexanes, respectively.

12.3 ANTIOXIDATIVE EFFECTS OF CITRUS PEELS

Phenolic compounds have been shown to be capable of scavenging free radicals, chelating metal catalysts, activating antioxidant enzymes, reducing α-tocopherol radicals, and inhibiting oxidases (Alia et al. 2003; Amic et al. 2003). The high antioxidative activities of these compounds are due to the redox properties of the hydroxyl groups of the molecules (Materska and Perucka 2005; Rice-Evans et al. 1997).

12.3.1 Methods for Determination of Peroxidation

DPPH is a well-known radical and a scavenger for other radicals (Sharma and Bhat 2009). As a result of this property, the rate of reduction of a chemical reaction upon addition of DPPH can be used as an indicator of the radical reaction. The DPPH radical has a deep violet color in solution, with a strong absorption band at about 520 nm, and turns pale yellow when neutralized. The DPPH assay has been widely used to measure free radical scavenging. Limitations include color interference and sample solubility, which can affect the results of this assay (Dastmalchi et al. 2007).

Recently, an alternative method using a moderately stable nitrogen-centered radical species, 2,2′-azino-bis(3-ethylbenzothiazoline-6-sulphonic) acid radical (ABTS*) was developed (Oboh and Ademosun 2012; Re et al. 1999). However, for unknown reasons, results from the ABTS* and DPPH* assays have not always been in agreement, even though it is known that ABTS* scavenging ability reflects the reducing power and OH* scavenging ability of phenolic compounds in citrus peels.

Antioxidants chelate and deactivate transition metals that participate in lipid peroxidation and oxidative stress (Oboh and Rocha 2007). For example, the Fe^{2+}-chelating ability of phenolic compounds can be attributed to the presence of two or more functional groups, such as -OH, -SH, -COOH, $-PO_3H_2$, C=O, $-NR_2$, -S-, and -O (Yuan et al. 2005). Therefore, the ferric reducing ability of plasma (FRAP) method has been used for an antioxidative assay. A simple, automated FRAP assay has been presented as a novel method for assessing "antioxidant power" by Benzie et al. (1996). This assay is based on the principle that ferrous ion reduction at low pH leads to the formation of a colored ferrous-tripyridyltriazine complex. FRAP values are calculated by comparing the absorbance change at 593 nm with standard solutions containing ferrous ions in known concentrations. Generally, it appears that the Fe^{2+}-chelating ability of phenolic antioxidants in some plant foods, such as leafy vegetables, peppers, and spices, as well as citrus peels, is higher than their corresponding OH* scavenging abilities (Oboh and Rocha 2007).

12.3.2 Parameters for Antioxidative Activity

The antioxidative activities and compositions of bioactive components differ for different citrus varieties. Song et al. (2001) studied essential oils from 30 kinds of citrus and evaluated the antioxidative effects of 14 components, including tocopherols, by a thiocyanate method using linoleic acid. Their results showed antioxidative activity in the essential oils of yuzu, lemon, hassaku, sudachi, mochiyu, yuko, and Tarocco orange. The antioxidative activities of β-pinene, myrcene, α-terpinene, γ-terpinene, and decanal in essential oils were greater or similar to that of δ-tocopherol. Tocopherols as natural antioxidants exist in the essential oils of limes, Valencia oranges, yuzu, and lemons. However, these investigators reported that there was little correlation between tocopherol content and antioxidative activity. The terpenes are the major compounds with antioxidative activity in essential oils of citrus fruits.

Gorinstein et al. (2001) also examined the antioxidative activity of lemons, oranges, and grapefruits in terms of the total radical-trapping antioxidative potential as well as the contents of dietary fiber, total polyphenols, essential phenolic compounds, ascorbic acid, and some trace elements. They reported that there were no significant differences in the content of total dietary fiber in the sarcocarp and peel part among citrus varieties, but the content of total dietary fiber in peels was greater than in the sarcocarp in some varieties. The content of total polyphenols in peeled lemons and their peels was significantly higher than in peeled oranges or grapefruits and their peels. In addition, the contents of total polyphenols, essential phenolic compounds, and ascorbic acid in the peels were significantly higher than in the peeled fruits. The content of Fe in peeled lemons and their peels were significantly higher than in other tested peeled fruits and their peels, respectively. The authors concluded that antioxidant content was significantly higher in peels than in sarcocarp and was significantly higher in lemons than oranges and grapefruits, suggesting that lemons have a greater antioxidant potential than the other two fruits and may be preferable functional food sources for dietary prevention of cardiovascular and other diseases.

Anagnostopoulou et al. (2008) also evaluated the antioxidative activities of extracts of navel sweet orange (*Citrus sinensis*) peel by the DPPH and luminol-induced chemiluminescence methods and compared the results with those for reference compounds Trolox, ascorbic acid, and quercetin. High antioxidant activity was found in the ethyl acetate fractions of *C. sinensis* with flavanone glycosides, flavones, and flavonols predominating. Su et al. (2008) also determined total phenolic content and antioxidative activity of four citrus herbal products, pericarpium citri reticulatae (PCR), pericarpium citri reticulatae viride (PCRV), *Aurantii immaturus fructus* (AIF) and *Aurantii fructus* (AF) extracts. They found that the EC_{50} (50% effective concentration) values in the DPPH assay ranged from 0.1 mg/mL (AF) to 1.59 mg/mL (AIF); EC_{50} values for hydrogen peroxide-scavenging activities ranged from 0.08 mg/mL (AF) to 0.9 mg/mL (PCR); finally, EC_{50} values for ferrous ion-chelating activity ranged from 0.8 mg/mL (AF) to 2.08 mg/mL (AIF), indicating that AF was the most effective antioxidant.

The antioxidant properties of phenolic compounds differ according to the parts of the plant (e.g., seed versus peel) and the molecular state (bound or free form). For example, Bocco et al. (1998) obtained extracts by methanol (free phenolic compounds) or by alkaline hydrolysis (bound phenolic compounds) and, using a model system based on accelerated citronellal oxidation, reported that extracts of seeds possessed greater antioxidant activity than peels. Phenolic acids in free or bound forms account for approximately one-third of the total phenolic compounds in plants. The bound forms are known to be covalently linked to various plant components through ester, ether, or acetyl bonds (Robbins 2003).

Oboh and Ademosun (2012) also determined the distribution of free and bound phenolic compounds in Nigerian citrus peels, such as orange (*Citrus sinensis*), grapefruit (*Citrus paradisi*), and shaddock (*Citrus maxima*), and they characterized their antioxidative effects after extraction with 80% acetone for free phenolic compounds and the alkaline- and acid-hydrolyzed residue with ethyl acetate for the bound phenolic compounds. They concluded that free phenolic extracts had significantly greater DPPH* scavenging ability than the bound phenolic extracts, except in orange peels.

The antioxidative effect of phenolic compounds differs after food processing, such as heat treatment. Jeong et al. (2004) studied the effect of heat treatment at 50°C, 100°C, and 150°C for 10–60 min on the antioxidative effects of extracts from *C. unshiu* peels. Heat treatment at 150°C for 60 min increased TPC, radical scavenging activity, and reducing power of the extracts, compared to non-heat-treated controls. In addition, they reported that 2,3-diacetyl-1-phenylnaphthalene, ferulic acid, *p*-hydroxybenzaldoxime, 5-hydroxyvaleric acid, 2,3-diacetyl-1-phenylnaphthalene, and vanillic acid, which are low-molecular-weight phenolic compounds, were newly produced by heat treatment, suggesting that the heating process can be applied as a technique for increasing the antioxidative effect of peel extract. Xu et al. (2007) also studied the antioxidative effects of heat treatment on Huyou (*Citrus paradisi* Changshanhuyou) peel. Their results showed that the free fraction of phenolic acids increased after heat treatment, but ester, glycoside, and ester-bound fractions decreased, as did the content of total flavanone glycosides. The antioxidant activity of methanol extracts of Huyou peel had increased total phenolic contents in the TPC assay, ABTS assay, and FRAP assay, showing high correlation coefficients. In addition, Xu et al. reported that levels of total cinnamics and benzoics in the free fraction were very high and contributed to increased antioxidative activities, but flavanone glycosides may have been destroyed when heated at 120°C for 90 min or at 150°C for 30 min. They suggested that a proper and reasonable heat treatment could be used to increase the antioxidative activity of citrus peel.

Furthermore, in citrus peels three tocopherol analogs, methoxy-tocopherol, α-tocopherol, and γ-tocopherol, were discovered as antioxidative agents. Seo et al. (2015) evaluated the protective effects of the isolated compounds, including the tocopherol derivatives extracted from the peels of *C. unshiu* Marcovich, against *tert*-butyl hydroperoxide-induced hepatotoxicity in human liver-derived HepG2 cells and glutamate-induced oxidative stress in HT22-immortalized hippocampal cells. Their results showed that three compounds had hepatoprotective and neuroprotective effects on oxidatively stressed HepG2 cells and HT22 cells.

12.4 ANTI-INFLAMMATORY EFFECTS OF CITRUS PEELS

Inflammation is a kind of complicated protective biological response of tissues to harmful stimuli such as pathogens, damaged cells, or some irritants. When there is inflammation, the initial cause of cell injury can be eliminated and necrotic cells and tissues damaged from the original insult can be cleared to initiate tissue repair. Therefore, the inflammatory response plays a crucial role in host defense against various stimuli and results in restoration to the normal cell structure and function.

12.4.1 Mechanisms of Inflammation

Inflammation can be classified as acute or chronic, according to the nature and duration of the response. Acute inflammation is caused by the increased movement of plasma and leukocytes from the blood into the injured tissues as an initial response. During processing, a series of biochemical reactions involving the local vascular system and the immune system within the injured tissue spreads and develops the inflammatory response. Chronic inflammation is a prolonged response that causes simultaneous destruction and healing of the tissue from the inflammatory process.

Acute inflammation is manifested by vascular changes, edema, and predominant neutrophil infiltration that terminates within a few days due to the short half-life of inflammatory mediators (such as lipoxins, resolvins, protectins, and neutrophils) and due to the liberation of anti-inflammatory cytokines, such as transforming growth factor β and interleukin-10 (IL-10) (Gosslau et al. 2011). Vasodilation of blood vessels and high vascular permeability can also be caused by inflammatory mediators such as nitric oxide (NO), histamine, serotonin, prostaglandins, and leukotrienes. Figure 12.2 shows the pathway of acute inflammation.

At an initial step of the response, NF-κB bound with inhibitor of κB is activated by proinflammatory cytokines, such as tumor necrosis factor alpha (TNF-α) and IL-1β, by bacterial endotoxins such as lipopolysaccharide (LPS), and by oxidative stress. The signal cascade is triggered through the phosphorylation of IκB by IκB

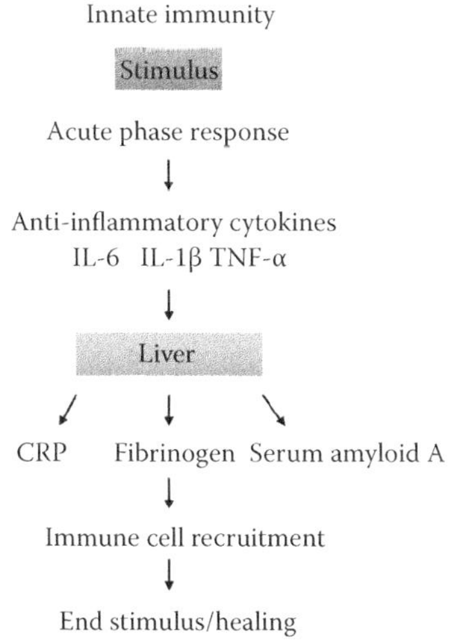

FIGURE 12.2 Pathway of acute inflammation. CRP, C-reactive protein; IL, interleukin; TNF, tumor necrosis factor.

kinase (IKK). Activated NF-κB translocates to the nucleus and leads to the transcription of target genes, such as the proinflammatory mediators cyclooxygenase 2 (COX-2) and inducible nitric oxide synthase (iNOS), various cytokines such as TNF-α, IL-1β, and IL-6, chemokines, and adhesion molecules (Lawrence et al. 2001). Anti-inflammatory agents mainly suppress the expression of proinflammatory genes by inhibiting the NF-κB activation pathway (Gilroy et al. 2004), and NF-κB inhibitors can be useful in the development of therapeutic drugs against inflammation associated with human diseases.

Acute inflammation facilitates rolling, adhesion, and endothelial transmigration of leukocytes toward the infected tissue site. Extravasation of neutrophils is coordinated by intercellular cell adhesion molecule-1 (ICAM-1) and vascular cell adhesion molecule-1 (VCAM-1), chemokines such as IL-8, and metabolites generated through the arachidonic acid pathway (Radi et al. 2001). Arachidonic acid produced by phospholipase A_2 is also a substrate for COX-2 and 5-lipoxygenase (5-LOX), which synthesize various prostaglandins and leukotrienes, respectively, which cause vasodilation (Robinson et al. 2010).

The COX-2 gene is highly inducible and regulated by different transcription factors, such as NF-κB, activator protein-1 (AP-1), cyclic AMP response element binding protein (CREB), and nuclear factor IL-6 (Smith et al. 2000). For the LOX pathway, leukotriene A_4 hydrolase produces leukotriene B_4, which is a potent neutrophil chemoattractant (Chen et al. 2004). Activation of neutrophils facilitates phagocytosis and intracellular degradation of the material ingested by lysosomal enzymes, accompanied by an oxidative burst that is caused by enzymatic generation of electrophilic species, including reactive oxygen species (ROS) and reactive nitrogen species (RNS). iNOS is also involved when an oxidative burst is caused by NO generation (Novo and Parola 2008).

Chronic inflammation is a prolonged response caused by persistent infections, immune-mediated inflammatory diseases, or prolonged exposure to toxic reagents; it results in simultaneous destruction caused predominantly by mononuclear macrophages and healing of the tissue from the inflammatory process. Figure 12.3 shows the pathway of chronic inflammation.

Chronic inflammation increases the level of proinflammatory mediators (such as iNOS, COX-2) and various cytokines (such as TNF-α, IL-1β, and IL-6). iNOS and COX-2 facilitate production of NO and prostaglandins, respectively. In chronic inflammation, macrophages have major cellular roles, with their life span of several months to years. Macrophages differentiate from monocytes when they cross the endothelium, which is governed by adhesion molecules and chemokines, such as monocyte chemotactic protein-1 and macrophage inflammatory protein-1. During chronic inflammation, there is a bidirectional positive interaction between lymphocytes and macrophages, including several cytokines (such as interferon-gamma, TNF-α, IL-1β, and IL-12) and chemokines (IL-8, monocyte chemotactic protein-1, and macrophage inflammatory protein-1) (Kundu and Surh 2008). The macrophages are activated mainly by Toll-like receptor signaling and IFN-γ released by T lymphocytes. In addition, IFN-γ expression can be increased by various ILs (IL-12, IL-15, and IL-18) and facilitates macrophage activation and an oxidative burst

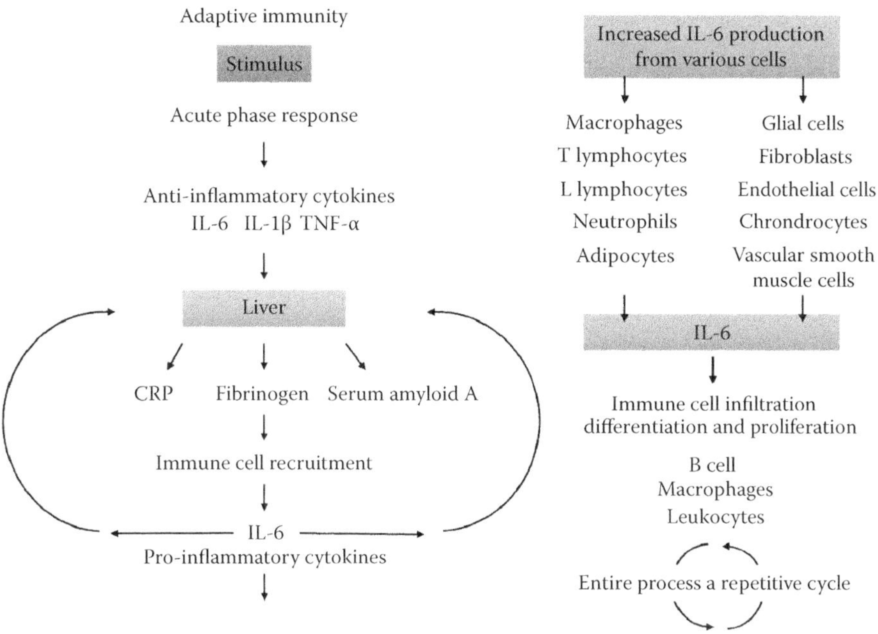

FIGURE 12.3 Pathway of chronic inflammation. CRP, C-reactive protein; IL, interleukin; TNF, tumor necrosis factor.

(Aggarwal et al. 2009; Novo and Parola 2008). During chronic inflammation, proinflammatory mediators, such as iNOS, COX-2, and various cytokines (TNF-α, IL-1β, and IL-6) are increased. iNOS and COX-2, which are responsible for increased levels of NO and prostaglandins, respectively, are active in the pathogenesis of various chronic inflammatory diseases, such as multiple sclerosis, Alzheimer's disease, and colon cancer (Heiss et al. 2001).

NO as a mediator of allergic inflammation exerts multiple modulating effects on inflammation, as well as being a major factor in the regulation of immune responses (Guzik et al. 2003); it affects every step of inflammation. Figure 12.1 shows effects of NO in the chronic inflammation process.

Low concentrations of NO inhibit adhesion molecule expression, synthesis of cytokines and chemokines, and adhesion and transmigration of leukocytes. Meanwhile, high concentrations of NO produced by iNOS can be toxic and proinflammatory. Actions of NO are dependent on the cellular context, NO concentration, and initial priming of immune cells. The effects of NO are exerted through multiple mechanisms in immune regulation. In addition, NO leads to modification of transcription factor activity, and thus modulates the expression of many other mediators of inflammation. Here, we discuss the mechanisms of NO- and superoxide-dependent modulation of inflammatory reactions in experimental animals and in humans.

In the immune system, cascades of mitogen-activated protein kinases (MAPKs), which are important factors for regulation of cell differentiation, cell growth, and cellular responses to cytokines, including p38 MAPK, extracellular signal-regulated kinase, and c-Jun N-terminal kinase (JNK) (Kang et al. 2011). MAPK activated by kinase cascades such as MAPK kinase can phosphorylate transcription factors such as NF-κB and activator protein-1. MAPKs can be activated by LPS and suppress the production of iNOS and COX-2 in macrophages (Chen et al. 1999). In addition, in LPS-induced macrophages, phosphorylation of MAPKs is related to proinflammatory cytokines as well as transcriptional activation of NF-κB, which is a transcriptional activator that has a major role in gene expression in immune and inflammatory responses (Carter et al. 1999).

12.4.2 Anti-Inflammatory Effects of Extracts of Citrus Peels

There are many kinds of phytochemicals, including flavonoids, in citrus peels. In particular, citrus plants have been known to be a major source of various flavonoids, such as hesperidin, naringin, nobletin, and other polymethoxylated flavones (Kang et al. 2011).

Kang et al. (2011) examined whether flavonoids (nobiletin, naringin, and hesperidin) in *Citrus aurantium* L. inhibited proinflammatory mediators by blocking NF-κB and MAPK signaling in LPS-stimulated RAW 264.7 macrophages. Flavonoids suppressed mRNA and protein expression of COX-2 and iNOS in LPS-induced macrophages, by inhibiting the degradation/phosphorylation of IκB-α and nuclear translocation of the NF-κB p65 as well as phosphorylation of MAPK, suggesting that flavonoids have good anti-inflammatory effects.

Mohanty et al. (2015) studied the anti-inflammatory effects of ethanol extracts from *Citrus limetta* fruit peels *in vivo*. Their study revealed that the extract was safe at high doses (500 mg/kg of body weight), that proinflammatory cytokine production was reduced, and malaria pathogenesis was attenuated, suggesting potential antimalarial activity by inhibiting the parasitemia and inflammatory mediators such as IFN-γ, TNF-α, and IL-6, which are related to malaria pathogenesis. In addition, it appeared that the extract was able to improve the hemoglobin and glucose levels and increase the survival time when used as a pharmaceutical agent against malaria pathogenesis.

Choi et al. (2007a) also studied the anti-inflammatory effects of posttreatment with nobiletin purified from the fruit peel of *Citrus sunki* Hort. ex Tanaka on LPS-activated RAW 264.7 cells. They reported that it significantly suppressed transcriptional activation of NF-κB, production of NO and PGE_2, and expression of iNOS and COX-2 protein. However, it appeared that nobiletin did not affect the phosphorylation/degradation of IκB-α or nuclear translocation of NF-κB in LPS-induced cell lines, but it did inhibit the DNA binding activity of NF-κB. The authors also reported that nobiletin inhibited the production of ROS as a regulator of NF-κB. On the basis of these results, they suggested that nobiletin exerts an anti-inflammatory effect through the interruption of NF-κB DNA binding activity as well as the suppression of ROS generation. As an anti-inflammatory agent, Lin et al. (2003) reported that nobiletin can inhibit matrix degradation of the articular cartilage and pannus formation in osteoarthritis and rheumatoid arthritis and also downregulate gene expression of other proinflammatory cytokines (such as IL-1α, IL-1β, TNF-α, and

IL-6). Ishiwa et al. (2000) also showed that nobiletin suppresses gene expression and also production of some matrix metalloproteinases (such as MMP-1, MMP-3, and MMP-9) in rabbit articular chondrocytes and synovial fibroblasts, and it inhibits pro-MMP-1 and pro-MMP-3. Many researchers have suggested that nobiletin might be employed as a novel anti-inflammatory and immunomodulatory drug.

Hesperidin also suppresses inflammatory cytokine production from mast cells. Choi et al. (2007b) studied the inhibitory effect of hesperidin on expression of hypoxia-inducible factor-1α and inflammatory cytokine production from HMC-1 cells. They suggested that hesperidin is one of several inhibitors of HIF-1a and cytokines on the mast cell-mediated inflammatory responses.

Naringinin, which is the glucoside of naringenin, also has an anti-inflammatory effect. Kanno et al. (2006) studied the inhibitory effect of naringin on LPS-induced endotoxin shock in mice and NO production in RAW 264.7 macrophages, and they reported that naringin inhibited inflammatory signals, such as COX-2, iNOS, and cytokines, in macrophage cells.

Kang et al. (2011) examined whether flavonoids (nobiletin, naringin, and hesperidin) isolated from *Citrus aurantium* L. inhibited proinflammatory mediators, including cytokines, by blocking NF-κB and MAPK signaling in LPS-stimulated RAW 264.7 macrophages. They found that the flavonoids suppressed mRNA and protein expression of COX-2 and iNOS in LPS-induced macrophages and inhibited the degradation/phosphorylation of IκB-α and nuclear translocation of the NF-κB p65 as well as phosphorylation of MAPK, suggesting that flavonoids block NF-κB and MAPK signaling and exert anti-inflammatory effects by suppressing expression of COX-2, iNOS, and cytokines.

In addition, auraptene, which is a natural bioactive monoterpene coumarin ether in citrus, has been known to have an anti-inflammatory effect. Okuyama et al. (2014) evaluated the anti-inflammatory effect of auraptene extracted from peels of *Citrus kawachiensis* in an animal model of systemic inflammation, and they showed that it had the ability to suppress loss of body weight and abnormal behavior in the open field, activation of microglia and astrocytes in the hippocampus, and expression of COX-2, which was coexpressed in astrocytes.

The correct heat treatment can increase anti-inflammatory activity of citrus peels. Ho and Lin (2008) studied heat treatment for enhancing the anti-inflammatory activity of *Citrus reticulata* peels by measuring their inhibitory effects on NO production by LPS-activated RAW 264.7 macrophages. Their results showed that the anti-inflammatory effect of citrus peel heat treated for 100°C increased significantly, and they suggested that major factors for anti-inflammatory activity in citrus peel extract were correlated with the levels of nobiletin and tangeretin.

Choi et al. (2007c) also studied the correlation between the content of flavonoids (naringin, naringenin, hesperidin, hesperetin, rutin, nobiletin, and tangeretin) and the NO production-inhibitory activity of peel extracts from various citrus fruits, using 20 citrus plants. They showed that citrus peel extracts inhibited LPS-induced NO production in RAW 264.7 cells with a positive and significant correlation ($r = 0.858$ to 0.879) and that the levels of nobiletin and tangeretin correlated with anti-inflammatory effects significantly and positively. In addition, they found that nobiletin showed a greater inhibitory activity on NO production than tangeretin (IC_{50}, 136.6 mM).

Recently, an extract blended with various citrus peels was studied in an effort to increase its biological activity. Lin et al. (2014) found that formulated extract from six citrus variety peels impaired dendritic cell function and attenuated allergic contact hypersensitivity more effectively, and also that the formulated extract interfered with LPS-induced MAPK–JNK, p38 phosphorylation, and nuclear translocation of NF-κB p65.

12.5 CONCLUSIONS

Phytochemicals have been a major component of traditional therapeutic approaches for wound healing across diverse cultures. These compounds act as natural biofunctional materials with health benefits that have been studied and increasingly applied to pharmaceutical and functional food products due to their fewer side effects compared to synthetic drugs. Recently, plants have been recognized as a good resource of thousands of new useful phytochemicals of great diversity. In particular, extracts of by-products or waste in agricultural food processing are relatively cheap and easily available and their economic value thus can be increased.

The additive and/or synergistic effects of the free or bound phenolic compounds of citrus peels as antioxidants could contribute to food preservatives and functional food products as well as medicinal products.

Flavonoid contents and inhibitory effects of the peel extracts on NO production of citrus fruit peels are highly correlated, providing various beneficial features of citrus fruit peels, such as the prevention of diseases and excessive NO release. Additionally, it has been shown that flavonoids in citrus peels mainly inhibit proinflammatory mediators, as well as various cytokines. Flavonoids are potential promising natural potential chemotherapeutic agents against inflammatory diseases.

Extracts naturally contain many kinds of phytochemicals. Therefore, they must be evaluated rigorously and chemically characterized to ensure sufficient consistency of their bioactivities. However, research in this field is complicated, because comparisons of results across studies are very difficult. The composition of every phytochemical also varies according to the source citrus variety, its cultivation conditions, and the extraction conditions. Therefore, it is necessary to accumulate more scientific information to confirm the consistency of performance of natural extracts for further practical applications.

REFERENCES

Aggarwal, B. B., S. Shishodia, S. K. Sandur, M. K. Pandey, and G. Sethi. 2006. Inflammation and cancer: How hot is the link? *Biochemical Pharmacology* 72:1605–1621.

Alam, P., A. Alam, M. K. Anwer, and S. I. Alqasoumi. 2014. Quantitative estimation of hesperidin by HPTLC in different varieties of citrus peels. *Asian Pacific Journal of Tropical Biomedicine* 4:262–266.

Alia, M., C. Horcajo, L. Bravo, and L. Goya. 2003. Effect of grape antioxidant dietary fiber on the total antioxidant capacity and the activity of liver antioxidant enzymes in rats. *Nutrition Research* 23:1251–1267.

Amic, D., D. Davidovic-Amic, D. Beslo, and N. Trinajstic. 2003. Structure related scavenging activity relationship of flavonoids. *Croatia Chemica Acta* 76:55–61.

Anagnostopoulou, M. A., P. Kefalas, V. P. Papageorgiou, A. N. Assimopoulou, and D. Boskou. 2006. Radical scavenging activity of various extracts and fractions of sweet orange peel (*Citrus sinensis*). *Food Chemistry* 94:19–25.

Arif, T., J. D. Bhosale, N. Kumar, T. K. Mandal, R. S. Bendre, G. S. Lavekar, and R. Dabur. 2009. Natural products: Antifungal agents derived from plants. *Journal of Asian Natural Products Research* 11:621–638.

Bagatela, B. S., A. P. Lopes, E. C. Cabral, F. F. Perazzo, and D. R. Ifa. 2015. High-performance thin-layer chromatography/desorption electrospray ionization mass spectrometry imaging of the crude extract from the peels of *Citrus aurantium* L. (Rutaceae). *Rapid Communications in Mass Spectrometry* 29:1530–1534.

Benavente-García, O., and J. Castillo. 2008. Update on uses and properties of citrus flavonoids: New findings in anticancer, cardiovascular, and anti-inflammatory activity. *Journal of Agricultural and Food Chemistry* 56:6185–6205.

Benzie, I. F., and J. J. Strain. 1996. The ferric reducing ability of plasma (FRAP) as a measure of "antioxidant power": The FRAP assay. *Analytical Biochemistry* 239:70–76.

Berkarda, B., H. Koyuncu, G. Soybir, and F. Baykut. 1998. Inhibitory effect of hesperidin on tumour initiation and promotion in mouse skin. *Research in Experimental Medicine* 198:93–99.

Bocco, A., M. E. Cuvelier, H. Richard, and C. Berset. 1998. Antioxidant activity and phenolic composition of citrus peel and seed extracts. *Journal of Agricultural and Food Chemistry* 46:2123–2129.

Brito, A., J. E. Ramirez, C. Areche, B. Sepulveda, and M. J. Simirgiotis. 2014. HPLC-UV-MS profiles of phenolic compounds and antioxidant activity of fruits from three *Citrus* species consumed in northern Chile. *Molecules* 19:17400–17421.

Carter, A. B., K. L. Knudtson, M. M. Monick, and G. W. Hunninghake. 1999. The p38 mitogen-activated protein kinase is required for NF-kappaB-dependent gene expression. The role of TATA-binding protein (TBP). *Journal of Biological Chemistry* 274:30858–30863.

Cheigh, C. I., E. Y. Chung, and M. S. Chung. 2012. Enhanced extraction of flavanones hesperidin and narirutin from *Citrus unshiu* peel using subcritical water. *Journal of Food Engineering* 110:472–477.

Chen, C., Y. H. Chen, and W. W. Lin. 1999. Involvement of p38 mitogen-activated protein kinase in lipopolysaccharide-induced iNOS and COX-2 expression in J774 macrophages. *Immunology* 97:124–129.

Chen, X., S. Wang, N. Wu, and C. Yang. 2004. Leukotriene A4 hydrolase as a target for cancer prevention and therapy. *Current Cancer Drug Targets* 4:267–283.

Choi, S. Y., J. H. Hwang, H. C. Ko, J. G. Park, and S. J. Kim. 2007a. Nobiletin from citrus fruit peel inhibits the DNA-binding activity of NF-κB and ROS production in LPS-activated RAW 264.7 cells. *Journal of Ethnopharmacology* 113:149–155.

Choi, I. Y., S. J. Kim, H. J. Jeong, S. H. Park, Y. S. Song, J. H. Lee et al. 2007b. Hesperidin inhibits expression of hypoxia inducible factor-1 alpha and inflammatory cytokine production from mast cells. *Molecular and Cellular Biochemistry* 305:153–161.

Choi, S. Y., H. C. Ko, S. Y. Ko, J. H. Hwang, J. G. Park, S. H. Kang, S. H. Han, S. H. Yun, and S. J. Kim. 2007c. Correlation between flavonoid content and the NO production inhibitory activity of peel extracts from various citrus fruits. *Biological and Pharmaceutical Bulletin* 30:772–778.

Dastmalchi, K., H. J. D. Dorman, M. Kosar, and R. Hiltunen. 2007. Chemical composition and *in vitro* antioxidant evaluation of a water soluble Moldavian balm (*Dracocephalum moldavica* L.) extract. *Lebensmittel Wissenschaft Technologie* 40:239–248.

de Carvalho, J. M., G. A. Maia, A. V. da Fonseca, P. H. de Sousa, and S. Rodrigues. 2015. Effect of processing on physicochemical composition, bioactive compounds and enzymatic activity of yellow mombin (*Spondias mombin* L.) tropical juice. *Journal of Food Science and Technology* 52:1182–1187.

de Rijke, E., P. Out, W. M. A. Niessen, F. Ariese, C. Gooijer, and U. Brinkman. 2006. Analytical separation and detection methods for flavonoids. *Journal of Chromatography A* 1112:31–63.

Gabor, M. 1988. Szent-Györgyi and the bioflavonoids: New results and perspectives of pharmacological research into benzo-pyrone derivatives, *Plant Flavonoids in Biology and Medicine II*, eds. V. Cody, E. Middleton, Jr., J. B. Harborne, and A. Beretz, pp. 1–15. New York: Alan R. Liss.

Gilroy, D. W., T. Lawrence, M. Perretti, and A. G. Rossi. 2004. Inflammatory resolution: New opportunities for drug discovery. *Nature Reviews Drug Discovery* 3:401–416.

Gorinstein, S., O. Martin-Belloso, Y. S. Park, R. Haruenkit, A. Lojek, M. Ciz, A. Caspi, I. Libman, and S. Trakhtenberg. 2001. Comparison of some biochemical characteristics of different citrus fruits. *Food Chemistry* 74:309–315.

Gosslau, A., S. Li, C. T. Ho, K. Y. Chen, and N. E. Rawson. 2011. The importance of natural product characterization in studies of their anti-inflammatory activity. *Molecular Nutrition and Food Research* 55:74–82.

Guzik, T. J., R. Korbut, and T. Adamek-Guzik. 2003. Nitric oxide and superoxide in inflammation and immune regulation. *Journal of Physiology and Pharmacology* 54:469–487.

Heiss, E., C. Herhaus, K. Klimo, H. Bartsch, and C. Gerhauser. 2001. Nuclear factor kappa B is a molecular target for sulforaphane-mediated anti-inflammatory mechanisms. *Journal of Biological Chemistry* 276:32008–32015.

Ho, S. C., and C. C. Lin. 2008. Investigation of heat treating conditions for enhancing the anti-inflammatory activity of citrus fruit (*Citrus reticulata*) peels. *Journal of Agricultural and Food Chemistry* 56:7976–7982.

Ishiwa, J., T. Sato, Y. Mimaki, Y. Sashida, M. Yano, and A. Ito. 2000. A citrus flavonoid, nobiletin, suppresses production and gene expression of matrix metalloproteinase 9/gelatinase B in rabbit synovial fibroblasts. *Journal of Rheumatology*. 27:20–25.

Jeong, S. M., S. Y. Kim, D. R. Kim, S. C. Jo, K. C. Nam, D. U. Ahn, and S. E. C. Lee. 2004. Effect of heat treatment on the antioxidant activity of extracts from citrus peels. *Journal of Agricultural and Food Chemistry* 52:3389–3393.

Jerkovic, I., J. Drulzic, Z. Marijanovic, M. Gugic, S. Jokic, and M. Roje. 2015. GC-FID/MS Profiling of supercritical CO_2 extracts of peels from *Citrus aurantium*, *C. sinensis* cv. Washington navel, *C. sinensis* cv. Tarocco and *C. sinensis* cv. Doppio Sanguigno from Dubrovnik Area (Croatia). *Natural Product Communications* 10:1315–1318.

Jiang, J., L. Shan, Z. Chen, H. Xu, J. Wang, Y. Liu, and Y. Xiong. 2014. Evaluation of antioxidant-associated efficacy of flavonoid extracts from a traditional Chinese medicine Hua Ju Hong (peels of *Citrus grandis* (L.) Osbeck). *Journal of Ethnopharmacology* 158(Part A):325–330.

Kang, S. R., K. I. Park, H. S. Park, D. H. Lee, J. A. Kim, A. Nagappan, E. H. Kim, W. S. Lee, S. C. Shin, M. K. Park, D. Y. Han, and G. S. Kim. 2011. Anti-inflammatory effect of flavonoids isolated from Korea *Citrus aurantium* L. on lipopolysaccharide-induced mouse macrophage RAW 264.7 cells by blocking of nuclear factor-kappa B (NF-κB) and mitogen-activated protein kinase (MAPK) signalling pathways. *Food Chemistry* 129:1721–1728.

Kanno, S., A. Shouji, A. Tomizawa, T. Hiura, Y. Osanai, M. Ujibe et al. 2006. Inhibitory effect of naringin on lipopolysaccharide (LPS)-induced endotoxin shock in mice and nitric oxide production in RAW 264.7 macrophages. *Life Sciences* 78:673–681.

Kawaii, S., Y. Tomono, E. Katase, K. Ogawa, and M. Yano. 1999. HL-60 differentiating activity and flavonoid content of the readily extractable fraction prepared from citrus juices. *Journal of Agricultural and Food Chemistry* 47:128–135.

Kundu, J. K., and Y. J. Surh. 2008. Inflammation: Gearing the journey to cancer. *Mutation Research* 659:15–30.

Lawrence, T., D. W. Gilroy, P. R. Colville-Nash, and D. A. Willoughby. 2001. Possible new role for NF-kappaB in the resolution of inflammation. *Nature Medicine* 7:1291–1297.

Lee, C. L., T. L. Hwang, C. Y. Peng, C. J. Chen, C. L. Kuo, W. Y. Chang, and Y. C. Wu. 2015. Anti-inflammatory and cytotoxic compounds from *Solanum macaonense. Natural Product Communications* 10:345–348.

Li, S., H. Yu, and C. T. Ho. 2006a. Nobiletin: Efficient and large quantity isolation from orange peel extract. *Biomedical Chromatography* 20:133–138.

Li, S., C.-Y. Lo, and C.-T. Ho. 2006b. Hydroxylated polymethoxyflavones and methylated flavonoids in sweet orange (*Citrus sinensis*) peel. *Journal of Agricultural and Food Chemistry* 54:4176–4185.

Li, S., Y. C. Lin, C. T. Ho, P. Y. Lin, M. Suzawa, H. C. Wang, C. L. Chu, D. Y. Chen, and C. C. Lin. 2014. Formulated extract from multiple citrus peels impairs dendritic cell functions and attenuates allergic contact hypersensitivity. *International Immunopharmacology* 20:12–23.

Lin, N., T. Satoa, Y. Takayama, Y. Mimaki, Y. Sashida, M. Yano, and A. Ito. 2003. Novel anti-inflammatory actions of nobiletin, a citrus polymethoxy flavonoid, on human synovial fibroblasts and mouse macrophages. *Biochemical Pharmacology* 65: 2065–2071.

Lu, Y., C. Zhang, P. Pbucheli, and D. Wei. 2006. Citrus flavonoids in fruit and traditional Chinese medicinal food ingredients in China. *Plant Foods for Human Nutrition* 61:57–65.

Ma, Y., X. Ye, Y. Hao, G. Xu, G. Xu, and D. Liu. 2008a. Ultrasound-assisted extraction of hesperidin from Penggan (*Citrus reticulata*) peel. *Ultrasonics Sonochemistry* 15: 227–232.

Ma, Y. Q., X. Q. Ye, Z. X. Fang, J. C. Chen, G. H. Xu, and D. H. Liu. 2008b. Phenolic compounds and antioxidant activity of extracts from ultrasonic treatment of Satsuma mandarin (*Citrus unshiu* Marc.) peels. *Journal of Agricultural and Food Chemistry* 56:5682–5690.

Ma, Y. Q., J. C. Chen, D. H. Liu, and X. Q. Ye. 2008c. Effect of ultrasonic treatment on the total phenolic and antioxidant activity of extracts from citrus peel. *Journal of Food Science* 73:T115–T120.

Manthey, J. A., and K. Grohmann. 2001. Phenols in citrus peel byproducts. Concentrations of hydroxycinnamates and polymethoxylated flavones in citrus peel molasses. *Journal of Agricultural and Food Chemistry* 49:3268–3273.

Materska, M., and I. Perucka. 2005. Antioxidant activity of the main phenolic compounds isolated from hot pepper fruit (*Capsicum annuum* L.). *Journal of Agricultural and Food Chemistry* 53:1750–1756.

Min, K. Y., K. A. Lee, H. J. Kim, K. T. Kim, M. S. Chung, P. S. Chang., H. Park, and H. D. Paik. 2014. Antioxidative and anti-inflammatory activities of *Citrus unshiu* peel extracts using a combined process of subcritical water extraction and acid hydrolysis. *Food Science and Biotechnology* 23:1441–1446.

Mohanty, S., A. K. Maurya, Jyotshna, A. Saxena, K. Shanker, A. Pal, and D. U. Bawankule. 2015. Flavonoids rich fraction of *Citrus limetta* fruit peels reduces proinflammatory cytokine production and attenuates malaria pathogenesis. *Current Pharmaceutical Biotechnology* 165:544–552.

Nayak, B., F. Dahmoune, K. Moussi, H. Remini, S. Dairi, O. Aoun, and M. Khodir. 2015. Comparison of microwave, ultrasound and accelerated-assisted solvent extraction for recovery of polyphenols from *Citrus sinensis* peels. *Food Chemistry* 187:507–516.

Nogata, Y., K. Sakamoto, H. Shiratsuchi, T. Ishii, M. Yano, and H. Ohta. 2006. Flavonoid composition of fruit tissues of *Citrus* species. *Bioscience Biotechnology and Biochemistry* 70:178–192.

Novo, E., and M. Parola. 2008. Redox mechanisms in hepatic chronic wound healing and fibrogenesis. *Fibrogenesis Tissue Repair* 1:5.
Oboh, G., and A. O. Ademosun. 2012. Characterization of the antioxidant properties of phenolic extracts from some citrus peels. *Journal of Food Science and Technology* 49:729–736.
Oboh, G., and J. B. T. Rocha. 2007. Polyphenols in red pepper (*Capsicum annuum* var *avicu-lare* [Tepin]) and their protective effect on some pro-oxidants induced lipid peroxidation in brain and liver. *European Food Research and Technology* 225:239–247.
Okuyama, S., K. Yamamoto, H. Mori, N. Toyoda, M. Yoshimura, Y. Amakura, T. Yoshida, K. Sugawara, M. Sudo, M. Nakajima, and Y. Furukawa. 2014. Auraptene in the peels of *Citrus kawachiensis* (Kawachi Bankan) ameliorates lipopolysaccharide-induced inflammation in the mouse brain. *Evidence-Based Complementary and Alternative Medicine* 2014:408503.
Radi, Z. A., M. E. Kehrli, Jr., and M. R. Ackermann. 2001. Cell adhesion molecules, leukocyte trafficking, and strategies to reduce leukocyte infiltration. *Journal of Veterinary Internal Medicine* 15:516–529.
Re, R., N. Pellegrini, A. Proteggente, A. Pannala, M. Yang, and C. Rice-Evans. 1999. Antioxidant activity applying an improved ABTS radical cation decolorisation assay. *Free Radical Biology and Medicine* 26:1231–1237.
Rice-Evans, C., N. J. Miller, and G. Paganga. 1997. Antioxidant properties of phenolic compounds. Trends in *Plant Science* 2:152–159.
Robbins, R. J. 2003. Phenolic acids in foods: An overview of analytical methodology. *Journal of Agricultural and Food Chemistry* 51:2866–2887.
Schieber, A., F. C. Stintzing, and R. Carle. 2001. By-products of plant food processing as a source of functional compounds: Recent developments. *Trends in Food Science and Technology* 12:401–413.
Seo, K. H., D. Y. Lee, D. S. Lee, H. C. Kang, H. S. Kim, Y. C. Kim, and N. I. Baek. 2015. Hepatoprotective and neuroprotective tocopherol analogues isolated from the peels of *Citrus unshiu* Marcovich. *Natural Product Researc*h 29:571–573.
Sharma, O. P., and T. K. Bhat. 2009. DPPH antioxidant assay revisited. *Food Chemistry* 113:1202–1205.
Smith, W. L., D. L. DeWitt, and R. M. Garavito. 2000. Cyclooxygenases: Structural, cellular, and molecular biology. *Annual Review of Biochemistry* 69:145–182.
Song, H.-S., H. Ukeda, and M. Sawamura. 2001. Antioxidative activities of citrus peel essential oils and their components against linoleic acid oxidation. *Food Science and Technology Research* 7:50–56.
Su, M. S., Y. T. Shyu, and P. J. Chien. 2008. Antioxidant activities of citrus herbal product extracts. *Food Chemistry* 111:892–896.
Tripoli, E., M. L. Guardia, S. Giammanco, D. D. Majo, and M. Giammanco. 2007. Citrus flavonoids: Molecular structure, biological activity and nutritional properties. A review. *Food Chemistry* 104:466–479.
Wang, Y. C., Y. C. Chuang, and H. W. Hsu. 2008. The flavonoid, carotenoid and pectin content in peels of citrus cultivated in Taiwan. *Food Chemistry* 106:277–284.
Whitman, S. C., E. M. Kurowska, J. A. Manthey, and A. Daugherty. 2005. Nobiletin, a citrus flavonoid isolated from tangerines, selectively inhibits class A scavenger receptor-mediated metabolism of acetylated LDL by mouse macrophages. *Atherosclerosis* 178:25–32.
Xu, G. H., J. C. Chen, D. H. Liu, Y. H. Zhang, P. Jiang, and X. Q. Ye. 2008. Minerals, phenolic compounds, and antioxidant capacity of citrus peel extract by hot water. *Journal of Food Science* 73:C11–C18.

Xu, G., X. Ye, J. Chen, and D. Liu. 2007. Effect of heat treatment on the phenolic compounds and antioxidant capacity of citrus peel extract. *Journal of Agricultural and Food Chemistry* 55:330–335.

Yuan, Y. V., D. E. Bone, and M. F. Carrington. 2005. Antioxidant activity of dulse (*Palmaria palmata*) extract evaluated *in vitro*. *Food Chemistry* 91:485–494.

13 Citrus Coumarins and Their Health Properties

Carlos A. Ledesma-Escobar
and María D. Luque de Castro

CONTENTS

13.1 INTRODUCTION

Coumarins, derivatives of cinnamic acid, are secondary metabolites that are widely distributed in the vegetal kingdom, both in the free form and as glycosides. More than 1300 coumarins have been isolated and reported from natural sources, particularly belonging to the Rutaceae, Apiaceae, Fabaceae, and Asteraceae families (Thuong et al. 2010). Coumarins show multiple bioactive properties. such as antioxidant, chemotherapeutic, antimicrobial, and anticoagulant activities, among others (Yu et al. 2005). The basic structure of coumarins, also known as benzo-2-pyrones, consists of the union of a benzene ring and a pyrane ring. Depending on the substituents, coumarins are structurally diverse, as they can yield hydroxylated, alkylated, and alkoxylated derivatives from the parent compound. They can even contain a furan ring fused to that of coumarin to generate furocoumarins, or they can be linked to another pyrane ring, generating pyranocoumarins (Smyth et al. 2011).

13.2 NATURAL OCCURRENCE OF COUMARINS IN CITRUS

Coumarins (or benzo-2-pyrones), as products from the phenylpropanoid pathway, can be classified into four categories: simple coumarins (benzo-α-pyrones), oxygenated coumarins (furanocoumarins), pyranocoumarins (benzodipyran-2-ones), and phenylcoumarins (benzo-benzopyrones); methoxylated and hydroxylated coumarins and furanocoumarins are the most abundant in citrus fruits (Bourgaud et al. 2006; Dugrand-Judek et al. 2015). Simple coumarins encompass the less complex coumarins and their hydroxylated, alkoxylated, alkylated, and glycosylated derivatives. Furanocoumarins have a typical molecular structure of a furan ring attached to the coumarin nucleus. According to the position of the furan ring, these metabolites are of either linear or angular types (Lin et al. 2014).

13.2.1 Biosynthesis of Coumarins

The biosynthesis of coumarins (Figure 13.1) proceeds from cinnamic acid, which is generated from primary metabolites through the shikimate and general phenylpropanoid pathways. This biosynthetic pathway has been widely discussed by several authors (Bourgaud et al. 2006; Lin et al. 2013; Talapatra and Talapatra 2015). In general terms, phenylalanine is converted into cinnamic acid by phenylalanine ammonia lyase; next, cinnamic acid can be converted into either *o*-coumaric acid or *p*-coumaric acid by cinnamic acid 2-hydroxylase or cinnamic acid 4-hydroxylase, respectively. This isomerization step is crucial for the formation of simple coumarin via lactonization of *o*-coumaric acid (Figure 13.1a), or for the production of 2,4-dihydroxycinnamic acid via 4-coumaric acid 2-hydroxylase. The next step entails conversion by lactonization into umbelliferone, the first hydroxylated coumarin (Figure 13.1b). Subsequent steps in the biosynthesis pathway from umbelliferone can produce other hydroxylated or methoxylated coumarins, and they can also be converted into furanocoumarins derivatives via prenyltransferase.

13.2.2 Main Coumarins in Citrus Fruits

Coumarin biosynthesis is mainly determined by the plant species. As mentioned above, furanocoumarins, methoxylated, and hydroxylated coumarins are the most abundant types in citrus; however, the citrus variety establishes the diversity of these metabolites. The main coumarins identified in citrus fruits are listed in Table 13.1. Also, the citrus variety is a determinant in the concentration of certain coumarins; thus, osthol is significantly more abundant in oranges, limettin is more abundant in lemons and limes, and bergamottin is more abundant in grapefruits. In dealing with the total content of coumarins, limes possess the maximum concentration of these metabolites, followed by grapefruits, with the lowest total concentration in mandarins.

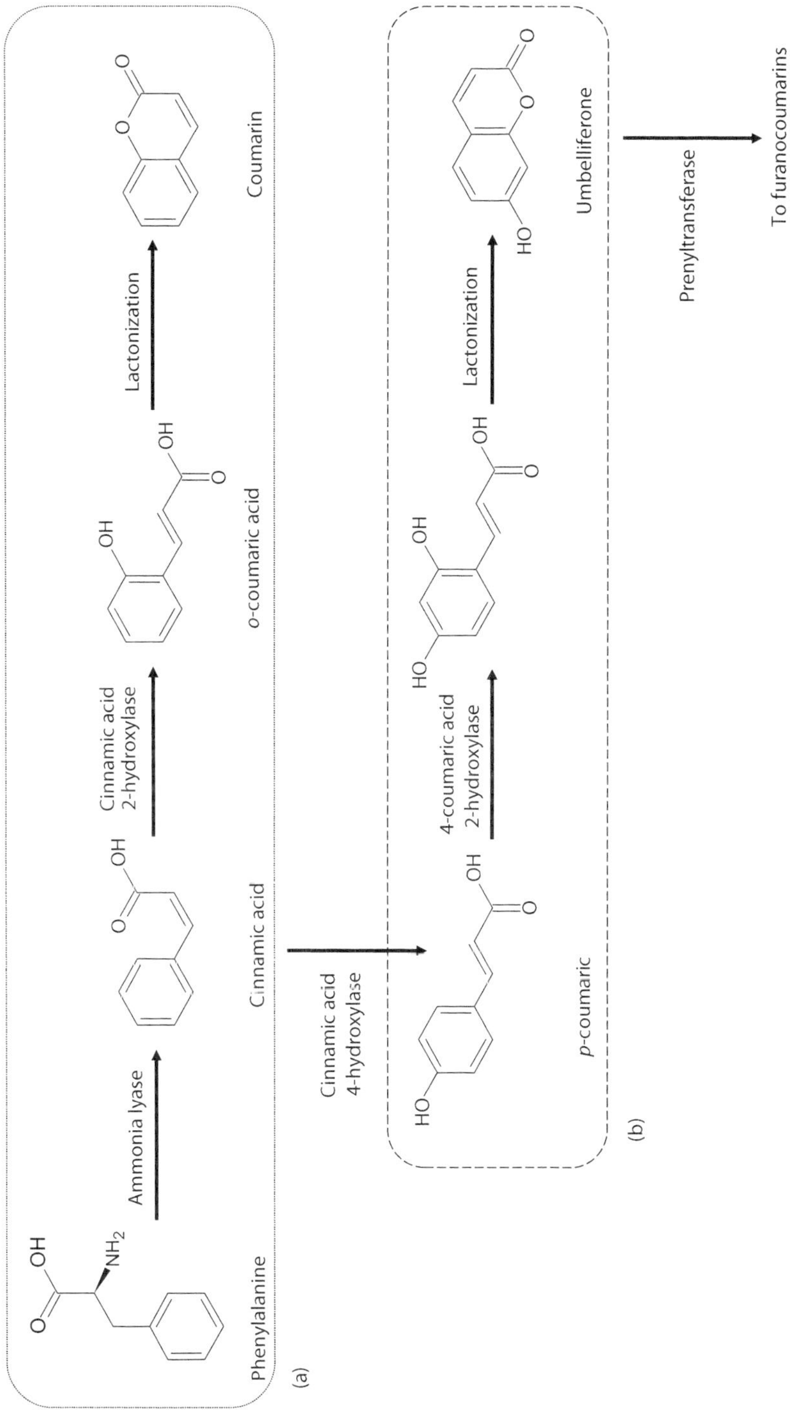

FIGURE 13.1 General metabolic pathways of natural occurring citrus coumarins. (a) Simple coumarins and (b) furanocoumarins.

TABLE 13.1
Main Coumarins Identified in Citrus Fruits

Compound	Oranges	Mandarins	Lemons	Grapefruits	Limes
		Coumarins			
Umbelliferone		x	x	x	x
Limettin	x	x	x	x	x
Osthol	x			x	
Scopoletin			x		x
Citropten			x		x
Aurapten		x	x	x	x
Epoxyaurapten	x			x	x
Esculetin	x	x			x
		Furanocoumarins			
Bergapten	x	x	x	x	x
Epoxybergamottin					
Oxypeucedanin	x	x	x	x	x
Bergamottin	x	x		x	x
Isopimpinellin	x	x			x
Bergaptol				x	
Epoxybergamottin	x			x	
Dihydroxybergamottin	x			x	
Psoralen				x	
Isoimperatorin	x		x		x
Heraclenin			x		x
Heraclenol			x		x
Imperatorin			x		x
Phellopterin			x		
Byakangelicin			x		
Byakangelicol			x		x
Xanthotoxin					x
Cnidilin					x

Sources: Barreca, D. et al., *Food Chemistry* 124(2):576–582, 2011a; Barreca, D. et al., *Food Chemistry* 129(4):1504–1512, 2011b; Dugrand-Judek, A. et al., *PLoS One* 10(11):e0142757, 2015; Ledesma-Escoba, C. A. et al., *Journal of Mass Spectrometry* 50(11):1196–1205, 2015b; Molina-Calle, M. et al., *Talanta* 141:150–157, 2015.

13.3 ISOLATION AND ANALYSIS OF COUMARINS FROM CITRUS

The growing demand for healthy and safe products has led to increasing interest of industries, mainly the food, pharmaceutical, and cosmetic industries, to replace synthetic additives with natural ones in both agricultural practices and processed products. In this sense, residues from the citrus industry represent a large source of

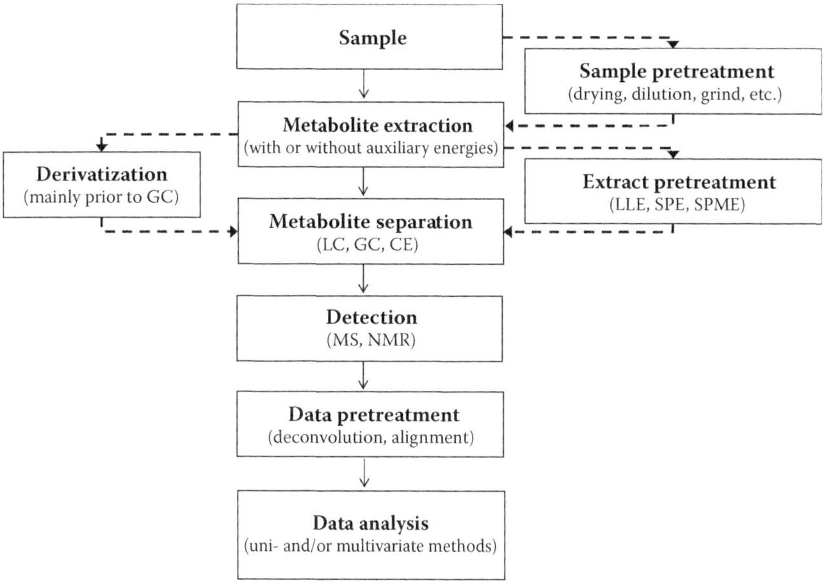

FIGURE 13.2 Scheme of the main steps involved in the study of bioactive compounds in citrus fruits.

natural bioactive compounds, mainly antioxidants as coumarins. In addition, the use of residues to obtain bioactive compounds can help to prevent serious contamination problems (Ledesma-Escobar and Luque de Castro 2014) and promote a comprehensive exploitation of citrus. A mandatory step for proper exploitation of these bioactive metabolites is the development of appropriate methods for sample pretreatment, extraction of the target compounds, and analysis of the extracts (Figure 13.2).

13.3.1 Sample Treatment Prior to the Extraction of Coumarins from Citrus

Sample pretreatment in studies of citrus is usually carried out by dehydration, with lyophilization or air drying as less common methods. Despite the large number of studies on extraction of citrus components, only in a few of them was the effect of sample pretreatment on the extractable metabolites evaluated. In fact, most of these studies dealt with the effect of different air drying temperatures, particularly with regard to phenolic compounds, and the results obtained with samples thus prepared were compared with those from fresh or lyophilized samples.

Lyophilization, also known as freeze-drying, consists of reducing the pressure on the drying chamber below the vapor pressure of ice, and drying is achieved by ice sublimation. In this way, it is possible to protect thermolabile compounds and minimize exposure to oxygen, which can produce undesirable oxidation reactions. Nevertheless, lyophilization is a relatively expensive dehydration process (Kasper and Friess 2011); while air drying is cheaper than lyophilization, the sample is exposed to heat and air for long intervals (Ledesma-Escobar and Luque de Castro

2014). This is why the common assumption is that lyophilization is the best sample pretreatment; however, in the case of citrus coumarins, this is not necessarily true. Recent studies conducted by the authors revealed that air drying increases both the concentration and the number of coumarins in polar lemon extracts. Thus, while in extracts from air-dried samples seven coumarins were identified, only five coumarins were identified in extracts from fresh samples; and the number was reduced to three coumarins in extracts from lyophilized samples. In addition, coumarins in extracts from air-dried samples were significantly more concentrated. The differences in the concentration of coumarins could be caused by the accelerated metabolism produced by increasing the temperature during air-drying treatment. The increased temperature could promote isomerization of coumarins, yielding others that naturally do not occur in lemons (Ledesma-Escobar et al. 2016b). These results suggest that air drying is the most suitable method for sample treatment prior to coumarin extraction.

13.3.2 Methods for Extraction of Coumarins from Citrus

The extraction method is of paramount importance in the quality of the final extract. Ideally, this step aims at (i) releasing the target metabolites from the sample in an efficient manner, (ii) avoiding the presence of undesirable metabolites, (iii) making the extract compatible with the analytical equipment, and (iv) concentrating trace metabolites, if necessary (Álvarez-Sánchez et al. 2010). Selection of the extraction step depends on the aim of the study; the solvents and extraction mode must be carefully chosen to obtain the expected results. In dealing with solid samples, the solid–liquid extraction step (properly known as leaching or lixiviation) can consist of maceration, Soxhlet extraction, ultrasound-assisted extraction (UAE), microwave assisted extraction (MAE), supercritical fluid extraction (SFE), or superheated liquid extraction (SHLE). Metabolites from liquids are mainly removed from the sample by liquid–liquid extraction (LLE), solid-phase extraction (SPE), or solid-phase microextraction (SPME).

In studies involving citrus coumarins, the raw material is usually a solid (dehydrated whole fruit, peels, or citrus residues from the juice industry); therefore, solid–liquid extraction based on shaking, UAE, or MAE is a common sample preparation step. Special attention must be paid to potential degradation of the target metabolites (particularly phenols) resulting from the effects of auxiliary energies, such as microwaves, ultrasound, or high temperatures (Ledesma-Escobar et al. 2016a; Qiao et al. 2014). Thanks to the polarity of coumarins, water, methanol, ethanol, or alcohol–water mixtures are commonly used as extractants (Dugrand-Judek et al. 2015; Ledesma-Escobar et al. 2015b; Molina-Calle et al. 2015). In dealing with liquid samples, such as cold-pressed grapefruit oil, LLE has been used to purify the oil, with hexane used to extract the lipid components, and coumarins have been determined in the residue (Chebrolu et al. 2013).

13.3.3 Overall and Individual Determination of Coumarins in Citrus Extracts

Determination of metabolites in a given extract can be aimed at the total content of target compounds endowed with a given characteristic, or at the concentration of individual compounds. An example of the former approach is the Folin–Ciocalteu (F–C) method, which is widely used for determination of total phenol compounds in citrus fruits extracts and provides fast information on the overall content of phenols. However, the F–C method provides no information about the given compounds; therefore, its use is limited to general determinations using a single reference standard, typically caffeic or gallic acid (Ledesma-Escobar and Luque de Castro 2014). Taking into account that this photometric method is based on the formation of characteristic blue complexes by interaction of the F–C reagent with the phenolic ring, a relative good estimator of the total content of coumarins in the extract is highly dependent on previous purification steps.

Individual determination requires prior separation of the target compounds. The most used separation methods for analysis of coumarins are based on liquid chromatography (LC), which allows good or acceptable separation of a given mixture into its individual components. Once separated in the chromatographic column, the metabolites reach a detector for individual analysis. The most common detector coupled to LC is of an ultraviolet (UV) molecular absorption nature, that is, the LC–UV configuration, which is simple to manage and relatively cheap and allows an acceptable accuracy for identification and quantitation of compounds (based on the absorption spectrum and retention time for each), providing that standards of the target compounds are available. The necessity for commercial standards makes it difficult (at most times impossible) to obtain the entire profile of the sample components (Ledesma-Escobar and Luque de Castro 2014), which in coumarin analyses is monitored in the wavelength range between 320 and 340 nm (Chu et al. 2012).

The use of mass spectrometry (MS) detectors for metabolites analysis is each time more extended because, depending on the type of instrument and method, this technique offers either high sensitivity for quantification or high mass resolution for identification. Hybrid MS equipment, like quadrupole time-of-flight (QTOF) instruments, produces tandem MS (MS/MS) spectra that provide information on structures and fragmentation patterns for subsequent tentative identifications supported by information in databases (Matsuda et al. 2009). Tentative identification based on MS/MS spectra is very dependent on the ability of the analyst to interpret the different data from neutral mass losses, common fragments, parent ions, possible adducts, and thus to solve the puzzle (Ledesma-Escobar et al. 2015b). Triple-quadrupole (QqQ) detectors provide a lower mass accuracy than other mass analyzers but they allow analysis with high sensitivity and selectivity; therefore, they are preferably used for quantitative analysis, as they monitor the precursor ion and fragments that provide greater sensitivity.

13.4 HEALTH PROPERTIES OF CITRUS COUMARINS

A number of studies have provided evidence that coumarins possess important biological activities, including antioxidant, anti-inflammatory, anticoagulant, antimicrobial, anticancer, and anti-AIDS properties (Azelmat et al. 2015; Kostova et al. 2011; Lin et al. 2013; Spino et al. 1998; Thakur et al. 2015). Also, the potential activities of this class of metabolites against neurodegenerative and cardiovascular diseases has been reported (Anand et al. 2012). These properties have promoted coumarins as metabolites with promising therapeutic applications.

13.4.1 Antioxidant Activities of Citrus Coumarins

Antioxidant activity is the most studied characteristic of citrus coumarins. This property is related to the capacity of some metabolites to react against reactive oxygen species (ROS) or reactive nitrogen species (RNS), which are commonly known as free radicals, either neutralizing them or preventing their formation. These radicals, produced during cellular metabolism, play a key role in cell signaling, apoptosis, gene expression, and ion transportation; nevertheless, excessive free radicals cause oxidative stress, which can damage DNA, RNA, proteins, and lipids and result in an increased risk for cardiovascular disease, cancer, neurodegenerative damage, autism, and other diseases (Fialkow et al. 2007; Lü et al. 2010). The consumption of dietary antioxidants such as coumarins may help to maintain a proper balance of free radicals in the body, thus preventing damages caused by the latter (Biesalski et al. 2009; Zou et al. 2016).

The antioxidant or antiradical activity of metabolites from natural products are most often exploited by using a raw extract without a previous separation–purification step, thus making it difficult (or even impossible) to obtain information on the antioxidant behavior of single metabolites in *in vitro* models. However, a few studies have used commercial standards of coumarins that occur naturally in citrus fruits and have produced evidence that these metabolites do not possess the same quality or strength of antioxidant activity as other do natural antioxidants, like flavonoids or phenolic acids (Foti et al. 1996). Based on a β-carotene–linoleic acid bleaching assay model, bergapten has been shown to possess less than 7% antioxidant activity, while flavonoids like scutellarein or kaempferol showed activities over 50%. Similar results were obtained in inhibiting the reduction of NitroBlue Tetrazolium, and in the α,α-diphenyl-β-picrylhydrazyl (DPPH) scavenging assay, as bergapten showed no activity against these free radicals under the working conditions (Yu et al. 2005). Nevertheless, at concentrations exceeding 1 mM, coumarins can be effective for inhibition of some enzymes, such as xanthine oxidase or guaiacol peroxidase (Payá et al. 1992).

Despite the lower activity of coumarins against ROS in *in vitro* models compared to other natural antioxidants, like flavonoids or phenolic acids, several reports indicate that these compounds are involved in the regulation of several cellular pathways and, therefore, they can be useful against some types of cancer and in reducing neurodegenerative damages, among other diseases.

13.4.2 Citrus Coumarins as Anticancer Agents

Naturally occurring coumarins, including those found in citrus fruits, are endowed with variable structures due to the various types of substitutions in their basic structures that can influence their biological activities. Thus, coumarins possess a range of anticarcinogenic activities, as shown in Figure 13.3, with minimum side effects, thanks to their ability to regulate diverse cellular pathways (Kostova 2005;

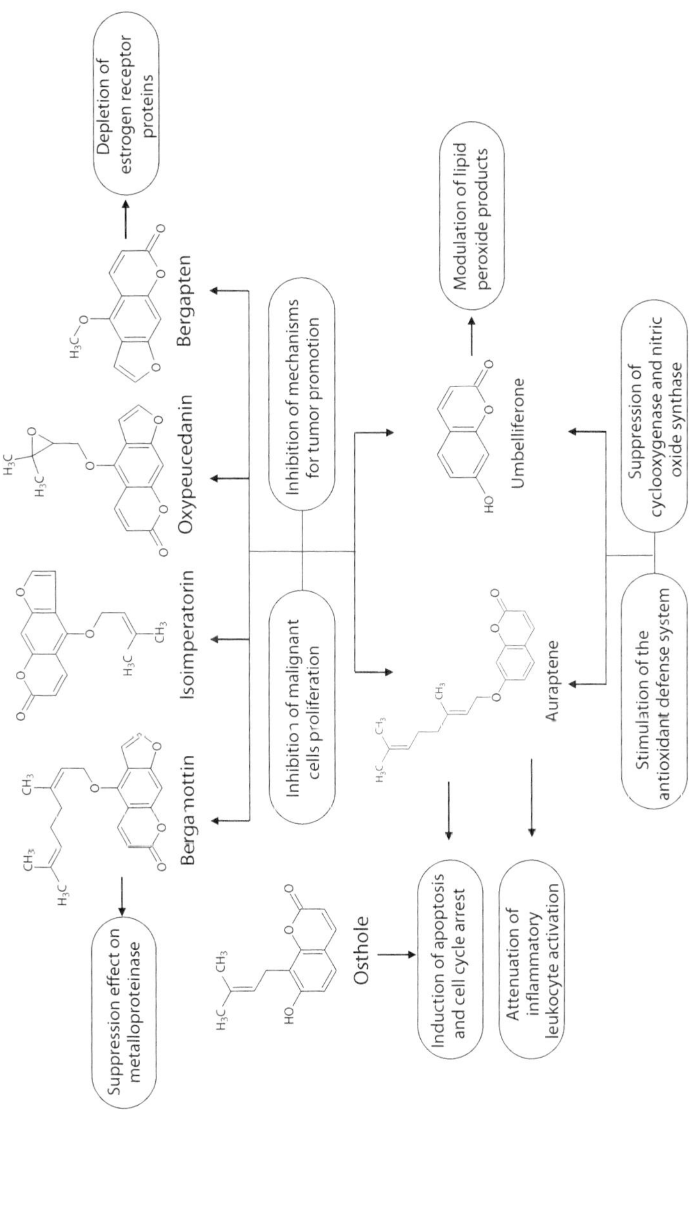

FIGURE 13.3 Main anticancer activities of natural occurring citrus coumarins.

Prince et al. 2009). In this sense, biological investigations have revealed that both natural and synthetic coumarins inhibit a number of pathways related to cancer, such as those of kinase, cell cycle arrest, angiogenesis, heat shock proteins, telomerase, antimitotic activity, carbonic anhydrase, monocarboxylate transporters, aromatases, and sulfatases, as the most important inhibitions these compounds can cause (Thakur et al. 2015).

In dealing with the given citrus coumarins, auraptene (AUR) is considered one of the promising chemopreventive agents against skin, tongue, esophagus, colon, and prostate cancer, based on tests in rodents (Murakami et al. 2000; Tang et al. 2007). The most important role of AUR against cancer is to prevent inflammation, which is a universal and physiological response in carcinogenesis processes, by attenuating inflammatory leukocyte activation. Thus, compared to a control group, AUR decreased by 43% the levels of edema formation, 85% H_2O_2 production, 92% leukocyte infiltration, and 84% proliferation of the cell nuclear antigen index in *in vivo* rodent models. Therefore, AUR might inhibit a mechanism for tumor promotion (Murakami et al. 1997, 2000; Tanaka et al. 1998). Also in *in vitro* models, both AUR and umbelliferone (UMB) have shown the potential to suppress the expression of cyclooxygenase and nitric oxide synthase (Kohno et al. 2006; Murakami et al. 2000).

Dietary administration of AUR in mice led to effects such as (i) significant inhibition of tongue carcinogenesis induced by 4-nitroquinoline-1-oxide during the initiation and postinitiation phases; (ii) reduction of the frequency of dysplastic lesions; (iii) inhibition of cell proliferation biomarker expression (Tanaka et al. 1998), and (iv) suppression of colitis-related colon carcinogenesis induced by dextran sodium sulfate and azoxymethane (Kohno et al. 2006). Also, UMB has shown promising effects against colon carcinogenesis in rats, as it modulates lipid peroxide products and stimulates the antioxidant defense system during 1,2-dimethylhydrazine-induced colon carcinogenesis (Muthu et al. 2013). In *in vitro* cancer cell line models, UMB induced the apoptosis of laryngeal cancer, reducing the viability and migration of malignant cells (Kiełbus et al. 2013).

As a chemopreventative agent against human prostate cancer, AUR induced apoptosis and cell cycle arrest, apparently by a non-androgen-mediated pathway (Tang et al. 2007), while osthole (OST) has shown cytotoxic and apoptotic induction effects against human lung carcinoma, neuroblastoma, and prostate cancer cell lines at low micromolar concentrations (Shokoohinia et al. 2014).

In dealing with furanocoumarins, bergapten has demonstrated both antiproliferative effects and proapoptotic responses in human breast cancer cells via depletion of estrogen receptor proteins (tamoxifen-sensitive and resistant cells), thereby preventing cross talk between the receptor and growth factor mitogenic signaling (Panno et al. 2012) or counteracting the stimulant effect of some growth factors on breast cancer cell growth and progression (Panno et al. 2009). On the other hand, bergamottin has shown the ability to inhibit both signal transducers and activators of transcription 3, which are closely related to growth, survival, proliferation, metastasis, and angiogenesis of various cancer cells, through the induction of tyrosine phosphatase, which makes bergamottin an effective suppressor of tumor cell survival proliferation and metastasis (Kim et al. 2014). This furanocoumarin also has a suppression effect on metalloproteinase, which plays a key role in the invasion and metastasis of cancer cells, and it thus contributes to antitumor activity (Hwang et al. 2010). The results obtained using oxypeucedanin in an *in vitro* human prostate carcinoma cell line DU145 model indicated that this furanocoumarin inhibits malignant cell growth by G_2/M cell cycle arrest, where repair might occur along with preparation for mitosis, and induces apoptotic cell death (Kang et al. 2009). The cytotoxic effects of oxypeucedanin and isoimperatorin against cultured human tumor cell lines, such as those for non-small cell lung, ovary, melanoma, central nervous system, and colon cancers, in *in vitro* models have also been reported (Kim et al. 2007).

Finally, several studies have provided evidence for the effects of both hydroxylated coumarins and furanocoumarins on promotion of the induction of phase II enzymes (detoxifying enzymes), such as glutathione *S*-transferase and quinone reductase, in the target organ, contributing to cancer chemopreventative properties (Muthu et al. 2013; Pokharel et al. 2006; Prince et al. 2009; Tanaka et al. 1997, 1998; Tang et al. 2007).

13.4.3 Citrus Coumarins as Neuroprotector Agents

Among the pharmacological properties of citrus coumarins, numerous studies have revealed that these compounds exert great activity on the central nervous system (CNS), acting both as preventive and therapeutic treatments of various neurodegenerative diseases, such as Alzheimer's disease (AD), Parkinson's disease (PD), cerebral ischemia, or even traumatic brain injury (TBI), thanks to their interaction with neurotransmitter receptors, inhibition of target oxidative enzymes, and anti-inflammatory properties (Skalicka-Wozniak et al. 2016), as summarized in Figure 13.4.

Amino acid-induced toxicity is one of the most frequently investigated targets for neuroprotection. In this sense, the activity of citrus coumarins against *N*-methyl-D-aspartate receptor (NMDA), which acts as a coagonist with glutamate, producing changes in glutamate transmission that are associated with a number of CNS pathologies (Trist 2000), has been investigated *in vitro* based on NMDA-induced toxicity in mixed cortical cell cultures that contain both neurons and astrocytes. The results of this study revealed that both AUR and isopentenyloxycoumarin possess a protective effect (83% and 71%, respectively, compared with the control) against NMDA-induced neurotoxicity, in particular at concentrations ranging from 1 to 10 μM. In addition, the anti-radical-scavenging capacities of both coumarins were tested in the DPPH assay, and the results revealed that these compounds do not exert this kind of activity; therefore, the authors concluded that the neuroprotective effect observed in their experiment was not due to a radical-scavenging effect (Epifano et al. 2008).

Concerning AD, several studies have revealed that both natural and synthetic coumarin derivatives have activity against this disease, which is a progressive degenerative disorder of the brain and the most common form of dementia. It is associated with a losses in the cholinergic system with decreased levels of acetylcholine in the brain areas that deal with learning, memory, behavior, and emotional responses (Anand et al. 2012). To prevent the metabolic hydrolysis of acetylcholine (Ach), most drugs are focused on inhibition of acetylcholinesterase (AChE) (Benzi and Moretti 1998). Regarding citrus coumarins, the potential of scopolin (SCN) and

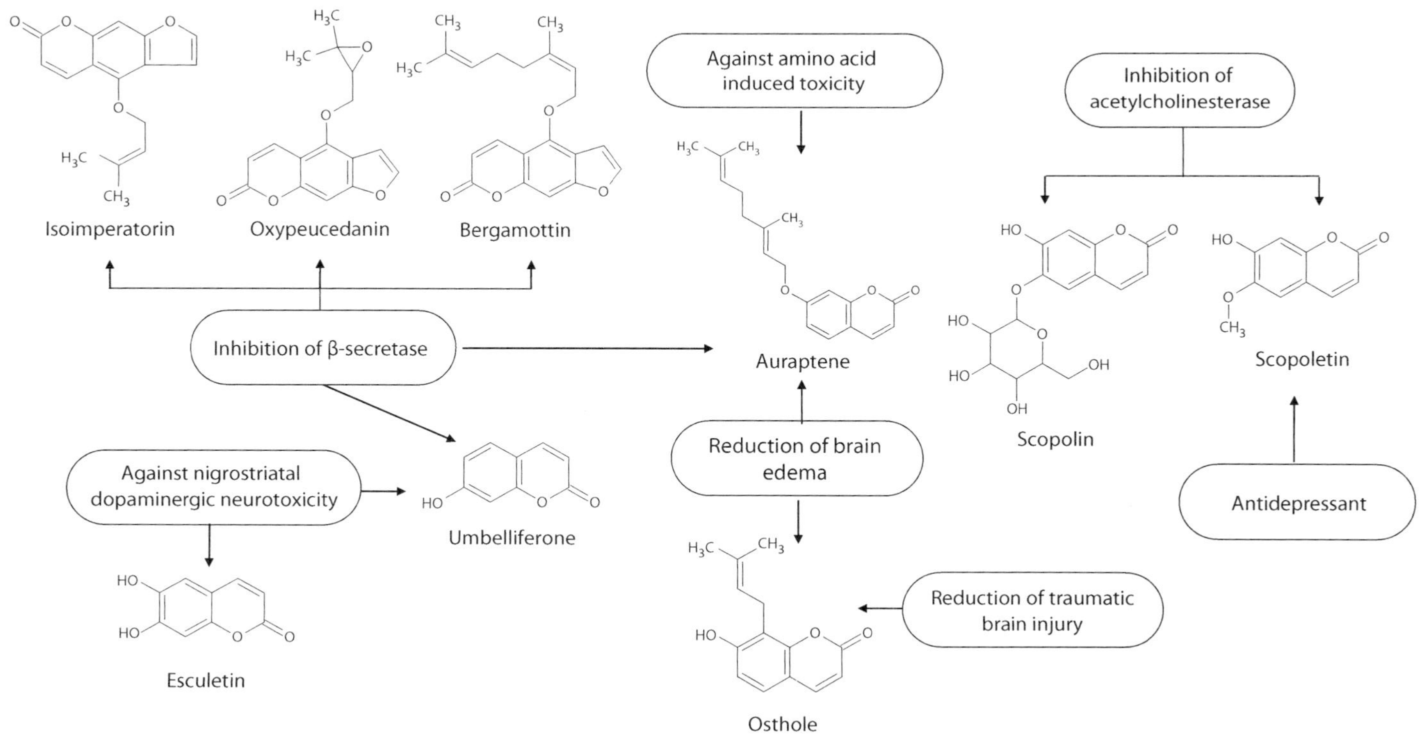

FIGURE 13.4 Main neuroprotective activities of natural occurring citrus coumarins.

its glucoside, scopoletin (SCT), as AChE inhibitors has been reported, based on a validated pharmacophore model. The results of this study suggested that both SCN and SCT are potential inhibitors of AChE by reduction of the extracellular concentration of ACh in the nucleus accumbens. In *in vivo* models, the activity of SCT was similar to that of galanthamine, one of the most potent alkaloids for treatment of AD (Rollinger et al. 2004).

On the other hand, given the role of the β-amyloid (Aβ) peptide as a central player in the pathogenesis of AD and the strong association between Aβ and AD, it is likely that therapeutic strategies to lower the levels of Aβ in the brain should prove beneficial for AD treatment. One such strategy could involve inhibition of β-secretase (BACE1), which is the key enzyme in Aβ generation (Vassar 2004). For this purpose, the potential of citrus coumarins as BACE1 inhibitors has been investigated in *in vitro* models, revealing a greater inhibitory activity by coumarins containing in their skeleton a geranyloxy group as a substituent. In fact, UMB (7-hydroxycoumarin) exhibited a lower inhibitory effect than AUR (7-geranyloxycoumarin), and similar results were observed with furanocoumarins; while bergapten (5-methoxypsoralene) showed no activity, bergamottin (5-geranyloxypsoralen) was the most potent BACE1 inhibitor among the 46 analyzed coumarins (Marumoto and Miyazawa 2012). Other furanocoumarins found in citrus, such as imperatorin, isoimperatorin, and oxypeucedanin, have also shown potential to inhibit BACE1 (Marumoto and Miyazawa 2010); however, all coumarins tested in both studies were less effective inhibitors of BACE1 than a statin-based synthetic peptidomimetic used as a reference. Despite the greater activity exhibited by a synthetic reference peptide, citrus coumarins could be useful for prevention or treatment of AD by inhibition of BACE1. Thanks to their low molecular weight and high lipophilicity, materials such as coumarins could easily reach the target action site in the brain following oral or transdermal administration, since the molecules cross the blood–brain barrier (Marumoto and Miyazawa 2012; Zhang et al. 2011).

The progressive degeneration of the dopaminergic nigrostriatal system in PD patients, which results in a deficiency of dopamine in the striatum and leads to the characteristic motor features of the disease (Kordower et al. 2013), could be avoided or mitigated by the effect of UMB and esculetin. These coumarins act against neurotoxins that cause degeneration of dopaminergic neurons, as evaluated *in vivo* using rodents exposed to the neurotoxin 1-methyl-4-phenyl-1,2,3,6-tetrahydro-pyridine (MPTP), which causes nigrostriatal dopaminergic neurotoxicity. The results showed that dietary administration of UMB and esculetin attenuated 75% of MPTP-induced neurotoxicity in the substantia nigra pars compacta compared to the control group fed butylated hydroxyanisole, which only attenuated 45% the MPTP-induced neurotoxicity under the studied conditions. The neuroprotective effect of these coumarins is linked to their ability to restore glutathione levels, reducing the progression of PD and preventing apoptosis (Subramaniam and Ellis 2013).

In addition to PD, dopaminergic neurotoxicity induced by oxidative stress can lead to depression disorders, which are among the top five leading causes of disability and disease burden throughout the world (Caspi et al. 2003). Animal models predictive of antidepressant action have been used for development of novel therapeutic compounds and for understanding the neural substrates underlying depressive behavior, as is the case for SCT, which has demonstrated a specific antidepressant-like effect in the tail suspension test, an animal model predictive of antidepressant activity in which a depressant-like behavior induced by acute immobility stress was able to be reversed. The results of SCT antidepressant activity were compared to those obtained with the commercial antidepressant fluoxetine, used as positive control, and the comparison revealed similar results, thanks to the interaction of SCT with the serotonergic, noradrenergic, and dopaminergic systems (Capra et al. 2010).

Citrus coumarins have also been used to prevent cerebral ischemia, which causes a focal or global insufficiency of blood flow to the brain, delaying neuronal cell death in the hippocampus and resulting in sequential cognitive impairments (Namura et al. 2013). In this sense, administration of OST (doses of 20 or 40 mg/kg of body weight) led to significantly reduced brain edema following acute ischemic stroke induced by middle cerebral artery occlusion (MCAO) in rats. The rats, pretreated with OST for 30 min before MCAO induction, exhibited a significant reduction in the infarct volume, cerebral edema, and neurological deficit scores by the decreased activity of both myeloperoxidase and inflammatory cytokines. The results of this study suggest that the mechanisms of the effect might be involved in its antioxidative action and anti-inflammatory property (Chao et al. 2010). Similarly, AUR (dose of 25 mg/kg to rats) was effective in suppressing neuronal cell death in the hippocampus, cyclooxygenase-2 expression, and microglial activation in rats, suggesting that AUR acts as a neuroprotective agent in the ischemic brain, which may be mediated by suppression of the inflammatory response (Okuyama et al. 2013).

Finally, OST demonstrated protective effects against TBI in adult rats. The animals pretreated with OST (20 or 40 mg/kg) 30 min before TBI showed a significant reduction of neurological deficits, cerebral edema, and hippocampal neuron loss 24 h after TBI compared with the control group. Also, OST reduced the level of oxidative stress and active caspase-3 expression, thus suggesting that OST may exert its neuroprotective effects via antioxidative and antiapoptotic properties against TBI in rats (He et al. 2012).

13.4.4 Other Health Effects of Coumarins

Some other bioactive properties of coumarins and derivatives have been reported and indicate that these compounds can act as antimicrobial agents, either inhibiting cell growth or reducing biofilm formation and virulence of some microorganisms (Ojala et al. 2000).

Citrus coumarins have been demonstrated to reduce biofilm formation and the virulence of *Escherichia coli* O157:H7. Thus, coumarin and UMB at 50 μg/mL was found to inhibit *E. coli* O157:H7 biofilm formation by more than 80% and 90%, respectively, compared with the control, without affecting bacterial growth; however, concentrations of 200 μg/mL had inhibitory cell growth effects. Unlike antibiotics, that aim to inhibit cell growth, biofilm inhibitors do not inhibit bacterial growth, but they may reduce the risk of drug resistance. The role of coumarins as inhibitors of biofilm formation seems to be due to antibiofilm activity and not to antimicrobial activity (Lee et al. 2014).

On the other hand, OST has been demonstrated effective to control a plant-pathogen, powdery mildew caused by *Sphaerotheca fuliginea*, on foliage. At concentration of 25 or 50 mg/mL, OST inhibits germination of *S. fuliginea* conidia by 85.7 or 100%, respectively. Compared to synthetic fungicides, OST was as effective as difenoconazole and more effective than triadimefon against *S. fuliginea*. This behavior suggests that OST is a promising natural fungicide to partially replace the common use of synthetic fungicides for powdery mildew control (Wang et al. 2009).

ACKNOWLEDGMENTS

We thank the Junta de Andalucía and FEDER program for financial support through Project FQM-1602. C.A.L.E. thanks the Mexican National Council for Science and Technology (CONACYT, No. 224905) for grants to support his research in both countries.

REFERENCES

Álvarez-Sánchez, B., F. Priego-Capote, and M. D. Luque de Castro. 2010. Metabolomics analysis. II. Preparation of biological samples prior to detection. *Trends in Analytical Chemistry* 29(2):120–127.

Anand, P., B. Singh, and N. Singh. 2012. A review on coumarins as acetylcholinesterase inhibitors for Alzheimer's disease. *Bioorganic and Medicinal Chemistry* 20(3):1175–1180.

Azelmat, J., S. Fiorito, V. A. Taddeo, S. Genovese, F. Epifano, and D. Grenier. 2015. Synthesis and evaluation of antibacterial and anti-inflammatory properties of naturally occurring coumarins. *Phytochemistry Letters* 13:399–405.

Barreca, D., E. Bellocco, C. Caristi, U. Leuzzi, and G. Gattuso. 2011a. Distribution of C- and O-glycosyl flavonoids, (3-hydroxy-3-methylglutaryl)glycosyl flavanones and furocoumarins in *Citrus aurantium* L. juice. *Food Chemistry* 124(2):576–582.

Barreca, D., E. Bellocco, C. Caristi, U. Leuzzi, and G. Gattuso. 2011b. Elucidation of the flavonoid and furocoumarin composition and radical-scavenging activity of green and ripe chinotto (*Citrus myrtifolia* Raf.) fruit tissues, leaves and seeds. *Food Chemistry* 129(4):1504–1512.

Benzi, G., and A. Moretti. 1998. Is there a rationale for the use of acetylcholinesterase inhibitors in the therapy of Alzheimer's disease? *European Journal of Pharmacology* 346(1):1–13.

Biesalski, H. K., L. O. Dragsted, I. Elmadfa, R. Grossklaus, M. Muller, D. Schrenk, and P. Weber. 2009. Bioactive compounds: Definition and assessment of activity. *Nutrition* 25(11–12):1202–1205.

Bourgaud, F., A. Hehn, R. Larbat, S. Doerper, E. Gontier, S. Kellner, and U. Matern. 2006. Biosynthesis of coumarins in plants: A major pathway still to be unravelled for cytochrome P450 enzymes. *Phytochemistry Reviews* 5(2–3):293–308.

Capra, J. C., M. P. Cunha, D. G. Machado, A. D. Zomkowski, B. G. Mendes, A. R. Santos, and A. L. Rodrigues. 2010. Antidepressant-like effect of scopoletin, a coumarin isolated from *Polygala sabulosa* (Polygalaceae) in mice: Evidence for the involvement of monoaminergic systems. *European Journal of Pharmacology* 643(2–3):232–238.

Caspi, A., K. Sugden, T. E. Moffitt, A. Taylor, I. W. Craig, H. Harrington, and R. Poulton. 2003. Influence of life stress on depression: Moderation by a polymorphism in the 5-HTT gene. *Science* 301(5631):386–389.

Chao, X., J. Zhou, T. Chen, W. Liu, W. Dong, Y. Qu, and Z. Fei. 2010. Neuroprotective effect of osthole against acute ischemic stroke on middle cerebral ischemia occlusion in rats. *Brain Research* 1363:206–211.

Chebrolu, K. K., G. K. Jayaprakasha, J. Jifon, and B. S. Patil. 2013. Purification of coumarins, including meranzin and pranferin, from grapefruit by solvent partitioning and a hyphenated chromatography. *Separation and Purification Technology* 116:137–144.

Chu, J., S. L. Li, Z. Q. Yin, W. C. Ye, and Q. W. Zhang. 2012. Simultaneous quantification of coumarins, flavonoids and limonoids in *Fructus citri sarcodactylis* by high performance liquid chromatography coupled with diode array detector. *Journal of Pharmaceutical and Biomedical Analysis* 66:170–175.

Dugrand-Judek, A., A. Olry, A. Hehn, G. Costantino, P. Ollitrault, Y. Froelicher, and F. Bourgaud. 2015. The distribution of coumarins and furanocoumarins in *Citrus* species closely matches citrus phylogeny and reflects the organization of biosynthetic pathways. *PLoS One* 10(11):e0142757.

Epifano, F., G. Molinaro, S. Genovese, R. T. Ngomba, F. Nicoletti, and M. Curini. 2008. Neuroprotective effect of prenyloxycoumarins from edible vegetables. *Neuroscience Letters* 443(2):57–60.

Fialkow, L., Y. Wang, and G. P. Downey. 2007. Reactive oxygen and nitrogen species as signaling molecules regulating neutrophil function. *Free Radical Biology and Medicine* 42(2):153–164.

Foti, M., M. Piattelli, M. T. Baratta, and G. Ruberto. 1996. Flavonoids, coumarins, and cinnamic acids as antioxidants in a micellar system. Structure–activity relationship. *Journal of Agricultural and Food Chemistry* 44(2):497–501.

He, Y., S. Qu, J. Wang, X. He, W. Lin, H. Zhen, and X. Zhang. 2012. Neuroprotective effects of osthole pretreatment against traumatic brain injury in rats. *Brain Research* 1433:127–136.

Hwang, Y. P., H. J. Yun, J. H. Choi, K. W. Kang, and H. G. Jeong. 2010. Suppression of phorbol-12-myristate-13-acetate-induced tumor cell invasion by bergamottin via the inhibition of protein kinase Cδ/p38 mitogen-activated protein kinase and JNK/nuclear factor-κB-dependent matrix metalloproteinase-9 expression. *Molecular Nutrition and Food Research* 54(7):977–990.

Kang, T. J., S. Y. Lee, R. P. Singh, R. Agarwal, and D. S. Yim. 2009. Anti-tumor activity of oxypeucedanin from Ostericum koreanum against human prostate carcinoma DU145 cells. *Acta Oncologica* 48(6):895–900.

Kasper, J. C., and W. Friess. 2011. The freezing step in lyophilization: Physico-chemical fundamentals, freezing methods and consequences on process performance and quality attributes of biopharmaceuticals. *European Journal of Pharmaceutics and Biopharmaceuticals* 78(2):248–263.

Kiełbus, M., K. Skalicka-Woźniak, A. Grabarska, W. Jeleniewicz, M. Dmoszyńska-Graniczka, A. Marston, and A. Stepulak. 2013. 7-Substituted coumarins inhibit proliferation and migration of laryngeal cancer cells in vitro. *Anticancer Research* 33(10):4347–4356.

Kim, S. M., J. H. Lee, G. Sethi, C. Kim, S. H. Baek, D. Nam, and K. S. Ahn. 2014. Bergamottin, a natural furanocoumarin obtained from grapefruit juice induces chemosensitization and apoptosis through the inhibition of STAT3 signaling pathway in tumor cells. *Cancer Letters* 354(1):153–163.

Kim, Y.-K., Y. S. Kim, and S. Y. Ryu. 2007. Antiproliferative effect of furanocoumarins from the root of *Angelica dahurica* on cultured human tumor cell lines. *Phytotherapy Research* 21(3):288–290.

Kohno, H., R. Suzuki, M. Curini, F. Epifano, F. Maltese, S. P. Gonzales, and T. Tanaka. 2006. Dietary administration with prenyloxycoumarins, auraptene and collinin, inhibits colitis-related colon carcinogenesis in mice. *International Journal of Cancer* 118(12):2936–2942.

Kordower, J. H., C. W. Olanow, H. B. Dodiya, Y. Chu, T. G. Beach, C. H. Adler, and R. T. Bartus. 2013. Disease duration and the integrity of the nigrostriatal system in Parkinson's disease. *Brain* 136(8):2419–2431.

Kostova, I. 2005. Synthetic and natural coumarins as cytotoxic agents. *Current Medicinal Chemistry Anti Cancer Agents* 5(1):29–46.

Kostova, I., S. Bhatia, P. Grigorov, S. S. Balkansky, V. K. Parmar, A. Prasad, and L. Saso. 2011. Coumarins as antioxidants. *Current Medicinal Chemistry* 18(25):3929–3951.

Ledesma-Escobar, C. A., and M. D. Luque de Castro. 2014. Towards a comprehensive exploitation of citrus. *Trends in Food Science and Technology* 39(1):63–75.

Ledesma-Escobar, C. A., F. Priego-Capote, and M. D. Luque de Castro. 2015a. Comparative study of the effect of auxiliary energies on the extraction of citrus fruit components. *Talanta* 144:522–528.

Ledesma-Escobar, C. A., F. Priego-Capote, and M. D. Luque de Castro. 2015b. Characterization of lemon (*Citrus limon*) polar extract by liquid chromatography–tandem mass spectrometry in high resolution mode. *Journal of Mass Spectrometry* 50(11):1196–1205.

Ledesma-Escobar, C. A., F. Priego-Capote, and M. D. Luque de Castro. 2016a. Comparative study of the effect of sample pretreatment and extraction on the determination of flavonoids from lemon (*Citrus limon*). *PLoS One* 11(1):e0148056.

Ledesma-Escobar, C. A., F. Priego-Capote, and M. D. Luque de Castro. 2016b. Effect of sample pretreatment on the extraction of lemon (*Citrus limon*) components. *Talanta* 153:386–391.

Lee, J. H., Y. G. Kim, H. S. Cho, S. Y. Ryu, M. H. Cho, and J. Lee. 2014. Coumarins reduce biofilm formation and the virulence of *Escherichia coli* O157:H7. *Phytomedicine* 21(8–9):1037–1042.

Lin, Y., R. Jain, and Y. Yan. 2014. Microbial production of antioxidant food ingredients via metabolic engineering. *Current Opinion in Biotechnology* 26:71–78.

Lin, Y., X. Sun, Q. Yuan, and Y. Yan. 2013. Combinatorial biosynthesis of plant-specific coumarins in bacteria. *Metabolic Engineering* 18:69–77.

Lü, J.-M., P. H. Lin, Q. Yao, and C. Chen. 2010. Chemical and molecular mechanisms of antioxidants: Experimental approaches and model systems. *Journal of Cellular and Molecular Medicine* 14(4):840–860.

Marumoto, S., and M. Miyazawa. 2010. β-Secretase inhibitory effects of furanocoumarins from the root of *Angelica dahurica*. *Phytotherapy Research* 24(4):510–513.

Marumoto, S., and M. Miyazawa. 2012. Structure-activity relationships for naturally occurring coumarins as beta-secretase inhibitor. *Bioorganic and Medicinal Chemistry* 20(2):784–788.

Matsuda, F., K. Yonekura-Sakakibara, R. Niida, T. Kuromori, K. Shinozaki, and K. Saito. 2009. MS/MS spectral tag-based annotation of non-targeted profile of plant secondary metabolites. *Plant Journal* 57(3):555–577.

Molina-Calle, M., F. Priego-Capote, and M. D. Luque de Castro. 2015. Ultrasound-assisted emulsification–extraction of orange peel metabolites prior to tentative identification by LC–QTOF MS/MS. *Talanta* 141:150–157.

Murakami, A., W. Kuki, Y. Takahashi, H. Yonei, Y. Nakamura, Y. Ohto, and K. Koshimizu. 1997. Auraptene, a citrus coumarin, inhibits 12-O-tetradecanoylphorbol-13-acetate-induced tumor promotion in ICR mouse skin, possibly through suppression of superoxide generation in leukocytes. *Japanese Journal of Cancer Research* 88(5):443–452.

Murakami, A., Y. Nakamura, T. Tanaka, K. Kawabata, D. Takahashi, K. Koshimizu, and H. Ohigashi. 2000. Suppression by citrus auraptene of phorbol ester- and endotoxin-induced inflammatory responses: Role of attenuation of leukocyte activation. *Carcinogenesis* 21(10):1843–1850.

Muthu, R., P. Thangavel, N. Selvaraj, R. Ramalingam, and M. Vaiyapuri. 2013. Synergistic and individual effects of umbelliferone with 5-flurouracil on the status of lipid peroxidation and antioxidant defense against 1,2-dimethylhydrazine induced rat colon carcinogenesis. *Biomedicine and Preventive Nutrition* 3(1):74–82.

Namura, S., H. Ooboshi, J. Liu, and M. A. Yenari. 2013. Neuroprotection after cerebral ischemia. *Annals of the New York Academy of Science* 1278(1):25–32.

Ojala, T., S. Remes, P. Haansuu, H. Vuorela, R. Hiltunen, K. Haahtela, and P. Vuorela. 2000. Antimicrobial activity of some coumarin containing herbal plants growing in Finland. *Journal of Ethnopharmacology* 73(1–2):299–305.

Okuyama, S., S. Minami, N. Shimada, N. Makihata, M. Nakajima, and Y. Furukawa. 2013. Anti-inflammatory and neuroprotective effects of auraptene, a citrus coumarin, following cerebral global ischemia in mice. *European Journal of Pharmacology* 699(1–3):118–123.

Panno, M. L., F. Giordano, M. G. Palma, V. Bartella, V. Rago, M. Maggiolini, and S. Ando. 2009. Evidence that bergapten, independently of its photoactivation, enhances p53 gene expression and induces apoptosis in human breast cancer cells. *Current Cancer Drug Targets* 9(4):469–481.

Panno, M. L., F. Giordano, P. Rizza, M. Pellegrino, D. Zito, C. Giordano, and S. Andò. 2012. Bergapten induces ER depletion in breast cancer cells through SMAD4-mediated ubiquitination. *Breast Cancer Research and Treatment* 136(2):443–455.

Payá, M., B. Halliwell, and J. R. S. Hoult. 1992. Interactions of a series of coumarins with reactive oxygen species. *Biochemical Pharmacology* 44(2):205–214.

Pokharel, Y. R., E. H. Han, J. Y. Kim, S. J. Oh, S. K. Kim, E. R. Woo, and K. W. Kang. 2006. Potent protective effect of isoimperatorin against aflatoxin B1-inducible cytotoxicity in H4IIE cells: Bifunctional effects on glutathione S-transferase and CYP1A. *Carcinogenesis* 27(12):2483–2490.

Prince, M., Y. Li, A. Childers, K. Itoh, M. Yamamoto, and H. E. Kleiner. 2009. Comparison of citrus coumarins on carcinogen-detoxifying enzymes in Nrf2 knockout mice. *Toxicology Letters* 185(3):180–186.

Qiao, L., Y. Sun, R. Chen, Y. Fu, W. Zhang, X. Li, and X. Ye. 2014. Sonochemical effects on 14 flavonoids common in citrus: Relation to stability. *PLoS One* 9(2):e87766.

Rollinger, J. M., A. Hornick, T. Langer, H. Stuppner, and H. Prast. 2004. Acetylcholinesterase inhibitory activity of scopolin and scopoletin discovered by virtual screening of natural products. *Journal of Medicinal Chemistry* 47(25):6248–6254.

Shokoohinia, Y., L. Hosseinzadeh, M. Alipour, A. Mostafaie, and H.-R. Mohammadi-Motlagh. 2014. Comparative evaluation of cytotoxic and apoptogenic effects of several coumarins on human cancer cell lines: Osthole induces apoptosis in p53-deficient H1299 cells. *Advances in Pharmacological Sciences* 2014:847574.

Skalicka-Wozniak, K., I. E. Orhan, G. A. Cordell, S. M. Nabavi, and B. Budzynska. 2016. Implication of coumarins towards central nervous system disorders. *Pharmacological Research* 103:188–203.

Smyth, W. F., J. L. Morgan, E. O'Kane, T. J. Millar, and V. N. Ramachandran. 2011. The characterisation of coumarins from selected structural classes by electrospray ionisation quadrupole time-of-flight tandem mass spectrometry. *Rapid Communications in Mass Spectrometry* 25(9):1308–1314.

Spino, C., M. Dodier, and S. Sotheeswaran. 1998. Anti-HIV coumarins from calophyllum seed oil. *Bioorganic and Medicinal Chemistry Letters* 8(24):3475–3478.

Subramaniam, S. R., and E. M. Ellis. 2013. Neuroprotective effects of umbelliferone and esculetin in a mouse model of Parkinson's disease. *Journal of Neuroscience Research* 91(3):453–461.

Talapatra, S., and B. Talapatra. 2015. Shikimic acid pathway, in *Chemistry of Plant Natural Products*, pp. 625–678. Heidelberg: Springer Verlag.

Tanaka, T., K. Kawabata, M. Kakumoto, H. Makita, A. Hara, H. Mori, and H. Ohigashi. 1997. Citrus auraptene inhibits chemically induced colonic aberrant crypt foci in male F344 rats. *Carcinogenesis* 18(11):2155–2161.

Tanaka, T., K. Kawabata, M. Kakumoto, K. Matsunaga, H. Mori, A. Murakami, and H. Ohigashi. 1998. Chemoprevention of 4-nitroquinoline 1-oxide-induced oral carcinogenesis by citrus auraptene in rats. *Carcinogenesis* 19(3):425–431.

Tang, M., K. Ogawa, M. Asamoto, N. Hokaiwado, A. Seeni, S. Suzuki, and T. Shirai. 2007. Protective effects of citrus nobiletin and auraptene in transgenic rats developing adenocarcinoma of the prostate (TRAP) and human prostate carcinoma cells. *Cancer Science* 98(4):471–477.

Thakur, A., R. Singla, and V. Jaitak. 2015. Coumarins as anticancer agents: A review on synthetic strategies, mechanism of action and SAR studies. *European Journal of Medicinal Chemistry* 101:476–495.

Thuong, P. T., T. M. Hung, T. M. Ngoc, D. T. Ha, B. S. Min, S. J. Kwack, and K. Bae. 2010. Antioxidant activities of coumarins from Korean medicinal plants and their structure–activity relationships. *Phytotherapy Research* 24(1):101–106.

Trist, D. G. 2000. Excitatory amino acid agonists and antagonists: Pharmacology and therapeutic applications. *Pharmaceutica Acta Helvetiae* 74(2–3):221–229.

Vassar, R. 2004. BACE 1. *Journal of Molecular Neuroscience* 23(1):105–113.

Wang, C. M., W. Zhou, C. X. Li, H. Chen, Z.-Q. Shi, and Y. J. Fan. 2009. Efficacy of osthol, a potent coumarin compound, in controlling powdery mildew caused by *Sphaerotheca fuliginea*. *Journal of Asian Natural Products Research* 11(9):783–791.

Yu, J., L. Wang, R. L. Walzem, E. G. Miller, L. N. Pike, and B. S. Patil. 2005. Antioxidant activity of citrus limonoids, flavonoids, and coumarins. *Journal of Agricultural and Food Chemistry* 53(6):2009–2014.

Zhang, X., Y. Xie, W. Cao, Q. Yang, S. Miao, and S. Wang. 2011. Brain distribution study of imperatorin in rats after oral administration assessed by HPLC. *Chromatographia* 74(3):259–265.

Zou, Z., W. Xi, Y. Hu, C. Nie, and Z. Zhou. 2016. Antioxidant activity of citrus fruits. *Food Chemistry* 196:885–896.

14 Main Industrial Citrus By-Products in Spain—Citrus Dietary Fiber

A Health-Promoting Functional Food Ingredient

Francisco Marin

CONTENTS

14.1 INTRODUCTION

Citrus is the most abundant fruit tree crop in the world. Its production and consumption have grown rapidly from an average harvest of 57 × 10^6 metric tons per year during the 1980s, up to in excess of 115 × 10^6 metric tons in the 2010/2011 season. Among the most important citrus are oranges (70.6 million tons), mandarins (25.5 million tons), lemons and limes (12.9 million tons), and grapefruits (6.4 million tons) (FAO 2012).

Gross fresh citrus consumption has grown over the last few decades. During the 1980s, 35.2 million tons per year was allocated for fresh fruits, reaching an amount of 86 million tons during the 2010/2011 season. In contrast, former figures indicate that 19.8 million tons of citrus was industrially processed per year during the 1980s, and 25 million tons during the 2010/2011 season, which means that the percentage of industrialized citrus has notably decreased from 36% to 25% of the annual citrus crop (FAO 2012). In any case, 25 million tons is an immense amount.

The citrus processing industry has focused for many years not only on the production of juices, but also on the canning industry, for example, production of marmalade and segments of mandarin, and the chemical industry, for example, extraction of flavonoids and essential oils (Ortuño et al. 1997; Izquierdo and Sendra 2003). Nonedible citrus residues represent between 50% and 65% of the total weight of the fruit and remain as the primary residue, resulting in large amounts of peels which are almost a quarter of the whole fruit mass, seeds, and fruit pulp (Braddock 1999; Mandalari et al. 2006), which may be estimated to exceed 15 × 10^6 metric tons per year (Marin et al. 2007). These solid residues, referred to as citrus waste, constitute a severe environmental problem (McNary et al. 1957; Laufenberg et al. 2003; Montgomery 2004) that has led to the generation of large quantities of putrefying waste in some regions, presenting a significant risk to local watercourses and in some cases leading to uncontrolled methane production. This major environmental problem associated with citrus peel is due to its highly fermentable carbohydrate content, which accelerates its degradation under the warm climate of the citrus belt (Lin et al. 2013; Mamma and Christakopoulus 2014).

Other conservative disposal routes for citrus waste have included utilizing it as raw material in the manufacture of cattle feed upon drying or combusting it (Vergamini et al. 2015; Volpe et al. 2015). Apart from the transport costs, cattle feed needs to be dried, decreasing moisture content from 80% to 10%, which is highly energy intensive and costly, to render a protein content usually lower than 10% in dry weight (Marin et al. 2007). In addition to the low nutritional value of citrus waste for cattle feed, it can cause diseases to the consuming animals (e.g., mycotoxicosis, ruminal parakeratosis), proving this end use to be of low profitability and with limited application (Duoss-Jennings et al. 2013; Ledesma-Escobar and De Castro 2014). In that respect, combustion would also generate global warming gases contributing to pollution, although some research approaches to using it as fuel have been carried out (Wilkins et al. 2007; Lohrasbi et al. 2010; Lin et al. 2013). Industrial citrus wastes have also been traditionally used as raw material in the chemical industry for the production of pectins (Laufenberg et al. 2003), often employed to increase the viscosity of liquid or semisolid foods, and for the extraction of flavonoids (Ortuño et al. 1997).

The aforementioned conservative routes for citrus do not account for the potentiality of recycling value compounds present in the food waste, which is in disagreement with the new tendencies of environmental protection worldwide (Kosseva 2013). At the same time, due to the huge amount of food materials discharged worldwide and the existing technologies able to recover and recycle high-added value compounds inside the food chain, these routes have generated interesting perspectives (Galanakis 2012). This convergence has stimulated recent significant interest in developing more responsible ways of dealing with waste products, ideally with the added benefit of yielding high-value products and establishing more responsible approaches.

This chapter reviews the state of the art of the use and the potential use of bioactive compounds and citrus by-products obtained from citrus waste.

14.2 CITRUS FRUIT AND ITS BIOACTIVE COMPONENTS

Oranges, lemons, limes, tangerines, and grapefruits, among others, are all *Citrus* spp., family Rutaceae, which bear fruits in the form of hesperidia. A hesperidium is a modified berry with a tough, leathery rind. The peel contains volatile oil glands in pits. The fleshy interior is composed of separate sections, called carpels, filled with fluid-filled vesicles that are actually specialized hair cells. The outer ovary wall becomes the thick spongy layer of the rind, while the inner ovary wall becomes very juicy with several seeds (Swingle and Reece 1967).

From a more practical point of view, the citrus fruit is clearly split in two parts: peel and flesh (Figure 14.1). The peel consists of flavedo and albedo. The flavedo (or exocarp) is the outermost layer of the fruit and bears oil glands and pigments. Flavedo is mostly composed of cellulosic material and also contains other components, such as essential oils, paraffin waxes, steroids and triterpenoids, fatty acids, pigments (carotenoids, chlorophylls), flavonoids, and bitter principles such as limonoids, among others. When ripe, flavedo cells contain carotenoids, mostly xanthophyll, inside chromoplasts, which in a previous developmental stage contain chlorophyll. The internal region of the flavedo is rich in multicellular bodies that have spherical or pyriform shapes and are full of essential oils. The albedo (or mesocarp) is contiguous with the flavedo. It is the inner part of the peel which is commonly removed before eating and is rich in dietary fiber, especially in pectin. The endocarp is separated into two sections, which are most commonly called segments, and the juicy pulp filling the segments is usually referred to as juice sacs or vesicles (Schneider 1968; Izquierdo and Sendra 2003).

Most of industrialized citrus are submitted to obtain juice, which leads to the bulk of citrus waste being constituted of peel and the extenuated pulp of mature fruits, with a minor contribution of immature whole fruits used in the chemical industry to extract bioactive compounds. The fruit developmental stage determines the chemical composition of wastes, what can be obtained from citrus wastes, and what can be done with them. The precise bioactive principle content may vary from one citrus waste to another, depending on the *Citrus* species (i.e., lemon, lime, mandarin, grapefruit, and so on), on the phenological stage of the plant (e.g., mature fruit vs. immature, ripeness), on the cultural conditions, or even on the industrial process undergone by the citrus fruit; this last element can affect waste composition in such a way that the type of squeezer machine can cause variations not only to the juice composition but also to the citrus waste (Marin et al. 2002a). Although averages for citrus wastes composition can be found elsewhere (Braddock 1999), shown in Table 14.1 are the effects of specie and industrial process on the chemical composition of

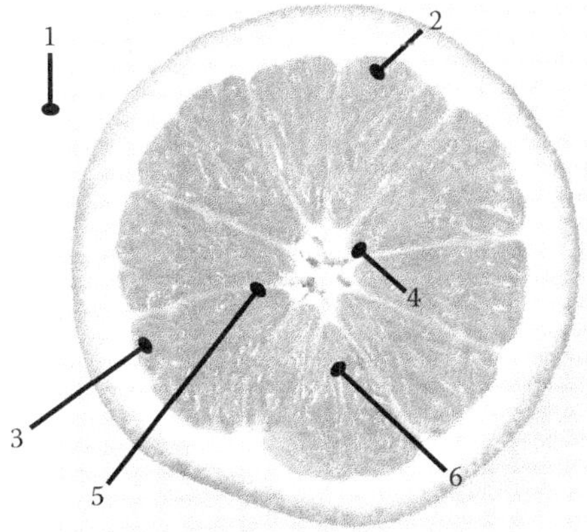

FIGURE 14.1 Structure of citrus fruit. Peel: *1*, flavedo or exocarp; *2*, albedo or mesocarp. Pulp or endocarp: *3*, pulp segment; *4*, segment wall; *5*, central wall; *6*, juice sac or vesicle.

TABLE 14.1
Chemical Compositions of Different Citrus By-Products

Industrial Process	*Citrus* Species (Fruit Part)	Ash (% DW)	Sugar (% DW)	Fats (% DW)	Proteins (% DW)	Flavonoids (% DW)	Fiber (% DW)
Extraction of flavonoids	Sour orange (whole mature fruit)	11.03 ± 0.29	8.01 ± 0.23	0.55 ± 0.03	12.87 ± 0.90	1.51 ± 0.02	48.58 ± 2.22
Extraction of flavonoids	Satsuma (peels)	10.03 ± 0.15	8.50 ± 0.20	1.57 ± 0.06	7.33 ± 0.33	3.06 ± 0.06	57.69 ± 3.35
Extraction of flavonoids	Grapefruit (whole mature fruit)	8.09 ± 0.41	8.02 ± 0.36	0.52 ± 0.02	12.51 ± 0.87	3.04 ± 0.03	52.25 ± 3.09
Canning industry	Satsuma (peels)	5.05 ± 0.22	10.07 ± 0.54	1.59 ± 0.09	7.50 ± 0.21	5.09 ± 0.05	53.12 ± 2.11
Juice industry	Lemon (peels)	2.52 ± 0.15	6.52 ± 0.48	1.51 ± 0.11	7.00 ± 0.44	12.54 ± 0.62	51.71 ± 1.92
Juice industry	Lemon (pulp)	2.54 ± 0.05	9.01 ± 0.87	3.09 ± 0.12	8.72 ± 0.36	4.52 ± 0.10	77.17 ± 3.24
Juice industry	Sweet orange (peels)	2.55 ± 0.09	6.04 ± 0.41	1.52 ± 0.05	6.55 ± 0.32	11.00 ± 0.54	51.67 ± 2.01
Juice industry	Sweet orange (pulp)	2.56 ± 0.10	9.57 ± 0.22	4.00 ± 0.15	9.06 ± 0.38	4.50 ± 0.15	78.66 ± 3.10

Note: The values are expressed as average ± effect size (ES) estimated as range ($n = 7$).

several kinds of citrus wastes from research work carried out by our group (Marin et al. 2007).

A large number of studies have been carried out to extract and identify bioactive components in different parts of citrus fruits, and therefore in citrus waste, to gain a deeper knowledge of their relationship with diet, health benefits or reduced risk of diseases. Natural products present in the peel such as fiber, for example, cellulose or pectin, flavonoids, vitamin C, carotenoids, folic acid, and essential oils, among others, have shown to be very useful both in the food industry and in the maintenance of human health (Patil et al. 2009; González-Molina et al. 2010).

Generally, these kinds of research have focused on specific families of chemicals. Thus, special attention has been paid to dietary fiber, in particular to pectin. In this way, the pectin found in citrus peel has been used not only in the food industry as a gelling agent but also in the pharmaceutical industry as an ingredient for the preparation of antidiarrheal drugs, as a detoxifying agent, and for preparation of suspensions for controlled drug release (Liu et al. 2003; Piriyaprasar and Sriamornsak 2011). Special attention will be paid to this component.

The first thing we notice in a citrus fruit is its color. Except for some rare varieties of oranges (blood oranges) that show a dark (as dark as venous blood) red color, mainly produced by cyanidin 3-*O*-glucoside (Felgines et al. 2008), the range of color from the pale yellow of some lemons, through the more subdued color of pummelo and grapefruit, to bright orange, are all caused by carotenoids. Carotenoids are well recognized as health-promoting chemicals, and information regarding this topic can be found elsewhere (see, e.g., Rao and Rao 2007; Britton et al. 2009). In the specific case of carotenoid from citrus, benefits such as the following have been described: reduction in the risk of breast and colon cancer (Fraser and Bramely 2004) and some help with sunburn and lipid peroxidation in human skin cells induced by UV radiation (Aust et al. 2001).

Another group of interesting bioactive compounds in citrus is that of essential oils. They represent between 2% and 3% of dry citrus peel with slight variations in composition depending on the *Citrus* species (Shaw 1979; Mustafa 2015). These oils, valued for their flavor and fragrance that allow for their use as natural food flavoring, have been shown to possess antibacterial and antifungal properties, and when these are added to food and subjected to thermal treatment for preservation, no alteration of food organoleptic properties is produced (Tyagi et al. 2014; Mustafa 2015). The former is the reason why citrus essential oils could be considered as suitable alternative to chemical additives for use in the food industry, attending to the needs of safety and satisfying the demand of consumers for natural food components (Viuda-Martos et al. 2008). Although antimicrobial and flavor properties cannot be considered health-promoting properties and, therefore, could not be used for functional food design, we will allocate some lines to them due to the general interest in it.

Unquestionably, special attention has been paid to phenolic compounds, and specifically to flavonoids, because many epidemiological and interventional studies have associated consumption of these compounds with lower risk for different types of cancer and cardiovascular diseases, showing that they possess antioxidant, anti-inflammatory, and radical scavenging activities, among others, as reported in the excellent reviews of Benavente-García et al. (1997) and Benavente-García and Castillo (2008).

14.3 CITRUS DIETARY FIBER AS THE MAIN CITRUS BY-PRODUCT: PROPERTIES AND USES

The consumption of dietary fiber plays an important role in the prevention of diseases, such as constipation, hemorrhoids, hypercholesterolemia, and colorectal cancer (Silalahi 2002; Brownlee 2011). Dietary fibers are desirable not only for their nutritional value but also for their functional and technological properties (Thebaudin et al. 1997; Elleuch et al. 2011). The main advantage of dietary fiber from citrus fruits, compared to that from other alternative sources such as cereals, is its higher proportion of soluble dietary fiber (Gorinstein et al. 2001). This is important, considering that the requirements for dietary fiber intake must be balanced, that is, the water-soluble fraction should represent between 30% and 50% of the total dietary fiber (Bingham et al. 2003; Aune et al. 2011; Brownlee 2011). Moreover, citrus fruits have better qualities than other sources of dietary fiber due to the presence of bioactive compounds (e.g., flavonoids and vitamin C) with antioxidant properties, which may provide additional health-promoting effects (Marin et al. 2002b).

Although citrus wastes possess a large variety of bioactive compounds that might be considered as potential sources of functional components under a chemical breakdown (Schieber et al. 2001), citrus wastes can be easily conditioned, as seen in the following, to obtain a by-product that we will call CDF (citrus dietary fiber). CDF, except for ascorbic acid, contains more bioactive compounds such as phenolic acids, flavonoids—for example, hesperidin, narirutin, naringin, and eriocitrin, among others (Fernández-López et al. 2004)–limonoids, and fibers than the juices do (Bocco et al. 1998; Manthey and Grohmann 2001; Gorinstein et al. 2001). These compounds have attracted more attention because of their properties related to human health, which are attributed to their antioxidant activities and free radical-scavenging abilities (Bocco et al. 1998; Imeh and Khokhar 2002). They also have antimicrobial, anti-inflammatory, and antiatherosclerotic activity, and they are chemopreventive or anticancer agents (Tripoli et al. 2007; Benavente-García and Castillo 2008). Fibers from citrus are being considered to be of higher quality than those from cereal, as mentioned earlier, due to a better balance between soluble and insoluble dietary fiber content and also due to their higher water and oil-holding capacities (Larrauri 1999). The additional advantage of citrus fiber is due to their content of associated bioactive compounds (flavonoids and vitamin C) with antioxidant properties, which may exert higher health-promoting effects than the dietary fiber itself (Lario et al. 2004). Lemon possesses the highest antioxidant potential among citrus fruits, and it is the most suitable fiber for dietary prevention of cardiovascular and other diseases (Gorinstein et al. 2001). CDF is desirable not only for its nutritional aspects but also for its functional and technological properties (Moure et al. 2001). Thus, the presence of all those citrus bioactive compounds together with the major component (i.e., fiber) gives citrus wastes a special and, possible, unique properties.

The biochemistry of citrus has been extensively studied, and values regarding the fiber content of citrus fruits can be found elsewhere (Braddock and Graumlich 1981; Braddock 1999; Gorinstein et al. 2001; Fernández-López et al. 2004). However, fiber content in citrus by-products not only differs from the whole fresh fruit but also differs according the industrial process undergone by citrus. Table 14.2 shows the

TABLE 14.2
Fiber Compositions of Different Citrus By-Products

Industrial Process	*Citrus* Species (Fruit Part)	Pectin (% DW)	Lignin (% DW)	Cellulose (% DW)	Hemicellulose (% DW)
Extraction of flavonoids	Sour orange (whole mature fruit)	6.54 ± 0.56	14.37 ± 1.32	20.74 ± 1.92	6.57 ± 0.56
Extraction of flavonoids	Satsuma (peels)	2.58 ± 0.22	13.54 ± 1.26	30.53 ± 2.35	11.04 ± 0.98
Extraction of flavonoids	Grapefruit (whole mature fruit)	8.53 ± 0.68	11.56 ± 0.98	26.57 ± 2.01	5.59 ± 0.42
Canning industry	Satsuma (peels)	16.01 ± 1.21	8.59 ± 0.76	22.55 ± 2.22	6.01 ± 0.56
Juice industry	Lemon (peels)	13.00 ± 1.06	7.56 ± 0.54	23.06 ± 2.11	8.09 ± 0.81
Juice industry	Lemon (pulp)	22.53 ± 1.95	7.55 ± 0.66	36.22 ± 3.25	11.05 ± 1.09
Juice industry	Sweet orange (peels)	12.07 ± 1.12	7.51 ± 0.62	24.52 ± 2.00	7.57 ± 0.66
Juice industry	Sweet orange (pulp)	23.02 ± 2.12	7.52 ± 0.59	37.08 ± 3.10	11.04 ± 1.05

Note: The values are expressed as an average ± ES ($n = 7$).

fiber compositions of different citrus by-products obtained after different industrial processes.

The fiber content of citrus by-products, and the portions of every kind of fiber (i.e., pectin, lignin, cellulose, hemicellulose), is more dependent on the industrial source than on the *Citrus* species. By-products from the chemical industry (i.e., extraction of flavonoids) show lower pectin content and higher lignin content, which is explained by the use of hydroalcholic solvents during the flavonoid extraction process. Thus, the pectin content of by-products generated by the food industry may be 2- to 10-fold higher than those from the chemical industry, while citrus by-products from the chemical industry have a higher content of insoluble fiber (cellulose, hemicelluloses, and lignin), reaching percentages of 80% of the total fiber. Moreover, citrus by-products undergo a process to turn wastes into fiber. This process comprises two critical steps: scalding (which includes washing) and drying, which also affects the chemical composition of the final product. In general terms, contents of ash, sugar, proteins, flavonoids, and pectin decreased, while contents of fats, lignin, cellulose, and hemicellulose increased after this treatment as shown in Figure 14.2. The changes alter many more parameters than the former ones. Thus, other parameters such as antioxidant activity, ascorbic acid content, or technological properties such as fiber water-binding capacity or fiber lipid-binding capacity are modified (Marin et al. 2007).

In this way, a different product other than just simple waste may be obtained. Several developments have happened (Lario et al. 2004; Kang et al. 2006). A good example of this, by its simplicity, is the approach carried out by Lario et al. (2004),

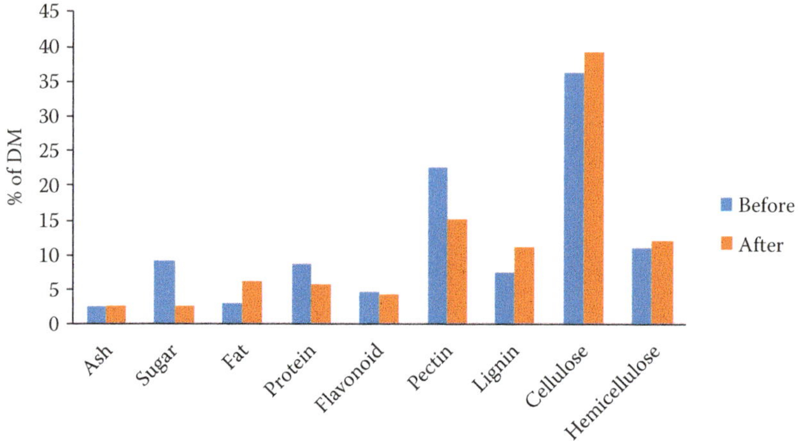

FIGURE 14.2 Changes in lemon wastes from the juice industry during the process of fiber setup. Before: chemical composition of citrus by-product. After: chemical composition after scalding and drying. (Reprinted from *Food Chemistry*, 100, Marin. F.R. et al., By-products from different citrus processes as a source of customized functional fibres, 736–741, Copyright (2007), with permission from Elsevier.)

in which a process for producing CDF from lemon waste is developed and the physical and chemical characteristics, as well as microbiological properties, are studied. Figure 14.3 shows the proposed flowchart for transforming citrus waste into CDF.

The aforementioned by-product, CDF, may be used in the formulation of several kinds of foodstuffs, supplying them with technological and health-promoting properties. Research efforts, and developments, have focused mainly on the use of CDF not only in meat products but also, although to a lesser extent, in dairy and bakery products and beverages.

CDF is suitable for addition to meat products and has been used in cooked meat products to increase the cooking yield due to its water-binding and fat-binding properties and to improve texture (Fernández-Ginés et al. 2005). Citrus by-products (lemon and orange fiber powder) have been added, in different concentrations, to cooked and dry-cured sausages (Aleson-Carbonell et al. 2003, 2004; Fernández-Ginés et al. 2003, 2004; Fernández-López et al. 2004), to typical Spanish dry-fermented products (i.e., *salchichón*) (Fernández-López et al. 2008) in burgers (Aleson-Carbonell et al. 2005) and in sucuk sausage (Yalınkılıç et al. 2012) among others, with excellent results.

Lemon CDF is usually added in different concentrations, ranging from 2.5% to 10%, in cooked sausages and dry-cured sausages. The addition of this fiber may have additional effects due to the presence of bioactive compounds that also induce a decrease in residual nitrite levels; its sensory properties are similar to those of conventional sausages (Fernandez-Lopez et al. 2005; Bhat and Bhat 2011). Other types of CDF such as orange fiber are added to lower concentrations (0.5%–2%), due to the extra color contribution, to different kinds of cooked sausages. The

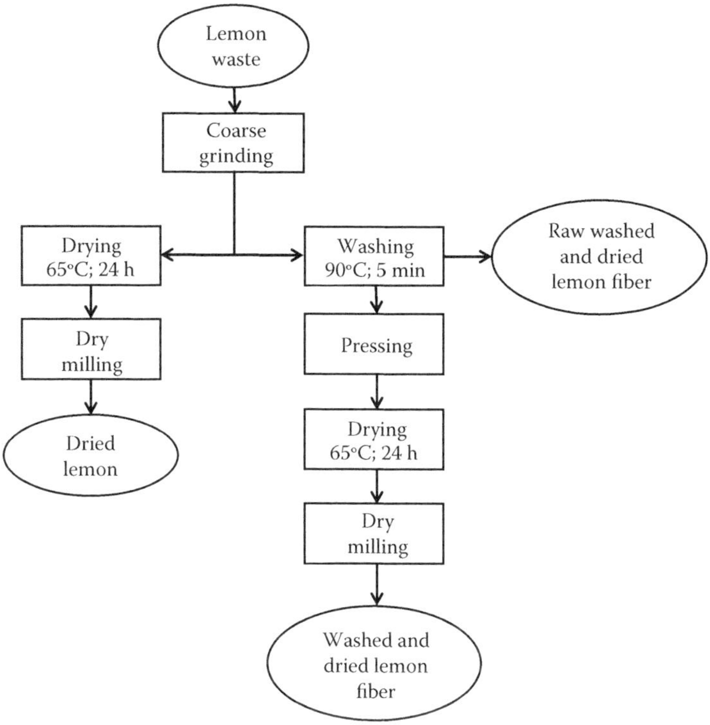

FIGURE 14.3 Flowchart of juice by-product processing. (Reprinted from *Innovative Food Science & Emerging Technologies*, 5, Lario, Y. et al., Preparation of high dietary fiber powder from lemon juice by-products, 113–117, Copyright (2004), with permission from Elsevier.)

results show that this addition improves the nutritional value, decreases the residual nitrite level, and delays the oxidation process as determined by 2-thiobarbituric acid (TBA) values and the red color. CDF in all concentrations made the products harder and less springy and chewy (Fernández-Ginés et al. 2003; Fernandez-Lopez et al. 2005; Bhat and Bhat 2011). Other strategies mix cereal and citrus fibers in any equal amounts (1.5%:1.5%). The addition of such amounts, after ripening, in the fiber content in sausages of about 2% improves their nutritional properties and retains the sensory profile. In the same way, mixtures of CDF, orange fiber at 1%, and essential oils have also been used, proving to extend the product shelf life with no relevant changes in sensory properties (Viuda-Martos et al. 2010). However, the upper limit for CDF addition seems to be about 10%; over this amount, an increase in hardness and lower sensory properties is observed (García et al. 2002).

In summary, the use of CDF in meat products has interesting advantages such as improving cooking yield, due to its water-binding and fat-binding properties, texture, or hardness while keeping the sensory property score. Regarding the health-promoting properties, increased fiber content by itself may be justification for better and healthier properties of the new food products. Furthermore, a very interesting secondary effect observed when meat products are produced with CDF is their ability to reduce residual nitrite levels, avoiding the possible formation of nitrosamines and nitrosamides. For instance, reductions in residual nitrite level of 60%, for CDF supplementation between 0.5% and 2%, in bologna sausages and 50%, with supplementations of 5%, in dry-cured sausages have been described by Viuda-Martos et al. (2009). Curiously, this effect is explained again by the presence in the citrus fiber of different bioactive chemicals such as flavonoids and ascorbic acid, that is, antioxidant properties, and not by the fiber content itself (Viuda-Martos et al. 2011).

However, to a lesser extent, as mentioned earlier, citrus fiber by-products have also been incorporated into other foods besides meat products. Thus, CDF has been added to dairy products such as ice cream in quantities ranging from 0.4% to 1.2%. Although authors conclude that a combination of citrus fiber and emulsifier produces desirable ice cream properties, they also report that CDF, by itself, cannot improve viscosity or overrun the sensory properties of an ice cream sample (Dervisoglu and Yazici 2006). Researchers with similar results have concluded that CDF is not suitable for yogurt elaboration, observing that the presence of particles alters yogurt structure (Sendra et al. 2010; Yi et al. 2014). Although, as mentioned earlier, CDF does not seem suitable for yogurt elaboration, there are still some opportunities to develop functional foods based on fermented milk and citrus fiber, as previously reported. This is because probiotic bacteria can survive and their growth is not restricted but enhanced (Sendra et al. 2008). In this way, approaches to the use of CDF for its satiating power and as a fat replacer have been done with more than acceptable results (Perriguee et al. 2009).

Some attempts have also been made to use CDF as an ingredient in bakery products such as muffins, where the addition of orange fiber, at 10%, decreased the predicted glycemic index without any changes in the sensory score (Romero-Lopez et al. 2011) and in biscuits, where the author concluded that citrus fiber could be added to biscuits at a replacement level of 5% without noticeable adverse effects on sensory evaluation (Kohajdova et al. 2011). Moreover, the use of low levels of citrus fiber (2.5%) in bread formula has been reported to be an effective way to increase bread yield, which is explained by increasing water absorption (Miller 2011). As bread undergoes baking and fermentation, some changes are produced during elaboration. However, authors report that phenolic components of citrus fiber are more affected than the different fiber fractions themselves (Shyu et al. 2014). So far, most research approaches aimed to resolve an environmental waste problem by incorporating a by-product in food, without taking into consideration the putative health-promoting properties associated with the by-product. However, some research approaches have directly pointed to their use in improving nutritional value of the food. Thus, some researchers have used CDF as a fat replacer as reported by Stoll et al. (2015), and Lim et al. (2014) used fiber from the less common citrus yuja (*Citrus junos*) has been introduced to replace fat in baked products with good results.

Apart from the research covered by the standard channels of science dissemination, there are a considerable number of patents that in one way or another show the relevance of CDF and its potential application to the food industry. Examples of this are inventions related to preparation of citrus fiber with hydrophobic vitamins to be added to beverages (Vanhemelrijck and McCare 2007), methods for preparing an edible emulsion consisting of fibers (Almeida-Rivera et al. 2015), development of new beverages based on soy proteins and citrus fiber (Beckmann et al. 2014), flavored beverages fortified with citrus fiber (Li and Xiong 2013), or industrial methods for producing water-soluble fiber (Zhong et al. 2010), which differs from that already published by Lario et al. (2004) in two steps: one hydrolysis step and one extraction step with CO_2, under supercritical conditions, to eliminate color-causing agents, carotenoids, and bitter agents such as limonoids.

Unquestionably, the huge number of patents covering the issue of citrus fiber shows the economic interest of the food industry. However, from a scientific point of view, it is difficult to reach accurate and critical conclusions on that basis, due to the nature of a patent content, the lack of discussion, the partial results, and the blend between what is scientifically demonstrated and what is a putative truth. For instance, patents can be found claiming antiallergic properties of CDF (Cheng et al. 2014), to which tea extract and polyphenols have been added, on theoretical speculation but with no experimental evidence. Claims based on such conjecture are wishes that do not achieve the standard demonstration of a cause–effect relationship between consumption and claimed effect in humans, as is usually requested by food agencies, as seen in the following. Also, patents of dubious novelty can be found, such as that submitted by Zhang and Wang (2011), entitled "Citrus fiber meat product and preparation method thereof," in which protection is claimed for a citrus fiber meat product and a preparation method, aimed to provide a more healthy meat product by using citrus fiber (at 0.1%–4% of citrus fiber in weight percentage) as a fat replacer. If the former application is legally approved as a patent title, its scientific contribution is uncertain, based upon the previous published research works.

Finally, nowadays CDF is commercialized worldwide under several trademarks. To avoid free advertising, it is suggested that the reader should search the web. However, all commercial uses in food are focused on technological properties, although some mention of the health-promoting properties is often made.

14.4 OTHER BIOACTIVE COMPOUNDS IN CITRUS DIETARY FIBER: PROPERTIES AND USES

CDF, as a citrus by-product, is not formed exclusively by different types of fibers, as explained earlier, but from many other chemical compounds that were initially present in the citrus waste (i.e., flavonoids, carotenoids, essential oils, and so on). Moreover, some finding such as the effect of its ability to reduce residual nitrite levels cannot be explained as an effect of the chemical fiber but is due to other compounds with antioxidant properties (Viuda-Martos et al. 2009).

The composition of several citrus wastes, including pulp and peel wastes from lemons and oranges generated during industrial squeezing, and changes in composition after setting up citrus wastes into CDF, respectively, are shown in Table 14.1 and Figure 14.2. Special attention should be paid to two relevant facts: First, there are noticeable differences in the amounts of relevant compounds such as flavonoids that can increase three times according to the anatomic source of the waste (e.g., flavonoids in pulp waste reach an amount of approximately 4.5%, while in peel wastes, they can reach about 12% of dry matter), as shown in Table 14.1. Second, water-soluble products drop their relative abundances (i.e., pectins, sugars, proteins, and so on), while the opposite happens for the less water-soluble ones (i.e., lignin, cellulose, lipids). Shown in Table 14.3 are the changes in phenolics, flavonoids, carotenoids, ascorbic acid, and antioxidant activity after lemon wastes are transformed into CDF. It is important to note that the most dramatic change is produced in ascorbic acid, which has practically disappeared, while approximately 80% of the initial antioxidant activity remains, measured as total antioxidant activity. In a similar way, total phenolics and flavonoids remain for more than 80% of the initial amount, while carotenoid content decreases, which may be due to the loss of carotenoids in water during extraction and oxidation under high temperature (90°C) (Rock et al. 1998). Additionally, it has been confirmed that leaks of these compounds are produced during the washing step (Viuda-Martos et al. 2011). Thus, these researchers have found phenolics such as caffeic acid, ferulic acid, or *p*-coumaric acids and flavonoids such as eriocitrin, narirutin, hesperidin, and neohesperidin at concentrations ranging from 1.5 to 3.5 mg/L for phenolic acids and from 2 to 39 mg/L for flavonoids when the water from the washing of the orange juice wastes was analyzed.

Four types of flavonoids (flavanones, flavones, flavonols, and anthocyanins, the last only in blood oranges) occur in *Citrus* spp. In this genus, flavanones accumulate in a greater quantity than flavones do. The concentration of these compounds depends upon the age of the fruit, and the highest levels are detected in immature fruits (Castillo et al. 1992). These compounds not only play an important physiological and ecological role but are also of commercial interest because of the multitude of their applications in the food and pharmaceutical industries (Benavente-García et al. 1997).

Significantly, much of the activity of citrus flavonoids appears to impact blood and microvascular endothelial cells, and the two main areas of research on the

TABLE 14.3
Contents of Relevant Compounds Present in Lemon Wastes and CDF from Lemon

	Pulp Lemon Waste (mg/g DM)	CDF from Lemon (mg/g DM)
Total phenolics[a]	63.84 ± 1.77	55.79 ± 1.15
Total flavanoids[b]	45.19 ± 1.03	40.22 ± 0.20
Carotenoids[c]	3.98 ± 0.03	2.17 ± 0.02
Ascorbic acid	5.64 ± 0.11	0.42 ± 0.01
TAA	60.07 ± 1.03	47.34 ± 0.66

Source: Reprinted from *Food Chemistry*, 100, Marin F. R. et al., By-products from different citrus processes as a source of customized functional fibres, 736–741, Copyright (2007), with permission from Elsevier.

Note: The values are expressed as average ± ES ($n = 7$).

Abbreviation: DM, dry matter; TAA, total antioxidant activity.

[a] Results expressed as gallic acid equivalent per gram of DM.

[b] Results expressed as quercetin equivalent per gram of DM.

[c] Total carotenoids were calculated according to Al-Farsi et al. (2005).

biological actions of citrus flavonoids have been inflammation and cancer (Manthey et al. 2001). Epidemiological and animal studies point to a possible protective effect of flavonoids against cardiovascular diseases and some types of cancer (Benavente-García and Castillo 2008).

Flavonoids have been studied for more than 50 years, and although the precise cellular mechanisms involved in their biological action are still not completely known (Gee et al. 2002), anticarcinogenic properties, through antiproliferative, anti-invasive, and antiangiogenic mechanisms, and cardio-protective effects, through antithrombotic, anti-ischemic, antioxidant, and vasorelaxant mechanisms, have been reported. Also, an anti-inflammatory activity with putative effect on degenerative diseases has been described (Benavente-García and Castillo 2008). However, although health-promoting properties have been attributed to flavonoids in *in vivo* or *in vitro* models, there is no evidence demonstrating a cause–effect relationship between consumption of a specific food containing flavonoids and the claimed effect in humans.

The two most relevant studies on phenolic and flavonoid content in industrial citrus wastes, in our opinion and up to now, are the ones carried out by Manthey and Grohmann (2001) and by Delpino-Rius et al. (2015). These authors described that the main phenolic compounds in citrus fruit fibers are flavones and flavanones. According to the source, that is, *Citrus* species, the profiles of tangerine and orange fibers are similar to those described for fresh fruit (Abad-García et al. 2012). Hesperetin-7-rutenoside (hesperidin) and naringenin-7-*O*-rutenoside (narirutin) account for more than 90% of the total phenolic content in these fibers. Other characteristic compounds of citrus fruits, such as isokuratenin-7-rutenoside (didymin), ferulic acid, and apigenin derivatives, were also found. Although flesh and peel fibers from orange and tangerine show the same phenolic profile, peel fibers show a much higher total phenolic compound content than flesh and the profile and content of these compounds in orange fibers are similar to those described for nonindustrial fibers processed in a pilot food plant (Fernández-López et al. 2009) and previous studies on industrial fibers (Coll et al. 1998; Marin et al. 2007). On the other hand, hesperidin and eriodictyol-7-rutenoside (eriocitrin) are the predominant phenolic compounds in lemon fibers, as they are in fresh lemon (Abad-García et al. 2012). Apigenin, diosmetin, and naringenin derivatives are found in minor amounts in this fiber.

Other compounds of a phenolic nature, such as hydroxycinnamic acids and polymethoxylated flavones, can also be found in citrus wastes and, therefore, in CDF. In this way, Manthey and Grohmann (2001) found that the highest concentrations of hydroxycinnamates occur in orange and tangerine wastes compared to grapefruit and lemon wastes. Moreover, these authors reported that hydroxycinnamic acids are not in free form but as amides and esters, describing the occurrence of *p*-coumaric, ferulic, sinapic, phlorin, coniferin, and feruloylputrescine, with ferulic acid being the most abundant. Regarding polymethoxylated flavones, the same authors reported the presence of sinsetin, isosinsetin, hexa-*o*-methylquercetagetin, nobiletin, tetramethylscutellarein, and desmethylnobiletin in orange, tangerine, and grapefruit CDF, while none of them were detected in lemon by-products. The former quantitative description of polymethoxylated flavone occurrence in citrus by-products is consistent with that reported by other authors (Del Río et al. 1998, 2004; Nogata et al. 2006).

Besides phenolic compounds, it has also been reported (Delpino-Rius et al. 2015) that in the particular case of CDF obtained from lemon pulp wastes, a considerable amount of 5-hydroxymethylfurfural is measured, about 800 µg/g of dry matter, but not in other CDF obtained from other *Citrus* species, where it remains 10-fold lower than in lemon CDF. This finding can be explained by the degradation caused by the temperature reached during processing (Del Caro et al. 2004). However, due to the lack of availability of other research reports, these figures should be treated with caution.

Beyond the presence in CDF of the most abundant flavonoids, such as flavanones and flavones, is the presence of minor compounds such as polymethoxylated flavones, which may provide by-products with extra benefits. Thus, it has been reported that diets containing 1% citrus polymethoxylated flavones reduced, in animal models, serum cholesterol and low-density lipoproteins up to 27% and 40%, respectively (Kurowska and Manthey 2004). If this effect is also produced in humans or as an additive to reduce cholesterol levels, as described for dietary fiber (Brown et al. 1999), further research is required to demonstrate it. Other studies have also reported other beneficial effects on health, such as better homeostasis of lipid and glucose metabolism in induced insulin resistance in animal models. However, unrealistic dosages were used in the former study. If this were translated into human use, it would imply intakes of approximately 10 g per day of these compounds (Li et al. 2006).

Regarding the industrial application of flavonoids, there is also tremendous commercial interest in them. However, and possibly because of flavonoids, they can be isolated and handled as pure compounds; their main area of exploitation is as drugs or additives. Paradigmatic examples of this are the flavone diosmin and a modified form of the flavanone hesperidin, neohesperidin dihydrochalcone.

For instance, diosmin is a flavone with distribution in citrus and aromatic plants (Marin et al. 1998; Marin and Del Río 2001) used to ameliorate varicose veins under the trademark of Daflon, Thus, hundreds if not several thousands, of patents on this topic can be found; a few examples are seen in the patents of Huet et al. (1996) and Zhang and Zhang (2006), in which a pharmaceutical composite of flavonoids extracted from Rutaceae and, especially diosmin, is claimed for the first one and a production process is claimed for the second one.

Neohesperidin dihydrochalcone is an artificial sweetener produced from the neohesperidin flavanone through an industrial process that involves several chemical reductions (Marin et al. 2002b; Wining et al. 2007). As seen for diosmin, hundreds of patents about neohesperidin dihydrochalcone can be found. Some good examples of these are those filed by Horowitz and Bruno (1961) and by Guo et al. (2014), as the first patent in which the taste properties of neohesperidin dihydrochalcone are claimed and as a method for producing a sweetener from a source different from the flavanone neohesperidin, respectively.

Much less attention has been paid to other bioactive compounds present in citrus by-products such as terpenoids, that is, limonene or carotenoids. Thus, for the latest ones, there were, at the time of writing this chapter, less than 700 indexed research papers, while more than 2300 were found for flavonoids. *Citrus* spp. display a wide range of fruit coloration due to differences in carotenoids. Thus, with qualitative and quantitative changes, depending on specie, variety, and ripening stage, citrus present a wide collection of carotenoids such as phytoene, phytofluene, α-carotene, β-carotene, ζ-carotene, β-cryptoxanthin, β-citraurin, zeaxanthin, violaxanthin, lycopene, neoxanthin, and lutein, among others (Lado et al. 2015). Although, total carotenoid can vary depending on the aforementioned factors, reasonable and accurate figures may range between 2.0 and 0.2 mg/g dry matter approximately (Zou et al. 2016), the highest amounts corresponding to *C. reticulata* (mandarin) and the lowest ones to *C. grandis* (pummelo) (Wang et al. 2008). On the other hand, sadly, only a small amount of research has been conducted on the content of carotenoids in citrus wastes and/or citrus by-products (El-Sharnouby et al. 2013). However, taking into consideration the chemical properties of these compounds and the process undergone by citrus wastes to become CDF, it seems reasonable to infer a slight increase in its carotenoid content.

Apart from the academic scientific interest in carotenoids, the industry is also very keen on them. Thus, as happens for other citrus bioactive compounds, a considerable number of patents on this topic can be found; most of them make claims on the extraction process to obtain carotenoids. Some examples are those claimed by Toulinim (1950) or by Johnson et al. (1977), where traditional methods are described, and others, where more advanced methods such as those based on supercritical fluid extraction are claimed (Chen et al. 2012).

Other classes of terpenoids, such as limonoids, are also present in citrus wastes and have shown health-promoting properties (Jacob et al. 2000; Manners et al. 2007). Limonoid content in different citrus parts and species can be found in the paper of Hasegawa et al. (1993). Limonoids are abundant in citrus fruits, where the highest concentrations occur in the seed with limonin, nomilin, and nomilin-17-β-D-glucose predominating, where about 300 mg/kg of fresh weight can be reached. In the peel and pulp, limonin and nomilin are the predominant limonoids bound to glucose, where amounts of 100 and 30 mg/kg of fresh weight can be easily found (Russo et al. 2014). Limonin and nomilin are presently concentrated an adsorption resin during debittering of grapefruit juice, orange juice, pulp wash, and peel wash. These limonoids can be recovered by extraction of spent resin material along with flavonoids. A large source of these limonoids is available from the peel and debittering waste streams which could be recovered. However, due to their insolubility in aqueous solutions and extreme bitter nature, limonoids probably have little value for use in foodstuffs (Widmer and Montanari 1994; Kuroyanagi et al. 2008; Russo et al. 2014). On the other hand, most patents on limonoids cover their use for cancer treatment, which is completely different from the topic covered in this chapter.

Finally, a few words should be dedicated to citrus essential oils. Although essential oils are relevant compounds present in citrus, their applications in the food industry are restricted in practice to use as flavoring agents and as potential antibacterial and antifungal agents, both properties unrelated to functionality of foods (Mustafa 2015). However, an extremely interesting application as a health-promoting agent is arising from citrus essential oils. Thus, research is currently being carried out regarding the potential use of essential oils to treat dysbiosis (Myers et al. 2009; Lang and Buchbauer 2012).

14.5 HEALTH-PROMOTING FUNCTIONAL FOODS BASED ON CITRUS DIETARY FIBER

CDF has been widely used, as outlined earlier, in several kinds of foods. For instance, sausages, CDF has been added to hamburgers, ice cream, yogurt, and muffins and an acceptable sensory score was achieved. Furthermore, technological properties better than those of the original food are claimed and some relevant improvements such as lower residual nitrites, in the case of meat products, have been made. Furthermore, CDF contains not only soluble and insoluble fibers but also many other bioactive compounds, such as flavanones, flavones, polymethoxylated flavones, carotenoids, and limonoids, among others, for which *in vitro* evidence, and in some cases *in vivo*, from using animal models has been reported everywhere and at every moment. But is it enough to make a claim for a functional food?

According to more advanced and strict legislations, that is not enough. It is necessary to demonstrate that a cause–effect relationship is established between consumption and claimed effect in humans, at the used dose and under an affordable food intake. So far, no clinical evidence proving in humans that the relationship between consumption of foods supplemented with CDF and health improvement has been found. Therefore, although scientific data may suggest that health-promoting properties may be attributed to any citrus by-product, the legality (at least in the European Union context) says that the evidence shall consist primarily of studies in humans; additionally, European Food Safety Agency (EFSA) guidance asks to demonstrate that a cause–effect relationship be established between consumption and claimed effect in humans (EFSA Panel on Dietetic Products, Nutrition and Allergies 2011d), which means that no commercial claim can be made on meat (or other) products elaborated with citrus fiber.

To date, more than 2000 health claims have been submitted to EFSA and the vast majority of them have been rejected, including 13 related to citrus. Shown in Table 14.4 are these 13 applications and the rejection reasons. The reason is simple: "Non-compliance with the Regulation because on the basis of the scientific evidence assessed, this claimed effect for this food has not been substantiated"; that is, not enough evidence has been supplied to demonstrate that a cause–effect relationship has been established between consumption and the claimed effect in humans, which must be shown in studies in humans. It is important to note that even a health claim on diosmin, which can be bought from any European chemist, was rejected by EFSA, because the lack of studies in humans in the form of food, not because of the active principle does not have pharmacological properties.

It is a real issue that papers contain the statement "more research is needed," and usually it is true. However, for future development of functional foods based on citrus by-products, it is absolutely necessary to go in depth into their effect on human physiology. The first steps have been made. Thus, recent research is focused on the effect *in vitro* of fibers, including citrus fibers, on human satiety (Logan et al. 2015) or on the effect of dietary fiber and polyphenols in the microbiota of healthy volunteers (Cuervo et al. 2014), both cases showing positive effects. These kinds of studies correlating chemical compounds and physiological effects seem promising for the future to demonstrate that there is a cause–effect relationship between citrus by-products and human health.

TABLE 14.4
Requested Health Claims Related to Citrus

Food, Nutrient, or Substance	Health Claim	EU Legal Status	Motive of Rejection	Reference
Hesperidin	Helps maintain healthy/strong bones	Nonauthorized	Noncompliance with the regulation because on the basis of the scientific evidence assessed, this claimed effect for this food has not been substantiated	EFSA Panel on Dietetic Products, Nutrition and Allergies (2010b)
Naringin	Helps maintain healthy/strong bones	Nonauthorized	Noncompliance with the regulation because on the basis of the scientific evidence assessed, this claimed effect for this food has not been substantiated	EFSA Panel on Dietetic Products, Nutrition and Allergies (2010b)
Citrus bioflavonoids	May help to keep joints healthy	Nonauthorized	Noncompliance with the regulation because on the basis of the scientific evidence assessed, this claimed effect for this food has not been substantiated	EFSA Panel on Dietetic Products, Nutrition and Allergies (2010b)
Lemon (*C. limon*): flavonoids	Acts as a natural antioxidant, helps to reduce oxidative stress, helps to reduce aging effects, necessary for cell protection, improves the antioxidant defensive system	Nonauthorized	Noncompliance with the regulation because on the basis of the scientific evidence assessed, this claimed effect for this food has not been substantiated	EFSA Panel on Dietetic Products, Nutrition and Allergies (2010a)
Diosmin (a component of citrus peel extract and precursor of diosmetin)	Helps maintain a good venous blood circulation, supports a normal venous function, helps maintain healthy venous circulation in the legs, protects veins from inflammatory reactions, supports the strength of blood vessels	Nonauthorized	Noncompliance with the regulation because on the basis of the scientific evidence assessed, this claimed effect for this food has not been substantiated	EFSA Panel on Dietetic Products, Nutrition and Allergies (2011a)

(Continued)

TABLE 14.4 (CONTINUED)
Requested Health Claims Related to Citrus

Food, Nutrient, or Substance	Health Claim	EU Legal Status	Motive of Rejection	Reference
Hesperidin (a component of citrus peel extract and precursor of hesperitin)	Helps maintain normal blood cholesterol levels/supports heart health	Nonauthorized	Noncompliance with the regulation because on the basis of the scientific evidence assessed, this claimed effect for this food has not been substantiated	EFSA Panel on Dietetic Products, Nutrition and Allergies (2011b)
Bioflavonoids from citrus	Helps to maintain healthy venous circulation	Nonauthorized	Noncompliance with the regulation because on the basis of the scientific evidence assessed, this food is not sufficiently characterized for a scientific assessment of this claimed effect and the claim could not therefore be substantiated	EFSA Panel on Dietetic Products, Nutrition and Allergies (2011a)
Grapefruit (*C. paradisi*)	Antioxidative properties/supports the body organs and tissues in case of oxidative damage	Nonauthorized	Noncompliance with the regulation because on the basis of the scientific evidence assessed, this claimed effect for this food has not been substantiated	EFSA Panel on Dietetic Products, Nutrition and Allergies (2010a)
Grapefruit (*C. paradisi*)	May help the detoxification process, possesses antioxidant activity, can be considered as a detoxifying/purifying agent due to its antioxidant properties, provides antioxidant protection	Nonauthorized	Noncompliance with the regulation because on the basis of the scientific evidence assessed, this claimed effect for this food is not a beneficial physiological effect as required by the regulation	EFSA Panel on Dietetic Products, Nutrition and Allergies (2010c)

(Continued)

TABLE 14.4 (CONTINUED)
Requested Health Claims Related to Citrus

Food, Nutrient, or Substance	Health Claim	EU Legal Status	Motive of Rejection	Reference
Grapefruit (*C. paradisi*)	Flavonoids contained within the grapefruit contribute to the microbial balance in the body organs and tissues	Nonauthorized	Noncompliance with the regulation because on the basis of the scientific evidence assessed, this food is not sufficiently characterized for a scientific assessment of this claimed effect and the claim could not therefore be substantiated	EFSA Panel on Dietetic Products, Nutrition and Allergies (2011a)
Grapefruit (*C. paradisi*)	Flavonoids contained within the grapefruit contribute to the microbial balance in the body organs and tissues	Nonauthorized	Noncompliance with the regulation because on the basis of the scientific evidence assessed, this food is not sufficiently characterized for a scientific assessment of this claimed effect and the claim could not therefore be substantiated	EFSA Panel on Dietetic Products, Nutrition and Allergies (2011a)
Lemon (*C. limon*): flavonoids	Helps to support digestion, contributes to the normal function of intestinal tract, helps support the digestive juice flow	Nonauthorized	Noncompliance with the regulation because on the basis of the scientific evidence assessed, this claimed effect for this food has not been substantiated	EFSA Panel on Dietetic Products, Nutrition and Allergies (2011b)
Sweet orange (*C. sinensis*)	Traditionally used for the good circulation of blood in microvessels/traditionally used to decrease the sensations of heavy legs	Nonauthorized	Noncompliance with the regulation because on the basis of the scientific evidence assessed, this claimed effect for this food is not sufficiently defined to be able to be assessed and the claim could not therefore be substantiated	EFSA Panel on Dietetic Products, Nutrition and Allergies (2011c)

In summary, an affordable by-product from the industrialization of citrus is available, undergoing an easy transformation; moreover, it is full of bioactive compounds that exert interesting pharmacological properties beyond the physiological ones of the fiber. The food industry needs only to design foods based on CDF and fund human research to demonstrate a cause–effect relationship between any new putative functional food based on CDF and human health. As always, what is required is a great deal of dedicated work and sufficient funding.

REFERENCES

Abad-García, B., S. Garmón-Lobato, L. A. Berrueta, B. Gallo, and F. Vicente. 2012. On line characterization of 58 phenolic compounds in *Citrus* fruit juices from Spanish cultivars by high-performance liquid chromatography with photodiode-array detection coupled to electrospray ionization triple quadrupole mass spectrometry. *Talanta* 99:213–224.

Al-Farsi, M, C. Alasalvar, A, Morris, M. Baron and F. Shahidi. 2005. Comparison of antioxidant activity, anthocyanins, carotenoids, and phenolics of three native fresh and sun-dried date (*Phoenix dactylifera* L.) varieties grown in Oman. *Journal of Agricultural and Food Chemistry* 53(6):7592–7599.

Aleson-Carbonell, L., J. Fernández-López, E. Sayas-Barberá, E. Sendra, and J. A. Pérez-Alvarez. 2003. Utilization of lemon albedo in dry-cured sausages. *Journal of Food Science* 68(5):1826–1830.

Aleson-Carbonell, L, J. Fernández-López, E. Sendra, E. Sayas-Barberá, and J. A. Pérez-Alvarez. 2004. Quality characteristics of a non-fermented dry-cured sausage formulated with lemon albedo. *Journal of the Science of Food and Agriculture* 84(15):2077–2084.

Aleson-Carbonell, L., J. Fernández-López, J. A. Pérez-Alvarez, and V. Kuri. 2005. Characteristics of beef burger as influenced by various types of lemon albedo. *Innovative Food Science & Emerging Technologies* 6(2):247–255.

Almeida-Rivera, C. P., J. Nijsse, and S. T. A. Regismond. 2015. Method for preparing a fibre containing emulsion. Patent US89456559B2.

Aune, D., D. S. Chan, R. Lau, R. Vieira, D. C. Greenwood, E. Kampman, and T. Norat. 2011. Dietary fibre, whole grains, and risk of colorectal cancer: Systematic review and dose–response meta-analysis of prospective studies. *BMJ* 343:d6617.

Aust, O., H. Sies, W. Stahl, and M. C. Polidori. 2001. Analysis of lipophilic antioxidants in human serum and tissues: Tocopherols and carotenoids. *Journal of Chromatography A* 936(1–2):83–93.

Beckmann, C. H., M. Mellema, and A. H. E. Ström. 2014. Beverage comprising soy proteins and citrus fibre. Patent EP2560498B1.

Benavente-García, O., J. Castillo, F. R. Marin, A. Ortuño, and J. A. Del Río. 1997. Uses and properties of citrus flavonoids. *Journal of Agricultural and Food Chemistry* 45(12):4505–4515.

Benavente-García, O., and J. Castillo. 2008. Update on uses and properties of citrus flavonoids: New findings in anticancer, cardiovascular, and anti-inflammatory activity. *Journal of Agricultural and Food Chemistry* 56(15):6185–6205.

Bhat, Z. F., and H. Bhat. 2011. Functional meat products: A review. *International Journal of Meat Science* 1(1):1–14.

Bingham, S. A., N. E. Day, R. Luben, P. Ferrari, N. Slimani, T. Norat, F. Clavel-Chapelon et al. 2003. Dietary fibre in food and protection against colorectal cancer in the European Prospective Investigation into Cancer and Nutrition (EPIC): An observational study. *The Lancet* 361(9368):1496–1501.

Bocco, A., M. E. Cuvelier, H. Richard, and C. Berset. 1998. Antioxidant activity and phenolic composition of citrus peel and seed extracts. *Journal of Agricultural and Food Chemistry* 46(6):2123–2129.

Braddock, R. J. 1999. In: Braddock, R. J., and E. Weiss (Eds.), Handbook of citrus by-products and processing technology. Wiley: New York. *The Citrus Industry* :1–11

Braddock, R. J., and T. R. Graumlich. 1981. Composition of fiber from citrus peel, membranes, juice vesicles and seeds. *Lebensmittel-Wissenschaft und Technologie—Food Science and Technology* 14:229–231.

Britton, G., S. Liaaen-Jensen, and H. Pfander (Eds.). 2009. *Carotenoids volume 5: Nutrition and health*. Basel, Switzerland: Springer.

Brown, L., B. Rosner, W. W. Willett, and F. M. Sacks. 1999. Cholesterol-lowering effects of dietary fiber: A meta-analysis. *American Journal of Clinical Nutrition* 69(1):30–42.

Brownlee, I. A. 2011. The physiological roles of dietary fibre. *Food Hydrocolloids* 25(2):238–250.

Castillo, J., O. Benavente-García, and J. A. Del Río. 1992. Naringin and neohesperidin levels during development of leaves, flower buds, and fruits of *Citrus aurantium*. *Plant Physiology* 99(1):67–73.

Chen, J., D. Liu, Y. Sun, D. Wu, X. Ye, and X. Zhang. 2012. Environment-friendly ultrasonic auxiliary extraction method of citrus carotenoid extraction. Patent CN101863811B.

Cheng, J., S. Lu, J. Xing, Q. Xia, and M. Zheng. 2014. Antiallergic citrus dietary fiber preparation method and product thereof. Patent CN103462041B.

Coll, M. D., L. Coll, J. Laencina, and F. A. Tomas-Barberan. 1998. Recovery of flavanones from wastes of industrially processed lemons. *Zeitschrift für Lebensmitteluntersuchung und-Forschung A* 206(6):404–407.

Cuervo, A., L. Valdés, N. Salazar, C. G. de los Reyes-Gavilán, P. Ruas-Madiedo, M. Gueimonde, and S. González. 2014. Pilot study of diet and microbiota: Interactive associations of fibers and polyphenols with human intestinal bacteria. *Journal of Agricultural and Food Chemistry* 62(23):5330–5336.

Del Caro, A., A. Piga, I. Pinna, P. M. Fenu, and M. Agabbio. 2004. Effect of drying conditions and storage period on polyphenolic content, antioxidant capacity, and ascorbic acid of prunes. *Journal of Agricultural and Food Chemistry* 52(15):4780–4784.

Del Río, J. A., M. C. Arcas, O. Benavente-García, and A. Ortuño. 1998. Citrus polymethoxylated flavones can confer resistance against *Phytophthora citrophthora*, *Penicillium digitatum*, and *Geotrichum* species. *Journal of Agricultural and Food Chemistry* 46(10):4423–4428.

Del Río, J. A., M. D. Fuster, P. Gómez, I. Porras, A. García-Lidón, and A. Ortuño. 2004. *Citrus limon*: A source of flavonoids of pharmaceutical interest. *Food Chemistry* 84(3):457–461.

Delpino-Rius, A., J. Eras, F. Vilaró, M. Á. Cubero, M. Balcells, and R. Canela-Garayoa. 2015. Characterisation of phenolic compounds in processed fibres from the juice industry. *Food Chemistry* 172:575–584.

Dervisoglu, M., and F. Yazici. 2006. Note: The effect of citrus fibre on the physical, chemical and sensory properties of ice cream. *Food Science and Technology International* 12(2):159–164.

Duoss-Jennings, H. A., T. B. Schmidt, T. R. Callaway, J.A. Carroll, J. M. Martin, S. A. Shields-Menard, P. R. Broadway, and J. R. Donaldson. 2013. Effect of citrus byproducts on Survival of O157:H7 and non-O157 *Escherichia coli* Serogroups within in vitro bovine ruminal microbial fermentations. *International Journal of Microbiology (ID 398320), Vol. 2013, 5 pages.* (http://dx.doi.org/10.1155/2013/398320).

EFSA Panel on Dietetic Products, Nutrition and Allergies. 2010a. Scientific Opinion on the substantiation of health claims related to various food(s)/food constituent(s) and protection of cells from premature aging, antioxidant activity, antioxidant content. and antioxidant properties, and protection of DNA, proteins and lipids from oxidative damage pursuant to Article 13(1) of Regulation (EC) No 1924/2006. *EFSA Journal* 8(2):1489.

EFSA Panel on Dietetic Products, Nutrition and Allergies. 2010b. Scientific Opinion on the substantiation of health claims related to various food(s)/food constituent(s) claiming maintenance of joints (ID 1799, 1973, 2022, 2178, 2202, 2254, 2255, 2311, 2394, 2417, 2418, 2458, 2649, 2794, 2798, 3119, 3144, 3274, 3283, 3318, 3339, 3495, 3511, 3523, 3555, 3624, 3699, 3748, 3770, 3835, 3884, 3892, 3904, 3943, 3978, 4012, 4020, 4056, 4137, 4175), maintenance of bone (ID 1764, 1907, 2418, 4012, 4020, 4056, 4175) and maintenance of muscles (ID 2254, 2311) pursuant to Article 13(1) of Regulation (EC) No 1924/200. *EFSA Journal* 8(2):1493.

EFSA Panel on Dietetic Products, Nutrition and Allergies. 2010c. Scientific Opinion on the substantiation of health claims related to various food(s)/food constituent(s) and protection of cells from premature ageing (ID 1668, 1917, 2515, 2527, 2530, 2575, 2580, 2591, 2620, 3178, 3179, 3180, 3181, 4329, 4415), antioxidant activity, antioxidant content, and antioxidant properties (ID 857, 1306, 2515, 2527, 2530, 2575, 2580, 2591, 2629, 2728, 4327, 4365, 4380, 4390, 4394, 4455, 4464, 4507, 4694, 4705), protection of DNA, proteins and lipids from oxidative damage (ID 1196, 1211, 1216, 1306, 1312, 1440, 1441, 1666, 1668, 1692, 1900, 1914, 1948, 2023, 2158, 2517, 2522, 2527, 2575, 2591, 2620, 2637, 2639, 2663, 2860, 3079, 3276, 3564, 3818, 4324, 4329, 4351, 4397, 4416, 4424, 4507, 4527, 4528, 4542, 4611, 4629, 4659), and bioavailability of anthocyanins in black currants (ID 4220) pursuant to Article 13(1) of Regulation (EC) No 1924/2006. *EFSA Journal* 8(10):1752.

EFSA Panel on Dietetic Products, Nutrition and Allergies. 2011a. Scientific Opinion on the substantiation of health claims related to: Flavonoids and ascorbic acid in fruit juices, including berry juices (ID 1186); flavonoids from citrus (ID 1471); flavonoids from *Citrus paradisi* Macfad. (ID 3324, 3325); flavonoids (ID 1470, 1693, 1920); flavonoids in cranberry juice (ID 1804); carotenoids (ID 1496, 1621, 1622, 1796); polyphenols (ID 1636, 1637, 1640, 1641, 1642, 1643); rye bread (ID 1179); protein hydrolysate (ID 1646); carbohydrates with a low/reduced glycaemic load (ID 476, 477, 478, 479, 602) and carbohydrates which induce a low/reduced glycaemic response (ID 727, 1122, 1171); alfalfa (ID 1361, 2585, 2722, 2793); caffeinated carbohydrate-containing energy drinks (ID 1272); and soups (ID 1132, 1133) pursuant to Article 13(1) of Regulation (EC) No 1924/2006. *EFSA Journal* 9(4):2082.

EFSA Panel on Dietetic Products, Nutrition and Allergies. 2011b. Scientific Opinion Part I on the substantiation of health claims related to various food(s)/food constituent(s) not supported by pertinent human data (ID 411, 559, 1174, 1184, 1197, 1380, 1409, 1656, 1667, 1670, 1763, 1767, 1806, 1884, 1908, 1997, 2141, 2159, 2243, 2244, 2325, 2331, 2333, 2336, 2652, 2717, 2727, 2752, 2788, 2861, 2870, 2885, 2894, 3077, 3101, 3516, 3595, 3726, 4252, 4288, 4290, 4406, 4509, 4709) pursuant to Article 13(1) of Regulation (EC) No 1924/2006. *EFSA Journal* 9(6):2246.

EFSA Panel on Dietetic Products, Nutrition and Allergies. 2011c. Scientific Opinion on the substantiation of health claims related to various food(s)/food constituent(s) and health relationships that are not sufficiently defined (ID 9, 377, 531, 555, 560, 569, 582, 583, 584, 585, 589, 590, 675, 692, 847, 1199, 1265, 1267, 1342, 1352, 1357, 1368, 1369, 1379, 1382, 1383, 1399, 1401, 1477, 1480, 1482, 1530, 1625, 1732, 1777, 1792, 1814, 1837, 1998, 2175, 2212, 2223, 2329, 2487, 2504, 2753, 2916, 3076, 3667, 3692, 4241, 4243, 4247, 4248, 4278, 4407); are not referring to a function of the body (ID 1233, 1385, 1406, 1746, 1879, 1950, 3131, 4202); are related to the prevention or treatment of a disease (ID 1859, 2552, 2743, 2773, 3087, 3164, 3687); are not referring to a beneficial physiological effect (ID 544, 545, 648, 773, 782, 1399, 1781, 1795, 1842, 1906, 1944, 2759, 2825, 2886, 3128, 3137, 3153, 3519, 3972, 4112, 4662, 4691) pursuant to Article 13(1) of Regulation (EC) No 1924/2006. *EFSA Journal* 9(6):2228.

EFSA Panel on Dietetic Products, Nutrition and Allergies. 2011d. Scientific and technical guidance for the preparation and presentation of an application for authorisation of a health claim (revision 1). *EFSA Journal* 9(5):2170–2206.

Elleuch, M., D. Bedigian, O. Roiseux, S. Besbes, C. Blecker, and H. Attia. 2011. Dietary fibre and fibre-rich by-products of food processing: Characterisation, technological functionality and commercial applications: A review. *Food Chemistry* 124(2):411–421.

El-Sharnouby, G. A., S. M. Aleid, and M. M. Al-Otaibi. 2013. Conversion of processed citrus wastes into nutritional components. *Journal of Food Processing & Technology* 4(8):259–264.

FAO (Food and Agriculture Organization of the United Nations). 2012. Citrus Fruit: Fresh and Processed: Annual Statistics 2012. (http://www.fao.org/fileadmin/est/COM_MARKETS_MONITORING/Citrus/Documents/CITRUS_BULLETIN_2012.pdf).

Felgines, C., O. Texier, C. Besson, P. Vitaglione, J. L. Lamaison, V. Fogliano, A. Scalbert, L. Luca Vanella, and F. Galvano. 2008. Influence of glucose on cyanidin 3-glucoside absorption in rats. *Molecular Nutrition & Food Research* 52(8):959–964.

Fernández-Ginés, J. M., J. Fernández-López, E. Sayas-Barberá, E. Sendra, and J. A. Pérez-Alvarez. 2003. Effect of storage conditions on quality characteristics of bologna sausages made with citrus fiber. *Journal of Food Science* 68(2):710–714.

Fernández-Ginés, J. M., J. Fernández-Lopez, E. Sayas-Barberá, E. Sendra, and J. A. Pérez-Alvarez. 2004. Lemon albedo as a new source of dietary fiber: Application to bologna sausages. *Meat Science* 67(1):7–13.

Fernández-Ginés, J. M., J. Fernández-López, E. Sayas-Barberá, and J. Pérez-Alvarez. 2005. Meat products as functional foods: A review. *Journal of Food Science* 70(2):R37–R43.

Fernandez-Lopez, J., Aleson-Carbonell, L., Perez-Alvarez, J. A., and Kuri, V. 2005. Antioxidant and antibacterial activities of natural extracts: Application in beef meatballs. *Meat Science* 69(3):371–380.

Fernández-López, J., J. M. Fernández-Ginés, L. Aleson-Carbonell, E. Sendra, E. Sayas-Barberá, and J. A. Pérez-Alvarez. 2004. Application of functional citrus by-products to meat products. *Trends in Food Science & Technology* 15(3):176–185.

Fernández-López, J., E. Sendra, E. Sayas-Barberá, C. Navarro, and J. A. Pérez-Alvarez. 2008. Physico-chemical and microbiological profiles of "salchichón" (Spanish dry-fermented sausage) enriched with orange fiber. *Meat Science* 80(2):410–417.

Fernández-López, J., E. Sendra-Nadal, C. Navarro, E. Sayas, M. Viuda-Martos, and J. A. Perez-Alvarez. 2009. Storage stability of a high dietary fibre powder from orange by-products. *International Journal of Food Science & Technology* 44(4):748–756.

Fraser, P. D., and P. M. Bramley. 2004. The biosynthesis and nutritional uses of carotenoids. *Progress in Lipid Research* 43(3):228–265.

Galanakis, C. M. 2012. Recovery of high added-value components from food wastes: Conventional, emerging technologies and commercialized applications. *Trends in Food Science & Technology* 26(2):68–87.

Garcia, M. L., R. Dominguez, M. D. Galvez, C. Casas, and M. D. Selgas. 2002. Utilization of cereal and fruit fibres in low fat dry fermented sausages. *Meat Science* 60(3):227–236.

Gee, J. M., G. Williamson, and I. T. Johnson. 2002. Evidence for consistent patterns between flavonoid structures and cellular activities. *Proceedings of the Nutrition Society* 61(1):97–103.

González-Molina, E., R. Domínguez-Perles, D. A. Moreno, and C. García-Viguera. 2010. Natural bioactive compounds of *Citrus limon* for food and health. *Journal of Pharmaceutical and Biomedical Analysis* 51(2):327–345.

Gorinstein, S., O. Martıń-Belloso, Y. S. Park, R. Haruenkit, A. Lojek, M. Ĉıž, A. Caspi, I. Libman, and S. Trakhtenberg. 2001. Comparison of some biochemical characteristics of different citrus fruits. *Food Chemistry* 74(3):309–315.

Guo, W., J. Zhao, and J. Xiaojin. 2014. Method for synthesizing neohesperidin dihydrochalcone from naringin. Patent CN104119408A.

Hasegawa, S., M. Miyake, and Y. Ozaki. 1993. Biochemistry of citrus limonoids and their anticarcinogenic activity. pp. 198–208. In: Haung, M. T., O. Osawa, C. T. Ho, and R. Rosen (Eds.), *Food phytochemicals for cancer prevention I: Fruits and vegetables.* ACS Symposium Series No. 546, pp. 198–208. Washington, DC: American Chemical Society.

Horowitz, R. H., and G. Bruno. 1961. Dihydrochalcone derivatives and their use as sweeting agents. Patent US3087821A.

Huet, D. B., N. Piot, and A. Cuine. 1996. Pharmaceutical composition for the oral administration of flavonoids. Patent EP0711560A1.

Imeh, U., and S. Khokhar. 2002. Distribution of conjugated and free phenols in fruits: Antioxidant activity and cultivar variations. *Journal of Agricultural and Food Chemistry* 50(22):6301–6306.

Izquierdo, L., and J M. Sendra. 2003. Citrus fruit composition and characterization. In: B. Caballero, L. Trugo, and P. Finglas (Eds.), *Encyclopedia of food sciences and nutrition*, pp. 1335–1341. Cambridge, MA: Academic Press.

Jacob, R., S. Hasegawa, and G. Manners. 2000. The potential of citrus limonoids as anticancer agents. *Perishables Handling Quarterly* 102:6–8.

Johnson, J. D., H. E. Viale, and D. M. Wait. 1977. Method for extracting carotenoid pigments from citrus oils. Patent US4126709A.

Kang, H. J., S. P. Chawla, C. Jo, J. H. Kwon, and M. W. Byun. 2006. Studies on the development of functional powder from citrus peel. *Bioresource Technology* 97(4):614–620.

Kohajdova, Z., J. Karovičová, M. Jurasova, and K. Kukurová, K. 2011. Application of citrus dietary fibre preparations in biscuit production. *Journal of Food and Nutrition Research* 50(3):182–190.

Kosseva, M. R. 2013. Introduction: Causes and challenges of food wastage. In: Kosseva, M., and C. Webb (Eds.), *Food industry wastes: Assessment and recuperation of commodities*, pp. xv–xxiv. Elsevier: Amsterdam.

Kurowska, E. M., and J. A. Manthey. 2004. Hypolipidemic effects and absorption of citrus polymethoxylated flavones in hamsters with diet-induced hypercholesterolemia. *Journal of Agricultural and Food Chemistry* 52(10):2879–2886.

Kuroyanagi, M., H. Ishii, N. Kawahara, H. Sugimoto, H. Yamada, K. Okihara, and O. Shirota. 2008. Flavonoid glycosides and limonoids from *Citrus* molasses. *Journal of Natural Medicines* 62(1):107–111.

Lado, J., L. Zacarías, A. Gurrea, A. Page, A. Stead, and M. J. Rodrigo. 2015. Exploring the diversity in *Citrus* fruit colouration to decipher the relationship between plastid ultrastructure and carotenoid composition. *Planta* 242(3):645–661.

Lang, G., and G. Buchbauer. 2012. A review on recent research results (2008–2010) on essential oils as antimicrobials and antifungals: A review. *Flavour and Fragrance Journal* 27(1):13–39.

Lario, Y., E. Sendra, J. Garcı́, C. Fuentes, E. Sayas-Barberá, J. Fernández-López, and J. A. Perez-Alvarez. 2004. Preparation of high dietary fiber powder from lemon juice by-products. *Innovative Food Science & Emerging Technologies* 5(1):113–117.

Larrauri, J. A. 1999. New approaches in the preparation of high dietary fibre powders from fruit by-products. *Trends in Food Science & Technology 3* 10(1):3–8.

Laufenberg, G., K. Kunz, and M. Nystroem. 2003. Transformation of vegetable waste into value added products: (A) The upgrading concept; (B) Practical implementations. *Bioresource Technology* 87:176–198.

Ledesma-Escobar, C. A., and M. D. L. de Castro. 2014. Towards a comprehensive exploitation of citrus. *Trends in Food Science & Technology* 39(1):63–75.

Li, R. W., A. G. Theriault, K. Au, T. D. Douglas, A. Casaschi, E. M. Kurowska, and R. Mukherjee. 2006. Citrus polymethoxylated flavones improve lipid and glucose homeostasis and modulate adipocytokines in fructose-induced insulin resistant hamsters. *Life Sciences* 79(4):365–373.

Li, Y., and B. Xiong. 2013. Fruit-flavored solid beverage. Patent CN102389147B.

Lim, J., S. Ko, and S. Lee. 2014. Use of Yuja (*Citrus junos*) pectin as a fat replacer in baked foods. *Food Science and Biotechnology* 23(6):1837–1841.

Lin, C. S. K., L. A. Pfaltzgraff, L. Herrero-Davila, E. B. Mubofu, S. Abderrahim, J. H. Clark, A. A. Koutinas et al. 2013. Food waste as valuable resource for the production of chemicals, materials and fuels: Current situation and global perspective. *Energy Environmental Sciences* 6:426–464.

Liu, L., M. L. Fishman, J. Kost, and K. B. Hicks. 2003. Pectin-based systems for colon-specific drug delivery via oral route. *Biomaterials* 24(19):3333–3343.

Logan, K., A. J. Wright, and H. D. Goff. 2015. Correlating the structure and in vitro digestion viscosities of different pectin fibers to in vivo human satiety. *Food & Function* 6(1):62–70.

Lohrasbi, M., M. Pourbafrani, C. Niklasson, and M. J. Taherzadeh. 2010. Process design and economic analysis of a citrus waste biorefinery with biofuels and limonene as products. *Bioresource Technology* 101(19):7382–7388.

Mamma, D., and P. Christakopoulus. 2014. Biotransformation of *Citrus* by-products into added value products. *Waste Biomass Valorization* 5:529–549.

Mandalari, G., R. N. Bennett, G. Bisignano, A. Saija, R. Lo Curto, C. B. Faulds, and K. W. Waldron. 2006. Characterization of flavonoids and pectins from bergamot (*Citrus bergamia* Risso) peel, a major byproduct of essential oil extraction. *Journal of Agricultural and Food Chemistry* 54(1):197–203.

Manners, G. D. 2007. Citrus limonoids: Analysis, bioactivity, and biomedical prospects. *Journal of Agricultural and Food Chemistry* 55(21):8285–8294.

Manthey, J. A., and K. Grohmann. 2001. Phenols in citrus peel byproducts: Concentrations of hydroxycinnamates and polymethoxylated flavones in citrus peel molasses. *Journal of Agricultural and Food Chemistry* 49(7):3268–3273.

Manthey, J. A., N. Guthrie, and K. Grohmann. 2001. Biological properties of citrus flavonoids pertaining to cancer and inflammation. *Current Medicinal Chemistry* 8(2):135–153.

Marin, F. R., A. Ortuno, O. Benavente-García, and J. A. Del Río. 1998. Distribution of flavone glycoside diosmin in *Hyssopus officinalis* plants: Changes during growth. *Planta Medica* 64(2):181–182.

Marín, F. R., and J. A. Del Río. 2001. Selection of hybrids and edible *Citrus* species with a high content in the diosmin functional compound: Modulating effect of plant growth regulators on contents. *Journal of Agricultural and Food Chemistry* 49(7):3356–3362.

Marin, F. R., M. J. Frutos, J. A. Pérez-Alvarez, F. Martinez-Sanchez, and J. A. Del Rio. 2002b. Flavonoids as nutraceuticals: Structural related antioxidant properties and their role on ascorbic acid preservation. In: E. Atta-Ur-Rahman (Ed.), Bioactive *Natural Products: Studies in natural products chemistry*. Vol. 26, pp. 741–778. Amsterdam: Elsevier.

Marin, F. R., M. Martinez, T. Uribesalgo, S. Castillo, and M. J. Frutos. 2002a. Changes in nutraceutical composition of lemon juices according to different industrial extraction systems. *Food Chemistry* 78(3):319–324.

Marin, F. R., C. Soler-Rivas, O. Benavente-García, J. Castillo, and J. A. Pérez-Álvarez. 2007. By-products from different citrus processes as a source of customized functional fibres. *Food Chemistry* 100:736–741.

McNary, R. R., M. H. Dougherty, and R. W. Wolford. 1957. Determination of the chemical oxygen demand of citrus waste waters. *Sewage and Industrial Wastes* 29 (8):894–900.

Miller, R. A. 2011. Increased yield of bread containing citrus peel fiber 1. *Cereal Chemistry* 88(2):174–178.

Montgomery, R. 2004. Development of biobased-products. *Bioresource Technology* 91: 1–24.

Moure, A., J. M. Cruz, D. Franco, J. M. Domínguez, J. Sineiro, H. Domínguez, J. Sineiro, H. Dominguez, M. J. Nuñez, and J. C. Parajó. 2001. Natural antioxidants from residual sources. *Food Chemistry* 72(2):145–171.

Mustafa, N. E. M. 2015. Citrus essential oils: Current and prospective uses in the food industry. *Recent Patents on Food, Nutrition and Agriculture* 7(2):115–127.

Myers, S. R., J. A. Hawrelak, and T. Cattley. 2009. Essential oils in the treatment of intestinal dysbiosis: A preliminary in vitro study. *Alternative Medicine Review* 14(4):380–384.

Nogata, Y., K. Sakamoto, H. Shiratsuchi, T. Ishii, M. Yano, and H. Ohta. 2006. Flavonoid composition of fruit tissues of *Citrus* species. *Bioscience, Biotechnology, and Biochemistry* 70(1):178–192.

Ortuño, A., I. Reynaldo, M. D. Fuster, J. Botía, D. García-Puig, F. Sabater, A. García-Lidón, I. Porras, and J. A. Del Río. 1997. Citrus cultivars with high flavonoid contents in the fruits. *Scientia Horticulturae* 68(1–4):231–236.

Patil, B. S., G. K. Jayaprakasha, M. N. Chidambara, and A. Vikram. 2009. Bioactive compounds: Historical perspectives, opportunities and challenges. *Journal of Agricultural and Food Chemistry* 57(18):8142–8160.

Perrigue, M. M., P. Monsivais, and A. Drewnowski. 2009. Added soluble fiber enhances the satiating power of low-energy-density liquid yogurts. *Journal of the American Dietetic Association* 109(11):1862–1868.

Piriyaprasarth, S., and Sriamonsak, P. 2011. Flocculating and suspending properties of commercial citrus pectin and pectin extracted from pomelo (*Citrus maxima*) peel. *Carbohydrate polymers* 83(2): 561–568.

Rao, A. V., and L. G. Rao. 2007. Carotenoids and human health. *Pharmacological Research* 55(3):207–216.

Rock, C. L., J. L. Lovalvo, C. Emenhiser, M. T. Ruffin, F. W. Flatt, and S. J. Schwartz. 1998. Bioavailability of beta-carotene is lower in raw than in processed carrots and spinach in women. *Journal of Nutrition* 128:913–916.

Romero-Lopez, M. R., P. Osorio-Diaz, L. A. Bello-Perez, J. Tovar, and A. Bernardino-Nicanor. 2011. Fiber concentrate from orange (*Citrus sinensis* L.) bagase: Characterization and application as bakery product ingredient. *International Journal of Molecular Sciences* 12(4):2174–2186.

Russo, M., I. Bonaccorsi, G. Torre, M. Sarò, P. Dugo, and L. Mondello. 2014. Underestimated sources of flavonoids, limonoids and dietary fibre: Availability in lemon's by-products. *Journal of Functional Foods* 9:18–26.

Schieber, A., F. C. Stintzing, and R. Carle. 2001. By-products of plant food processing as a source of functional compounds—Recent developments. *Trends in Food Science & Technology* 12(11):401–413.

Schneider, H. 1968. The anatomy of *Citrus*. In: Ruther, W., H. J. Weber, and L. D. Batchelor (Eds.), *The citrus industry*. Vol. II, p. 398. Berkeley, CA: University of California Press.

Sendra, E., P. Fayos, Y. Lario, J. Fernandez-Lopez, E. Sayas-Barbera, and J. A. Perez-Alvarez. 2008. Incorporation of citrus fibers in fermented milk containing probiotic bacteria. *Food Microbiology* 25(1):13–21.

Sendra, E., V. Kuri, J. Fernandez-Lopez, E. Sayas-Barbera, C. Navarro, and J. A. Perez-Alvarez. 2010. Viscoelastic properties of orange fiber enriched yogurt as a function of fiber dose, size and thermal treatment. *LWT—Food Science and Technology* 43(4):708–714.

Shaw, P. E. 1979. Review of quantitative analyses of citrus essential oils. *Journal of Agricultural and Food Chemistry* 27(2):246–257.

Shyu, Y. S., T. C. Lu, and C. C. Lin. 2014. Functional analysis of unfermented and fermented citrus peels and physical properties of citrus peel-added doughs for bread making. *Journal of Food Science and Technology* 51(12):3803–3811.

Silalahi, J. 2002. Anticancer and health protective properties of citrus fruit components. *Asia Pacific Journal of Clinical Nutrition* 11(1):79–84.

Stoll, L., S. H. Flôres, and R. C. S. Thys. 2015. Citrus peel fiber and its application as a fat substitute in loaf brad. *Ciência Rural* 45(3):567–573.

Swingle, W. T., and P. C. Reece. 1967. The botany of Citrus and its wild relatives. In Reuther, W., Webber, H. J., and Batchelor, L. D. (Eds.). University of California Press. Berkeley, CA, USA. *The Citrus Industry* 2(1):190–430.

Thebaudin, J. Y., A. C. Lefebvre, M. Harrington, and C. M. Bourgeois. 1997. Dietary fibres: Nutritional and technological interest. *Trends in Food Science & Technology* 8(2):41–48.

Toulinim, H. A. 1950. Method for extracting peel oils and other waste products. Patent US2708627A.

Tripoli, E., M. La Guardia, S. Giammanco, D. Di Majo, and M. Giammanco. 2007. Citrus flavonoids: Molecular structure, biological activity and nutritional properties: A review. *Food Chemistry* 104(2):466–479.

Tyagi, A. K., D. Gottardi, A. Malik, and M. E. Guerzoni. 2014. Chemical composition, in vitro anti-yeast activity and fruit juice preservation potential of lemon grass oil. *LWT—Food Science and Technology* 57(2):731–737.

Vanhemelrijck, J. G. R., and C. H. McCare. 2007. Citrus fruit fibre in emulsions. Patent WO 2007003391A1.

Vergamini, D., D. Cuming, and D. Viaggi. 2015. The integrated management of food processing waste: The use of the full cost method for planning and pricing Mediterranean citrus by-products. *International Food and Agribusiness Management Review* 18:153–172.

Viuda-Martos, M., Y. Ruiz-Navajas, J. Fernández-López, and J. Pérez-Álvarez. 2008. Antifungal activity of lemon (*Citrus lemon* L.), mandarin (*Citrus reticulata* L.), grapefruit (*Citrus paradisi* L.) and orange (*Citrus sinensis* L.) essential oils. *Food Control* 19(12):1130–1138.

Viuda-Martos, M., J. Fernández-López, E. Sayas-Barbera, E. Sendra, C. Navarro, and J. A. Pérez-Álvarez. 2009. Citrus co-products as technological strategy to reduce residual nitrite content in meat products. *Journal of Food Science* 74(8):R93–R100.

Viuda-Martos, M., Y. Ruiz-Navajas, J. Fernández-López, and J. A. Pérez-Álvarez. 2010. Effect of orange dietary fibre, oregano essential oil and packaging conditions on shelf-life of bologna sausages. *Food Control* 21(4):436–443.

Viuda-Martos, M., J. Fernandez-Lopez, E. Sayas-Barbera, E. Sendra, and J. A. Perez-Alvarez. 2011. Physicochemical characterization of the orange juice waste water of a citrus by-product. *Journal of Food Processing and Preservation* 35(2):264–271.

Volpe, M., D. Panno, R. Volpe, and A. Messineo. 2015. Upgrade of citrus waste as a biofuel via slow pyrolysis. *Journal Analytical and Applied Pyrolysis* 115:66–76.

Wang, Y. C., Y. C. Chuang, and H. W. Hsu. 2008. The flavonoid, carotenoid and pectin content in peels of citrus cultivated in Taiwan. *Food Chemistry* 106(1):277–284.

Widmer, W. W., and A. M. Montanari. 1994. Citrus waste streams as a source for phytochemicals. In *Proceedings—Florida State Horticultural Society* 107:284–288.

Wilkins, M. R., W. W. Widmer, and K. Grohmann. 2007. Simultaneous saccharification and fermentation of citrus peel waste by *Saccharomyces cerevisiae* to produce ethanol. *Process Biochemistry* 42(12):1614–1619.

Winnig, M., B. Bufe, N. A. Kratochwil, J. P. Slack, and W. Meyerhof. 2007. The binding site for neohesperidin dihydrochalcone at the human sweet taste receptor. *BMC Structural Biology* 7(1):66.

Yalınkılıç, B., G. Kaban, and M. Kaya. 2012. The effects of different levels of orange fiber and fat on microbiological, physical, chemical and sensorial properties of sucuk. *Food Microbiology* 29(2):255–259.

Yi, T., X. Huang, S. Pan, and L. Wang. 2014. Physicochemical and functional properties of micronized jincheng orange by-products (*Citrus sinensis* Osbeck) dietary fiber and its application as a fat replacer in yogurt. *International Journal of Food Sciences and Nutrition* 65(5):565–572.

Zhang, G., and Q. Zhang. 2006. Diosming producing process. Patent CN101089009A.

Zhang, K., and J. Wang. 2011. Citrus fibre meat product and preparation method thereof. Patent CN101946901A.

Zhong, Z., L. Weu, H. Dachuan, and C. Yan. 2010. Method for producing water-soluble citrus peel fibres. Patent CN101797037.

Zou, Z., W. Xi, Y. Hu, C. Nie, and Z. Zhou. 2016. Antioxidant activity of *Citrus* fruits. *Food Chemistry* 196:885–896.

15 Grapefruit-Like Varieties with Low Furanocoumarin Content

Eran Raveh, Yoram Eyal, Ron Porat, and Nir Carmi

CONTENTS

15.1 INTRODUCTION

Furanocoumarins (FCs) are plant defense compounds that function as toxins against pathogens, insects, and other plant pests (Berenbaum 2003; Widmer 2006). Insects exposed to linear FCs suffer decreased larval weight, extended generation time, and increased mortality (Diawara et al. 1993). High doses of some FCs are also toxic to humans; among the maladies caused in humans by FCs are "celery picker's itch" and "bartender's itch," caused by contact with FC-containing celery plants or oil of lime peels, respectively. Also, FCs can cause temporary or even permanent blindness if introduced into the eyes. Some FCs are also carcinogenic and teratogenic; that is, they can cause cancer and birth defects, respectively (Diawara et al. 1993; Gou et al. 2000; Pan et al. 2004).

FCs are found in several plant families, including edible plant species such as the *Citrus* species pummelo and grapefruit, but the levels of mutagenic FCs (such as psoralen) in pummelo and grapefruits are rather low; therefore, there is no risk associated with consumption of their fruit and juice (Diawara et al. 1993). Citrus is one of the most important fruit crops worldwide, with a total annual production of 115.5 million tons, to which grapefruit contributes about 5.5% (USDA-FAO 2014). However, in recent years, there has been a decline in grapefruit consumption, especially in the United States, as a result of the "grapefruit juice effect," attributed to FCs: FC-containing grapefruit juice negatively affects the body's degradation of medicines, leading to overdose effects.

The mechanism of the grapefruit juice effect is based on FCs that inhibit the intestinal drug catabolic enzyme CYP3A, thereby leading to accumulation of medicines, which in turn can lead to harmful effects ranging from relatively mild hypotension and dizziness in the case of some calcium channel blockers, to potentially severe nephrotoxicity in the case of some immunosuppressant drugs (Bailey et al. 1998, 2004; De Castro et al. 2006; Dugrand et al. 2013; Ohnishi et al. 2000; Pan et al. 2004; Paine et al. 2005, 2006; VanderMolen et al. 2014; Wen et al. 2002; Widmer and Haun 2005).

Drugs that interact with FCs (Bailey et al. 2013) include the following:

- Anticancer
  - Dasatinib (leukemia)
  - Erlotinib (lung cancer and pancreatic cancer)
  - Everolimus (kidney cancer)
  - Lapatinib (breast cancer)
  - Nilotinib (leukemia)
  - Pazopanib (kidney cancer)
  - Sunitinib (kidney/gastrointestinal cancer)
  - Vandetanib (thyroid cancer)
  - Venurafenib (skin cancer)
- Anti-infective
  - Erythromycin (antibiotic)
  - Halofantrine (antimalaria)
  - Maraviroc (anti-human immunodeficiency virus [anti-HIV])
  - Primaquine (antimalaria)
  - Quinine (antimalaria)
  - Rilpivirine (anti-HIV)
- Anticholesterol
  - Atorvastatin
  - Lovastatin
  - Simvastatin
- Cardiovascular
  - Amiodarone (heart rhythm disorders)
  - Apixaban (anticlotting)
  - Dronedarone (heart rhythm disorders)
  - Eplerenone (heart failure)
  - Felodipine (high blood pressure/angina)
  - Nifedipine (high blood pressure/angina)
  - Quinidine (heart rhythm disorders)
  - Rivaroxaban (blood anticlotting)
  - Ticagrelor (blood anticlotting after heart attack)

- Central nervous system
    - Oral alfentanil (painkiller)
    - Oral fentanyl (painkiller)
    - Oral ketamine (painkiller, sedative)
    - Lurasidone (schizophrenia/mental health problems)
    - Oxycodone (painkiller)
    - Pimozide (schizophrenia/other mental health problems)
    - Ziprasidone (schizophrenia, mania, bipolar disorder)
- Gastrointestinal
    - Domperidone (antinausea)
- Immunosuppressants
    - Cyclosporine (postorgan transplant, rheumatoid arthritis, psoriasis)
    - Sirolimus (postorgan transplant)
    - Tacrolimus (postorgan transplant)
- Urinary tract
    - Solifenacin (frequent urination/incontinence)
    - Silodosin (enlarged prostate)
    - Tamsulosin (enlarged prostate)

The FC biosynthesis pathway starts with the amino acid phenylalanine (Figure 15.1) via umbelliferone to obtain linear or angular FCs. In citrus, only the linear FC-biosynthesis pathway is present which leads to formation of demethylsuberosin by alkylation of umbelliferone (Figure 15.1). Three FCs—6,7-dihydroxybergamottin (6,7-DHB), bergamottin, and paradisin C—appear to cosegregate genetically, which suggests that these three compounds are tightly linked metabolically as coproducts of the bergamottin pathway (Chen et al. 2011). Ranked in descending order of their inhibitory potency, the FCs found in grapefruit are paradisin C > 6,7-DHB > bergamottin > isoimperatorin > bergapten > bergaptol (Ohnishi et al. 2000; Row et al. 2006).

In this chapter, we report on variations in the biosynthesis levels of various FCs in diverse varieties of grapefruit and pummelo. Twelve citrus cultivars (Table 15.1), including four pummelos, two grapefruits, one mandarin, one orange, and three newly selected grapefruit-like varieties, were tested for juice and leaf levels of five FCs—bergamottin, bergaptol, epoxybergamottin, 6,7-DHB, and isoimperatorin—and two FC precursors—psoralen and umbelliferone. The results showed that the orange and the mandarin do not contain FCs; moreover, the lack of FC precursor compounds suggests that the biosynthesis pathway is entirely not active in mandarins and oranges. We also observed great diversity in the accumulation of FCs and their precursor compounds in various grapefruit and pummelo cultivars. Finally, we describe two new low-FC and seedless grapefruit-like cultivars, "Aliza" and "Cookie," which were selected in our breeding program by using a mutation-inducing procedure (Vardi et al. 2008).

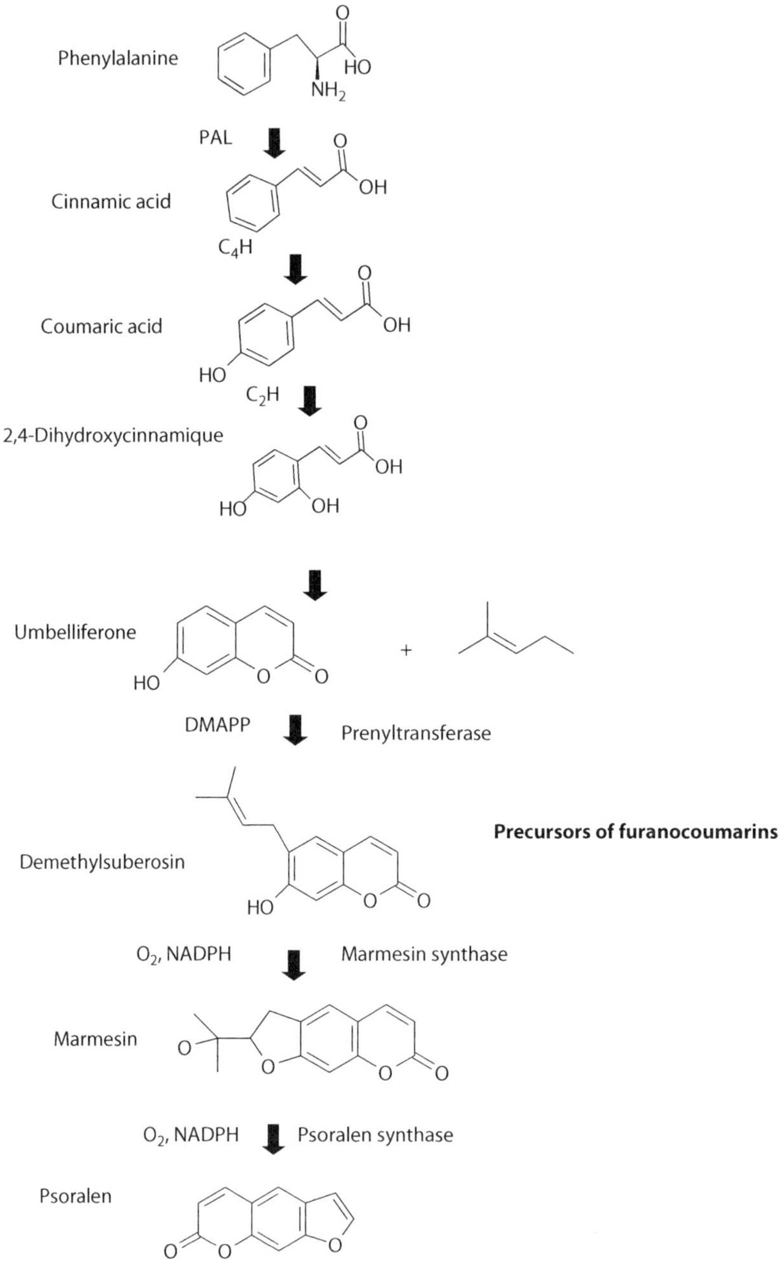

FIGURE 15.1 Linear FC biosynthesis pathway from phenylalanine to 6,7-dihydroxybergamottin. The chemical structures were taken from the PubChem database (Wang et al. 2009).

(Continued)

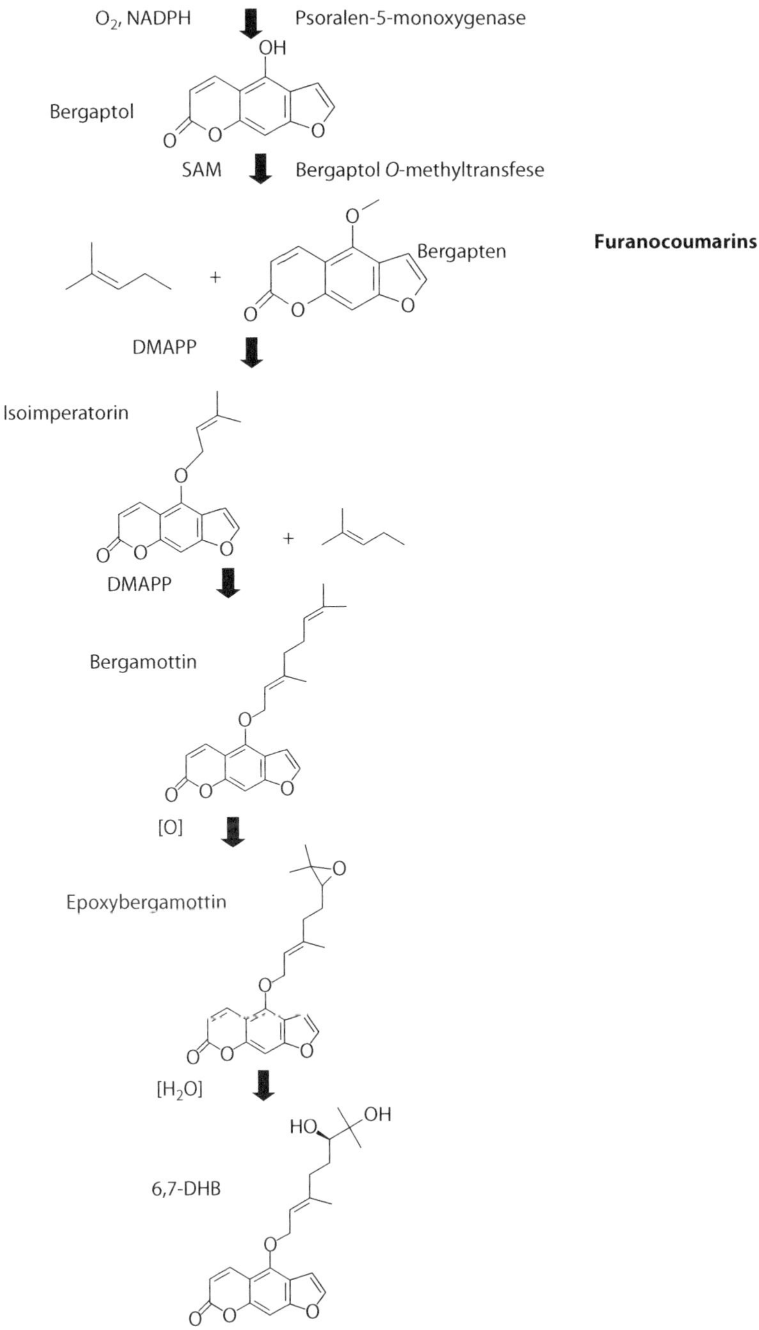

FIGURE 15.1 (CONTINUED) Linear FC biosynthesis pathway from phenylalanine to 6,7-dihydroxybergamottin. The chemical structures were taken from the PubChem database (Wang et al. 2009).

TABLE 15.1
List of Tested Cultivars

Cultivar	Group	♂	♀
Tahitian	Pummelo (*Citrus maxima* (Burm.) Merr.)	ND	ND
Chandler	Pummelo (*C. maxima* (Burm.) Merr.)	"Siamese Pink" pummelo (*C. maxima* (Burm.) Merr.)	"Siamese Sweet" pummelo (*C. maxima* (Burm.) Merr.)
Marsh	Grapefruit (*C. paradisi* Macf.)	Orange (*C. sinensis* (L.) Osbeck)	Pummelo (*C. maxima* (Burm.) Merr.)
Duncan	Grapefruit (*C. paradisi* Macf.)	Orange (*C. sinensis* (L.) Osbeck)	Pummelo (*C. maxima* (Burm.) Merr.)
Valencia	Orange (*C. sinensis* (L.) Osbeck)	*C. reticulata*	Pummelo (*C. maxima* (Burm.) Merr.)
Orah	Mandarin (*C. reticulata*)	"Temple" (*C. reticulata* × *C. sinensis*)	"Kinnow" mandarin (*C. reticulata* Blanco)
Flamingo	Pummelo (*C. maxima* (Burm.) Merr.)	"Chandler" pummelo (*C. maxima* (Burm.) Merr.)	"Tahitian" pummelo (*C. maxima* (Burm.) Merr.)
Hanna	Pummelo (*C. maxima* (Burm.) Merr.)	"Chandler" pummelo (*C. maxima* (Burm.) Merr.)	"Tahitian" pummelo (*C. maxima* (Burm.) Merr.)
Einat	A red triploid "Oroblanco"-like citrus fruit	Tetraploid "Hudson" (*C. paradisi* Macf.)	Acid-free pummelo (*C. maxima* (Burm.) Merr.)
Aliza	Grapefruit-like	"Orah" mandarin (*C. reticulata* Blanco)	"Chandler" pummelo (*C. maxima* (Burm.) Merr.)
Dany	Grapefruit-like	"Michal" mandarin (*C. reticulata* Blanco)	"Duncan" grapefruit (*C. paradisi* Macf.)
Cookie (seedless mandelo)	Grapefruit-like	"Frua" mandarin (*C. reticulata*)	"Siamese Sweet" pummelo (*C. maxima* (Burm.) Merr.)

Source: From Fidel, Y. et al., *Food and Nutrition Sciences*, 7, 90–101, 2016.
Abbreviation: ND, not determined.

15.2 CHARACTERIZATION OF FURANOCOUMARIN COMPONENTS

15.2.1 Extraction of Furanocoumarins

FCs were extracted from fruit juice and leaves. For the juice analysis, 10 mL of juice of each cultivar was mixed with an equal volume of ethyl acetate. For the leaf analysis, 10 g of leaf tissue was ground in liquid nitrogen and mixed with ethyl acetate at a volume equivalent to the weight, that is, 10 mL. The extraction was performed by vortex mixing for 5 min at 25°C followed by centrifugation at 3220 g for 25 min at 25°C. The organic phase was collected and reextracted with an equivalent volume of double-distilled water. The mixture was then remixed and centrifuged as described earlier. A 2 mL aliquot of the organic phase was evaporated under airflow and dissolved in 1 mL of acetonitrile for liquid chromatography–mass spectrometry (LC-MS) analysis.

15.2.2 Extraction of Naringin

The fruit slices were mixed in a blender for 1 min, and the blended material was mixed with butanol (1:1, vol/vol), centrifuged at 4°C at 10,000 g for 15 min, and the butanol fraction was centrifuged again at 10,000 g for 10 min. A 3 mL aliquot of the upper phase was dried in a speed-vacuum centrifuge, and the precipitate was dissolved in 5 mL of 100% methanol and filtered through a 0.45 μm filter. LC-MS analyses were conducted with an ultraperformance liquid chromatography (UPLC)–triple quadrupole–MS instrument (Waters Xevo TQ MS). The samples were again filtered, through a 0.22 μm Millex-HV Durapore (polyvinylidene fluoride [PVDF]) membrane, before injection into the LC-MS apparatus. Separation was performed on a 2.1 m × 50 mm inner diameter (i.d.), 1.7 μm UPLC BEH C18 column. Chromatographic and MS parameters were as follows. The mobile phase consisted of water (phase A) and 0.1% formic acid in acetonitrile (phase B). The linear gradient program was as follows: 100%–95% A over 0.1 min, 95%–5% A over 5 min, held at 5% A for 2 min, and then back to the initial conditions (95% A) for 3 min. The flow rate was 0.3 mL/min, and the column temperature was kept at 35°C. All of the analyses were performed with the electrospray ionization (ESI) source in positive-ion mode, with a capillary voltage of 3.2 kV, a cone voltage of 30 V, a desolvation temperature of 350°C, a desolvation gas flow rate of 850 L/h, and a source temperature of 150°C. Quantization was performed with multiple reaction monitoring (MRM) acquisition by monitoring the 581/152, 581/273 transitions, retention time (RT) = 6.07 (dwell time of 161 ms for each transition) for naringin.

15.2.3 LC-MS Analyses

LC-MS analyses were conducted with a UPLC Triple Quadrupole-MS (Waters Xevo TQ MS). Samples were filtered through a 0.22 μm Millex-HV Durapore (PVDF) membrane before being injected into the LC-MS apparatus. Separation was performed with a 2.1 m × 50 mm i.d., 1.7 μm UPLC BEH C18 column. The chromatographic and MS parameters were as follows: the mobile phase consisted of water (phase A) and 0.1% formic acid in acetonitrile (phase B). The linear-gradient program for bergamottin, epoxybergamottin, 6,7-DHB, bergaptol, and isoimperatorin was as follows: 100%–95% A over 0.1 min, 95%–5% A over 8 min, held at 5% A for 3 min, and then back to the initial conditions (95% A) for 3 min. The linear-gradient program for psoralen and umbelliferone was as follows: 100%–95% A over 0.1 min, 95%–5% A over 4 min, held at 5% A for 3 min, and then back to the initial conditions (95% A) for 3 min. The flow rate was 0.3 mL/min, and the column temperature was kept at 35°C. All of the analyses were performed with the ESI source used in positive-ion mode, with a capillary voltage of 3.2 kV, a cone voltage of 30 V, a desolvation temperature of 350°C, a desolvation gas flow of 650 L/h, and a source temperature of 150°C. Quantitation was performed with MRM acquisition by monitoring the following transitions: 339/147, 339/203 (RT = 7.1, dwell time of 161 ms) for bergamottin; 355/153, 355/203 (RT = 5.6, dwell time of 78 ms) for epoxybergamottin; 373/153, 373/203 (RT = 4.2, dwell time of 78 ms) for 6,7-DHB; 202/131, 202/146 (RT = 3.1, dwell time of 161 ms) for bergaptol; 271/146, 271/203 (RT = 5.5, dwell time of 161 ms) for isoimperatorin; 187/115, 187/13 (RT = 2.46, dwell time of 78 ms) for psoralen; and 163/91, 163/107 (RT = 1.8, dwell time of 78 ms) for umbelliferone.

15.3 FURANOCOUMARIN CONTENTS IN DIFFERENT VARIETIES

FCs were found at varied concentrations in different cultivars, but were completely absent from mandarins and oranges and were close to absent in the new grapefruit-like varieties "Aliza" and "Cookie" (Figure 15.2). In grapefruit and pummelo, bergamottin and 6,7-DHB were found at higher concentrations than other FCs, probably because they are final products that accumulate over time. In correlation, the FC precursors are present in very low concentrations in grapefruit and pummelo (Figure 15.2), as they are utilized for biosynthesis of the final products. Interestingly, although the FC biosynthesis pathway of grapefruit (Figure 15.1) was inherited from pummelo, the accumulated concentrations of bergamottin and 6,7-DHB in the grapefruit cultivars "Duncan" and "Marsh" (a seedless mutation of "Duncan") were much higher than those observed in the pummelo cultivars that we examined (Figure 15.2).

The complete absence of FCs and FC precursors from umbelliferone to psoralen in oranges and mandarins (Figure 15.2) suggests that the entire pathway for biosynthesis of FC is absent or inactive in these species and that only a null allele for FC biosynthesis can be inherited from these species. The new selection "Aliza," a hybrid of "Chandler" pummelo and "Orah" mandarin, is almost completely void of FCs, suggesting that the pummelo parent is heterozygous for FC biosynthesis (i.e., contains one active copy and one inactive copy, presumably inherited by "Aliza"). Furthermore, the finding that the pummelo hybrid "Hanna" (a cross between "Chandler" and "Tahitian" pummelos) has no FCs (Figure 15.2) indicates that "Chandler" probably also contributed its inactive allele for FC biosynthesis to "Hanna." In contrast, "Chandler" contributed its active allele for FC biosynthesis to the pummelo "Flamingo" (a different cross between "Chandler" and "Tahitian" pummelos) (Figure 15.2); the rather higher level of 6,7-DHB in "Flamingo" might be a result of the release of the isoimperatorin that accumulates in "Tahitian," by the active allele of "Chandler" (Fidel et al. 2016).

The concentrations of bergamottin and 6,7-DHB observed in 4- to 5-month-old leaves were extremely low relative to those observed in the fruits. Pummelo cultivars have much lower concentrations of epoxybergamottin in their fruits versus other FC compounds (Figure 15.2a), but the accumulated amounts of this FC in their leaves were equivalent to the highest concentrations observed among all of the examined cultivars (Figure 15.2). During commercial processing, epoxybergamottin may be hydrolyzed to 6,7-DHB, which has been identified as a potentially important inhibitor of CYP3A4 (Bailey et al. 2013; Wangensteen 2003).

In all of the fruit juices of the examined cultivars, bergaptol and isoimperatorin were found at significantly lower concentrations than the other FCs, reflecting the fact that they are intermediate products in the pathway. However, pummelo "Tahitian"

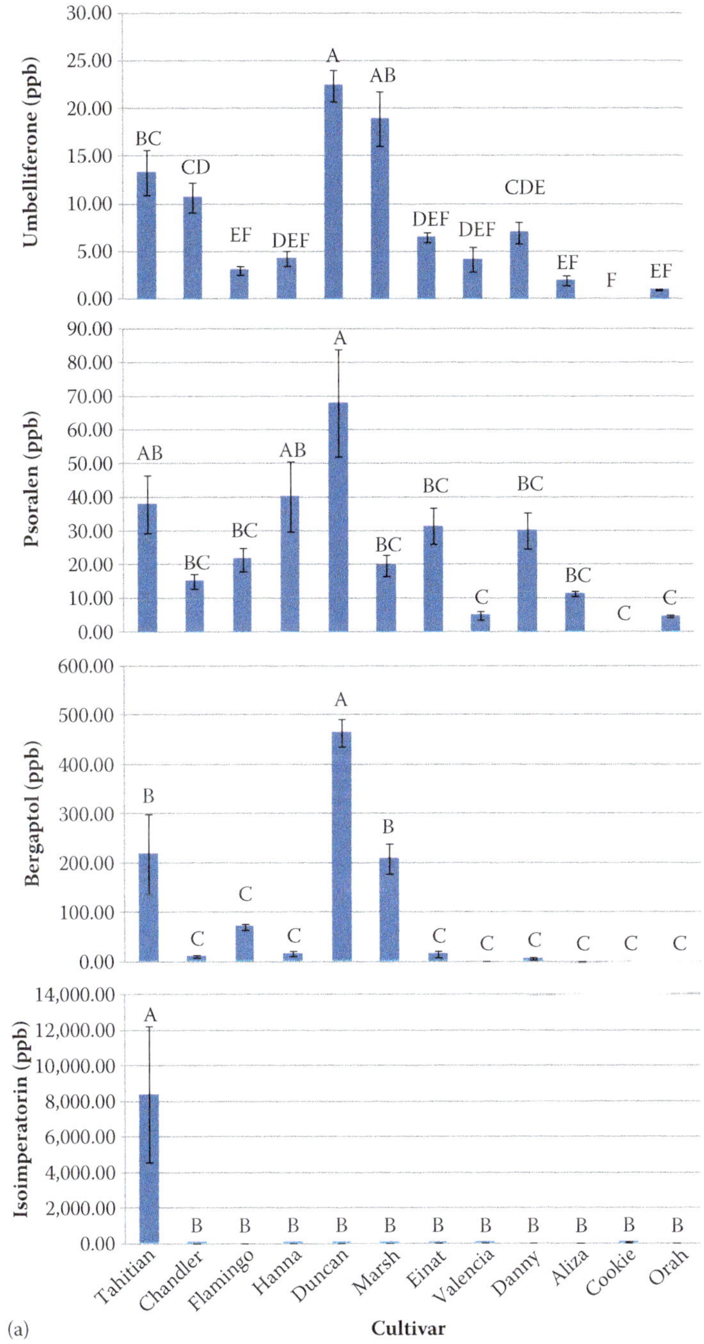

FIGURE 15.2 Analysis of FC levels in 12 citrus cultivars. FCs were analyzed in (a) fruit tissues. Data are means of three replications. Different letters above bars indicate significant differences at $p \leq 0.05$. *(Continued)*

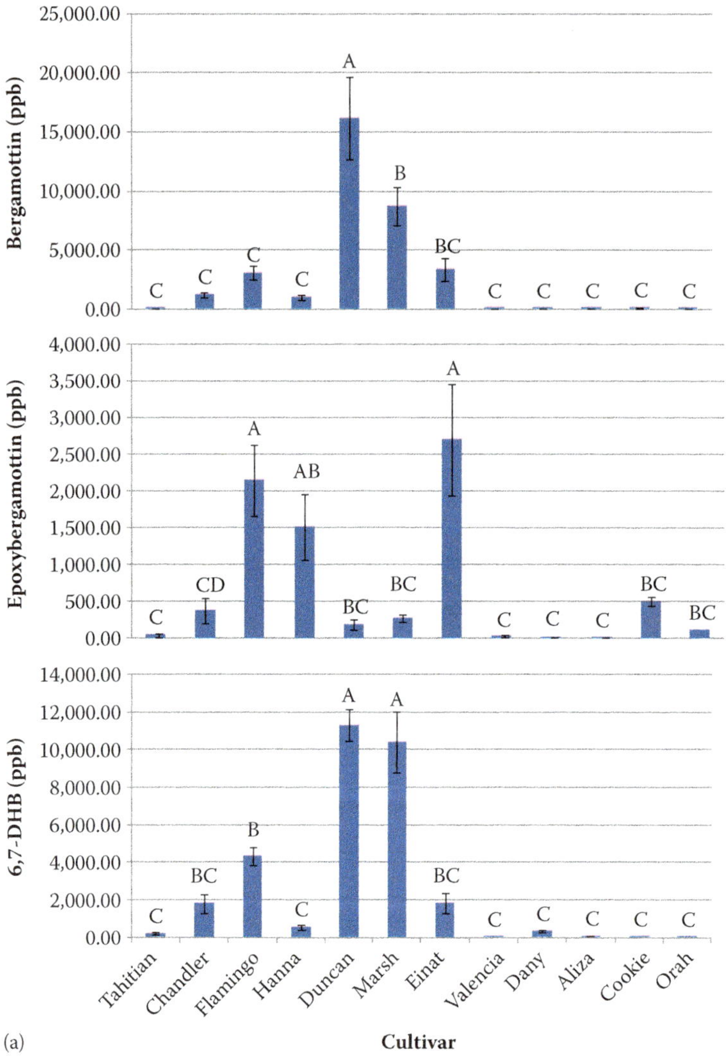

FIGURE 15.2 (CONTINUED) Analysis of FC levels in 12 citrus cultivars. FCs were analyzed in (a) fruit tissues. Data are means of three replications. Different letters above bars indicate significant differences at $p \leq 0.05$. *(Continued)*

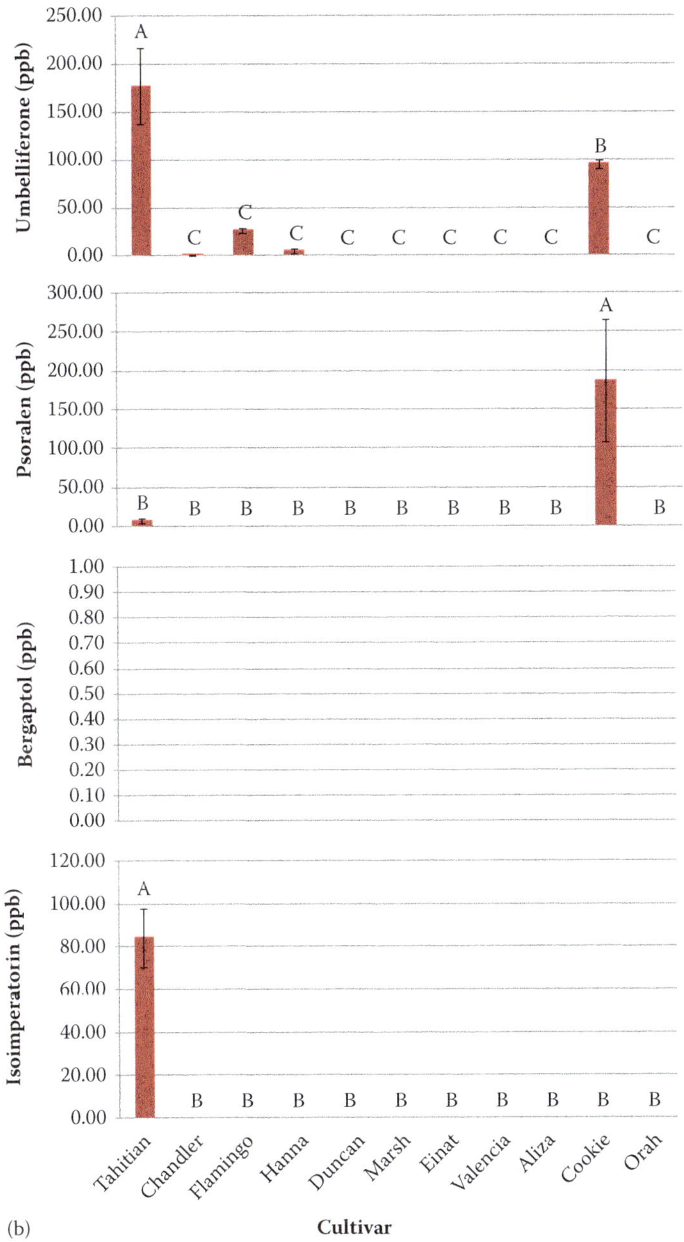

FIGURE 15.2 (CONTINUED) Analysis of FC levels in 12 citrus cultivars. FCs were analyzed in (b) leaf tissues. Data are means of three replications. Different letters above bars indicate significant differences at $p \leq 0.05$. (From Fidel, Y. et al., *Food and Nutrition Sciences*, 7, 90–101, 2016.) (*Continued*)

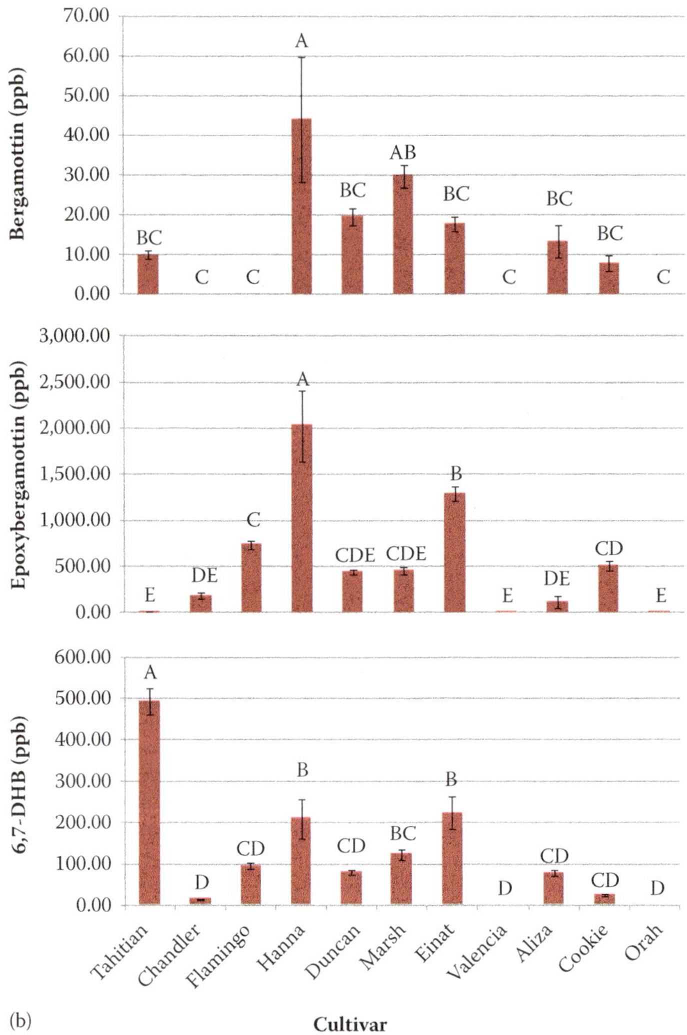

FIGURE 15.2 (CONTINUED) Analysis of FC levels in 12 citrus cultivars. FCs were analyzed in (b) leaf tissues. Data are means of three replications. Different letters above bars indicate significant differences at $p \leq 0.05$. (From Fidel, Y. et al., *Food and Nutrition Sciences*, 7, 90–101, 2016.)

contained an extraordinarily high concentration of isoimperatorin; this pummelo also had very low concentrations of bergamottin and 6,7-DHB, suggesting that the pathway in this variety is partially blocked and it accumulates isoimperatorin instead of bergamottin and 6,7-DHB.

We have selected two new seedless grapefruit-like cultivars, "Aliza" and "Cookie" (Table 15.1), which do not accumulate FCs. At the start of ripening,

"Aliza" fruits have green skin and yellow flesh, but the former turns yellow later in the season, and when fully ripe, these fruits have orange skin and orange flesh (Figure 15.3). By inducing mutations, we selected seedless types of "Aliza" and "Cookie" (Figures 15.3 and 15.4). Both of these selections have unique and favorable characteristics: they are juicy and, because they contain naringin (Table 15.2), they have been classified as grapefruit-like cultivars; "Aliza" and "Cookie" are also easy to peel (Fidel et al. 2016).

(a)

(b)

FIGURE 15.3 Photographs of "Aliza," a new grapefruit-like cultivar, taken in (a) November and (b) February. (From Fidel, Y. et al., *Food and Nutrition Sciences*, 7, 90–101, 2016.)

FIGURE 15.4 Photographs of "Cookie," a new grapefruit-like cultivar. (From Fidel, Y. et al., *Food and Nutrition Sciences*, 7, 90–101, 2016.)

15.4 SUMMARY

FCs are a group of related plant defense metabolites found in several plant families, including some species in the genus *Citrus*, such as grapefruit and pummelo. FCs function as toxins against pathogens, insects, and other plant pests, and some are toxic to humans at high levels. Although the levels of FCs in grapefruits are nontoxic to humans, they inhibit the intestinal enzyme CYP3A, thus preventing degradation of medicines, such as statins, and causing dangerous overdose effects. This overdosing could cause devastating side effects, ranging from stomach bleeding to kidney problems, muscle aches, and irregular heartbeats. The levels of FC pathway intermediates and end products in 12 citrus cultivars, including mandarin (*C. reticulata*), orange (*C. sinensis* (L.) Osbeck), pummelo (*C. maxima* (Burm.) Merr.), grapefruit (*C. paradisi* Macf.), and two newly selected grapefruit-like varieties ((*C. reticulate*) × (*C. maxima* (Burm.) Merr) were analyzed. The orange and mandarin varieties do not contain FCs or FC precursor compounds, suggesting that this biosynthetic pathway is absent or inactive in mandarins and oranges and therefore a good genetic source for null alleles to FC biosynthesis. The two new low-FC and seedless grapefruit-like varieties, "Aliza" and "Cookie," are developed as a cross between pummelo and mandarin. The fruits of these varieties resemble grapefruit and contain high levels of the flavanone naringin, typical of grapefruit, but contain only trace amounts of FCs. Based on the variability of FC content and inheritance in *Citrus* species, the results suggest that future development of new low-FC grapefruit varieties is an achievable objective. FC compositions and concentrations are highly variable in citrus, suggesting that selection for new grapefruit like cultivars with low FCs is a feasible objective. The selection of two new low FC grapefruit-like varieties is potentially safe for consumption by humans using statins and other drugs. The correlation of FC accumulation between leaves and fruits provides a potential marker for screening young plantlets for additional FC-free grapefruit-like varieties.

TABLE 15.2
Quality Traits and Naringin Contents of Juices of 12 Different Citrus Cultivars

Cultivar	Seeds (No. ± SE)	Sugar (°Bx ± SE)	Acid (% ± SE)	Juice (% ± SE)	Size (g ± SE)	Naringin (ppb ± SE)	Ripening Season
Tahitian	120.00 (±6.00)	12.00 (±0.07)	1.50 (±0.04)	30.50 (±2.50)	542.00 (±28.00)	66,690.56 (B) (±3,072.85)	October–November
Chandler	84.00 (±6.00)	12.00 (±0.04)	0.30 (±0.02)	9.00 (±0.80)	684.00 (±34.00)	144,550.72 (A) (±6,871.74)	October–November
Flamingo	0.30 (±0.20)	10.70 (±0.10)	0.90 (±0.5)	19.10 (±1.50)	1,218.00 (±60.00)	75,495.57 (B) (±4,382.96)	November–January
Hanna	26.00 (±4.00)	9.50 (±0.23)	0.60 (±0.07)	23.80 (±1.50)	917.00 (±147.00)	86,314.11 (B) (±11,235.48)	November–February
Duncan	55.00 (±5)	11.80 (±0.15)	2.30 (±0.04)	33.80 (±0.90)	339.00 (±16.00)	88,589.18 (B) (±12,461.56)	October–April
Marsh	3.00 (±1.00)	11.20 (±0.12)	2.00 (±0.04)	35.60 (±1.10)	372.00 (±9.00)	42,496.21 (C) (±3,789.91)	October–April
Einat	2.70 (±1.30)	9.50 (±0.3)	1.80 (±0.05)	23.70 (±1.10)	687.00 (±30.00)	67,598.29 (B) (±5,105.92)	November–February
Valencia	4.00 (±0.70)	9.60 (±0.25)	1.40 (±0.05)	45.90 (±0.80)	263.00 (±8)	0.00 (E) (±0.00)	February–May
Dany	9.00 (±1.40)	11.00 (±0.09)	1.70 (±0.06)	22.10 (±1.40)	458.00 (±21.00)	0.00 (E) (±0.00)	January
Aliza	0.00	10.40 (±0.20)	0.70 (±0.03)	42.80 (±1.60)	621.00 (±58.00)	31,124.00 (CD) (±3,000.33)	November–February
Cookie	0.00	11.00 (±0.20)	0.50 (±0.04)	38.50 (±2.00)	486.00 (±25.00)	13,865.00 (DE) (±1,839.04)	October–November
Orah	16.00 (±1.60)	12.10 (±0.18)	1.10 (±0.04)	35.70 (±1.10)	155.00 (±5.00)	0.00 (E) (±0.00)	January–March

Source: From Fidel, Y. et al., *Food and Nutrition Sciences*, 7, 90–101, 2016.
Note: Data are means ± standard error (SE) of three replications. Letters in parenthesis indicate significant differences at $p \leq 0.05$.

REFERENCES

Bailey, D. G., J. Malcolm, O. Arnold, and J. D. Spence. 1998. Grapefruit juice–drug interactions. *British Journal of Clinical Pharmacology* 46:101–110.

Bailey, D. G., J. Malcolm, O. Arnold, and J. D. Spence. 2004. Grapefruit juice–drug interactions. *British Journal of Clinical Pharmacology* 58:S831–S843.

Bailey, D. G., G. Dresser, and J. M. Arnold. 2013. Grapefruit–medication interactions: Forbidden fruit or avoidable consequences? *CMAJ* 185:309–316.

Berenbaum, M. R. 2003. Characterization of furanocoumarin metabolites in parsnip webworm, *Depressaria pastinacella. Journal of Chemical Ecology* 29:671–682.

Chen, C., P. Cancalon, C. Haun, and F. Gmitter Jr. 2011. Characterization of furanocoumarin profile and inheritance toward selection of low-furanocoumarin seedless grapefruit cultivars. *Journal of the American Society for Horticultural Science* 136:358–363.

De Castro, W. V., S. Mertens-Talcott, A. Rubner, V. Butterweck, and H. Derendorf. 2006. Inhibitory effect on monooxygenases involved in drug metabolism and potential toxicity. *Journal of Agricultural and Food Chemistry* 54:249–255.

Diawara, M. M., J. T. Trumble, K. K. White, W. G. Carson, and L. A. Martinez. 1993. Toxicity of linear furanocoumarins to *Spodoptera exigua*: Evidence for antagonistic interactions. *Journal of Chemical Ecology* 19:2473–2484.

Dugrand, A., A. Olry, T. Duval, A. Hehn, Y. Froelicher, and F. Bourgaud. 2013. Coumarin and furanocoumarin quantitation in citrus peel via ultraperformance liquid chromatography coupled with mass spectrometry (UPLC-MS). *Journal of Agricultural and Food Chemistry* 61:10677–10684.

Fidel, Y., M. Carmeli-Weissberg, Y. Yaniv, F. Shaya, N. Dai, E. Raveh, Y. Eyal, R. Porat, and N. Carmi. 2016. Breeding and analysis of two new grapefruit-like varieties with low furanocoumarin content. *Food and Nutrition Sciences* 7:90–101.

Gou, L. Q., K. Fukuda, T. Ohta, and Y. Yamazoe. 2000. Role of furanocoumarin derivatives in grapefruit juice-mediated inhibition of human CYP3A activity. *Drug Metabolism and Disposition* 28:766–771.

Ohnishi, A., H. Matsuo, S. Yamada, H. Takanaga, S. Morimoto, Y. Shoyama, H. Ohtani, and Y. Sawada. 2000. Effect of furanocoumarin derivatives in grapefruit juice on the uptake of vinblastine by *Caco*-2 cells and on the activity of cytochrome P450 3A4. *British Journal of Pharmacology* 130:1369–1377.

Paine, M. F., A. B. Criss, and P. B. Watkins. 2005. Two major grapefruit juice components differ in time to onset of intestinal CYP3A4 inhibition. *Journal of Pharmacology and Experimental Therapeutics* 312:1151–1160.

Paine, M. F., W. W. Widmer, H. L. Hart, S. N. Pusek, K. L. Beavers, A. B. Criss, S. S. Brown, B. F. Thomas, and P. B. Watkins. 2006. A furanocoumarin-free grapefruit juice establishes furanocoumarins as the mediators of the grapefruit juice-felodipine interaction. *The American Journal of Clinical Nutrition* 83:1097–1105.

Pan, L., Z. Wen, J. Baudry, M. R. Berenbaum, and M. A. Schuler. 2004. Identification of variable amino acids in the SRS1 region of CYP6B1 modulating furanocoumarin metabolism. *Archives of Biochemistry and Biophysics* 422:31–41.

Row, E. C., S. A. Brown, A. V. Staculski, and M. S. Lennard. 2006. Design, synthesis and evaluation of furanocoumarin monomers as inhibitors of CYP3A4. *Organic & Biomolecular Chemistry* 4:1604–1610.

USDA-FAO (US Department of Agriculture Foreign Agricultural Service). 2014. *Citrus: World Markets and Trade*, pp. 1–9. http://apps.fas.usda.gov/psdonline/circulars/citrus.pdf.

VanderMolen, K. M., G. R. Ainslie, M. F. Paine, and N. H. Oberlies. 2014. Labeled content of two furanocoumarins in dietary supplements correlates with neither actual content nor CYP3A inhibitory activity. *Journal of Pharmaceutical and Biomedical Analysis* 98:260–265.

Vardi, A., I. Levin, and N. Carmi. 2008. Induction of seedlesness in *Citrus*: From classical to emerging biotechnological approaches. *Journal of the American Society for Horticultural Science* 133:117–126.

Wang, Y., J. Xiao, T. O. Suzek, J. Zhang, J. Wang, and S. H. Bryant. 2009. PubChem: A public information system for analyzing bioactivities of small molecules. *Nucleic Acids Research* 37:623–633.

Wangensteen, H., E. Molden, H. Christensen, and K. Malterud. 2003. Identification of epoxybergamottin as a CYP3A4 inhibitor in grapefruit peel. *European Journal of Clinical Pharmacology* 58:663–668.

Wen, Y. H., J. Sahi, E. Urda, S. Kulkarni, K. Rose, X. Zheng, J. F. Sinclair, H. Cai, S. C. Strom, and V. E. Kostrubsky. 2002. Effects of bergamottin on human and monkey drug-metabolizing enzymes in primary cultured hepatocytes. *Drug Metabolism and Disposition* 30:977–984.

Widmer, W. 2006. One tangerine/grapefruit hybrid (tangelo) contains trace amounts of furanocoumarins at a level too low to be associated with grapefruit/drug interactions. *Journal of Food Science* 70:419–422.

Widmer, W., and C. Haun. 2005. Variation in furanocoumarin content and new furanocoumarin dimers in commercial grapefruit (*Citrus paradisi* Macf.) juices. *Journal of Food Science* 70: 307–312.

16 Young Citrus Fruits as By-Products and Their Chemical Properties as Affected by Food Processing

Xingqian Ye, Jinhu Tian, Yujing Sun, Jiang Ping, and Qianying (Sophia) Ye

CONTENTS

16.1 INTRODUCTION TO CITRUS YOUNG FRUITS

During the cultivation process, a large number of young citrus fruits will drop from their trees. This physiological dropping usually occurs three times from anthesis to maturity. Typically, during the second physiological dropping, several young citrus fruits will drop and numerous bioactive compounds will accumulate in the fruits. These fruits are commonly used as the starting material for the extraction of bioactive compounds.

Young citrus fruits are used in traditional Chinese medicine and are commonly known as *Zhishi* (*Fructus Aurantii Immaturus*). Zhishi has been used in China for a long time as a traditional medicine and was recorded in almost all of the famous ancient medical books such as *Agricultural Huang*. For instance, in one of the chapters from the *Compendium of Materia Medica*, the basic properties of Zhishi are described and its differences with trifoliate orange are also discussed. In another ancient Chinese horticulture book (*Julu*), Zhishi is also fully described and *C. aurantium* L. var. decumana Bonar was considered to be the raw materials (Xie; Cai et al. 1999). In the early days, the raw materials were mainly from *Poncirus trifoliata* L. but recently the raw materials changed to sour oranges (Xie).

According to the *Chinese Pharmacopoeia* (2005), Zhishi is made with the young dry fruit of *C. aurantium* L. or *C. sinensis* Osbeck. The raw materials are collected from the dropped young fruits during May and June. After discarding the impurities, the young fruits are cut in half and processed by sun-drying or low-temperature drying.

As one of the traditional Chinese medicines, the raw materials of Zhishi are widely distributed in China and the main producing area are Sichuan, Jiangxi, Hunan, Fujian, and so on. In particular, the sour oranges from Jiangxi are considered to have the highest quality (Zhang et al. 1989). According to the *Chinese Pharmacopoeia* (2005), the raw materials of Zhishi are mainly *C. aurantium* "Huangpi," *C. aurantium* "Daidai," *C. aurantium* "Chuluan," and *C. aurantium* "Tangcheng" (Lu et al. 2006).

As an herbal medicine, with its bitter, pungent, sour, and slightly cold properties, Zhishi is generally used to treat phlegm, bloating, splanchnic prolapse, and so on (Chinese Pharmacopoeia Commission 2010). Generally, the structure of Zhishi is hemispherical, but some are spherical with diameters from 0.5 to 2.5 cm. The exocarp is dark green or dark brown, granular, protuberant, and wrinkled. The stigma remnants or stem scar are clearly visible. The mesocarps are slightly uplifted and yellowish white or yellow-brown and have 0.3–1.2 cm thick oil chamber columns (one to two) in the edges. The flesh capsule looks brown and hard with a delicate fragrance and bitter, sour taste. Due to the provisions of the Pharmacopoeia, the identification method for Zhishi is its feeling indexes as well as the contents of aurantiamarin and synephrine. Thus, many young citrus fruits from different areas are also considered as raw materials of Zhishi.

16.2 MAJOR BIOACTIVE COMPOUNDS IN YOUNG CITRUS FRUITS

In China, many young citrus fruits (*C. aurantium*) can be used as raw material for traditional Chinese medicine. Several young citrus fruits drop from trees during the rainy season and end up as waste material. Extensive research has been conducted on this waste material to produce valued-added citrus products and to promote the development of the citrus industry (Ye et al. 2011).

16.2.1 Content and Composition of the Main Flavonoids in Different Citrus Varieties

Four flavanone glycosides and two polymethoxylated flavones are detected in many different varieties of young citrus fruits. The four flavanone glycosides and the two polymethoxylated flavones are also the major flavonoids of the seven citrus fruit categories. In addition, some young citrus fruits, such as lemons, contain other flavonoids (Table 16.1).

The main types of flavonoids are different in the citrus fruit varieties. Hesperidin is the main flavanone glycoside in the young fruits of *mandarins*, *sweet oranges*, and *lemons*. In these three categories, no naringin or neohesperidin is detected. In young fruits of *pummelo* and *Huyou*, naringin is the main flavanone glycoside; while in the hybrid of pummelo and orange (*Gaocheng*), the naringin content is significant higher than that of other flavonoids ($p < 0.05$).

When taking species into consideration, young mandarin fruits have high levels of polymethoxylated flavones, and *Zhuhong* fruits have the highest nobiletin (13.66 mg/g DW). This is flavanone.

The young fruits of pummelo, Huyou, and lemon have very little polymethoxylated flavone content. Among the 16 varieties of young fruits, Huyou and Gaocheng are two varieties that have flavonoids; Foyou grapefruit has a high naringin content of 61.60 mg/g DW.

Orange is one of the few high flavonoids species and Huyou are detected 98.75% for six of the total flavonoids in six varieties, while content of other flavonoids is extremely low.

The content of the main flavonoids in the different varieties of citrus is significantly different, even though they belong to the same species of mandarins or oranges. Also, the different varieties of citrus differ in their flavonoid composition, which can determine their end use. For example, the young citrus fruits of Owari Satsuma could be used as the raw material for hesperidin (261.63 mg/g DW) extraction, while small red-orange citrus fruits could be used as the raw material for narirutin (119.11 mg/g DW) extraction.

TABLE 16.1
Composition of Flavanone Glycosides and Polymethoxylated Flavones in Young Citrus Fruits (mg/g DW)

Variety	Species	Narirutin	Naringin	Hesperidin	Neohesperidin	Nobiletin	Tangeretin	Total
Citrus poonensis Hort. ex Tanaka (Penggan)	Mandarin	10.34	ND	188.62	ND	8.28	7.07	214.31
Citrus unshiu var. praecox Tanaka cv Nichinan No.1 (Southday1 satsuma)	Mandarin	38.09	ND	141.90	ND	0.45	0.07	180.51
Citrus unshiu var. praecox Tanka cv Miyagawa wase (Miyagawa satsuma)	Mandarin	46.86	ND	151.14	ND	0.63	0.19	198.82
Citrus unshiu Marc. cv Yamada (Yamada satsuma)	Mandarin	38.43	ND	141.97	ND	0.64	0.20	181.24
Citrus unshiu Marc. cv Owari (Owari satsuma)	Mandarin	43.90	ND	261.63	ND	0.92	0.36	306.82
Citrus erythrosa Hort. ex Tanka (*C. erythrosa)*	Mandarin	5.89	ND	191.81	ND	13.66	11.95	223.31
Citrus tardiferax Hort. ex Tanka (Manju mandarin)	Mandarin	8.28	ND	153.41	ND	7.10	8.27	177.07
Citrus succosa Hort. ex Tanaka (Bendizao)	Mandarin	5.81	ND	255.00	ND	3.42	2.91	267.13
Citrus suavissima Hort. ex Tanaka (Ougan)	Mandarin	11.00	10.54	163.71	ND	11.58	10.29	207.12
Jin orange	Sweet orange	17.33	ND	191.53	ND	3.00	0.26	212.13
Citrus Sinensis var. brasiliensis Tanaka (Navel orange)	Sweet orange	26.70	ND	131.54	ND	1.40	0.01	159.64
Citrus aurantium Linn cv Xiaohongcheng (Xiaohongcheng)	Sour orange	119.11	ND	68.68	ND	0.85	0.85	189.49
Citurs limon (L.) Burm.f. cv Eureka (Eureka)	Lemon	1.62	ND	42.05	ND	0.03	ND	43.70
Citrus grandis (L.) Osbeck cv Foyou (Foyou)	Pummelo	ND	61.60	ND	ND	ND	0.78	62.38
Citrus paradise Macf. Changshanhuyou (Huyou)	Huyou	6.20	54.59	3.92	41.00	0.24	0.09	106.03
Citrus gradis × *Citrus sinensis* (Gao orange)	Hybrid	2.25	67.50	3.04	100.03	1.86	1.30	175.98

Abbreviation: ND, not detected.

Citrus fruits are rich in flavonoids. Manthey and Grohmann (1996) reported that the content of hesperidin was 19.17–31.75 mg/g DW in the peel of mature oranges; Lu et al. (2006) reported the content of hesperidin and naringin in the peel of mature Huyou were 6.25 mg/g DW and 32.5 mg/g DW, respectively; Xu et al. (2007) measured the content of rutin, naringin, hesperidin, and neohesperidin in the peel of mature Huyou and reported that the contents were 2.81, 31.57, 2.04, and 24.09 mg/g DW, respectively. Comparing the values reported in the literature, the content of flavonoids in young citrus fruit was much higher than that in mature citrus fruit peels according to these results, young citrus fruits could be used as the raw material for the extraction of flavonoids.

16.2.2 Main Content and Composition of Phenolic Acids in Different Varieties of Citrus Fruit

16.2.2.1 Content and Composition of Extractable Phenolic Acids in Different Varieties of Young Citrus Fruits

Table 16.2 lists the content of seven main phenolic acids in young citrus fruits, along with the statistical results that indicate which citrus varieties have statistically different phenolic acid contents. No gallic acid was detected in any of the citrus fruits. Four kinds of cinnamic acids (caffeic acid, coumaric acid, ferulic acid, erucic acid) and three kinds of benzoic acids (protocatechuic acid, *p*-hydroxybenzoic acid, and vanillic acid) were detected.

The results in Table 16.2 indicate that 14 citrus fruit varieties have the following phenolic acids: caffeic acid, coumaric acid, ferulic acid, protocatechuic acid, and *p*-hydroxybenzoic acid. Two varieties (Manju mandarin and Ougan) do not contain vanillic acid and one variety, *C. erythrosa*, lacks sinapic acid. Among the various varieties of citrus fruits, the proportions of the phenolic acids are very different, especially with the methanol-soluble cinnamic-based phenolic acids. In most cases, ferulic acid is the major phenolic acid, followed by coumaric and caffeic acids. However, in eureka lemon, the content of coumaric (1784.26 μg/g DW) is five times greater than ferulic acid content (321.23 μg/g DW). For the benzoic-based phenolic acids, hydroxybenzoic acid is present in the highest amounts in the young mandarin fruits, sweet oranges, and lemons, while the levels of vanillic acid in sour oranges, pummelos, and Gao oranges are greater than those for *p*-hydroxybenzoic acid. Overall, it appears that the phenolic acid content in the young citrus fruit varieties, except for Buddha pummelo, are high. The content of each phenolic acid in the different varieties of citrus fruit also exhibits very large ranges. For example, the caffeic acid content in the young mandarin fruit (*Zhu Hong*) can reach as high as 7087.70 μg/g DW, but is only 16.47 μg/g DW in Buddha pummelo. The content of ferulic acid reaches a maximum (8424.04 DW (g/g)) and minimum (38.86 DW (g/g)) in the young mandarin fruit (*Zhu Hong*) and pummelo (*Foyou*), respectively.

16.2.2.2 Content and Composition of Bound Phenolic Acids in Different Varieties of Young Citrus Fruits

Table 16.3 lists the content and composition of bound phenolic acids in different varieties of young citrus fruits. The results for bound phenolic acids were

TABLE 16.2
Composition of Extractable Phenolic Acids in Young Citrus Fruits (μg/g DW)

		Cinnamics (μg/g DW)				Benzoics (μg/g DW)		
Variety	Species	Caffeic Acid	*p*-Coumaric Acid	Ferulic Acid	Sinapic Acid	Protocatechuic Acid	*p*-Hydroxybenzoic Acid	Vanillic Acid
Penggan	Mandarin	2848.45	1090.50	2330.95	58.05	29.64	127.91	12.88
Southday1 satsuma	Mandarin	567.20	599.61	1708.19	66.11	26.47	159.86	12.40
Miyagawa satsuma	Mandarin	120.40	494.91	333.65	14.37	35.75	271.57	26.77
Yamada satsuma	Mandarin	126.66	475.43	473.72	17.61	35.33	211.09	22.05
Owari satsuma	Mandarin	325.35	610.13	713.19	10.45	34.88	208.76	14.71
C. erythrosa	Mandarin	7087.70	1574.45	8424.04	ND	30.80	309.41	42.48
Manju mandarin	Mandarin	833.16	815.31	1540.31	15.14	23.06	122.64	ND
Bendizao	Mandarin	3028.04	1014.56	2806.65	20.43	80.33	151.17	55.17
Ougan	Mandarin	2677.86	762.50	5178.54	97.55	40.22	105.83	ND
Jin orange	Sweet orange	529.95	800.40	1767.32	45.86	24.41	206.91	39.97
Navel orange	Sweet orange	650.10	1131.92	1940.27	55.16	33.52	193.15	17.05
Xiaohongcheng	Sour orange	469.83	78.53	1484.28	81.95	25.00	253.28	691.35
Eureka	Lemon	37.05	1784.26	321.23	66.72	19.48	206.33	75.61
Foyou	Pummelo	16.47	86.63	38.86	69.91	7.79	50.74	137.92
Huyou	Huyou	328.33	216.35	515.48	46.90	11.63	125.19	386.35
Gao orange	Hybrid	775.63	282.86	1877.60	52.36	17.34	190.90	347.41

Abbreviation: ND, not detected.

TABLE 16.3
Composition of Bound Phenolic Acids in Young Citrus Fruits (μg/g DW)

		Cinnamics (μg/g DW)				Benzoics (μg/g DW)		
Variety	Species	Caffeic Acid	*p*-Coumaric Acid	Ferulic Acid	Sinapic Acid	Protocatechuic Acid	*p*-Hydroxybenzoic Acid	Vanillic Acid
Penggan	Mandarin	945.38	130.70	419.56	ND	6.74	35.23	4.03
Southday1 satsuma	Mandarin	110.13	63.68	167.02	ND	ND	39.12	6.07
Miyagawa satsuma	Mandarin	92.12	53 59	83.85	ND	14.32	56.34	18.58
Yamada satsuma	Mandarin	91.05	57.66	107.09	10.94	14.62	49.57	21.21
Owari satsuma	Mandarin	176.17	72.26	145.65	15.37	20.43	65.04	16.22
C. erythrosa	Mandarin	1089.20	279.27	1336.53	ND	6.53	128.13	24.28
Manju mandarin	Mandarin	459.51	269.50	778.98	ND	7.51	65.94	23.69
Bendizao	Mandarin	556.87	96.61	280.38	16.79	22.39	75.27	25.80
Ougan	Mandarin	459.71	71.11	513.81	ND	ND	30.40	ND
Jin orange	Sweet orange	142.65	116.48	273.01	15.37	4.02	45.25	6.39
Navel orange	Sweet orange	126.63	119.39	225.94	10.98	ND	35.99	3.76
Xiaohongcheng	Sour orange	50.05	43.88	108.64	8.89	ND	36.83	19.94
Eureka	Lemon	16.60	156.84	35.85	6.01	5.44	17.31	10.09
Foyou	Pummelo	6.35	25.93	7.84	9.79	5.80	7.32	8.59
Huyou	Huyou	88.14	48.24	98.80	5.49	ND	31.65	75.56
Gao orange	Hybrid	141.67	59.55	199.0	8.34	ND	22.08	11.92

Abbreviation: ND, not detected.

similar to the results for the extractable phenolic acids, given that no gallic acid was detected in the samples. The content of the bound (cinnamic-based) phenolic acids in the 16 citrus young fruit varieties is, for the most part, greater than the content of the bound (benzoic-based) phenolic acids. In the young mandarins, pummelos, and Gao oranges, the main bound phenolic acid was caffeic acid; in sweet orange and sour orange, it was mainly ferulic acid; and in eureka lemon, *p*-coumaric acid content was the highest. Sinapic acid content was the lowest (6.01–16.79 μg/g DW) among all the bound (cinnamic-based) phenolic acids in the 16 varieties of young citrus fruits. Benzoic-based phenolic acids are not major contributors of phenolic acids due to their low content in young citrus fruits. The concentration of *p*-hydroxybenzoic acid ranged from 7.32 to 128.13 μg/g DW.

16.2.2.3 Comparison of Total Phenolic Acids

The total amount of extractable phenolic acids and bound phenolic acids are shown in Figure 16.1. In all 16 varieties, the extractible phenolic acids account for almost all of the phenolic acids in young citrus fruits. In sour oranges (*Xiaohongcheng*), the percentage of extractable phenolic acids to total phenolic acids is the greatest (92%). The different varieties of young citrus fruit also have significantly different ($p < 0.05$) total phenolic acids content. The total phenolic acid content ranges from 479.94 to 20,332.83 μg/g DW. There are big differences in the content of total phenolic acid even in the same species of young citrus fruits. For example, among mandarin species, the *Zhu Hong* variety has the highest content of total phenolic acids

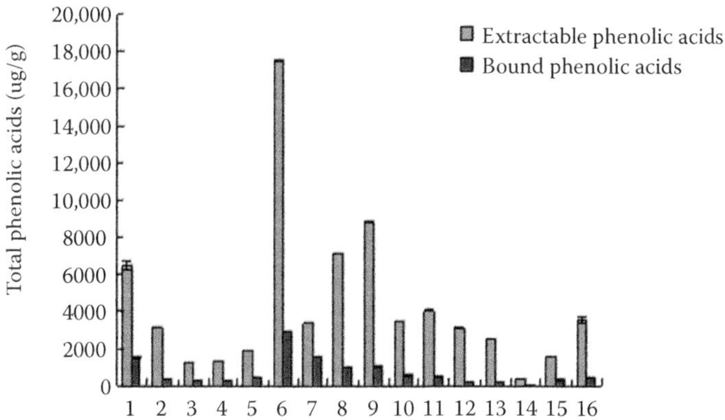

FIGURE 16.1 Comparison of extractable and bound total phenolic acids in young citrus fruits of different varieties. *1*, Penggan; *2*, Southday1 satsuma; *3*, Miyagawa satsuma; *4*, Yamada satsuma; *5*, Owari satsuma; *6*, *C. erythrosa*; *7*, Manju mandarin; *8*, Bendizao; *9*, Ougan; *10*, Jin orange; *11*, navel orange; *12*, Xiaohongcheng; *13*, eureka; *14*, Foyou; *15*, Huyou; *16*, Gao orange.

(20,332.83 μg/g DW), while in the Miyagawa variety, the content of total phenolic acid is only 1616.23 μg/g DW.

16.2.3 Content of Limonin and Nomilin in Different Young Citrus Fruits

Table 16.4 lists the composition and content of limonin and nomilin in 16 varieties of young citrus fruits. Limonin is found in all 16 varieties, while nomilin appears in only 14 varieties. The content of limonin and nomilin is closely associated with the citrus varieties. For most young citrus fruits, the content of limonin is higher than that of nomilin. These results are similar to those reported by Hashinaga and Itoo (1983). As shown in Table 16.4, the content of limonin achieves a maximum value (4851.26 μg/g DW) in the young fruits of pummelos (Foyou), and a minimum value (88.84 μg/g DW) in the young fruits of mandarins (Miyagawa). For nomilin, the maximum and minimum values are 790.91 and 3.65 μg/g DW for mandarins (Manju) and mandarins (Bendizao), respectively.

Limonin and nomilin are the aglycone of limonoids. The aglycones are the major source of the bitterness in citrus fruits, considering that their glycosides are only mildly bitter. This would indicate that the content of limonin and nomilin may have some impact on the sensory quality of different citrus varieties and remains to be further investigated.

TABLE 16.4
Composition of Limonin and Nomilin in Young Citrus Fruits (μg/g DW)

Variety	Species	Limonin	Nomilin	Total
Penggan	Mandarin	711.04	219.25	930.29
Southday1 satsuma	Mandarin	112.49	19.27	131.76
Miyagawa satsuma	Mandarin	88.84	15.70	104.54
Yamada satsuma	Mandarin	380.84	164.42	545.26
Owari satsuma	Mandarin	97.83	24.15	121.97
C. erythrosa	Mandarin	267.11	60.45	327.56
Manju mandarin	Mandarin	337.22	790.91	1128.13
Bendizao	Mandarin	192.67	3.65	196.32
Ougan	Mandarin	291.88	544.52	836.39
Jin orange	Sweet orange	327.49	245.28	572.77
Navel orange	Sweet orange	398.30	198.75	597.05
Xiaohongcheng	Sour orange	560.08	92.98	653.06
Eureka	Lemon	2251.97	614.84	2866.81
Foyou	Pummelo	4851.26	ND	4851.26
Huyou	Huyou	504.92	147.90	652.82
Gao orange	Hybrid	2,622.65	ND	2622.65

Abbreviation: ND, not detected.

TABLE 16.5
Composition of Synephrine in Young Citrus Fruits (mg/g DW)

Variety	Species	Synephrine
Penggan	Mandarin	13.51
Southday1 satsuma	Mandarin	7.58
Miyagawa satsuma	Mandarin	8.99
Yamada satsuma	Mandarin	9.02
Owari satsuma	Mandarin	10.79
C. erythrosa	Mandarin	22.47
Manju mandarin	Mandarin	17.64
Bendizao	Mandarin	12.01
Ougan	Mandarin	8.49
Jin orange	Sweet orange	5.40
Navel orange	Sweet orange	2.81
Xiaohongcheng	Sour orange	2.25
Eureka	Lemon	ND
Foyou	Pummelo	ND
Huyou	Huyou	ND
Gao orange	Hybrid	3.59

Abbreviation: ND, not detected.

16.2.4 Synephrine Content in Young Citrus Fruits

The content of synephrine in different varieties of young citrus fruits is given in Table 16.5. The results show no synephrine in the young fruits of lemon, pummelo, and Huyou. Generally, the levels of synephrine are high in young mandarin fruits, with the highest level (22.47 mg/g DW) in the young mandarin fruits (*Zhu Hong*), followed by lower levels in the sweet oranges and the lowest level (2.25 mg/g DW) in sour oranges (Xiahongcheng). These results are consistent with the results published by Chen et al. (2005), with the synephrine content, based on species, following the same order: mandarins > sweet oranges > sour oranges.

Xiao et al. (2000) reported the synephrine content in young citrus fruits harvested from four different locations ranged from 0.501 to 1.209 mg/g DW. Wang (2004) measured the synephrine content in young *C. aurantium* L. fruits from different areas around Shanghai and during different picking periods and reported values that ranged from 8.0 to 21.8 mg/g DW.

16.2.5 Total Phenolic Content and Antioxidant Capacity in Different Citrus Varieties

The total phenolic content was reported as gallic acid equivalent (GAE). The antioxidant capacity of young citrus fruits using three methods (ferric ion reducing

TABLE 16.6
Total Phenolic Content and Antioxidant Capacity of Extracts of Young Citrus Fruits

Variety	Species	Total Phenolic (GAE, mg/g DW)	FRAP (TEAC, mg/g DW)	DPPH (TEAC, mg/g DW)	ABTS (TEAC, mg/g DW)
Penggan	Mandarin	67.42	96.35	21.85	196.35
Southday1 satsuma	Mandarin	47.12	64.30	13.39	188.31
Miyagawa satsuma	Mandarin	50.02	66.46	12.42	165.12
Yamada satsuma	Mandarin	47.26	60.69	11.88	214.22
Owari satsuma	Mandarin	68.91	100.70	17.91	269.40
C. erythrosa	Mandarin	78.71	126.75	33.22	327.56
Manju mandarin	Mandarin	56.21	83.15	15.49	214.37
Bendizao	Mandarin	72.80	125.04	27.79	283.92
Ougan	Mandarin	51.93	98.40	18.58	199.24
Jin orange	Sweet orange	54.14	93.73	14.51	172.58
Navel orange	Sweet orange	43.51	72.28	14.16	188.52
Xiaohongcheng	Sour orange	52.07	46.77	13.55	202.32
Eureka	Lemon	21.17	37.90	8.26	104.64
Foyou	Pummelo	34.16	14.51	8.77	113.09
Huyou	Huyou	37.47	29.65	11.99	156.93
Gao orange	Hybrid	59.11	66.48	18.78	225.43

antioxidant powder [FRAP], 2,2-diphenyl-1-picrylhydrazyl [DPPH], 2,2′-azino-bis(3-ethylbenzothiazoline-6-sulfonic acid [ABTS]) was reported as Trolox equivalent antioxidant capacity (TEAC), and the experimental data are shown in Table 16.6. The total phenolic contents in the young fruits of mandarin, sweet orange, and sour orange are higher than that in Huyou, pummelo, and lemon. Among all citrus varieties, the young fruits of *C. erythrosa* has the highest total phenolic content (78.71 mg GAE/g DW), and it also has the highest TEAC in all three assay methods, which reflects the relatively high antioxidant capacity of the young fruits of *C. erythrosa*. In general, the young fruits of mandarin and sweet orange also have relatively high antioxidant capacities, making these young citrus fruits potential candidates for use in foods as an antioxidant functional component.

16.3 UTILIZATION OF YOUNG FRUITS THROUGH FOOD PROCESSING

In China, there are many formulations of Chinese medicines (Pharmacopoeia records) that utilize citrus fruits such as *C. aurantium* L. and *Fructus aurantii* as the raw material. When preparing these formulations of medicines, the most important processing step is the drying method. In traditional Chinese medicine, natural drying and storage are the main methods used. However, natural drying is influenced by the weather, climate, and air humidity, and the effects of these factors on the activity of functional ingredients and on antioxidant capacity are not entirely understood.

To better understand the drying process, four varieties of young citrus fruit (Penggan, Foyou, Gao orange, and Huyou) were subjected to three drying methods: sun-drying, 60°C hot air-drying, and freeze-drying. The dried samples were analyzed, and the effects of drying on the contents of flavonoids, phenolic acids, limonoids, and synephrine and on antioxidant capacity were studied to make a comprehensive evaluation of the capability and suitability of the three drying methods (Jiang 2008; Sun et al. 2015).

16.3.1 Impact of Different Drying Methods on the Content of Flavonoids in Young Citrus Fruits

The impact of different drying methods on the four flavanone glycosides and two polymethoxylated flavones in young citrus fruits is shown in Table 16.7. The total flavonoid content in the freeze-dried citrus fruit is the highest, while in the sun-dried samples the flavonoid content is the lowest. For example, the freeze-dried Penggan citrus fruit has a hesperidin content of 197.13 mg/g DW, while the 60°C hot air-dried and sun-dried samples have hesperidin levels of 193.08 and 188.62 mg/g DW, respectively.

The influence of different drying methods on the total content of six flavonoids in young citrus fruits is shown in Figure 16.2. Except for Penggan (*C. poonensis* Hort. ex Tanaka) ($p > 0.05$), the different drying methods significantly affected the total content of citrus flavonoids ($p < 0.05$). As can be seen from the experimental results, the freeze-drying method gives the best results. The 60°C hot air-drying method also maintains a high level of flavonoids in the young citrus fruits. However, the traditional sun-drying method used for preparing traditional Chinese medicines shows a significant loss of flavonoids. For example, the freeze-dried sample of Gaocheng has 196.65 mg/g DW of total flavonoids, while the sun-dried sample has only 175.98 mg/g DW. The sun-dried sample is 10.50% lower than the freeze-dried sample. Compared to the freeze-dried samples, the sun-dried samples of Penggan, Foyou, and Huyou show losses in total flavonoid content of 4.69%, 8.18%, and 8.90%, respectively.

16.3.2 Impact of Different Drying Methods on Total Phenolic Acid Content in Young Citrus Fruits

The impact of different drying methods on the total phenolic acid content in young citrus fruits is shown in Table 16.8. Using Duncan's multiple range test on four citrus varieties (*Penggan*, Gao orange, Foyou, Huyou), the effect of three different drying methods on total phenolic acid content were shown to be extremely significant ($p < 0.05$). The young fruits of *Penggan,* Gao orange, Foyou *and* Huyou that were freeze-dried have the highest total phenolic acid content. The total phenolic acid content in this four varieties after hot air-drying are 7.4%, 2.6%, 7.5% and 5.8% lower than in the freeze-dried samples, respectively. After sun-drying, the total

TABLE 16.7
Effect of Drying Methods on Flavanoid Content in Young Citrus Fruits (mg/g DW)

Variety	Drying Method	Narirutin	Naringin	Aurantiamarin	Neohesperidin	Nobiletin	Tangeretin
Pengan	Sun-drying	10.33	ND	188.62	ND	8.28	7.07
	60°C hot air-drying	11.12	ND	193.08	ND	9.33	8.01
	Freeze-drying	11.71	ND	197.13	ND	7.75	7.70
Gao orange	Sun-drying	2.2	67.50	3.04	100.03	1.86	1.30
	60°C hot air-drying	2.95	73.11	3.45	114.97	2.03	1.40
	Freeze-drying	3.45	71.39	3.50	114.70	2.19	1.43
Foyou	Sun-drying	ND	61.60	ND	ND	ND	0.78
	60°C hot air-drying	ND	64.36	ND	ND	ND	0.81
	Freeze-drying	ND	67.08	ND	ND	ND	0.87
Huyou	Sun-drying	6.20	54.59	3.92	41.00	0.24	0.09
	60°C hot air-drying	7.67	56.48	4.46	44.21	0.30	0.10
	Freeze-drying	7.55	57.93	4.25	46.33	0.24	0.10

Abbreviation: ND, not detected.

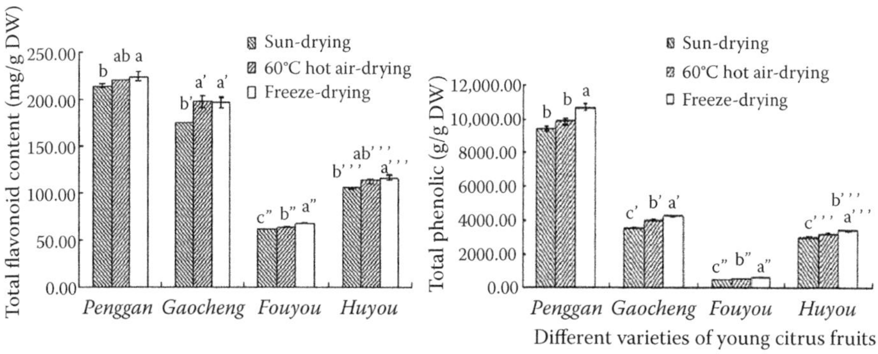

FIGURE 16.2 Effect of drying methods on total flavonoid and phenolic content in young citrus fruits (mg/g DW).

phenolic acid content are 1.7%, 14%, 14.2%, and 11% lower than the freeze-dried samples, respectively. As for individual phenolic acids, the result are consistent with the total phenolic acid results. For example, ferulic acid content in *Penggan* after freeze-drying was 3815.62 μg/g DW, while after 60°C hot air-drying and sun-drying, the content of ferulic acid was 3533.78 μg/g DW and 3399.32 μg/g DW, respectively, which represents a decrease of 7.4% and 10.9%, respectively. The results show that the heating and drying process can result in a significant decrease in the total contents of phenolic acid in young citrus fruits. This could be caused by the high temperatures used during hot air-drying, which results in phenolic acid oxidation and other damaging reactions. In addition, the exposure to UV radiation under sun-drying conditions, could also lead to phenolic acid loss.

16.3.3 Impact of Different Drying Methods on Limonin and Nomilin Content in Young Citrus Fruits

The impact of three different drying methods on limonin and nomilin contents in young citrus fruits are presented in Table 16.9. The four varieties, Penggan, Gao orange, Foyou, and Huyou, after freeze-drying have the highest limonin content (718.36, 2687.76, 4965.29, and 530.07 μg/g DW, respectively); after 60°C hot air-drying, the limonin contents are 716.59, 2664.08, 4895.55, and 518.02 μg/g DW, respectively. The effects of different drying procedures on nomilin content show similar trends: freeze-drying > 60°C hot air-drying > sun-drying. However, the Duncan multiple range test shows no significant difference in limonin and nomilin content among these three different drying methods at $p < 0.05$.

16.3.4 Impact of Different Drying Methods on Synephrine Content in Young Citrus Fruits

Among the four varieties (Penggan, Gao orange, Foyou, and Huyou), only Penggan and Gao oranges had detectable synephrine levels. Table 16.10 lists the results of these

TABLE 16.8
Effect of Drying Methods on Phenolic Acid Content in Young Citrus Fruits (μg/g DW)

Variety	Drying Method	Caffeic Acid	*p*-Coumaric Acid	Ferulic Acid	Sinapic Acid	Protocatechuic Acid	*p*-Hydroxybenzoic Acid	Vanillic Acid
Penggan	Sun-drying	4306.92	1368.72	3399.32	58.34	32.04	232.25	25.15
	60°C hot air-drying	4508.15	1425.55	3533.78	68.65	37.92	273.52	29.69
	Freeze-drying	4869.00	1538.84	3815.62	70.05	41.09	300.85	31.74
Gao orange	Sun-drying	772.83	298.55	1874.14	57.02	17.34	200.28	305.63
	60°C hot air-drying	808.46	337.38	2204.80	67.05	19.32	212.73	346.26
	Freeze-drying	873.11	364.35	2280.48	72.15	19.68	229.99	374.14
Foyou	Sun-drying	23.39	132.18	46.58	76.88	12.94	55.92	155.68
	60°C hot air-drying	27.60	137.39	54.58	78.80	15.16	60.92	168.95
	Freeze-drying	29.71	147.77	58.96	85.74	16.37	65.86	183.16
Huyou	Sun-drying	842.01	427.99	859.15	52.42	11.63	200.93	618.77
	60°C hot air-drying	880.42	456.69	906.03	61.77	12.23	220.99	649.29
	Freeze-drying	942.33	491.62	944.06	63.91	12.59	237.67	691.31

TABLE 16.9
Effect of Drying Method on Limonin and Nomilin Content in Young Citrus Fruits (μg/g DW)

Variety	Drying Method	Limonin	Nomilin	Total
Penggan	Sun-drying	711.04	219.25	930.29
	60°C hot air-drying	716.59	223.47	940.06
	Freeze-drying	718.36	223.47	941.83
Gao orange	Sun-drying	2622.65	ND	2622.65
	60°C hot air-drying	2664.08	ND	2664.08
	Freeze-drying	2687.76	ND	2687.76
Foyou	Sun-drying	4851.26	ND	4851.26
	60°C hot air-drying	4895.55	ND	4895.55
	Freeze-drying	4965.29	ND	4965.29
Huyou	Sun-drying	504.92	147.90	652.82
	60°C hot air-drying	518.02	148.15	666.17
	Freeze-drying	530.07	150.94	681.01

Abbreviation: ND, not detected.

TABLE 16.10
Effect of Drying Methods on Synephrine Content in Young Citrus Fruits (mg/g DW)

Variety	Drying Method	Synephrine
Penggan	Sun-drying	13.51
	60°C hot air-drying	13.64
	Freeze-drying	15.00
Gao orange	Sun-drying	3.59
	60°C hot air-drying	3.70
	Freeze-drying	3.93

three drying methods on the synephrine content. Statistical analysis revealed that the drying methods had a significant effect ($p < 0.05$) on the content of synephrine in young citrus fruits. As shown in Table 16.9, the content of synephrine is significantly higher in the freeze-dried samples. The values for Penggan and Gao oranges are 15.00 and 3.93 mg/g DW, respectively. Compared with the freeze-dried samples, the content of synephrine in the hot air-dried samples (13.64 mg/g DW Penggan and 3.70 mg/g DW Gao orange) and the sun-dried samples (13.51 mg/g DW Penggan and 3.59 mg/g DW Gao orange) are significantly lower ($p < 0.05$). However, the synephrine contents in the hot air-dried and sun-dried samples are not significantly different ($p > 0.05$).

16.3.5 Impact of Different Drying Methods on Total Phenolic Content and Antioxidant Capacity

The Folin–Ciocalteu method was used to determine the total phenol content and the FRAP, DPPH, and ABTS methods were applied to evaluate the antioxidant capacity of young citrus fruits. The results are shown in Figure 16.3. By statistical analysis, the effects of different drying methods on the total phenolic content in young citrus fruits show significant differences ($p < 0.05$), specifically in the four varieties of young citrus fruits (Penggan, Gao orange, Foyou, and Huyou). The highest total phenol content is in the freeze-dried samples, followed by the hot-air-dried samples and last by the sun-dried samples. For example, the total phenol content in the Foyou freeze-dried sample is 40.02 GAE mg/g DW, the hot-air-dried sample is 37.25 GAE mg/g DW, and the sun-dried sample is only 4.16 GAE mg/g DW.

In conclusion, the contents of flavonoids, phenolic acids, synephrine, limonin, and nomilin in these four varieties of young citrus fruits were the highest in the freeze-dried samples, while in the 60°C hot-air-dried samples and sun-dried samples, the contents were reduced.

Chism and Haard (1996) pointed out that fruits and vegetables have a high concentration of phenolic compounds, and this would suggest that all phenolic substances, including some intermediate metabolites, will accumulate in the vacuoles. In this chapter, the samples used were all fresh young citrus fruits and their cellular structures would have been intact. These samples, before the vacuum freeze-drying step, would undergo a prefreezing process at −80°C and render most intracellular

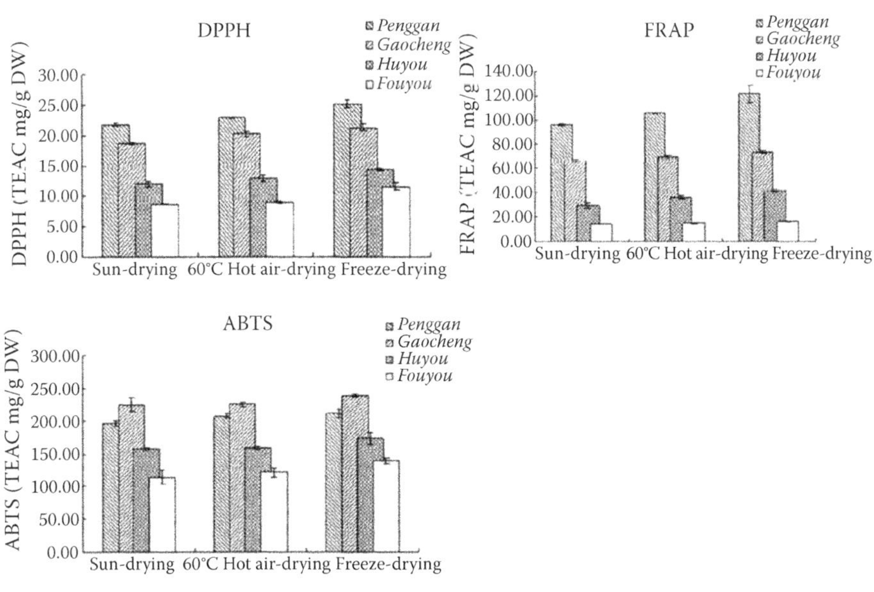

FIGURE 16.3 Effect of drying methods on free radical-scavenging (DPPH•, FRAP•, ABTS•+) activities of young citrus fruits (mg/g DW).

enzymes inactive. The temperature during vacuum freeze-drying is usually under 0°C and would stop metabolic activity within the cell and thus reduce the loss of phenolic compounds and other functional components. On the other hand, during the 60°C hot air-drying and sun-drying process, biological metabolism within the cell can remain active for a significant period. These dehydration procedures can cause the rupture of vacuoles and other cellular organelles, triggering a massive release of oxidative and hydrolytic enzymes such as polyphenol oxidase, leading to oxidation and hydrolysis of phenolic compounds. These destructive phenomena are more likely to occur under favorable environmental conditions such as suitable temperatures, available oxygen, and long drying times which are usually associated with sun-drying. Hence, we suspect that the oxidation and hydrolysis reactions performed by enzymes are responsible for the degradation of phenolic compounds and other functional ingredients during the 60°C hot air-drying and sun-drying processes. However, the real reasons are still not definitively known and will require additional scientific investigations.

16.4 SUMMARY

The composition and content of the main functional components in different varieties of citrus fruit were investigated. The main functional components included four flavanone glycosides (narirutin, naringin, hesperidin, neohesperidin), two polymethoxylated flavones (nobiletin, tangeretin), eight phenolic acids (gallic acid, protocatechuic acid, *p*-hydroxybenzoic acid, caffeic acid, vanillic acid, *p*-coumaric acid, ferulic acid, erucic acid), limonin, nomilin, and synephrine. The results show that flavanone glycosides were abundant in citrus young fruits and that there was a big difference in flavonoid content in the different varieties; polymethoxylated flavones were rare in citrus young fruits; the main phenolic acid in young citrus fruits was cinnamic acid; and in the majority of young citrus fruits, the content of limonin was higher than the content of nomilin. Among different species, the content of synephrine followed the order: mandarins > sweet oranges > sour oranges.

The antioxidant capacity of the extracted functional components that were determined by the FRAP, DPPH, and ABTS methods showed a significant correlation with total phenolics, total flavonoids, and total extractable phenolic acids ($p < 0.01$).

Different drying temperatures had a significant effect on flavonoid content, phenolic acid content, and synephrine content ($p < 0.05$), with the highest content in 60°C hot-air-dried sample. Using a lower or higher temperature caused a decrease in content, with the lowest content occurring at 120°C. Different drying temperatures also significantly affected the limonin and nomilin contents in young citrus fruits ($p < 0.05$).

Using a high temperature (120°C) during hot air-drying of young citrus fruits increased the antioxidant capacity of the methanol extracts but the content of flavonoids, phenolic acids, synephrine, and other functional ingredients decreased and this method is not considered a suitable drying procedure. Drying at 45°C and 30°C caused a decrease in the content of phenolic acids and other functional ingredients and also lowered the antioxidant capacity of the methanol extracts. Taking into consideration all the effects that temperature had on the content of functional ingredients and antioxidant capacity, the use of 60°C hot-air-drying of young citrus fruits was judged to be the most appropriate.

A large amount of flavanone glycosides is present in young citrus fruits, and the difference in content between species is significant. Polymethoxylated flavone is found in low concentration in young citrus fruits. The main phenolic acid in young citrus fruits is cinnamic acid. In the majority of young citrus fruits, the content of limonin is higher than that of nomilin. The order of synephrine content in different species is mandarins > sweet oranges > sour oranges.

Citrus fruit and its processed juice are one of the most popular fruits and drinks on the market. Citrus fruits have also been used as herbal medicine, as raw material in crude drugs, and as raw material in traditional Chinese medicine. The ongoing research on the efficacy of the bioactive components in *C. aurantium* remains unfinished, and the influence of the drying process on functional ingredients and antioxidant capacity is incomplete. Therefore, our research is focused on using young citrus fruits to study the composition and content of functional components in different *Citrus* species and to study the influence of hot air-drying at different temperatures on the content and antioxidant activity of their functional component.

ACKNOWLEDGMENTS

Financial supports from China Nature Sciences Foundation, project C200501 (2014–2017).

REFERENCES

Cai, Y. P., Y. G. Chen, and C. S. Fan. 1999. Studies on Chinese drugs zhiqiao and zhishi origin investigation and merchandise identification. *China Journal of Chinese Materia Medica* 24(5):259–262.

Chen, Z. 2015. Study on medicinal citrus species and their active constituents. *Chinese Wild Plant Resources* 24(4): 38–40.

Chinese Pharmacopoeia. 2005. Beijing: Chemical Publishing House.

Chism, G. W., and N. F. Haard. 1996. Characteristics of edible plant tissues. In: O. R. Fennema (Ed.), *Food chemistry*, pp. 943–1011. New York: Marcel Dekker.

Hashinaga, F., and S. Itoo. 1983. Seasonal changes of limonoids in the component parts of several citrus fruits. *Proceedings of the International Society of Citriculture* 2:901–905.

Jiang, P. 2008. Study on the functional components and antioxidant capacities of young fruits of citrus. Master dissertation, Zhejiang University, Hangzhou, pp. 33–98.

Lu, Y., C. Zhang, P. Bucheli, and D. Wei. 2006. Citrus flavonoids in fruit and traditional Chinese medicinal food ingredients in China. *Plant Foods for Human Nutrition* 61(2):55–63.

Manthey, A., and K. Grohmann. 1996. Concentrations of hesperidin and other orange peel flavonoids on citrus processing byproducts. *Journal of Agricultural and Food Chemistry* 44(3):811–814.

Sun, Y., Y. Shen, D. Liu, and X. Ye. 2015. Effects of drying methods on phytochemical compounds and antioxidant activity of physiologically dropped un-matured citrus fruits. *LWT–Food Science and Technology* 60:1269–1275.

Wang, H., and Hou T. 2004. Synephrine content of immature bitter orange around Shanghai with different mature time. *Journal of Shandong University of TCM* 28(5):377–379.

Xiao, M., Y. Wu, L. Zhu, X. Zeng, and S. Hu. 2000. Naringin and synephrine content of Zhi Ke in four different places of origin. *Journal of Chinese Medicinal Materials* 23(5):268–270.

Xu, G. H., X. Q. Ye, J. C. Chen, and D. H. Liu. 2007. Effect of heat treatment on the phenolic compounds and antioxidant capacity of citrus peel extract. *Journal of Agricultural and Food Chemistry* 55(2):330–335.

Ye, X. Q., J. C. Chen, D. H. Liu, P. Jiang, J. Shi, S. Xue, D. Wu, J. G. Xu, and Y. Kakuda. 2011. Identification of bioactive composition and antioxidant activity in young mandarin fruits. *Food Chemistry* 124:1561–1566.

Zhang, Z. Z., C. S. Fan, and A. H. Huang. 1989. The quality of immature bitter orange and *Hovenia dulcis* in Jiangxi. *China Journal of Chinese Materia Medica* 14(9):520–522.

Index

Page numbers followed by f and t indicate figures and tables, respectively.

B

N

U

V

W

X

Y

Z

* 9 7 8 1 4 9 8 7 4 2 7 2 6 *
#0005 - 230817 - C50 - 229/152/28 [30] - CB - 9781498742726